Dynamics of Fluids in Hierarchical Porous Media

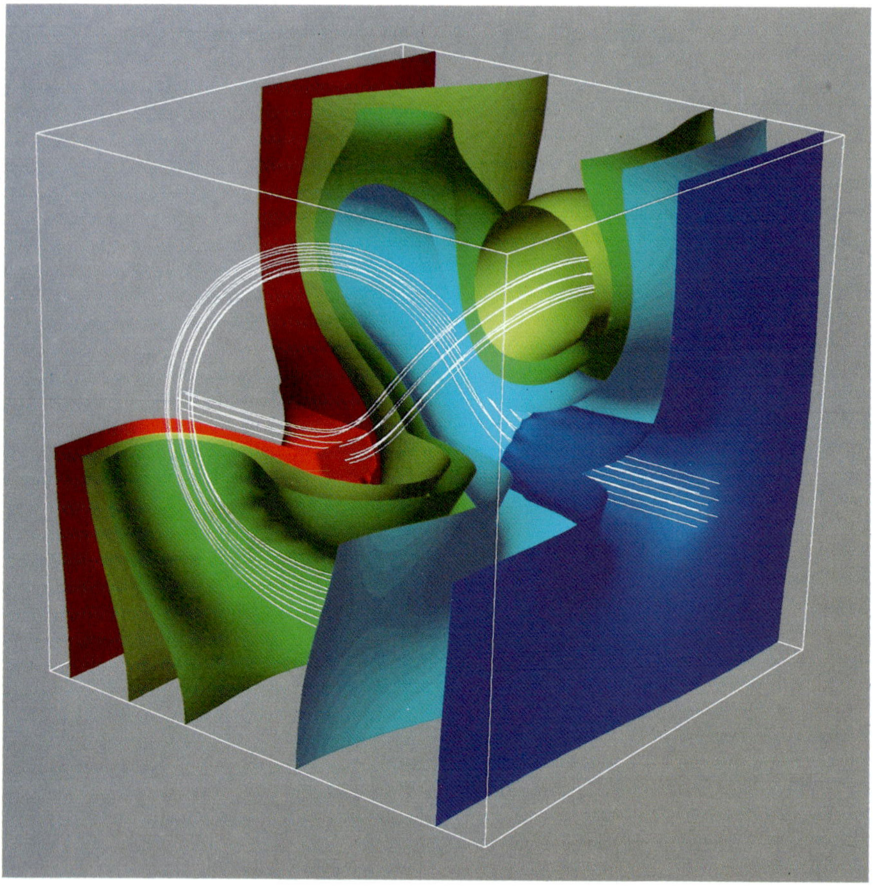

Visualization of complex three-dimensional flow resulting from a knot shaped, high conductivity channel embedded in a cube. Flow in this system is illustrated by streamlines superimposed on a cutaway view of the potential field depicted by shaded, colour-coded equal potential surfaces.

Dynamics of Fluids in Hierarchical Porous Media

Edited by

J.H. Cushman
Lilly Hall of Life Sciences
Purdue University
Indiana
USA

ACADEMIC PRESS
Harcourt Brace Jovanovich, Publishers
London San Diego New York
Boston Sydney Tokyo Toronto

ACADEMIC PRESS LIMITED
24/28 Oval Road
LONDON NW1 7DX

United States Edition published by
ACADEMIC PRESS INC.
San Diego, CA 92101

This book is printed on acid-free paper

British Library Cataloguing in Publication Data
Is available.

ISBN 0-12-200260-1

Printed in Great Britain by St Edmundsbury Press Ltd, Bury St Edmunds, Suffolk
Typeset by P&R Typesetters Ltd, Salisbury, Wiltshire, UK

Contributors

R. Ababou Department of Civil Engineering, Princeton University, Princeton, NJ 08544, USA
(Present address: Southwest Research Institute (Div. 20), San Antonio, TX 78228, USA)

T. Arbogast Department of Mathematics, Purdue University, West Lafayette, IN 47907, USA

R. Bhattacharya Department of Mathematics, Indiana University, Bloomington, IN 47405, USA

C. Cardone Department of Civil Engineering, Auburn University, Auburn, AL 36830, USA

M.A. Celia R.M. Parsons Laboratory Bldg. 48, Department of Civil Engineering, Massachusetts Institute of Technology, Cambridge, MA 02139, USA

C.R. Cole Pacific Northwest Laboratories, P.O. Box 999, Richland, WA 99352, USA

J.H. Cushman Lilly Hall of Life Sciences, Purdue University, West Lafayette, IN 47907-7899, USA

J.R. Douglas, Jr Department of Mathematics, Purdue University, West Lafayette, IN 47907, USA

H.I. Ene Department of Mathematics, The National Institute for Scientific and Technical Creation, Bd. Pacii 220, 79622 Bucharest, Romania
(Present address: University of Bucharest, Faculty of Mathematics, Mechanics Department, St. Academici 14, 70109 Bucharest, Romania)

L.A. Ferrand R.M. Parsons Laboratory Bldg. 48, Department of Civil Engineering, Massachusetts Institute of Technology, Cambridge, MA 02139, USA

H.P. Foote Pacific Northwest Laboratories, P.O. Box 999, Richland, WA 99352, USA

L.W. Gelhar Department of Civil Engineering, Massachusetts Institute of Technology, Cambridge, MA 02139, USA

V.K. Gupta Center for the Study of Earth from Space/CIRES and Department of Geological Sciences, University of Colorado, Boulder, CO 80309, USA

O. Güven Department of Civil Engineering, Auburn University, Auburn, AL 36830, USA

O. Levin Department of Hydrology and Water Resources, University of Arizona, Tucson, AZ 85721, USA

K.T.B. MacQuarrie Waterloo Centre for Groundwater Research, Waterloo, Ontario, Canada N2L 3G1
J.G. Melville Department of Civil Engineering, Auburn University, Auburn, AL 36830, USA
D.E. Moissis Department of Mathematics, University of Houston, Houston, TX 77005, USA
(Present address: 10434 Athens, Greece)
G. Moltyaner Environmental Research Branch, Chalk River Nuclear Laboratory, Chalk River, Ontario, Canada K0J 1J0
F.J. Molz Department of Civil Engineering, Auburn University, Auburn, AL 36830, USA
S.P. Neuman Department of Hydrology & Water Resources, University of Arizona, Tucson, AZ 85721, USA
O.A. Plumb Department of Mechanical and Materials Engineering, Washington State University, Pullman, WA 99164, USA
S.L. Schellenberg Waterloo Centre for Groundwater Research, Waterloo, Ontario, Canada N2L 3G1
G.A. Sharp Desert Research Institute, P.O. Box 60220, Reno, NV 89506-0220, USA
W.E. Soll R.M. Parsons Laboratory Bldg. 48, Department of Civil Engineering, Massachusetts Institute of Technology, Cambridge, MA 02139, USA
G. Sposito Department of Soil Science, University of California, Berkeley, CA 94720, USA
E.A. Sudicky Waterloo Centre for Groundwater Research, Waterloo, Ontario, Canada N2L 3G1
S.W. Tyler Desert Research Institute, P.O. Box 60220, Reno, NV 89506-0220, USA
S. Wheatcraft Desert Research Institute, P.O. Box 60220, Reno, NV 89506-0220, USA
M.F. Wheeler Department of Mathematics, University of Houston, Houston, TX 77005, USA
S. Whitaker Department of Chemical Engineering, University of California, Davis, CA 95616, USA
Y.-K. Zhang Department of Hydrology & Water Resources, University of Arizona, Tucson, AZ 85721, USA

Preface

Recent years have produced an explosion of research on the transport of fluid mixtures in natural, heterogeneous, porous media. During this period, it has become apparent that many, if not most, porous media possess a hierarchical structure. The purpose of this text is to tie together the most recent and powerful theoretical, computational, and experimental methods for studying such media. It is hoped that this text will provide the stimulus necessary to bring about a successful solution(s) to many of the complex problems associated with transport in natural porous media.

This text is an outgrowth of lectures presented at a symposium on hierarchical porous media, which was held in the fall of 1988 in San Francisco. I am grateful to the participants and the attentive and inquisitive audience for providing valuable feedback on the lectures. Several important topics which were omitted during the symposium have also been included in this text.

The chapters within the text have been peer reviewed, and I wish to thank the authors and publishers for putting up with the resulting delay in publication. I would also like to thank the Office of Health and Environmental Research of the US Department of Energy for providing financial support while I was editing the text.

West Lafayette, Indiana *John H. Cushman*

Contents

Chapter I

An Introduction to Hierarchical Porous Media

John H. Cushman
Lilly Hall of Life Sciences, Purdue University, West Lafayette, IN 47907-7899, USA

1. Prologue

Over the last two decades, researchers who have been trying to understand transport processes in porous media have begun to recognize that we do not know nearly as much as we once thought we did. Natural porous media are generally not homogeneous, nor are they uniformly random. Porous media may contain multiple, nested, natural length and time scales or continuously evolving scales. Measurements of constitutive variables on one scale may appear to have little relevance to any other scale, or for that matter to any known theory of transport. Moreover, standard theories of transport that have been in existence for several decades have often proven to be quite inadequate when tested on natural formations.

It is becoming more and more evident that to alleviate the problems mentioned above, scientists from many disciplines must band together in interdisciplinary research efforts. No longer can the experimentalist go one way and the theorist another. No longer can the mathematicians, physicists, chemical, civil, mechanical, and environmental engineers, soil scientists, chemists, and geologists work solely within their own disciplines and their own literature. Transport in porous media has become a truly multidisciplinary science that requires tools and concepts from many fields.

In this text we have attempted to put together lectures by authors from many different disciplines who possess many different types of skills and knowledge but all of whom are interested in transport in natural porous systems. The underlying theme for this text and unifying point for the lectures is the concept of natural hierarchy in porous media. We do not pretend these lectures present a complete, unified, coherent theory of transport in natural porous media. No such theory currently exists. Rather, we hope this series of lectures provides a starting point and catalyst for further research into the complex topic of

DYNAMICS OF FLUIDS IN HIERARCHICAL POROUS MEDIA
ISBN 0-12-200260-1

hierarchical porous media, and further hope the text illustrates the need for multidisciplinary research efforts and that it stimulates deeper thought into this very important and complex topic.

2. Hierarchy

As the title of the text indicates, the porous media which are of interest to us possess some sort of hierarchy. Hence, before we proceed, it is wise to clarify what we mean by such a system. The term hierarchy may be used in several different contexts. For example, it may imply either functional or structural hierarchy. The meaning of structural hierarchy should be clear; a porous medium possesses structural hierarchy if it can be decomposed into successively nested, interacting, physical subunits. On the other hand, functional hierarchy relates to hierarchical transport processes within a porous medium. Thus, functional deals with processes whereas structural deals with physical units. More often than not the two types of hierarchy, structural and functional, are intimately connected.

It is also possible to distinguish two types of structural (functional) hierarchy, discrete and continuous. In a discrete hierarchical medium there are a finite number of nested structural subunits or functional subprocesses, whereas in a continuous hierarchical medium the number of subunits (subprocesses) goes to infinity and no clear cut decomposition is in evidence. In a discrete hierarchical medium, every hierarchical level has a specific state space of variables of relevance to that level. The set of variables at a specific level consists of statistical moments, convolutions, or more general filters of the variables on the level immediately below. Hence, as we successively move up the discrete hierarchical system, we successively reduce the number of degrees of freedom which control the system, i.e. only portions of the information possessed by the system at the lowest level are passed to the next lowest and so on. The loss of information associated with the movement up the hierarchy is manifested in the appearance of new constitutive variables at each scale which are derived from local or non-local theories Provided the windows associated with experimental apparatus correspond to the natural discrete hierarchy of scales in a porous medium, we may decouple the scales and hence study information propagation over scales.

Continuous hierarchy is associated with extensive 'self-organization' which is evident in long-range correlation among dynamic or static properties. As an example, fractal porous media are extremely self-organized in that many static and dynamic properties have infinite correlation lengths. Information transfer in such systems can be compressed (scaled), but not decomposed as in discrete hierarchical systems. Constitutive variables in such a system are, in general, wave-vector and frequency dependent and result from non-local theories.

3. Local and non-local theories

In the previous section, local and non-local theories were mentioned. It is important for the reader to comprehend fully what is meant by this terminology. But before proceeding to define this terminology, a brief discussion of constitutive variables is in order.

In essence, as we move up the scales in a hierarchical system, constitutive variables are place holders in the hierarchical equations. They replace lost information associated with the reduction in the number of useful degrees of freedom. It is important to note that we never measure constitutive variables, we always compute them as a function of measured field variables, or rates of change of field variables on the scale(s) at which the field variables are defined. That is, constitutive variables result from the solution to an inverse problem. Basically, constitutive variables are parameters that show our ignorance of, or lack of need for, detailed information on the many degrees of freedom of relevance to lower scales within the system.

A primary goal of theories of transport in hierarchical media is to develop functional forms for constitutive variables at each scale in a discrete hierarchy, or over a continuum of scales in a continuous hierarchy, and to know the regions in space–time that are important to their definition. If the constitutive variable depends on what is happening at a point (here, as will be discussed in the next section, a 'point' has physical significance only with respect to the window and scale of resolution in the measurement process) in space–time or a very small neighbourhood of the point, the variable is said to be local and derived from a local theory. Such constitutive variables may still vary as a function of space and time.

On the other hand, if information is needed to define the constitutive variables from regions of space and time distinct from a neighbourhood of the space–time point where evaluation of the variables is to be made, then the theories and constitutive variables are said to be non-local in character. Non-local theories may result in space–time dependent constitutive variables that have two or more additional degrees of freedom. Possible manifestations of two additional degrees of freedom are lags in space and time. For example, a non-local dispersion tensor might be of the form $\mathbf{D}(\mathbf{x}, t, \mathbf{x} - \mathbf{x}_0, t - t_0)$ where $(\mathbf{x}_0, t_0)$ is the position and time of injection of a contaminant into the porous medium. A more general concept of non-locality, which we will not delve into here, allows the constitutive parameters to be functionals. As is evident from the porous media literature, the concept of non-locality is a point of considerable confusion. For example, it is not well understood that when one says a constitutive variable is wave-vector and frequency dependent, what is meant is that the wave-vector $\mathbf{k}$ dual to $\mathbf{x} - \mathbf{x}_0$ and the frequency ω dual to $t - t_0$ affect the constitutive parameter. One is not referring here to a wave-vector dual to position $(\mathbf{x})$ and a frequency dual to time (t). This is a very important point and a lack of appreciation for it has led to considerable misunderstanding of the literature.

4. Measurement

4.1 Sampling windows and instrument resolution

If the scales of measurement associated with experimental methods are not accounted for in theories of transport in hierarchical porous media, then the theories are metaphysical and of little practical consequence. This is because a change in the window of observation, or the resolution of observation, may radically alter the measured value of the field variable being examined—due to spatial or temporal heterogeneity. To illustrate this point, consider the following Gedanken experiment.

Suppose we are trying to determine the areal porosity of a natural porous formation; i.e. we are attempting to determine the ratio of void area to total area at a point x. Consider our experimental apparatus to be a camera with a fixed window of observation, but with variable resolution. For example, suppose the camera has a square window of area l^2 (where l is a unit of length) and a resolution of $(l/n)^2$ where n is an integer. For simplicity, assume that if the void space intersects a pixel of area $(l/n)^2$, then the pixel will be considered totally void, otherwise it will be considered as solid (this may be called a pixel weighting scheme). The porosity is the sum of the areas of the pixels containing void space divided by the window area l^2. Clearly, for fixed window size, the porosity is a function of the resolution and the pixel weighting scheme. If the porosity goes to zero as the number of pixels (n^2) goes to infinity (infinite resolution), then the void space is generally fractal. On the other hand, if the porosity goes to one, then the solid is generally fractal. Otherwise, the porosity will go monotonically to some constant which lies between zero and one. Thus, by letting the resolution become progressively greater, we can check to see if the medium displays fractal characteristics. That is, whether or not a porous medium's void space appears fractal is determined by adjusting instrument resolution. It should also be noted that the fractal dimension of a property of the porous medium may change with position or window size. Fractal porous media are discussed in more detail in Chapters XI and XIII.

Consider another Gedanken experiment in which the camera has fixed resolution, l_0^2, but l^2 (the window size) is allowed to vary. Maintain a common focal point and let l increase from l_0. If a plot of porosity against l^2 has one or more plateaux (note that the porosity is now varying with window size rather than window resolution), we say the porous medium has one or more natural scales (asymptotic limits, REAs, etc.) with respect to a given resolution of measurement, i.e. it possesses a discrete hierarchy. Such media are discussed in Chapters II–IV, V–X, and XIV–XVI. On the other hand, if there are no plateaux (or in regions between plateaux), we say there are evolving heterogeneities with respect to the resolution of the instrument. Media with evolving heterogeneities are discussed in Chapters XI, XIII–XV, and XVII. To ask if a porous medium possesses one or more natural scales at a point is a 'local' question. This is because to answer the question we must look at nested neighbourhoods at the point. On the other hand, to ask if a portion (e.g. the set consisting of the instrument window) of a porous medium is fractal is a global question; that is, rather than looking at nested neighbourhoods, one looks at refined partitions of the set. Thus, comparing the concept of an REA (REV) to the concept of a fractal is like comparing apples to oranges. It should be clear that a porous medium may possess both fractal and REA (REV) characteristics, one or the other, or neither. This point has not been well understood in the literature on porous media.

As mentioned earlier, the pixel weighting scheme for a given instrument need not be constant or uniform. In such cases, the instrument window is represented by a weight function, an abstract measure, or a generalized function. Methods for obtaining such weighting functions are discussed in Chapters II, III, and XVII.

4.2 Heterogeneity

From the discussion in the previous section, it should be apparent that in a hierarchical porous formation, heterogeneity will be a function of instrument window size and resolution. 'Explicit' heterogeneity is also a function of sampling location and sampling frequency. Explicit heterogeneity is an operational concept wherein heterogeneity is observed via multiple experimental observations made over a field with a specific instrument having a fixed window size and scale of resolution. This should be contrasted with 'implicit' heterogeneity. Implicit heterogeneity is heterogeneity which exists, but may not be measurable with an instrument having a fixed window size and resolution, or because of the sample locations chosen; i.e. heterogeneity may be implicitly present but not explicitly observed. Implicit heterogeneity exists because of scale constraints on experimental equipment. Implicit heterogeneity is generally accounted for with stochastic theories (see Chapters III, IX, XIII–XV, and XVII).

5. Outline of the text

Chapters II and III deal with experimental methods related to multiscale porous media. Asymptotic dispersion coefficients and tracer tests and their relation to the measurement process are the focus of Chapter II. Hydraulic conductivities, impeller meter tests, and tracer tests over different scales are the focus of Chapter III.

There are at least four basically equivalent methods for deriving transport equations for porous media which possess discrete hierarchy: Markov theories (Brownian dynamics), space–time averaging theories, homogenization, and classical continuum (mixture) theories. All involve the same basic physical assumptions and the existence of discrete scale hierarchies. Markovian theories are covered in Chapter IV, averaging theories are covered in Chapters V and VI, and homogenization is discussed in Chapters VII and VIII. The theory of mixtures is not covered in this text, but an excellent chapter (which is limited to two-scale hierarchy) on the topic by Ray Bowen may be found in Eringen (1976a).

Percolation theory is a pore-scale model of particular usefulness in deriving laboratory-scale constitutive equations. Percolation models are discussed in Chapter XVI.

When a porous medium possesses evolving heterogeneities (continuous hierarchy) theories other than those cited above should be applied. These may be either stochastic or deterministic. In general, theories for dealing with continuous hierarchy are much less developed than those for discrete hierarchy. As mentioned earlier, constitutive variables for systems possessing continuous hierarchy will be frequency and wave-vector dependent, consequently constitutive variables are often kernel functions in integrodifferential equations (see Chapter XVII). Stochastic theories of transport in media with evolving heterogeneities are discussed in Chapters XIII–XV, and XVII. A deterministic method which is based on fractal concepts, for systems with evolving heterogeneities is discussed in Chapter XI. An alternative deterministic method is non-local continuum mechanics. This method is not discussed in this text; however, an excellent discussion by D. Edelin can be found in Eringen (1976b).

Information compression via scaling is discussed periodically throughout the text. Of special reference in this regard are Chapters XI–XIV. Chapter XII is devoted solely to the use of continuous groups (Lie groups) for developing admissible scaling relationships for Richards' equation of unsaturated water transport.

Also included in the text are a number of different numerical techniques for dealing with problems which have hierarchical structure. Because information must be propagated between various scales, numerical techniques for hierarchical media must be very robust, i.e. they must be capable of developing information on fast and slow time scales simultaneously. Schemes of special importance in this regard are operator splitting and multigrid methods. In many cases, special numerical algorithms must be derived for each specific problem under investigation. Chapters which concentrate heavily on numerical methods include Chapters VII, IX, X, and XIII.

References

Eringen, A.C. (1976a). 'Continuum Physics', Vol. 3. Academic Press, New York.
Eringen, A.C. (1976b). 'Continuum Physics', Vol. 4. Academic Press, New York.

Chapter II

Field Studies of Dispersion: Radioactive Tracer Experiments on Scales of 20, 40 and 260 Metres

G.L. Moltyaner

Environmental Research Branch, Chalk River Nuclear Laboratories, Chalk River, Ontario, Canada K0J 1P0

1. Introduction

The varying capacity of the subsurface environment for storing and diluting contaminants is a matter of practical importance in operations or events that may result in the release or escape of such materials. Problems arising in the field have been among the greatest stimuli to experimental studies of dispersion processes in the subsurface environment. Undoubtedly, the most rewarding experimental studies of dispersion in groundwater systems have been those in which some easily detected tracer is introduced into the system and the concentrations occurring at various times and positions down-gradient from the source are measured by a precise chemical or physical technique. The tracer methods have the obvious advantage of enabling the dimensions and shape of a dispersing plume to be directly measured and the dispersion of the introduced material to be accurately described. They also have the attraction of being most directly relevant to the assessment of hazards arising from groundwater contamination.

A significant contribution to the understanding of dispersion processes has resulted from the application of radioactive tracers in field experimental studies (see, for instance, Killey and Moltyaner, 1988). Tritium is the most commonly used radioactive tracer. Although its chemical characteristics make it nearly ideal, its emissions of low-energy beta-radiation cannot be measured adequately *in situ*. The chemical analysis of samples taken is performed in the laboratory.

DYNAMICS OF FLUIDS IN HIERARCHICAL POROUS MEDIA
ISBN 0-12-200260-1

The tritium measurement process is referred to as the 'mixing-cup' measurement process (Moltyaner, 1987). Probably the second in the list of commonly used tracers is radioiodine. Radioiodine has been used extensively for tracing groundwater for relatively short transit times. It is a high-energy gamma-emitting tracer and is detectable at low concentration *in situ*; gamma-radiation detection with the aid of dry observation wells is referred to as the 'through-the-wall' measurement process (Moltyaner, 1987).

Only discrete samples of concentration are obtainable at monitoring points using the 'mixing-cup' measurement process but continuous samples of concentration are obtainable using the 'through-the-wall' measurement process. For this reason tracer techniques with the through-the-wall measurement device are more profitable; continuous through-the-wall measurements of tracer distribution are ideal for analysing the dispersion process. This chapter details this analysis by describing three field experiments that were designed specifically for this purpose. The experiments were carried out with tritium and radioiodine at a pre-selected site at Chalk River, Ontario (Killey and Moltyaner, 1988).

The dispersion of the tracer results from fluctuations in groundwater velocity. It is on these fluctuations of flow velocity, which vary slowly in space, that the scale of the tracer dispersion depends. A brief consideration of the typical tracer plume extending down-gradient from the source will serve to introduce the range of scales over which the velocity varies.

The spreading of the plume can be seen to occur in two distinct ways. First, any given section of the plume spreads as it travels, and this spreading action is clearly a result of differences of velocity within a relatively small volume (say, within an averaging volume associated with the measurement device) of the porous system occupied by the tracer. In addition, the plume as a whole possesses irregularities arising from variations in the bodily displacement of its component sections. These variations in bodily displacement are evidently associated with variations in the mean velocity of the larger volumes of groundwater (say, associated with individual layers) that transport sections of plume without exerting any important effect on their rate of growth until a larger or smaller scale heterogeneity is encountered. The combined effect of the two processes may be described at the scale of the whole plume by taking the simplest view that the mean flow is constant in velocity and by defining the groundwater flow velocity as the sum of mean velocity (plume-scale velocity) and fluctuations from this mean value. The effects of velocity fluctuations with a wide range of scales (heterogeneities with a wide range of sizes) are, thereby, directly demonstrated.

Most of our task of analysing the dispersion process observed at the Twin Lake aquifer consists of defining and describing these fluctuations and evaluating their action in redistributing and dispersing the injected tracers with the aid of the advection-dispersion model.

2. Theoretical treatment of dispersion

In general, in quantifying transport of a radioactive tracer through a porous medium one has to deal with the behaviour of a combined system of particles and fields. To avoid this we shall use a crude model in which the identities of individual particles are neglected and macroscopic-scale motion is considered.

For macroscopic motion we must consider an elementary volume with dimensions larger than the mean free path of a gamma-ray photon, but large enough for most of the particles within it at a time t also to be present at time $t + \mathrm{d}t$; in this consideration the particles leave and enter the elementary volume but the identity of an individual element persists. In fluid dynamics, this persistence is assured by collisions; particles are prevented from leaving a fluid element by collisions with neighbouring particles. This implies that the characteristic dimension of the element is larger than the mean free path of a fluid molecule. For a porous medium irradiated by a high-energy radioactive tracer this argument is less clear-cut, since the nature of scattering is different from that in fluid dynamics; scattering is accomplished mostly by different paths randomly selected by different tracer particles. Nevertheless, an essential requirement for applying infinitesimal calculus, without introducing any error that may be picked up by the measuring device (radiation detector) at the macroscopic scale, is still that the characteristic dimension of an elementary volume of porous medium is larger than the mean free path of the gamma-ray energy photon (during the measurement process the photons with the highest probability of reaching the detector are emitted by the tracer particles situated within a sphere of one mean-free-path radius: Moltyaner, 1989). At the local scale of continuum representation, at the scale of approximately three mean free paths, the tracer concentration distribution may be described by the advection-dispersion equation. In what follows the use of this equation and related issues are discussed.

2.1 Microscopic and macroscopic points of view

In principle, the transport of an ideal tracer in a porous medium can be described by the appropriate microscopic transport equation. However, because the geometry of the interface boundaries within a porous medium is too complicated to permit the direct solution of the microscopic transport equation, a macroscopic description of the transport problem is introduced. Generally speaking, we regard a macroscopic transport problem as one in which the details of the tracer distribution at the microscopic scale are inaccessible to us. There are different approaches to solving the problem. For example, macroscopic variables used in the macroscopic transport equations may be obtained by averaging the corresponding microscopic variables over a representative elementary volume (Bear, 1972). As an alternative, the weighting function method may be employed. Baveye and Sposito (1984) suggested defining the macroscopic variable by averaging microscopic values by means of the weighting function associated with a sampling device. They also suggested using a 'table-type' weighting function to define all macroscopic field variables in order to derive the macroscopic transport equation in a traditional manner. Cushman (1984) has shown that the weighting function can be regarded as a filter function of an instrument and has raised the question of how to determine the scale of the volume over which averaging should be performed; in fact, each macroscopic problem, due to variability in aquifer properties, has its own appropriate lower limit of averaging scale and this sets the scale of support for the weighting function to be used in the analysis.

To answer the question, we take the attitude that the averaged quantity can be always characterized by a finite scale of length (see Cushman, 1984, 1986). In a fluid mechanical description of a microscopic continuum the scale is the mean free path of a fluid molecule; the fluid continuum is introduced because of our inability to describe mathematically complicated trajectories of individual molecules. The fluid continuum description breaks down when the mean free path becomes comparable with the scale of support of the measurement device.

In porous media the macroscopic continuum description may be introduced when: (1) the size of the measurement device is much larger than the size of a mean pore; (2) it includes sufficient numbers of pores to make the statistical averaging required in the continuum concept meaningful (for details, see Cushman, 1983); and (3) it is small enough compared with the dimensions of the overall system to make the averaging physically interesting. If, therefore, we are to define the macroscopic properties in a manner similar to that in fluid mechanics, we should define a finite scale of length satisfying the stated conditions. This can be done with the aid of a high-energy gamma-emitting tracer injected into a system. We observe that gamma-ray photons move about in the porous system in a complicated path as the result of nuclear collisions. The extent to which photons interact with a porous medium is described in terms of quantities known as cross-sections (Lamarsh, 1983) which include both fluids and solids. By tradition, a related property known as a linear attenuation coefficient is used to characterize the interaction process. The linear attenuation coefficient defines the probability per unit path that a gamma-ray will have a collision in the medium. The average distance that a gamma-ray photon moves between collisions is called the mean free path and it is equal to the reciprocal of the linear attenuation coefficient. An essential requirement for applying infinitesimal calculus is that the characteristic dimension of an elementary volume of a porous medium is larger than a mean free path of the gamma-ray photon. This quantity is fundamental in describing the macroscopic motion of tracer particles in a porous medium. In describing the macroscopic motion, we must consider an elementary volume with dimensions larger than the mean free path of a gamma-ray photon. In this way the identity of the individual particle is neglected and only motion of the elementary volume is taken into account. Laboratory or field measurements of the tracer concentration performed at a scale larger than the mean free path may be treated as a continuous function in space which varies smoothly enough to be analysed by differential calculus. The measurements represent averages taken over many pores.

2.2 Discussion of spatial averaging

The conventional definition of a spatial macroscopic average of a microscopic function $A(x, t)$ with respect to a weighting function $g(x)$ is defined as (Anderson and Jackson, 1967; Baveye and Sposito, 1984; de Marsily, 1986) the convolution integral

$$B(x, t) = \int A(|x - x'|)g(x')\,d^3x' \tag{2.1}$$

where $d^3x' = dx'_1\,dx'_2\,dx'_3$ is a three-dimensional volume element, and x stands for the coordinates in space (x_1, x_2, x_3) and denotes a point where the macroscopic field is being evaluated (see Chapter I for more detail). The notation $|x - x'|$ has been used to indicate that only the interval between x and x' matters. With this in mind, Eq. 2.1 indicates a way to find the spatial average of A, as a sum of contributions at the point x from each point x' in the interval $|x - x'|$. In Eq. 2.1 $g(x)$ is a real positive function, non-zero in some neighbourhood of $x = 0$. It is normalized to unity and has dimensions large compared with the size of a large pore. Examples of the weighting function for a spherical averaging volume with an abrupt discontinuity at the surface of the sphere or a smooth weighting function with an unbounded support, usually exemplified by the Gaussian, are given elsewhere in this text.

In general, Eq. 2.1 does not specify the weighting function uniquely. In order to determine the macroscopic variables uniquely one has to specify the form chosen for the weighting function. Baveye and Sposito (1984) suggested the use of a 'table'-type function to determine uniquely the macroscopic mean variables. They presented an example of the 'table'-type function associated with an ideal cylindrical soil sampler. Moltyaner (1989) used the same type of weighting function for the description of a 'mixing-cup' solution sampler used in three natural-gradient dispersion experiments performed at Chalk River, Ontario. The essential feature of the mixing-cup device is that the tracer solution is extracted from the aquifer system during the sampling procedure and the chemical analysis of the collected sample is performed in the laboratory. The procedure is used for conventional tracers and/or radioactive tracers with low energy. In the case of a radioactive tracer with high energy, measurements of the tracer concentration are performed *in situ*. A measurement device used for sampling tracer concentration *in situ*, which includes a radiation detector and a fully penetrating dry observation well, is referred to as the 'through-the-wall' measurement device (Moltyaner 1987). The weighting function associated with the 'through-the-wall' measurement device is defined by considering the process of interaction of gamma-radiation emitted by a point source with aquifer materials and is found to follow an exponential attenuation law. The weighting function associated with a point source is introduced in the following form

$$g(x) = \mu \exp(-\mu|x|) \quad \forall x \tag{2.2}$$

and it is normalized to unity.

In field situations, the radiation source has an extended counting geometry. Assuming a uniform activity distribution in the interior of the source, a spherical averaging volume of $r = E\mu^{-1}$, where E is a given constant, can be used. The weighting function, therefore, is defined as (Moltyaner, 1989)

$$g(x) = \begin{cases} 3/(4\pi r^3), & |x| \leqslant r \\ 0, & |x| > r \end{cases} \tag{2.3a}$$

The weighting function associated with the mixing-cup measurement device is defined in the form

$$g(x) = \begin{cases} (\pi r^2 H)^{-1}, & |x| \in V \\ 0, & |x| \notin V \end{cases} \tag{2.3b}$$

where V is assumed to be a cylindrical volume element with height H and radius r containing solution extracted for measurement of tracer concentration. As a result of the measurement process all the microscopic fluctuations are averaged out, giving smooth and slowly varying macroscopic properties defined by (Eq. 2.1) with $g(x)$ defined by (Eq. 2.3). It is important to note here that although the weighting process is similar for both types of measurement devices, there is a difference in mean values of tracer concentration. In the case of the mixing-cup measurements the measured concentration represents an average over the liquid phase of the averaging volume (the microscopic concentration is discontinuous at the interfaces). In order to make the convolution procedure consistent with the measurement procedure the mean concentration value should be corrected by multiplying it by porosity. In the case of 'through-the-wall' measurements the measured concentration represents a value averaged over the entire domain (the high-energy photons penetrate both liquid and solid phases). The convolution procedure results in concentration values consistent with the measured ones.

In order to determine the size of the weighting function to be used for analysis and thus the size of the sampling device to be used in a field experiment, Moltyaner (1989) refers to a laboratory experiment using a point source of gamma-radiation. The size of the region over which the spatial averaging effectively takes place during the field measurement procedure is determined by the mean free path of the gamma-ray photon. Upon establishing the size, and thus the characteristic dimension of the porous medium continuum, it becomes immaterial what type of measurement device is used provided that it has an appropriate 'support'. To delimit the domain where we expect the introduced continuum description of a porous medium to work we observe that the description breaks down when the mean free path becomes comparable with the size of a large pore. The mean free path establishes an absolute lower limit to the macroscopic domain. The upper limit is determined by the properties of the exponential weighting function; the averaging effect will be virtually zero at three to five mean free paths. This sets the upper limit for the size of the measurement device to be used in a field experiment.

2.3 The dispersion equation

In a subsequent chapter Whitaker derives the dispersion equation in the form

$$\frac{\partial c}{\partial t} = \frac{\partial}{\partial x_i}\left(D_{ij}\frac{\partial c}{\partial x_j}\right) - v_i\frac{\partial c}{\partial x_i}, \quad D_{ij} = D^* + D'_{ij} \tag{2.4}$$

Here, all macroscopic variables are defined in terms of the averaging presented in the previous section. While assuming D_{ij} constant within the characteristic length scale of the averaging volume, Eq. 2.4 still allows for variation of D_{ij} with x. The reasoning supporting the assumption of gradient-dependent mixing used in Eq. 2.4 is based on the analogy with diffusion phenomena. The combination of the changing pore velocity coupled with molecular diffusion gives rise to a mixing process at the macroscopic scale which is characterized by the dispersion tensor, D_{ij}, equal to the sum of the molecular diffusion D^* and mechanical dispersion tensors D'_{ij}. If the dispersion tensor is expressed in its principal direction of anisotropy, it is limited to three components: D_{L} in the direction of flow and D_{T} in two perpendicular directions. The anisotropy of the dispersion tensor stems from the fact that the scattering of tracer particles is larger in the direction of flow velocity than in the transverse direction. In the calculation of the mechanical dispersion coefficient the following relations are generally used

$$D_{\mathrm{L}} = \alpha_{\mathrm{L}} v \tag{2.5}$$

$$D_{\mathrm{T}} - \alpha_{\mathrm{T}} v \tag{2.6}$$

where α_L and α_T are the longitudinal and transverse dispersivities. Based on our experimental data we express the dispersivity as

$$\alpha_L = \frac{\mu^{-1}}{3} \tag{2.7a}$$

The laboratory experiment showed that $\mu^{-1} = 2.3$ cm (Moltyaner, 1989) hence, $\alpha_L = 0.8$ cm which compares well with the dispersivities calculated from the field tests. According to the experimental results, we shall use an additional multiple of $1/10$ for calculating the transverse dispersivity

$$\alpha_T \simeq \frac{\alpha_L}{10} \tag{2.7b}$$

The three-dimensional advection–dispersion equation (Eq 2.4) is generally used for practical applications. In some problems, however, the one-dimensional approximation is quite sufficient. For the one-dimensional case equation Eq. 2.4 reduces to

$$\frac{\partial c}{\partial t} = \frac{\partial}{\partial x}\left(D_L \frac{\partial c}{\partial x}\right) - v\frac{\partial c}{\partial x} \tag{2.8a}$$

For the transport of a conservative tracer in field experiments, the three-dimensional form of the advection–dispersion equation with constant coefficients is often useful for analysis

$$\frac{\partial c}{\partial t} = D_L \frac{\partial^2 c}{\partial x^2} + D_{TL} \frac{\partial^2 c}{\partial y^2} + D_{TV} \frac{\partial^2 c}{\partial z^2} - v\frac{\partial c}{\partial x} \tag{2.8b}$$

where D_L, D_{TL}, D_{TV} are the longitudinal, lateral–transverse and vertical–transverse hydrodynamic dispersion coefficients.

2.4 Physical aspects of dispersivity at the local and plume scales

The concept of dispersivity as introduced in the earlier part of this chapter permits a phenomenological description of transport through porous media. An actual understanding of the phenomena, however, can be obtained only if the concept of dispersivity can be reduced to more fundamental physical properties. It seems reasonable to suppose that the property of dispersivity can be linked with other fundamental properties of a porous medium such as porosity, hydraulic conductivity and/or grain size distribution since all such properties are the manifestation of the geometrical structure of the medium.

One approach to linking the various properties is to establish empirical correlations between them. Very limited attempts have been made to establish whether any relationships exist between the various properties of porous media. Harleman and Rumer (1963) have found, based on the analysis of laboratory results, that the longitudinal dispersivity is proportional to the square root of hydraulic conductivity. This relationship was used recently by Gelhar *et al.* (1979). Our attempts to verify this relationship using results of field experiments were not successful; no correlation was observed between longitudinal dispersivity and hydraulic conductivity (Taylor *et al.*, 1987).

In view of the fact that a simple relationship between the dispersivity and various properties of the porous medium does not seem to exist, attempts have been made to attach a physical significance to dispersivity through theoretical considerations. The simplest way to establish correlations theoretically is to construct a mathematically tractable model of the porous medium. If such a model can be constructed it can be used to calculate dispersivity. Bear (1972) gives an account of various models which have been put forward to calculate the dispersivity of a porous medium. In the commonly accepted classical dispersion theory the porous medium is regarded as a random distribution in space of the small interconnected channels forming the pathways through the pores of the porous medium. According to the theory, the longitudinal and transverse dispersion coefficients are proportional to the average flow velocity (Eqs 2.5 and 2.6). The coefficient of proportionality, the dispersivity, is approximately an order of magnitude smaller in the transverse direction than in the longitudinal direction. The dispersion coefficients are measured experimentally and dispersivities are then calculated on the basis of Eq. 2.6. In the present work, we use a phenomenological approach according to which the Fickian type of law is assumed to be valid.

As was discussed in previous sections, the dispersion coefficient is then represented as a sum of molecular diffusion and mechanical dispersion coefficients and we may use the simple expression for the dispersion coefficient (Eq. 2.6) in the treatment of a radiotracer transport in groundwater. The coefficient of mechanical dispersion is proportional to the mean flow velocity and dispersivity (Eq. 2.5). The latter in turn is taken to be proportional to the mean free path of the gamma-ray photon. The mean free path is used here as a definition of the 'average size of dispersion' and is referred to as the scale of dispersion or dispersion length. The quantity provides a physically relevant specification for describing a scattering or mixing quality of the porous medium.

Up to this point, in the discussion of the dispersion phenomenon, we have considered only the local-scale dispersion as distinct from the plume-scale dispersion. Both local- and plume-scale dispersion concerns the spreading of tracer particles but a significant difference exists in the size of the particles and the type of heterogeneity that causes the spreading. The term local-scale dispersion is applied to processes occurring at the pore scale, that is, to processes of diffusion and pore-scale advection. The spreading out is accomplished by different paths randomly selected by the different particles. Randomness here is affected by the presence of two distinct phases, the solid and the fluid. In contrast, at the plume scale, the element of randomness exist in the variability of local-scale velocity across the section perpendicular to the mean direction of flow. The instantaneous distribution of the local-scale particles across the plume is affected at different horizontal levels by different velocities because of the layered structure of sedimentary sequences. If it is possible to assume that at different horizontal levels different parts of the plume are subject to the action of the same scale of dispersion, and that the instantaneous distribution of the particles across the plume changes randomly with time, the plume-scale longitudinal dispersion may be qualified as Fickian. Physically, it is evident that in a realistic model of the dispersion process the local-scale particles cannot be regarded as travelling consistently along a single line in the direction of mean flow due to the discontinuous nature of sedimentary layers. The random mechanism of scattering may be a reasonable assumption at distances significantly larger than the mean extent of the layers. In general, the local-scale velocity variations are far more effective at spreading out the plume than are the pore-scale velocity variations.

2.5 Models of transport process

A conceptual model is necessarily a simplification of the real physical process but should incorporate the important features of the physical system. The conceptual model adopted in this chapter to describe the transport process at the Twin Lake aquifer is represented by Eq. 2.8. It is assumed that the migration of the tracer slug can be described as the transport of a parallelepiped instantaneously injected into the aquifer. The corresponding solution to the Eq. 2.8a is (Carslaw and Jaeger, 1959, p. 56; Domenico and Robbins, 1984; Moltyaner, 1987):

$$c(x, y, z, t) = \frac{c_0}{8}\left[\operatorname{erf}\frac{(x + x_0/2 - vt)}{2(D_L t)^{1/2}} - \operatorname{erf}\frac{(x + x_0/2 - vt)}{2(D_L t)^{1/2}}\right]$$

$$\times\left[\operatorname{erf}\frac{(y + y_0/2)}{2(D_{TL} t)^{1/2}} - \operatorname{erf}\frac{(y - y_0/2)}{2(D_{TL} t)^{1/2}}\right]\left[\operatorname{erf}\frac{(z + z_0/2)}{2(D_{TV} t)^{1/2}} - \operatorname{erf}\frac{(z - z_0/2)}{2(D_{TV} t)^{1/2}}\right] \quad (2.9)$$

where c_0 is the tracer concentration at $t = 0$; and x_0, y_0 and z_0 are the initial dimensions of the parallelepiped. The one-dimensional form of Eq. 2.9 is

$$c(x, y) = \frac{c_0}{8}\left[\operatorname{erf}\frac{(x + x_0/2 - vt)}{2(D_L t)^{1/2}} - \operatorname{erf}\frac{(x - x_0/2 - vt)}{2(D_L t)^{1/2}}\right] \tag{2.10}$$

Equation 2.10 was used to estimate longitudinal dispersivity of the medium at the local scale. Equation 2.9 was used to estimate transverse dispersivity. At large distances from the source the relationship

$$D_L = \frac{v^2}{2}\sigma^2 t \tag{2.11}$$

can be used, where σ^2 is the variance of the concentration distribution.

For the instantaneously introduced plane source

$$c(x, t) = \frac{c_0}{2(\pi D_L t)^{1/2}} \exp\left[-(x - vt)^2/4D_L t\right] \tag{2.12}$$

3. Experimental studies of dispersion

The dispersion of a radioactive tracer artificially introduced into flowing groundwater is described by the dimensions and shapes of the distributions which are developed down-gradient from the source. In this section are discussed the details of these features, as revealed by numerous laboratory and 20-, 40- and 260-m field tracer experiments which were designed specially for this purpose and carried out using radioactive tracers at Chalk River, Ontario.

3.1 Principles of technique and analysis

Only over the last decade have larger-scale natural gradient dispersion tests been used to quantify dispersive properties of an aquifer of interest (Mackay *et al.*, 1986; Garabedian, 1987; Killey and Moltyaner, 1988). In a typical natural gradient dispersion experiment, a pulse of tracer-labelled groundwater is introduced into the system and its evolution in time is monitored in an array of observation wells situated down-gradient from the injection well. The main interest is to assess *in situ* the effects of aquifer heterogeneities on the dispersion process and to quantify the effect at the scale of support of the measurement device used in a particular experiment. Most of the field studies of the dispersion process performed using ideal tracers fall into one or another of two groups in which different measurement techniques are adopted for describing and measuring the effect of dispersion. These groups are associated with the through-the-wall and mixing-cup measurements of the concentration of a tracer material. Undoubtedly, the natural-gradient tracer experiments in which a pulse of tracer-labelled groundwater is monitored without any disturbance is most appropriate to the assessment of the hazards related to groundwater contamination. By sampling the concentrations occurring at various times and positions down-gradient from the injection point using such a monitoring technique (which is referred to as the 'through-the-wall' measurement technique) one follows the evolution of the tracer distribution in time and space in a most precise manner. Dry monitoring wells used for sampling the concentration field fully penetrate the aquifer and the measurements are performed using a radiation detector placed inside the dry well. The most frequently used measurement technique is the mixing-cup technique in which wells having different lengths of screened intervals are used to sample the concentration field by means of a pumping procedure. The chemical analysis of a collected sample performed in laboratory conditions leads to an average estimate of the mass of tracer, M, in a given volume of tracer solution, Qt, where Q is the known rate at which solution is drawn and t is the duration of the sampling process. In the case of through-the-wall sampling the effective value of Q is given by ua, where u is the flow velocity relative to the detector and a is the area normal to flow. Nearly all field tracer experiments designed to quantify the dispersion process in an aquifer make use of the mixing-cup measurement technique. Concentration measurements collected with the aid of this technique are used to describe the spread of a tracer plume. To measure this spread and thus to evaluate the dispersion coefficients a large number of closely spaced sampling points with the smallest possible size of the screened interval is required. Complete description of the spatial distribution of tracer material released at a 'point' requires frequent sampling by a team of observers at boreholes down-gradient from the source. Construction of a dense monitoring network requires significant financial and technical efforts, and in practice the monitoring is performed on a coarse spatial scale. The samples represent mixtures of waters taken from screened intervals of wells, and the number of such screened intervals is relatively small. In such cases, in order to obtain an adequate description of the rate at which the tracer breaks through, it is necessary to sample each screened interval with a sufficient frequency. It seems likely, based on the observations from the 260-m tracer test, that the process of dispersion can be studied in this way beyond a few hundreds of metres of travel. The extension of range should require merely an appropriate increase in the amount of tracer introduced.

A more convenient and accurate description of the dispersion process and a more precise description of the plume spread is obtained using high-energy radioactive tracers. In this case continuous vertical profiles of tracer distribution can be obtained using a novel through-the-wall measurement technique in conjunction with a gamma-emitting tracer (Killey and Moltyaner, 1988). The profiles provide an opportunity to draw an outline of the tracer plume and so to specify its shape and size. Furthermore, the vertical profiles provide real information about the vertical boundaries of the plume and describe with a high degree of accuracy those contributions to spreading which arise from the variations of flow velocity. For both types of measurement techniques, tracer migration is monitored in a fixed array of observation wells. It would be preferable to follow the trajectories of individual elements; in a qualitative way, at least, it is obvious that the dispersion of a tracer plume depends on the development of the particular elements of the plume as they are carried along by the flowing groundwater. To date, making true Lagrangian observations does not seem to be feasible for the subsurface environment. Under laboratory conditions visual examination of the development of the size and shape of a tracer plume is possible. The inherent smoothness of dispersion process can be shown by taking a number of photographs and by making necessary simulations to characterize the development of the plume.

3.2 Twin Lake tracer tests

Three natural-gradient tracer tests were performed at the Twin Lake aquifer at Chalk River, Ontario. In the autumn of 1982 a pulse of groundwater labelled with ^{131}I was injected into a well screened over the saturated thickness of the unconfined aquifer. The three-dimensional tracking of ^{131}I was performed over a 20-m distance. In the autumn of 1983, ^{131}I and ^{3}H were injected under conditions identical to those in 1982 but were monitored over 40 m down-gradient from the injection well. In 1987 we changed both the monitoring well and the injection conditions. This time a pulse of ^{3}H-labelled groundwater was injected for approximately three days rather than seven hours and monitored over a 260-m distance of subsurface flow path, and over approximately 30 m in the discharge zone. This time the injection well was screened over 4 m of saturated thickness of the aquifer. A large number of small-scale experiments were performed under laboratory conditions. For these experiments, undisturbed cores of Twin Lake aquifer sands were cut into sections from 8 to 15 cm length and prepared as laboratory columns. ^{3}H and NaCl were used as the ideal tracers. The breakthrough curves obtained from the laboratory columns were used to quantify the dispersion process at the small scale. The field programme which included borehole dilution and single well response tests was complemented by numerous laboratory experiments such as permeameter tests, porosity measurements and grain-size composition analysis described in detail elsewhere (Taylor *et al.*, 1987; Killey and Moltyaner, 1988). All ^{3}H, NaCl and some of the ^{131}I concentration data were measured using 'mixing-cup' measurement devices. The vertical dimension of the screened interval of multilevel samplers used at the Twin Lake site is 7 cm; the transverse spacing between screened intervals is 1 m. The samplers are spaced at 5 m in the direction of flow over the first 50 m of the flow paths; further down-gradient the spacing is irregular from 10 to 20 m. Most of the ^{131}I measurements at the site were performed using the 'through-the-wall' measurement technique. The fully penetrating dry observation wells used for this purpose are spaced longitudinally and laterally in the same manner as the multilevel samplers. The vertical resolution of collected data is 1 cm. The ^{131}I measurements were collected using a NaI detector and a Geiger tube. Most of the measurements were collected using the NaI detector. The effective radius of the sampling device over which the ^{131}I concentration was integrated is approximately 11 cm. The obtained measurements, therefore, refer to a support scale of the order of centimetres. In the course of the monitoring programme, two types of measurements were obtained at the support scale; by the 'through-the-wall' measurement technique a continuous record of the distribution of ^{131}I activity versus time at fixed locations, and repeated continuous records of the distribution of ^{131}I activity versus depth at each monitor location. One type of measurement was obtained using the 'mixing-cup' technique; discrete measurements of ^{131}I and ^{3}H over time at a pre-selected location. Vertical measurements were used for constructing the discrete concentration–time profiles at fixed elevation with a vertical spacing of 1 cm. The discrete concentration–time profiles were used to calculate the local-scale velocity field. The obtained velocities were used to convert the concentration–time data to concentration–distance data. To summarize, three types of measurements were actually obtained in the field: discrete measurements of concentration distribution over time using multilevel samplers, and continuous measurements of concentration distribution over time and over depth using dry-access wells. Two other types of results were derived from these measurements; discrete measurements of concentration distribution over time at fixed locations and instantaneous distribution of concentration in space at fixed times. The observed data were also averaged over the saturated thickness of the aquifer.

3.3 Time-dependent concentration data

As was discussed earlier, the observation of the trajectories of the individual particles is theoretically an ideal way of directly observing the basic processes of dispersion, but making true Lagrangian observations in the subsurface environment does not seem to be possible at this time. The measurement of instantaneous values of concentration simultaneously at a sufficient number of observation points, true Eulerian observations, also seems to be a formidable task. The observations of dispersion, therefore, have tended to rely mainly on discrete sampling of tracer concentrations at fixed points in the monitoring network. At the support scale, owing to the averaging effect of the device, dispersion is considered as occurring in a locally homogeneous and isotropic system. This implies that the local-scale advection–dispersion equation may be used for determining 'effective' properties of the aquifer (with the aid of discrete breakthrough curves) over regions associated with the monitoring network and separated by well-defined boundaries. The net effect is to view the aquifer as a set of subregions, each homogeneous, that behave in a manner analogous to the actual system. The accuracy of this approach depends entirely upon the degree of heterogeneity in the system and the density of monitoring network used to characterize the effect of heterogeneity on dispersive properties of the system. It is recognized that complete characterization of the system heterogeneity is impossible since the detail of any real process would not be sufficiently well known. However, significant progress in characterizing the system heterogeneity and spread of the plume has been made using the 'through-the-wall' measurement technique in conjunction with a gamma-emitting tracer; by scanning a dry observation well at 1 cm vertical intervals and by constructing discrete breakthrough curves at 1 cm intervals, the dispersive properties of the aquifer may be characterized with 1 cm spacing. Moreover, the vertical scans allow one to position the detector at the elevation of first arrival of the tracer at a borehole located along the main direction of flow. These observations possess the obvious advantage of enabling the distributions within a dispersing plume to be directly explored using continuous breakthrough curves, a feature which is not provided by the 'mixing-cup' measurement technique. A typical set of time–concentration observations obtained using the 'through-the-wall' measurement device in the 40-m test is shown in Fig. 1A. Notice the decrease in concentration and the increase in time between the rising and falling limbs of the curves as the tracer plume moves downstream. The same observation can be made with regard to breakthrough curves obtained at a monitoring well at different elevations due to the difference in the arrival time of the tracer (Fig. 1B). The continuous measurements of the concentration distribution using the 'through-the-wall' measurement technique showed that the basic shape of breakthrough curves observed at various distances down-gradient from the injection well seemed to conform to the Gaussian form; it should be mentioned that the theoretical treatment of Fickian type dispersion leads directly to this form of distribution. These observations may be regarded as supporting the adoption of a Gaussian distribution when the distance down-gradient from the source is large compared with the scale of dispersion. However, when the initial dimensions of injected pulse are large compared with the scale of dispersion, these considerations make sense only at a distance which is significantly larger than the size of the injected plume. On the other hand, at large distances the heterogeneous nature of the aquifer can cause deviations from the Gaussian form. In the Twin Lake experiment, the appropriateness of the Fickian approximation has been inferred from numerous observations of the tracer breakthroughs. There was a good reason to use the Fickian-dispersion model to describe the observed distributions. Quantification of longitudinal dispersivity with the Fickian dispersion model is performed in a manner similar to that used in a laboratory column. The main difference is that in field experiments the dispersion process occurs in an open system. The size of the

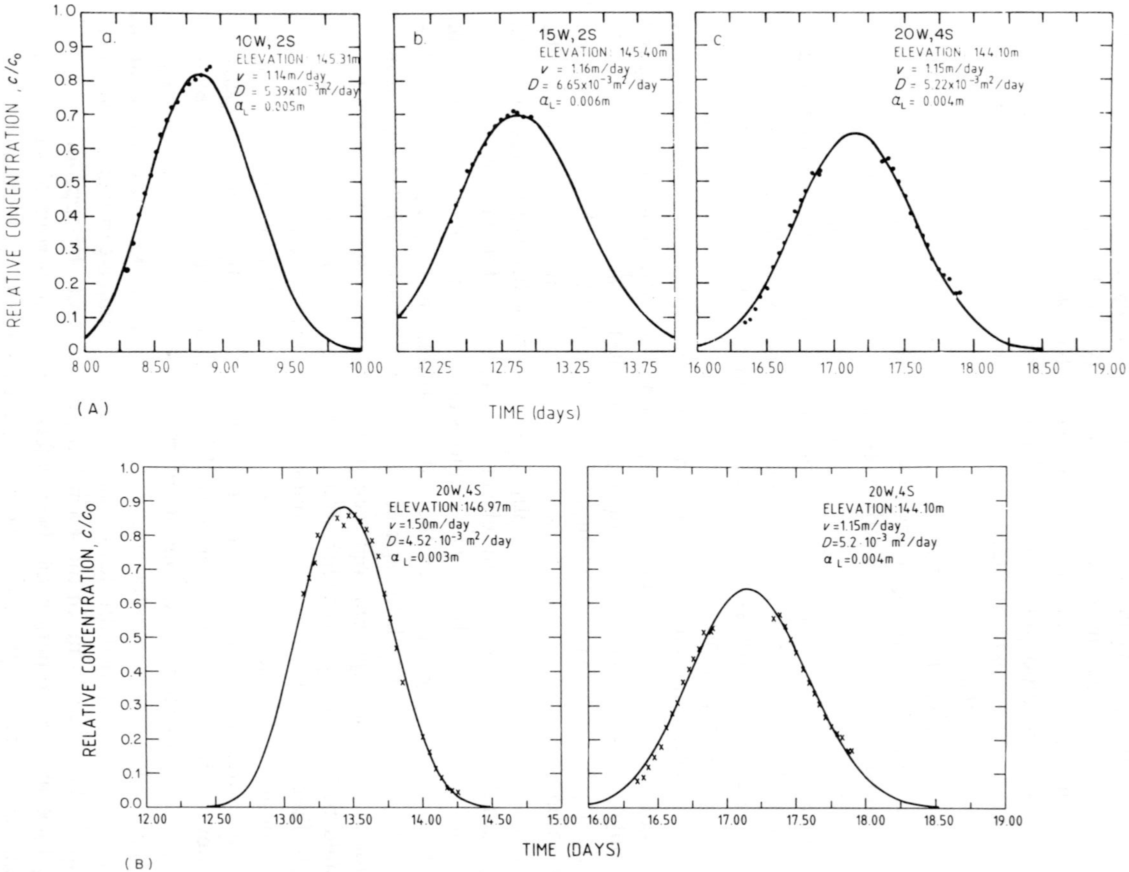

Figure 1. Observed (dots) and calculated (solid lines) breakthrough curves (A) at different distances and (B) at different elevations.

averaging volume associated with the sampling device 'fixes' the extent of the measured spreading; the 'walls' of the averaging volume act in a way similar to the walls of a laboratory column and keep the extent of the measured spreading at the local scale during the entire course of the monitoring programme. In rigorous terms, the measured concentration is defined by Eq. 2.1 and Eq. 2.3 and is associated with the centroid of the measurement device. The volume of aquifer contributing counts to the detector is characterized in terms of a mean free path. The distance is covered over time interval μ^{-1}/v, where μ^{-1} is the mean free path and v is the average velocity of the water particles with regard to the measurement device. From here it follows that the dispersion coefficient has meaning only for the volume element having radius of approximately μ^{-1} and time μ^{-1}/v required by a particle to traverse the volume element.

The same approach was used for analysing numerous discrete breakthrough curves obtained using the mixing-cup measurement device. It was stated earlier that concentration data were collected from multilevel samplers having vertical dimensions of screened intervals of approximately 7 cm. The radius of the cylindrical sampling volume is estimated from the volume of solution samples extracted for analysis as 5.9 cm. The radius is less than three mean free paths and larger than one mean free path of the gamma-ray photon. The measured concentration, therefore, is averaged over a volume similar to that in the case of the through-the-wall measurement device. Dispersion coefficients are similar, therefore, to those estimated using through-the-wall measurements. A thorough discussion of the issue is given by Moltyaner (1989).

3.4 Time-series of concentration profiles

The previous section was concerned with the longitudinal dispersion indicated by variations of concentration with time at a point as measured by an instrument fixed at that point. For quantifying the transverse dispersion, concentration measurements characterizing the tracer distribution in space are required. In practical cases, however, the measurements of concentration are performed along a line or across a plane over the entire period of the plume migration. Usually, these are time-series of instantaneous distributions across a reference plane or along a reference line. At the Twin Lake aquifer discrete measurements were made together with 'continuous' profiling of tracer distribution versus depth (the measurements performed at the Twin Lake aquifer permitted sensing the tracer activity over the full saturated thickness of the aquifer at 1 cm intervals). The time-series profiling provides information on the vertical variability in the concentration distribution, the shape of the transverse-to-flow distribution of concentration at any given time across the reference plane (the manner in which the concentration changes across the plane), the transverse dimension of the tracer cloud, the rate of change and the development of transverse dispersion at the plane of reference. The time-series of vertical profiles collected at a dry well are illustrated in Fig. 2. The problem of the vertical spread of a tracer may be considered in two categories according to whether the tracer is injected over a short interval above the bedrock surface or over the saturated thickness of the aquifer. In the latter case the vertical spreading is influenced from the very beginning by the variability in the bedrock surface. Both types of tracer injection were performed at the Twin Lake aquifer. The dimensions of the dispersing plume in the vertical transverse-to-flow direction is estimated at the time of arrival of the maximum-concentration vertical profile; the borehole with regard to which the time was estimated is determined by comparing the tracer arrival time at the boreholes located in the reference plane. The horizontal transverse-to-flow dimension of the plume is conveniently represented by the width at which the concentration falls to a given fraction of the peak value or, alternatively, by the standard deviation of the transverse displacement of the tracer. Corresponding definitions apply to the longitudinal and vertical dimensions, but when the tracer is injected over the saturated thickness, the plume height is defined as the height at which concentration falls to a particular fraction of the peak value. The growth and dissipation of the tracer plume can be observed by measuring the width of each vertical profile and by plotting it as a function of time. This gives a rough estimate of the relevant dimension for the dispersive spread. For the accurate estimation of the growth of the plume, the use of analytical solution (Eq. 2.9) is preferred. The maximum vertical profiles are special in the sense that as the tracer plume

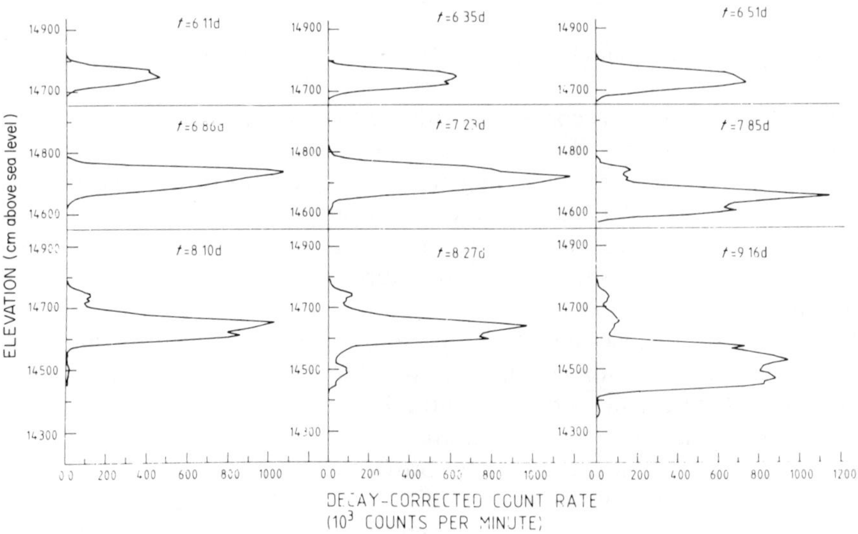

Figure 2. Concentration versus depth profiles collected 10 m down-gradient from the injection well. Data taken from the Twin Lake tracer test of 1983. Location 10W, 1.5S.

develops it is affected at different horizontal levels by different scales of dispersion determined by the heterogeneous properties of the materials. This heterogeneity is reflected in the irregular shape of the maximum vertical profiles (Fig. 3). The maximum vertical profiles give information on the relative separation of the individual zones of the dispersing plume. It is clear that the vertical extent of the maximum-concentration vertical profile is substantially larger than that of any other and the individual zone vertical profile may be used for quantifying the transverse dispersion at a borehole located along the mean direction of flow. The records of concentration distribution versus depth may be used to quantify dispersion at the plume scale by averaging the observed data. The averaging is performed in two ways: over the tracer-occupied zones of the macroscopic concentrations weighted by their respective macroscopic flow rates and of the macroscopic concentrations weighted by their respective macroscopic depth intervals. The corresponding composite time–concentration profiles are called depth-weighted distributions of mean concentration. The distributions may be used for estimating integrated plume-scale dispersivity (Moltyaner, 1987). Continuous vertical measurements can be conveniently used for indirect characterization of the aquifer heterogeneity with regard to the hydraulic conductivity; combined with borehole logs and sediment samples the information serves as a basis for constructing hydrostratigraphic scale models of the aquifer. Heterogeneity contrasts are identified using maximum-concentration profiles depicted in Fig. 3. The profiles are obtained along the boreholes of mean flow direction by superposition of the time sequence of vertical profiles available at

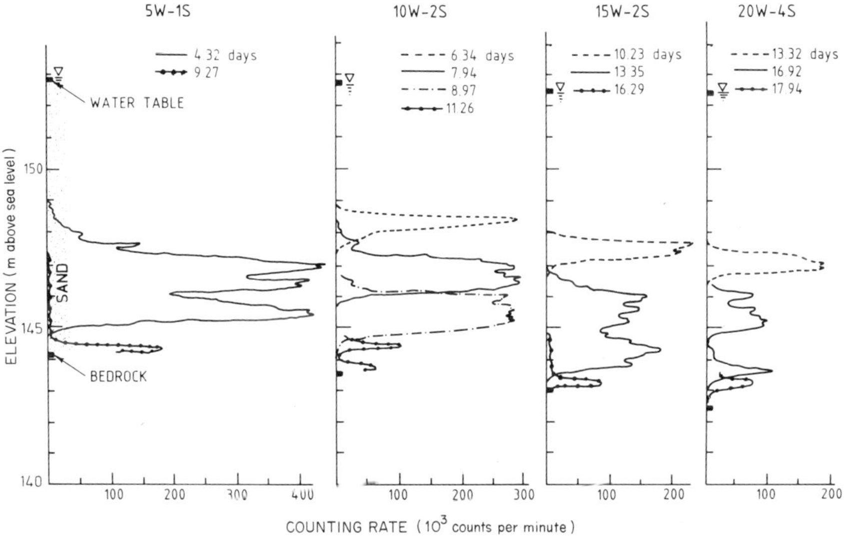

Figure 3. Maximum concentration vertical profiles collected at different distances from the injection well.

the borehole. The longitudinal variations in concentration can be observed in these figures by comparing concentration profiles at various locations. One can see that the top part of the injected slug was moving at a rate significantly higher than that in the underlying part of the aquifer. Figure 3 shows a significant degree of correlation in the arrival times of maximum-concentration profiles. There are three distinct zones which can be easily traced over the first 20 m of tracer migration. The maximum concentration profiles are given in Fig. 3 in the form of three separate profiles with different arrival times. The arrival time of the maximum-concentration profiles is in a very good agreement with the arrival time of peak concentration estimated from the concentration–time profiles constructed at the centroid of the maximum-concentration profile. The maximum-concentration profiles can be used to estimate the coefficient of transverse dispersion under the assumption of a homogeneous advection. The assumption is an essential feature of the analysis based on Eqs 2.9 and 2.10 and, as is clear from observations (Fig. 3), is quite reasonable.

3.5 Spatial distributions of the tracer plume

When the longitudinal dimension of the test slug is small in comparison with the longitudinal spacing of observation wells, the concentration measurements cannot be used to determine the instantaneous longitudinal distribution of tracer material. In order to provide a three-dimensional picture of the movement and spreading of the tracer and to make a detailed investigation at the scale of the tracer plume, the observed concentration–time data were transformed to concentration–distance data. The transformation was performed by making use of the local-scale velocities calculated on the basis of breakthrough curves constructed from vertical profiles at monitors along the mean direction of the tracer flow (Moltyaner and Wills, 1987). If we denote as $x(L, y_c, d, T)$ the longitudinal displacement at which the concentration value c, recorded at the observation well L, at elevation d, and at time T, the relation describing the translation along the longitudinal cross-section, with preliminary rotation in the x and z planes, is given as

$$x(L, y_c, d, T) = L - (t - T)v(y_c, d) \tag{3.1}$$

In this expression v represents the velocity of a local-scale solute particle at $z = d$, and y_c is the coordinate of the trajectory of the centre of mass. Since the concentration of the particle is supposed to remain constant during the transformation procedure, the reconstructed concentration field is represented in the term $c(x, y_c, z, T)$. By contouring the reconstructed concentration field one can obtain a spatial distribution of the tracer plume. Further quantitative information on the plume structure and movement may be obtained by vertically averaging the observed data. The analysis of transport mechanisms based on vertically averaged data is of interest in applications at the scale of full saturated thickness of the aquifer. The coordinates of a tracer particle in the reconstructed concentration field are obtained first by rotation in the plane $z = z_c$ through the angle that the mean direction of flow constitutes with the z axis and then by translation in the longitudinal direction on the basis of Eq. 3.1. In this case, v in Eq. 3.1 is a constant value; the implicit assumption is that velocity is uniform across a vertical direction. In Fig. 4 the evolution of the tracer plume along the mean direction of flow is illustrated. The main observation to make is the significant variability in plume movement; the initial irregularly shaped cylindrical plume spreads out along the mean flow direction. Contour plots of depth-averaged data are shown in Fig. 4B. The cross-sectional and vertically averaged contours of ^3H distribution observed in the 260-m tracer test are shown in Fig. 5 for comparison. The longitudinal extent of the plume is much less than in previous tests due to the fact that ^3H was injected only in the intermediate velocity zone.

4. Estimation of dispersivity

This section deals with the application of the ideas and results we have already discussed to the problem of estimating dispersion parameters using data collected from laboratory and field experiments. The methods used are valid only in the somewhat idealized conditions of a piecewise homogeneous advection. In the following section it is shown, however, that the methods give reliable estimates of dispersion parameters.

4.1 Local-scale longitudinal dispersivity

In the previous section it was seen that experimental data in the form of distribution of concentration versus time produced by tracer plumes show a tendency towards a Gaussian distribution. The assumption of Gaussian distribution leads to the use of Eq. 2.12 according to which concentration falls off with distance from the centre of the cloud (or axis of the plume) as an exponential function. As the mathematical form adopted in Eq. 2.12 allows concentration to fall completely to zero only at infinite values of coordinates, the dimensions of the plume are a matter of arbitrary definition. They may be conveniently defined by the root-mean-square deviations of the displacement of the material particles and estimated using moments of the breakthrough curves. For practical purposes we are concerned with injections performed over some finite interval of time and in general it is necessary to consider the effect of this on the dimensions of the plume. Equation 2.12 will apply only for large periods of time. More complex mathematical expressions Eqs 2.9–2.10 can be used. Application of Eq. 2.9 and/or Eq. 2.10 for estimation of dispersivity and

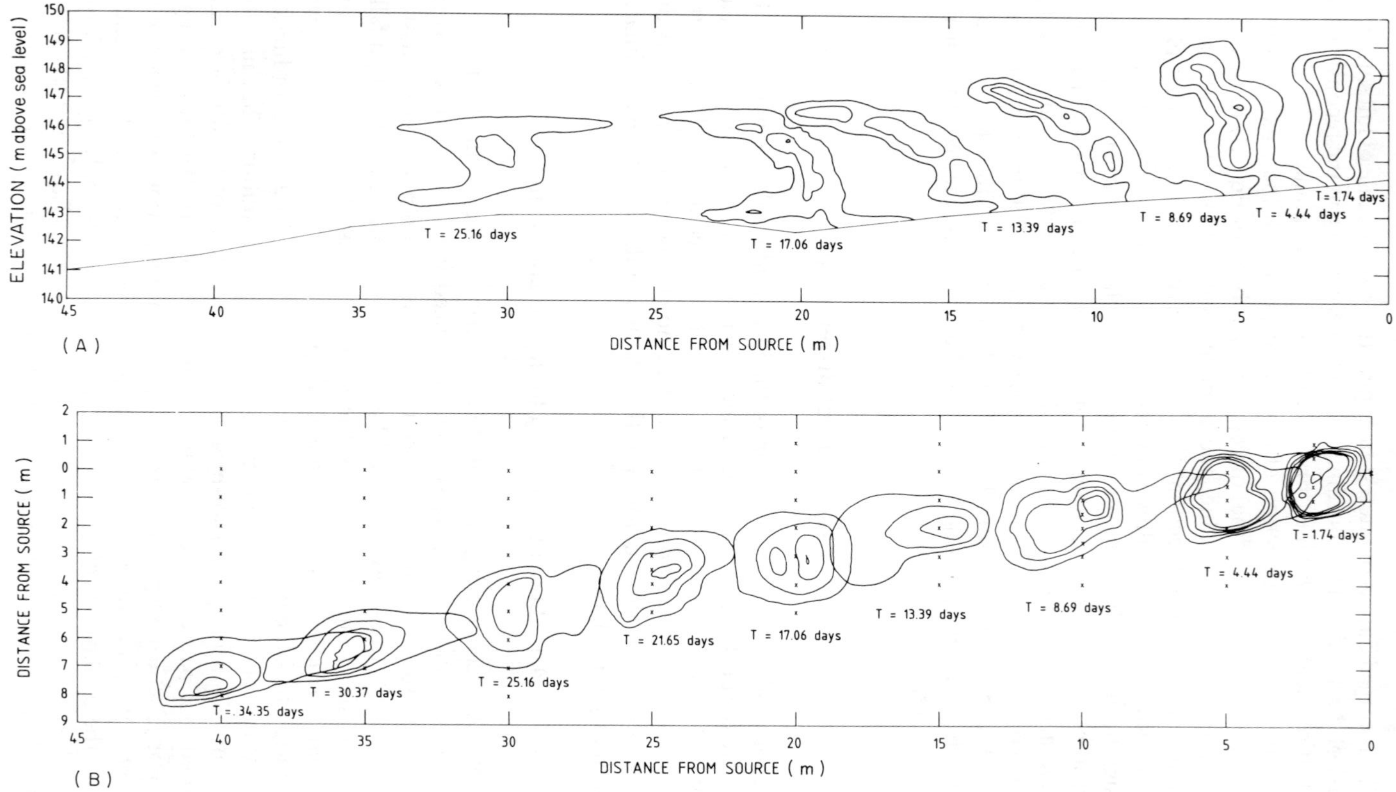

Figure 4. Concentration contours (A) along the mean direction of flow and (B) of vertically averaged data (40-m tracer test).

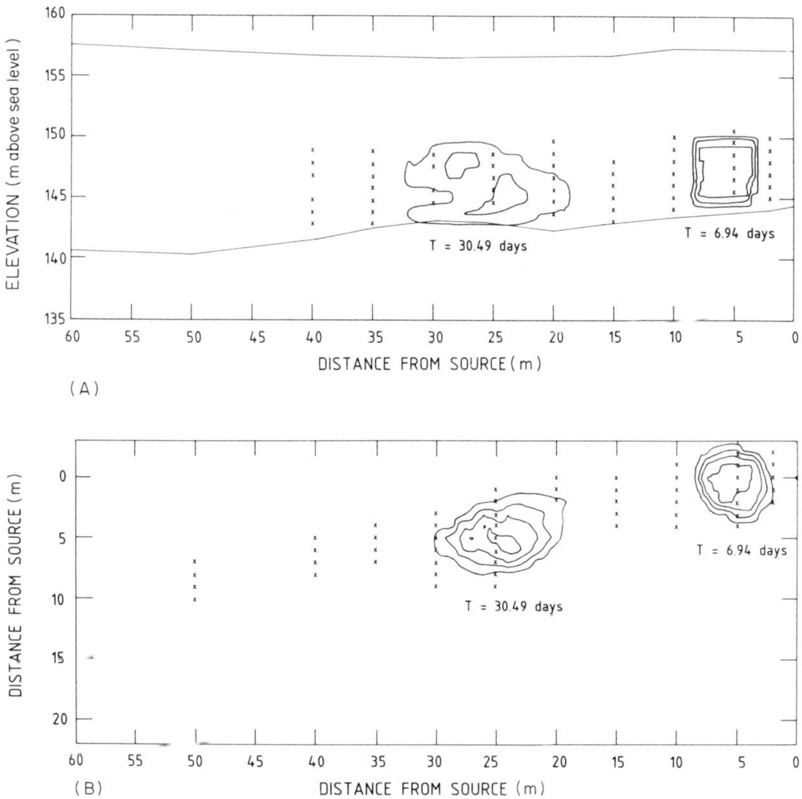

Figure 5. Concentration contours (A) along the mean direction of flow and (B) of vertically averaged data (260-m tracer test).

tracer velocity by means of the least-squares fitting procedure is illustrated in Fig. 1. The estimates of the local-scale dispersion parameters determined by fitting 86 discrete concentration–time profiles observed in the 40-m test is given by Killey and Moltyaner (1988), and Moltyaner and Killey (1988a, b). A close agreement between dispersivities calculated from time–concentration profiles observed at monitors located next to the trajectory of the plume centroid in the 260-m test and dispersivities calculated for individual layers from the 40-m test data was observed.

4.2 Determination of the local-scale dispersivity from laboratory tests

Laboratory experiments have been used traditionally to determine dispersive properties of an aquifer of interest. The commonly accepted procedure is to inject a chemically inert tracer into a laboratory column and to measure the tracer concentration as a function of time at selected location(s). In large-scale laboratory models, the dispersion equation might be used for local regions of the laboratory system to determine local values of dispersion parameters; the equation then can be integrated throughout the system to account for the detailed flow and mixing patterns. But in practice such a scheme seems to be time-consuming (especially when dealing with reactive contaminants) and laboratory experiments are generally limited to small columns when a tracer concentration is measured as a function of time only at the outlet of the column. The concentration versus time data obtained represent values averaged over the column cross-section and are used to determine dispersion parameters either by a fitting procedure or by using moments analysis. The obtained 'effective' parameters represent the aquifer properties at the scale of the length of the laboratory column. The advection-dispersion equation used to simulate contaminant migration in the aquifer is applicable then at the scale of the column length. Implicit in this application of the dispersion equation is the assumption that other aquifer parameters are calculated at the same scale (Cushman, 1986). Following the described methodology, laboratory studies were carried out with undisturbed cores of Twin Lake aquifer in order to obtain estimates of dispersion parameters (Moltyaner, 1986; Taylor *et al.*, 1987). The undisturbed cores were taken at the tracer test site at distances of 15 and 20 m from the injection well in the tracer-occupied zone. The core tubes were subdivided into sections varying from 8 to 16 cm in length in order to correspond to the 10–12 layers identified from the tracer migration and in order to evaluate effects of sample volume on the hydraulic and dispersive properties of the aquifer at the scale of several mean free paths of a gamma-ray photon. The internal diameter of aluminum tubing used to extrude the cores was 4.6 cm. A total of 23 column tracer tests were performed with the core sections using NaCl as a non-reactive tracer. The dispersive parameters of the column sediments were determined by fitting observed continuous breakthrough curves to the one-dimensional solution of the dispersion equation. The results of the experiments are given in Taylor *et al.* (1987). For all the tracer experiments velocities were such that spreading due to diffusion was negligibly small in comparison with that due to mechanical dispersion. Laboratory values of longitudinal dispersivity range from 0.06–0.32 cm with a mean of 0.14 cm. No trends in dispersivity value with column length were observed and it was concluded that a 15 cm core is appropriate for the characterization of local-scale dispersive properties of the Twin Lake aquifer (Moltyaner, 1986). It is important to note here that the extruded cores were sectioned in a way that made it possible to sample the whole range of the permeability contrasts in the tracer-occupied zone. Comparisons were made between dispersivity values obtained for a particular location in the laboratory tests with the dispersivity value obtained at the same location in the field test. Although the field-derived dispersivity values are within the range of those obtained in laboratory conditions, no correlations between field and laboratory dispersivities were observed.

4.3 Local-scale transverse dispersivity

The distributions shown in Fig. 2 indicate that a point-source approximation to the analysis of observed vertical spread of the tracer plume is inapplicable. For this reason analytical Eq. 2.9 is used to allow for the finite dimensions of the injected slug. The effective horizontal separation of maximum vertical profiles, owing to the vertical variability in aquifer properties, was observed in the arrival times of the plume at vertical positions. This means that it is more logical to treat the data as pertaining to three sources of smaller dimensions. It is, therefore, appropriate to use Eq. 2.9 for each of the smaller sources to quantify the magnitude of the vertical spread. The equation holds strictly for a homogeneous system and the application of this expression implies a homogeneity in the aquifer properties. Although individual profiles display considerable irregularities and distortion from the simple form due to an inherent lack of homogeneity over the sampling interval involved, there are also indications that the irregularities are random. Application of Eq. 2.9 to fit the maximum vertical profiles means that the relatively small variability within the dimensions of the profile is neglected. The fitting procedure acts as a homogenization procedure by filtering out noise in the system. The results of analysis based on this simplifying assumption are given by Moltyaner and Killey (1988a, b). The 'best-fit' vertical dispersivity is on average 0.1 cm. The data show remarkably little dispersion in the vertical direction with distance.

4.4 Plume-scale longitudinal dispersivity

One important result that emerged from field observations is that concentration variation in the mean direction of flow is significantly larger than in the transverse direction. Based on this observation, the concentration measurements were averaged over the cross-section perpendicular to the mean direction of flow and the obtained plume-scale concentration–time profiles were used for calculating the coefficient of longitudinal dispersion. Experimental data collected during the 260-m tracer test were used for this purpose. Since the averaging relates longitudinal dispersion to the variation of velocity over the plume transverse to the mean direction of flow, the concentration–time curve is also used for calculating plume-scale mean velocity. The plume-scale dispersivity is calculated as the ratio of the longitudinal dispersion coefficient to velocity. The mean velocity and coefficient of longitudinal dispersion obtained by fitting plume-scale Gaussian type breakthrough curves to those simulated by the one-dimensional solution to the advection-dispersion equation are given by Moltyaner and Wills (1987). The plume-scale value of dispersivity is approximately one order of magnitude larger than the local-scale dispersivity. There is probably no correlation between local- and plume-scale dispersivities. Nevertheless, there is some kind of association between them. The increase in the characteristic dimension of the averaging device over which the measured concentration is integrated, 0.1 m for the local-scale device versus 5 m for the hypothetical plume-scale device, is associated with an increase in dispersivity from 0.03 m to 0.38 m. The dispersivity was found to be equal to the average thickness of individual layers (Moltyaner, 1986).

5. Numerical simulation of groundwater flow and tracer transport

The hydrostratigraphic cross-sectional scale model of the aquifer system along the mean direction of tracer flow was used to construct a steady-state two-dimensional mathematical model for flow. The boundary conditions were selected to represent the real hydrologic boundaries and to approximate as closely as possible the artificially imposed boundary conditions. The right- and the left-hand boundaries were drawn approximately as equipotential lines with specified water fluxes. The water table elevations are refined by applying a zero flux boundary condition in the series of computer iterations and the values thus obtained were kept constant. The bedrock surface was simulated as a no-flow boundary. The finite element discretization of the hydrostratigraphic scale model was performed using three-node triangular elements. The two-dimensional linear interpolation functions were used to approximate the unknown hydraulic head distribution. The same interpolation functions were used in the weighted residuals procedure. The model was first calibrated against the hydraulic head only. The calibrated model showed a relatively poor agreement between simulated velocities and those estimated from the tracer test. A significant improvement in the model performance was gained from calibrating the model against both hydraulic head and tracer velocity measurements.

The velocity field was generated using the calculated hydraulic head and Darcy's law. The generated velocities were used in the transport part of the finite element model. The two-dimensional advection-dispersion equation describing tracer transport was solved under specified initial and boundary conditions using Galerkin finite-element approximation in space and finite-difference approximation in time. The model of tracer transport simulated the changes in the tracer concentration injected into the system along the right-hand boundary of the specified mass flux. The distribution of concentration observed 15 h after the injection was taken as the initial concentration distribution. A scenario using the observed 4.44 days concentration distribution as the initial one was also considered. The top and left-hand boundaries were treated through Dirichlet boundary condition. The bottom boundary conditions were Neumann type with a zero concentration gradient.

For the advection-dominated problem with small dispersivities, the finite-element model must meet a number of requirements to avoid oscillation, numerical dispersion and negativity of concentration. The effect of this phenomena on the solute transport may be controlled with a proper choice of grid Peclet and Curant numbers (Frind and Hokkanen, 1987). Unfortunately refinement of the grid spacing is limited by the restrictive size of computer memory and by the high computational costs. The optimum irregular grid accommodated by our site computer consisted of 452 nodes and 819 triangles. The average horizontal spacing between nodal points is approximately 1.25 m and the vertical spacing is 0.68 m. The time step is 0.1 days. The model is run until the plume reaches the boundary. After the first iteration, the model immediately produces small values of negative concentration. With the evolution of time this numerical error accumulates and eventually results in the creation of discontinuous small plumes. In the vicinity of the bedrock surface this leads to the development of a long dense tail. The effect of numerical dispersion was significant. The simulations based on the calibration of a flow model against both hydraulic head and velocity measurements show that the finite element model produces less spreading and the centroids of time concentration profiles are much better predicted as compared with the simulations based on the calibration against hydraulic head only. The results can be anticipated if one recalls that the velocity field affects the advective transport through the advective term and through the coefficients of hydrodynamic dispersion. A sensitivity analysis pertaining to the dispersivity showed that increase in dispersivity causes a decrease in negative values of concentration and in concentration gradients. The same quantitative effect was produced by increasing time steps. Increasing the value of transverse dispersivity causes the tracer to move towards the water table. A sensitivity analysis of the model with respect to space discretization of the domain was performed by refining the finite-element grid to 7128 nodes and 13 920 elements. This finite-element code was executed on a super minicomputer by Floating Point Systems Incorporated. The computer simulations begin with an initial tracer concentration distribution at 4.44 days after injection. The results are given in Moltyaner (1988). Numerical dispersion is reduced significantly and the observed variations in the tracer plume are predicted. The capability of the finite-element model to simulate the observed data is greatly enhanced.

6. Concluding remarks

The dispersive properties of the Twin Lake aquifer were studied by means of three natural-gradient radiotracer experiments and a large number of laboratory experiments. The use of a high energy gamma-emitting tracer in conjunction with the dry access well monitoring technique in the first two experiments allowed *in situ* measurement of vertical and temporal tracer distribution with a high degree of resolution, which in turn provided a means for quantifying the dispersion process at various scales of interest essential to understanding and predicting the fate of contaminants in a natural geological environment. The local-scale parameters were used for describing the dispersion process in a rigorous numerical model of the Twin Lake aquifer. The developed fine-mesh finite-element model of the radioiodine transport describes reasonably well the overall behaviour of the plume and finger-like spreading of its various parts, provided that an accurate simulation of the velocity field is performed. The dispersivity values required for these models may be calculated from laboratory experiments.

In general, transport through the Twin Lake aquifer is of a three-dimensional nature and the local-scale advection-dispersion equation with variable parameters adequately fits the observed data. However, since flow geometry is such that the lengthening of the plume in the longitudinal direction is significantly larger than in the transverse direction, a simpler approach may be used. According to this approach, it has been assumed that variation of dispersive flux in the transverse-to-flow direction can be incorporated into the one-dimensional model. The tracer migration was treated within the framework of the classical dispersion model at the plume scale. The predicted and measured concentration values are in close agreement.

The macroscopic transport equation (the advection-dispersion equation) predicts the average behaviour of a tracer. The equation may be derived from the microscopic equation by a suitable averaging procedure. We have argued that the spatial averaging with regard to the weighting function associated with the measurement device is the most appropriate one. The choice of the weighting function determines the region over which averaging takes place. Analysis of interaction of the gamma radiation with aquifer materials provides a means for choosing the weighting function and determines the size of the averaging region through the use of the mean free path of the gamma-ray photon. Each macroscopic transport problem has its own mean free path determined by the heterogeneous properties of aquifer materials. This sets the lower limit for the scale at which the transport problem may be considered as occurring in an equivalent homogeneous formation. The upper limit for the scale is determined by the properties of the exponential weighting function. The availability of tables for the mass attenuation coefficient for different materials makes the use of the mean free path of the gamma-ray photon for determining a characteristic dimension of the porous medium continuum a simple exercise. Using the tables or performing a laboratory experiment analogous to that described in this chapter, one determines the 'size' of the weighting function to be used for analysing experimental data and characteristic dimensions of the measurement device to be used for collecting the data. In this way the procedure generally used for determining the size of a measurement device and based primarily on intuitive considerations is replaced by a physically sound one.

Acknowledgements

This chapter was written during the author's sabbatical leave at the Institute of Wasserbau at the University of Stuttgart in West Germany; the hospitality received there is acknowledged with gratitude.

The financial support from the German Science Foundation provided through the Institute of Wasserbau (Dr H. Kobus, Director) is also acknowledged with gratitude.

References

Anderson, T.B. and Jackson, R. (1967). Fluid mechanical description of fluidized beds, equations of motion, *Ind. Eng. Chem. Fundam.* **6**, 527–539.

Baveye, P. and Sposito, G. (1984). The operational significance of continuum hypothesis in the theory of water movement through soils and sand aquifers, *Water Resour. Res.* **20**, 521–530.

Bear, J. (1972). 'Dynamics of Fluids in Porous Media', 796pp. Elsevier Scientific, New York.

Carslaw, H.S. and Jaeger, J.C. (1953). 'Conduction of Heat in Solids', 510pp. Clarendon Press, Oxford.

Cushman, J.H. (1983). Volume averaging, probabilistic averaging, and ergodicity, *Adv. Water Resourc.* **6**, 182–186.

Cushman, J.H. (1984). On unifying the concepts of scale, instrumentation and stochastics in the development of multiphase transport theory, *Water Resour. Res.* **20**, 1668–1676.

Cushman, J.H. (1986). On measurement, scale and scaling, *Water Resour. Res.* **22**, 129–134.

Domenico, P.A. and Robbins, G.A. (1984). A dispersion scale effect in model calibrations and field tracer experiments, *J. Hydrol.* **70**, 123–132.

Frind, E.O. and Hokkanen, G.E. (1987). Simulation of the Borden plume using the alternating direction Galerkin technique, *Water Resour. Res.* **23**, 918–930.

Garabedian, S.P. (1987). 'Local-Scale Dispersive Transport in Aquifers', Ph.D. Thesis, Massachusetts Institute of Technology.

Gelhar, L.W., Gutjahr, A.L. and Naff, R.L. (1979). Stochastic analysis of macrodispersion in a stratified aquifer, *Water Resour. Res.* **15**, 1387–1397.

Harleman, D.R.F. and Rumer, R.R. (1963). Longitudinal and lateral dispersion in an isotropic porous medium, *Fluid Mech.* **16**, 385–394.

Guven, O., Molz, F.J. and Melville, J.G. (1984). An analysis of dispersion in a stratified aquifer, *Water Resour. Res.* **20**, 1337–1354.

Killey, R.W.D. and Moltyaner, G.L. (1988). Twin Lake tracer tests: methods and permeabilities, *Water Resour. Res.* **24**, 1585–1612.

Lamarsh, J.R. (1983). 'Introduction to Nuclear Engineering'. Addison-Wesley, Reading, MA.

MacKay, D.M., Freyberg, D.L., Roberts, P.V. and Cherry, J.A. (1986). A natural gradient experiment on solute transport in a sand aquifer, *Water Resour. Res.* **22**, 2017–2030.

deMarsily, G. (1986). 'Quantitative Hydrogeology'. Academic Press, New York.

Moltyaner, G.L. (1986). Stochastic vs deterministic: a case study, *Hydrogeology, France* **2**, 183–196.

Moltyaner, G.L. (1987). Mixing-cup and through-the-wall measurements in field-scale tracer tests and their related scales of averaging, *J. Hydrol.* **89**, 281–302.

Moltyaner, G.L. (1988). Dispersion of contaminants in saturated porous media: validation of a finite-element model, *Dev. Water Sci.* **36**, 201–206.

Moltyaner, G.L. (1989). Hydrodynamic dispersion at the local scale of continuum representation, *Water Resour. Res.* **25**, 1041–1049.

Moltyaner, G.L. and Killey, R.W.D. (1988a). The Twin Lake tracer tests: longitudinal dispersion, *Water Resour. Res.* **24**, 1613–1627.

Moltyaner, G.L. and Killey, R.W.D. (1988b). The Twin Lake tracer tests: transverse dispersion, *Water Resour. Res.* **24**, 1628–1637.

Moltyaner, G.L. and Wills, C.A. (1987). 'Method of Moments Analysis of the Twin Lake Tracer Test Data'. AECL-9521, Atomic Energy of Canada Limited, Chalk River, Ontario.

Taylor, S.R., Moltyaner, G.L., Howard, K.W.F. and Killey, R.W.D. (1987). A comparison of field and laboratory methods for determining contaminant flow parameters, *Ground Water* **25**, 321–330.

Whitaker, S. (1966). The equation of motion in porous media, *Chem. Engng. Sci.* **21**, 291–300.

Chapter III

Hydraulic Conductivity Measurement at Different Scales and Contaminant Transport Modelling

Fred J. Molz, Oktay Güven, Joel G. Melville and Clay Cardone
Department of Civil Engineering, Auburn University, Auburn, AL 36849, USA

1. Introduction

While not well defined in a quantitative sense, the so-called 'scale problem' in subsurface hydrology is well known. Perhaps the most widely known manifestation of this problem is the scale-dependence of apparent full-aquifer dispersivity that has been reported by many investigators (Pickens and Grisak, 1981; Molz *et al.*, 1983; Anderson, 1984; Güven *et al.*, 1984; Güven and Molz, 1986, 1988; Taylor and Howard, 1987). In a physical sense, the scale problem arises mainly because of the pervasive heterogeneity of hydraulic conductivity in most natural aquifers. It is an artifact of the application of the dispersion model at a scale where the model is invalid. Other heterogeneous properties such as porosity and bulk density play a role also, but heterogeneity in hydraulic conductivity seems to dominate.

When a natural formation exhibits something approaching a continuous variation of its physical properties with length, one is forced to question whether a unique value for a given physical property, in the Darcy sense, can be assigned to the centroid of each representative elementary volume (REV), or even if REVs with the necessary invariance properties can be defined (Baveye and Sposito, 1984). In addition, the physical size and operating characteristics of any given instrument used to measure an aquifer property become important because different instruments have different sampling volumes, and thus average their measurements, generally in some weighted sense, over different volumes of aquifer. An extreme example would be hydraulic conductivity measured with a full-scale pumping test versus that resulting from a permeameter measurement in a laboratory.

As pointed out by Bear (1972) and by Baveye and Sposito (1984), adoption of the REV concept as a basis for deriving fundamental transport equations

DYNAMICS OF FLUIDS IN HIERARCHICAL POROUS MEDIA
ISBN 0-12-200260-1

implies that only measurements performed with instruments having a sampling volume equal to the REV of a porous medium will have a precise theoretical meaning. In order to overcome this logical limitation, Baveye and Sposito (1984) introduced the 'relativist concept' which defines measurement values, including the effect of different measuring instruments, in terms of weighted averages taken over the sampling volumes of the various instruments. Related ideas have been presented by Cushman (1984, 1986) where the scale of measurement is emphasized.

Although these ideas and concepts are relevant to much of the information presented here, they will not be discussed further. The major purpose of this chapter is to present measurements of horizontal hydraulic conductivity in a confined aquifer made at different lateral scales and at different vertical scales ranging from 1 ft (0.3048 m) up to the entire aquifer thickness. This will be followed by a discussion of the implications of such measurements for contaminant transport modelling. Most measurements were made using a borehole impeller meter (Molz *et al.*, 1989) for vertical scale information and fully penetrating pumping tests for obtaining information at various lateral scales. Such information is needed if one is going to make progress towards understanding the pervasive heterogeneity and scale problems that characterize natural porous media. The results will be relevant also to individuals working with modern stochastic continuum theories of transport in porous media because certain statistical properties of the measured hydraulic conductivity distributions become apparent.

2. Performance of experiments

All experiments were performed in a soil borrow area at the Barry Steam Plant of the Alabama Power Company, about 32 km (20 miles) north of Mobile, Alabama. The surface zone is composed of a low-terrace formation of Quaternary age consisting of interbedded sands and clays deposited along the western edge of the Mobile River. These sand and clay deposits extend to a depth of approximately 61 m (200 ft) where the contact between the Tertiary and Quarternary deposits is located. Below the contact, Miocene deposits are found that consist of undifferentiated sands, silty clays and thin-bedded limestones extending to an approximate depth of 305 m (1000 ft). The study formation is a sandy confined aquifer approximately 21.3 m (70 ft) thick which rests on the Tertiary–Quaternary contact as illustrated in Fig. 1 (Molz *et al.*, 1988).

The idea and methodology behind the impeller meter test are illustrated in Fig. 2. A small pump is placed in a well and operated at a constant flow rate, *QP*. After pseudo-steady-state behaviour is obtained, the flow meter, which measures vertical flow within the screen, is lowered to near the bottom of the well and a measurement of velocity is obtained. The meter is then raised a few feet,

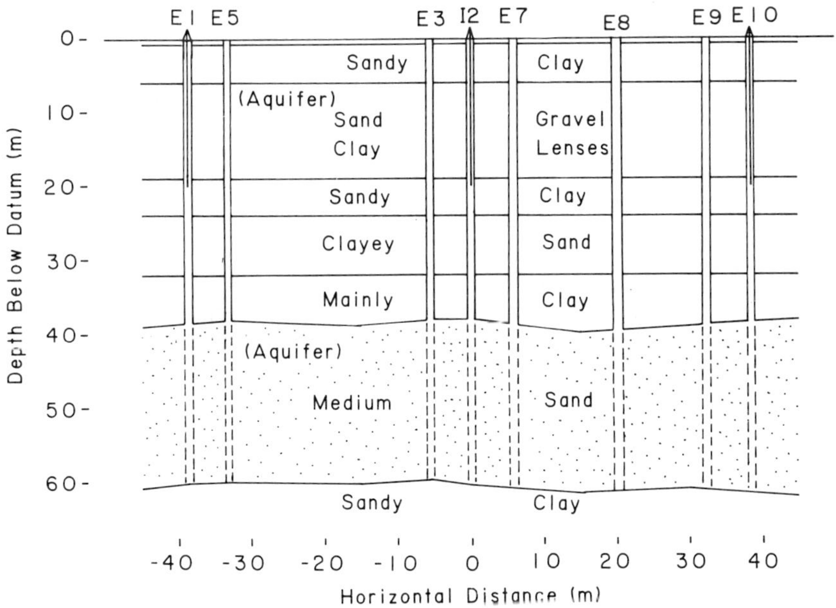

Figure 1. Vertical cross-sectional view through the centre of the study area at the Mobile site.

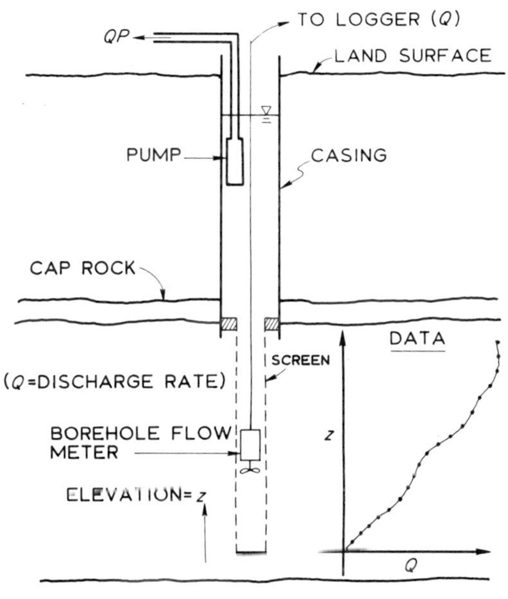

Figure 2. Apparatus and geometry associated with a borehole flow meter test.

another reading taken, raised another few feet—and so on. As illustrated in the lower portion of Fig. 2, the result is a series of data points giving vertical discharge, Q, within the well screen as a function of vertical position z. Just above the top of the screen the meter reading should be equal to QP, the steady pumping rate that is measured independently on the surface. The procedure may be repeated several times to ascertain that readings are stable.

As shown in Fig. 3, the basic data analysis procedure is quite easy. One assumes that the aquifer is composed of a series of n horizontal layers and takes the difference between two successive meter readings, which yields the net flow, ΔQ_i, entering the screen segment between the elevations z_i and z_{i+1} which are assumed to bound layer i ($i = 1, 2, \ldots, n$). One then employs the Cooper–Jacob (1946) formula for horizontal flow to a well from a layer, i, of thickness Δz_i, given by

$$\Delta H_i(r_w, t) = \frac{\Delta Q_i}{2\pi K_i \Delta z_i} \ln\left[\frac{1.5}{r_w}\left(\frac{K_i(\Delta z_i)t}{S_i}\right)^{\frac{1}{2}}\right] \tag{2.1}$$

where ΔH_i = drawdown in ith layer, ΔQ_i = flow from ith layer into well, K = horizontal hydraulic conductivity of ith layer, Δz_i = ith layer thickness, r_w = effective well radius, t = time since pumping started, and S_i = storage coefficient for ith layer. Solving Eq. 2.1 for the K_i outside the logarithmic term yields

$$K_i = \frac{\Delta Q_i}{2\pi \Delta H_i \Delta z_i} \ln\left[\frac{1.5}{r_w}\left(\frac{K_i(\Delta z_i)t}{S_i}\right)^{\frac{1}{2}}\right] \tag{2.2}$$

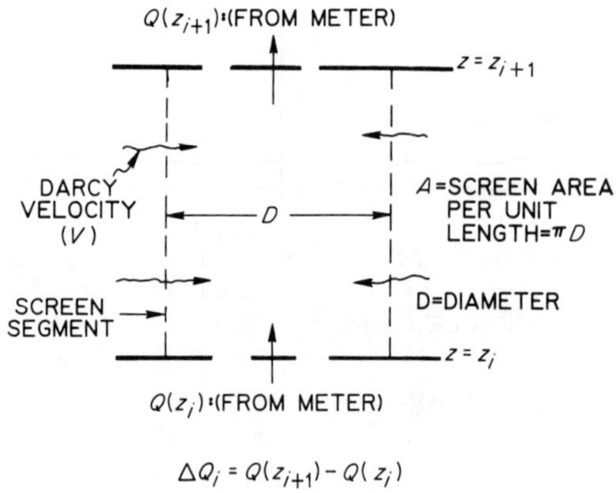

Figure 3. Illustration of the data collection and analysis procedures to determine the net flow, ΔQ_i, entering the segment of well between the elevations z_i and z_{i+1}.

which can be solved iteratively to obtain a value for K_i. Further details may be found in Morin *et al.* (1989) or Rehfeldt *et al.* (1989).

A convenient alternative method for obtaining a K distribution is based on the study of flow in a layered, stratified aquifer by Javandel and Witherspoon (1969). Their work showed that in idealized, layered aquifers, flow at the well bore radius, r_w, rapidly becomes horizontal even for relatively large permeability contrasts between layers. As the writers point out, under such conditions the radial gradients along the well bore are constant and uniform, and flow into the well from a given layer is proportional to the transmissivity of that layer, that is

$$\Delta Q_i = \alpha \Delta z_i K_i \tag{2.3}$$

where α is a constant of proportionality. This condition occurs when the dimensionless time $t_D = \bar{K}t/S_s r_w^2$ is $\geqslant 100$. (In this expression $\bar{K}$ is the average horizontal aquifer hydraulic conductivity defined as $\Sigma K_i \Delta z_i/B$, B is aquifer thickness, S_s is the aquifer specific storage, t is time since pumping started and r_w is well bore radius.)

To solve for α, sum the ΔQ_i over the aquifer thickness, to get

$$\sum_{i=1}^{n} \Delta Q_i = QP = \alpha \sum_{i=1}^{n} \Delta z_i K_i \tag{2.4}$$

Multiplying the right-hand side of Eq. 2.4 by B/B and solving for α yields

$$\alpha = \frac{QP}{B\bar{K}} \tag{2.5}$$

Finally, substituting for α in Eq. 2.3 and solving for $K_i/\bar{K}$ gives

$$\frac{K_i}{\bar{K}} = \frac{\Delta Q_i/\Delta z_i}{QP/B}; \quad i = 1, 2, \ldots, n \tag{2.6}$$

To obtain Eq. 2.6 it was assumed that ΔQ_i and QP do not change with time (i.e. pseudo-steady-state conditions apply). This will occur when $r_w^2 S/4Tt < 0.01$, where S and T are aquifer storage coefficient and transmissivity, respectively. Thus, from the basic data it is quite easy to get a plot of $K_i/\bar{K}$ versus elevation. If one then has the value of $\bar{K}$ from a fully penetrating pumping test, one can easily obtain dimensional values for K_i. At the Mobile site, the theoretical time for $t_D \geqslant 100$ is a fraction of a second, and after about 3 min of pumping, pseudo-steady-state conditions are reached. The $K_i/\bar{K}$ approach has practical appeal because one does not have to know values for r_w or S_i, which are difficult to specify precisely. Also, multiplicative errors in flow meter readings are cancelled out. However, a fully penetrating pumping test must be performed along with each flow meter test. Such pumping tests were performed at the Mobile site and the data analysed using standard procedures.

For a more detailed description of the impeller meter methodology as well as comparison of the results with other types of multilevel hydraulic conductivity tests, see Molz *et al.* (1989).

3. Results of tests

Shown in Fig. 4 is a plan view of the Mobile study site where the pumping tests and impeller meter tests were performed. The various wells are designated as I2, E6, A3, etc. The number in parentheses next to each well designation is the vertically averaged hydraulic conductivity in metres per day, $\bar{K}(x, y)$, that resulted from one or more small-scale pumping tests in which the designated well was the pumping well. Arrows indicate the pattern of the observation well/pumping-well arrangement in Fig. 4. They point from the observation well

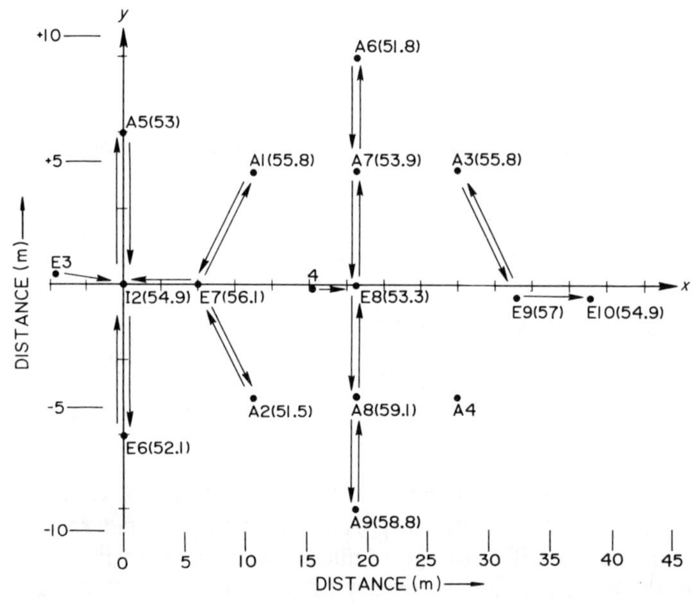

Figure 4. Plan view of the field site where small-scale pumping tests were performed. The numbers next to the dots are well designations, while the values in parentheses are the average hydraulic conductivities (m/day) assigned to the vicinity of each well. Each arrow represents a test and points from the observation well to the pumping well. Wells with more than one arrow pointing towards them were assigned average values.

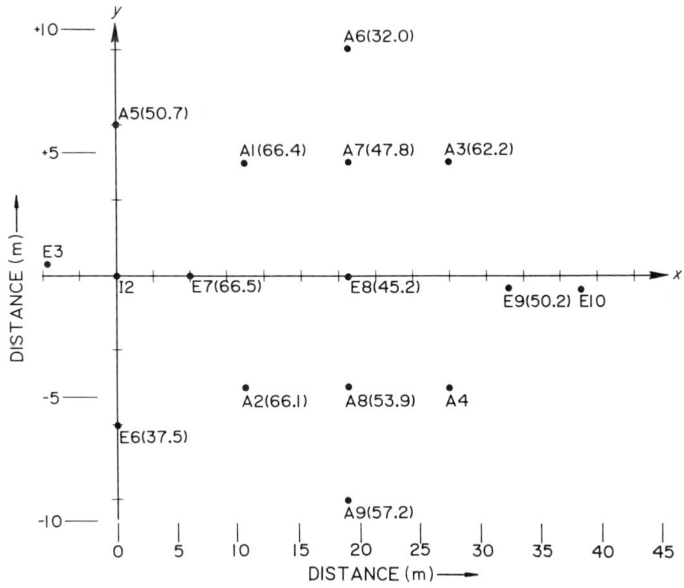

Figure 5. Results of small-scale pumping tests (m/day) in which the pumping wells were used as observation wells.

towards the pumping well. Each arrow represents one test with a pumping rate of about 0.22 m^3/min (58 gallons/min). The repeatability of any one test was good with the calculated values of $\bar{K}$ falling within less than 5% of each other.

A series of pumping tests in which the pumped wells were used as observation wells were performed also. Once again the pumping rate was approximately 0.22 m^3/min (58 gallons/min). The results of these tests are shown in Fig. 5.

$K_i/\bar{K}$ distributions based on impeller meter measurements at different scales in wells E7, E8, E9, A6, A7, A8 and A9 are shown in Figs. 6–12, respectively. Each figure was obtained with the use of Eq. 2.6 applied to impeller meter data from measurement intervals of 0.3 m (1 ft), 0.91 m (3 ft), 1.52 m (5 ft), 3.08 m (10 ft), and 4.27 m (14 ft).

As with the fully penetrating pumping tests, repeatability of the impeller meter tests was good. Evidence for this is shown in Fig. 13 which documents the results of repeated impeller meter tests in well E7.

4. Discussion of results

The vertically averaged hydraulic conductivity, $\bar{K}(x, y)$, shown in Fig. 4, seems to imply that the study aquifer is fairly homogeneous. The mean value

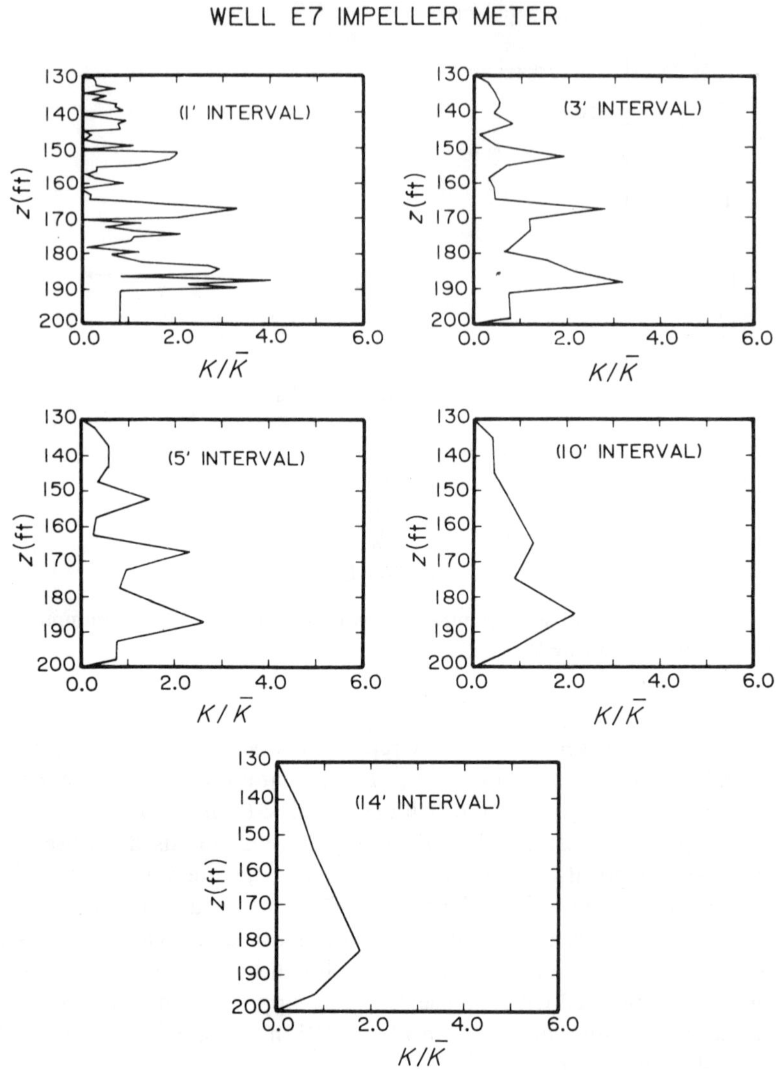

Figure 6. Dimensionless horizontal hydraulic conductivity distributions based on impeller meter readings taken at the various measurement intervals indicated on the figure.

of hydraulic conductivity is 54.9 m/day with a standard deviation of only 2.4 m/day. The mean value agrees well with the result of a large-scale pumping test (53.4 m/day) performed previously using I2 as the pumping well and pumping at the rate of 1.48 m^3/min (390 gallons/min) (Parr *et al.*, 1983).

As one would expect, the results shown in Fig. 5 are more variable because

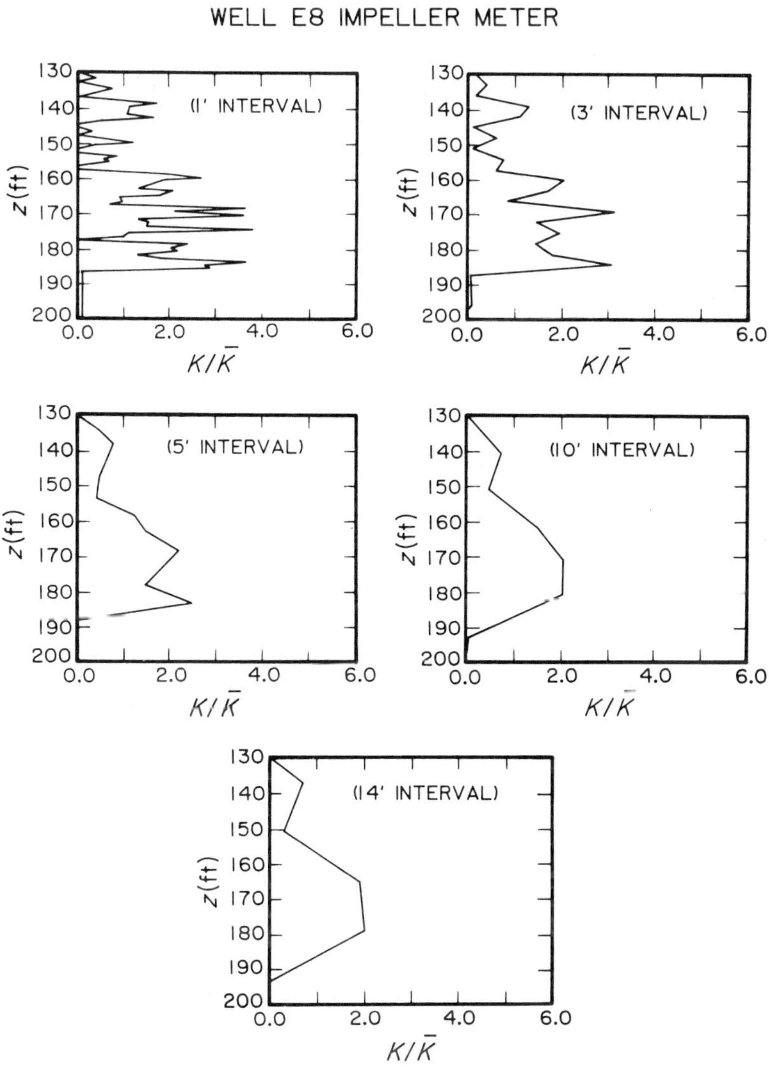

Figure 7. Dimensionless horizontal hydraulic conductivity distributions based on impeller meter readings taken at the various measurement intervals indicated on the figure.

a pumping test using the pumping well as an observation well will sample a smaller volume of the aquifer. Here the mean value is only 3.5% smaller at 53.0 m/day, but the standard deviation has increased by a factor of 4.75 to 11.4 m/day.

In the authors' judgement, no distinct pattern emerges from Fig. 4 or Fig. 5.

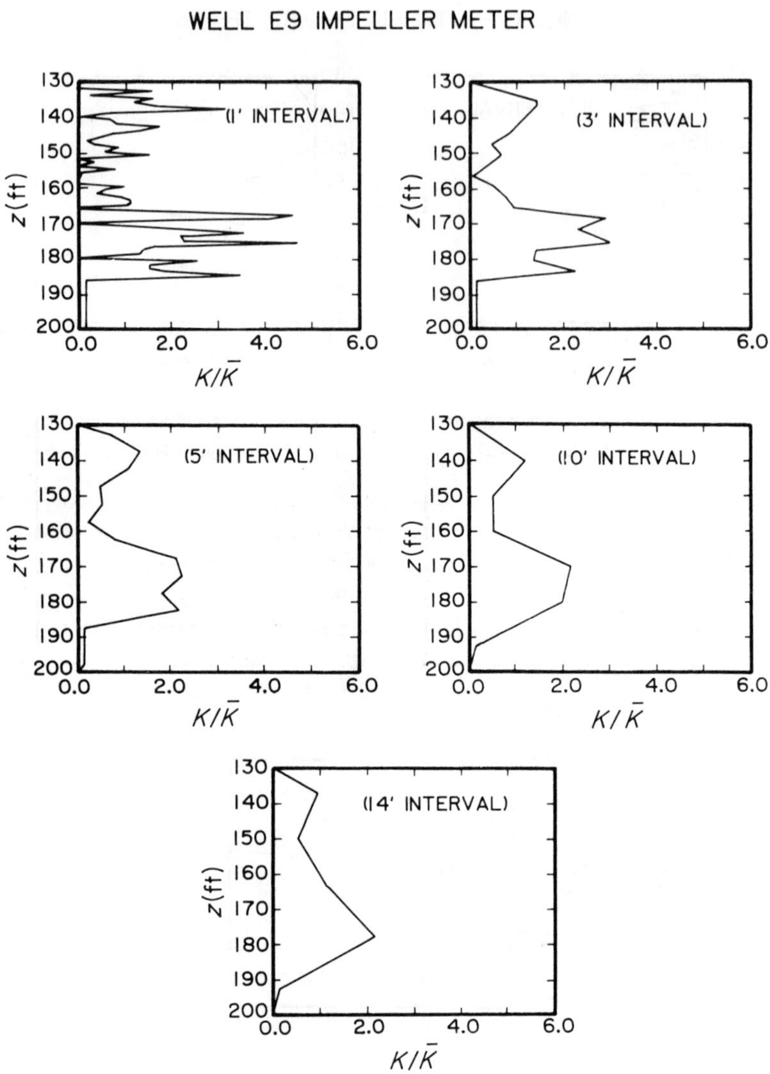

Figure 8. Dimensionless horizontal hydraulic conductivity distributions based on impeller meter readings taken at the various measurement intervals indicated on the figure.

It is probable that $\bar{K}(x, y)$ will show lateral trends over distances in excess of 38 m, which is the approximate distance between wells I2 and E10. However, over the lateral distances representative of Figs 4 and 5 the $\bar{K}(x, y)$ variations appear more or less random.

Given the generally layered nature of geological deposits in a fluvial

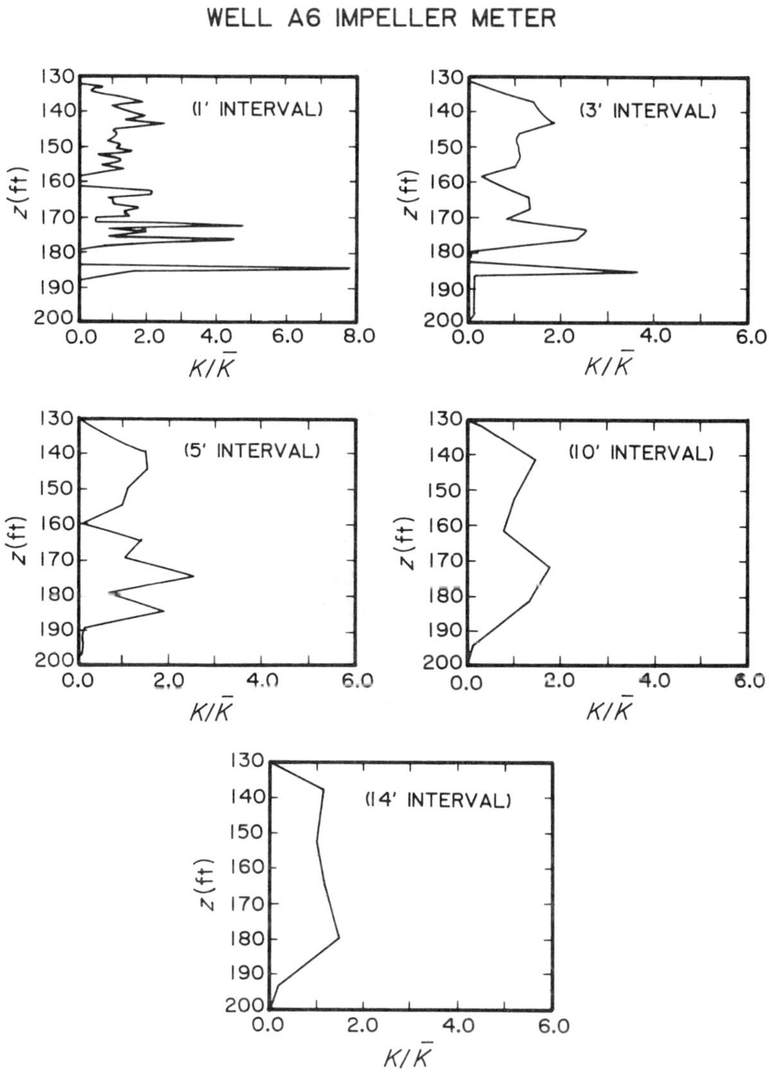

Figure 9. Dimensionless horizontal hydraulic conductivity distributions based on impeller meter readings taken at the various measurement intervals indicated on the figure.

environment, one would expect much more variability of horizontal hydraulic conductivity as a function of vertical position, $K(z)$, than of vertically averaged horizontal hydraulic conductivity as a function of lateral position, $\bar{K}(x, y)$. Examination of Figs 6–12 shows this to be the case. (Note that $K(z)$ at any particular z is still averaged over the 360° polar angle, so that the impeller

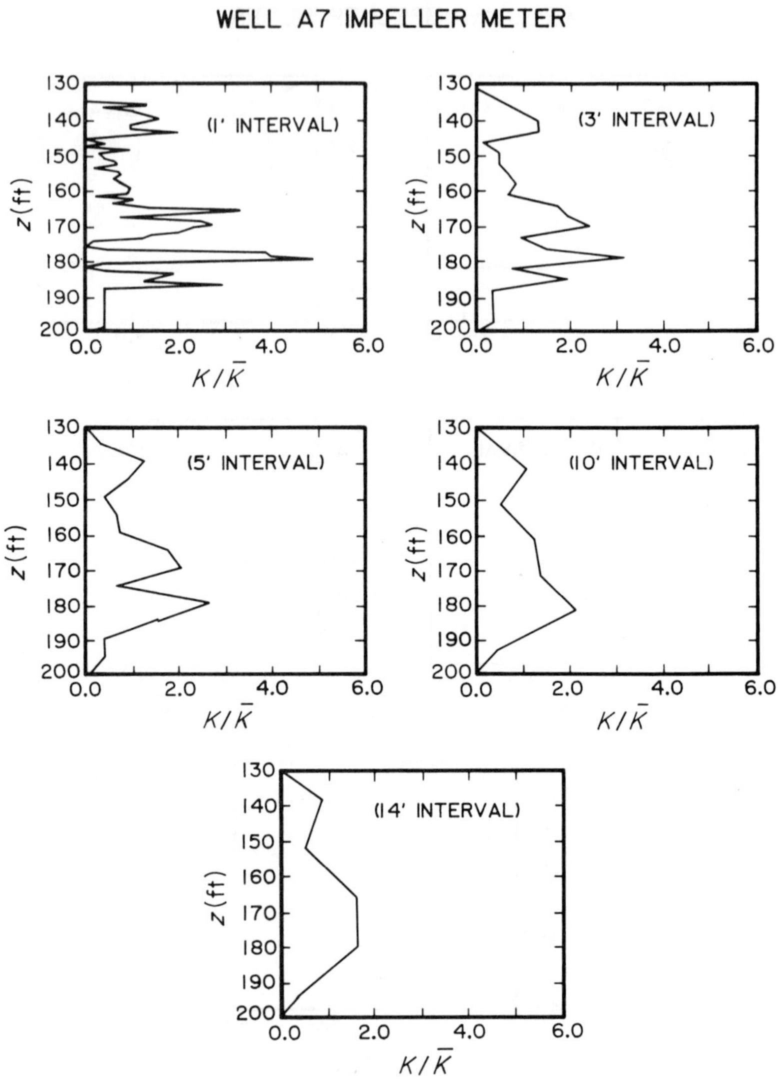

Figure 10. Dimensionless horizontal hydraulic conductivity distributions based on impeller meter readings taken at the various measurement intervals indicated on the figure.

meter test gives no information about lateral heterogeneity or anisotropy around a given well.) Different degrees of heterogeneity are apparent at the various measurement scales common to each figure. As the measurement interval decreases from 14 ft (4.27 m) to 1 ft (0.3 m), the variability of $K(z)$ increases, and there is every reason to expect that it would increase further if the

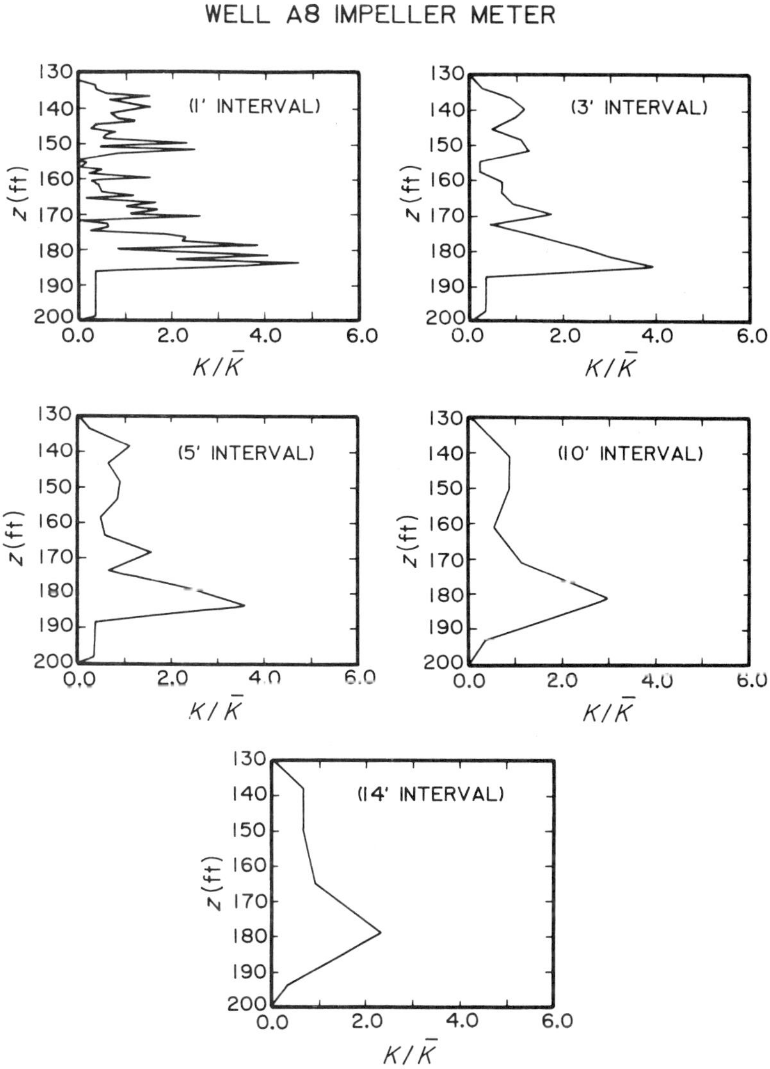

Figure 11. Dimensionless horizontal hydraulic conductivity distributions based on impeller meter readings taken at the various measurement intervals indicated on the figure.

measurement interval were decreased to 0.5 ft, 0.25 ft, etc. Obviously, this type of heterogeneity is not reflected in the results of fully penetrating pumping tests.

As shown in Fig. 5, wells E7, E8 and E9 are located along a line, which runs approximately in an east–west direction. As the measurement scale gets larger, all three wells begin to approximate something like a triangular distribution

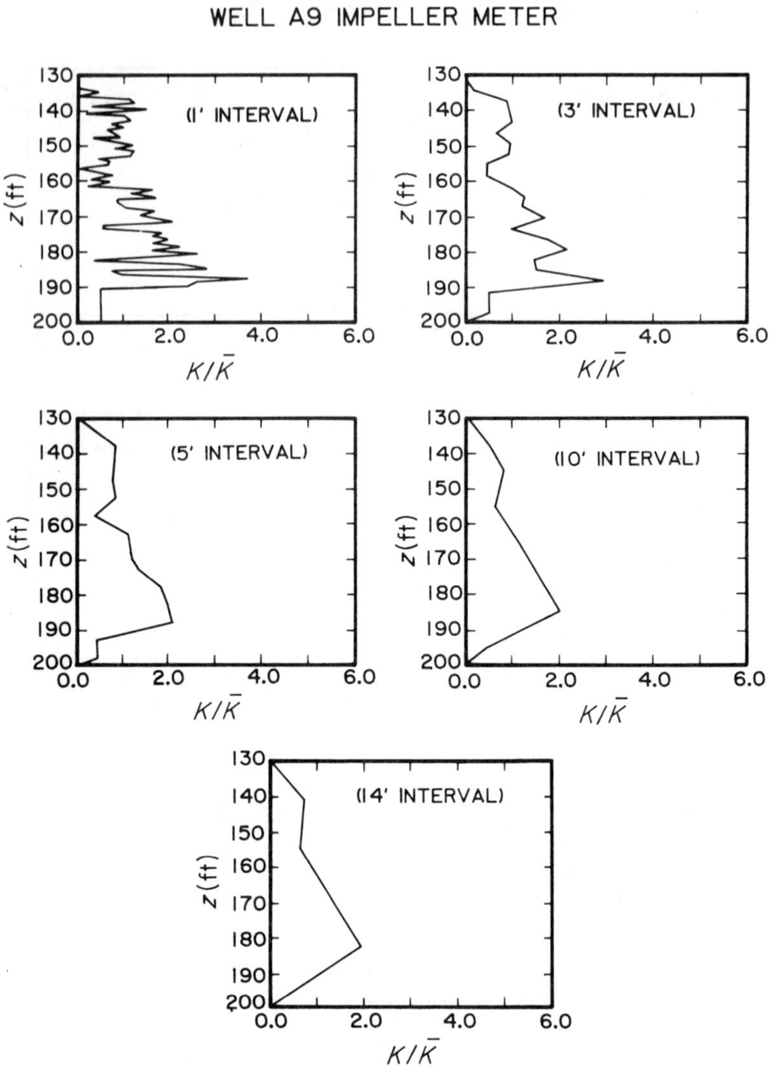

Figure 12. Dimensionless horizontal hydraulic conductivity distributions based on impeller meter readings taken at the various measurement intervals indicated on the figure.

with a peak in the bottom third of the aquifer. In the perpendicular direction, the same general behaviour is demonstrated by wells A7, A8, A9 and to a lesser extent by A6. Thus, with respect to the vertical coordinate the mean horizontal hydraulic conductivity is not constant with depth. There is a definite trend toward a hydraulic conductivity peak in the bottom third of the aquifer, and

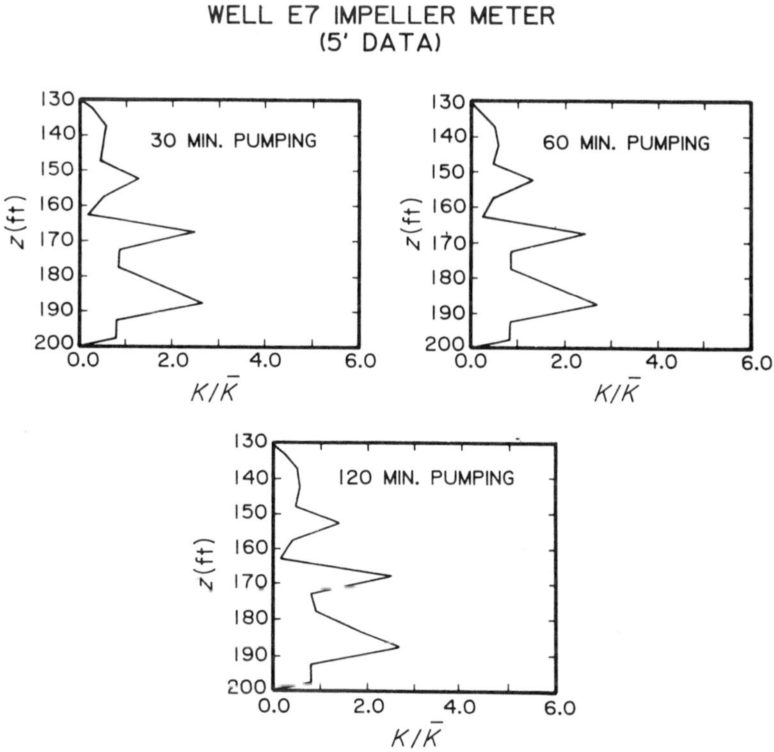

Figure 13. Dimensionless hydraulic conductivity distributions at 5 ft intervals in well E7 taken 30 min, 60 min and 120 min after the start of pumping. The results show good repeatability of the impeller meter method.

a smaller, less consistent peak in the top third. In the language of stochastic hydrology, one would say that the $K(z)$ distribution is non-stationary.

As mentioned previously, the impeller meter test averages over the 360° polar angle, and thus provides no information about K variations with respect to this variable. A limited amount of such information was obtained previously at the Mobile site using single-well tracer tests (Molz *et al.*, 1985, 1988). Shown in Fig. 14 are the results of three tracer tests conducted using I2 as the injection/recovery well, and E3, E6, and E7 as multi-level sampling wells. The $K_i/\bar{K}$ distributions in the three different coordinate directions (I2 → E3), (I2 → E6) and (I2 → E7) (see Fig. 5) all have a peak in the bottom third of the aquifer and somewhat similar overall trends. However, the magnitude of the hydraulic conductivity is larger at almost all levels in the I2 → E7 direction. Thus there are variations in $K(z)$ at a given z with respect to the polar angle.

Given the variability of the data displayed in Figs 6–12 at changing scales of measurement, a natural question that arises is 'at what scale should hydraulic

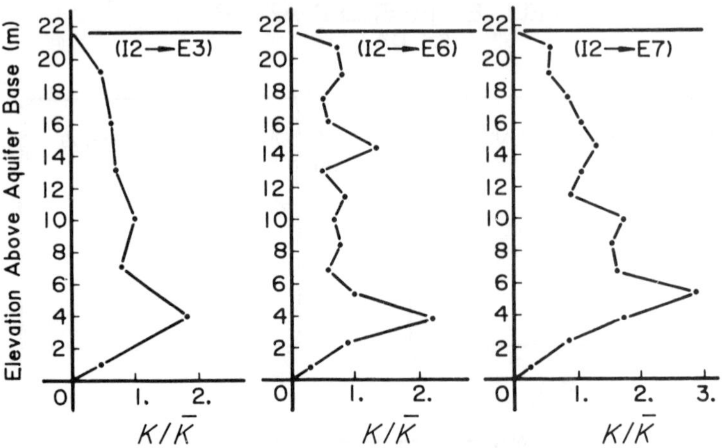

Figure 14. Apparent dimensionless horizontal hydraulic conductivity distributions based on single well tracer tests using I2 as the injection/withdrawal well (Molz *et al.*, 1988).

conductivity be measured for a particular application?' The authors are not aware of a general and precise answer to this question. See, however, Cushman (1984, 1986, 1987). However, we can present examples of previous tracer travel and concentration versus time predictions that were made at the Mobile site and agreed quite well with experiment. These predictions were based on the I2 → E3 and I2 → E7 hydraulic conductivity distributions, displayed in Fig. 14. With the relatively large number of $K(z)/\bar{K}$ distributions that we now have, it is possible to offer at least a partial explanation for the good agreement between theory and experiment that was achieved.

5. Tracer test performance and analysis

During the summers of 1984 and 1985, a series of single-well and two-well tracer tests were performed at the Mobile site (Molz *et al.*, 1985, 1986a, 1988).

Shown in Fig. 15 is a typical configuration for a single-well tracer test. The term 'single-well' represents the fact that only one pumping well is required in order to perform the test. As detailed in Güven *et al.* (1985), an observation well with multi-level samplers is required in order to obtain tracer travel time data at several vertical positions in the aquifer. One or more such observation/ sampling wells may be used in any particular tracer test. Actual test performance involves the injection of water having a known concentration of tracer in a well that is fully penetrating and screened over the entire thickness of the aquifer (Fig. 15). Concentration versus time measurements are then made at the various

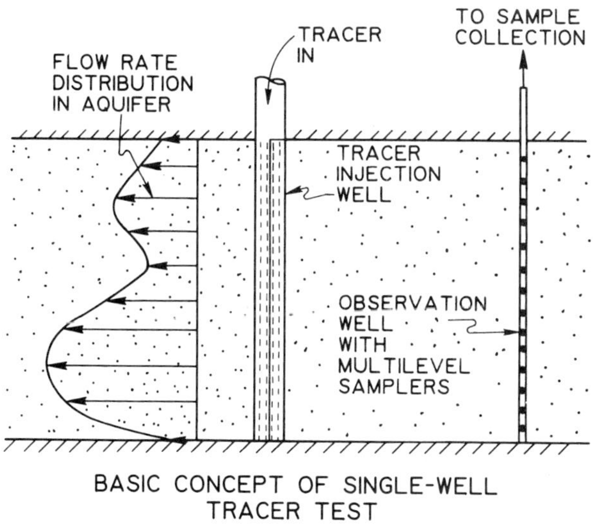

Figure 15. Vertical cross-sectional diagram illustrating the geometry associated with a single-well tracer test.

isolated sampling points in each multi-level sampling well during the experiment. The resulting tracer travel time information may be used to infer vertical profiles of horizontal hydraulic conductivity (Molz *et al.*, 1986b).

A typical configuration and flow pattern for a two-well tracer test is illustrated in Fig. 16. The experiment involves the simultaneous operation of an injection well and a withdrawal well, both of which are fully penetrating. Water is pumped into the injection well at a steady flow rate, Q, and is removed from the withdrawal well at the same rate. A conservative tracer of known concentration is added at the injection well for a period of time and the concentration of tracer in the water leaving the withdrawal well is measured and recorded as a function of time to give a concentration versus time breakthrough curve. The tracer injection period is usually short compared to the total time of the experiment.

Two-well tests may be carried out in either a recirculating or non-recirculating mode. In the recirculating mode, the water pumped from the withdrawal well is piped to the injection well, where it is injected back into the aquifer. In the non-recirculating mode, water produced from the withdrawal well is wasted at a safe distance from the test area. At the Mobile site, all two-well tests were conducted in the non-recirculating mode, and multilevel sampling wells were installed at various locations along the line connecting the injection and withdrawal wells (Molz *et al.*, 1986a, 1988).

As discussed by Pickens and Grisak (1981) and Güven *et al.* (1985, 1986), both single-well and two-well tracer tests can be used for a variety of purposes.

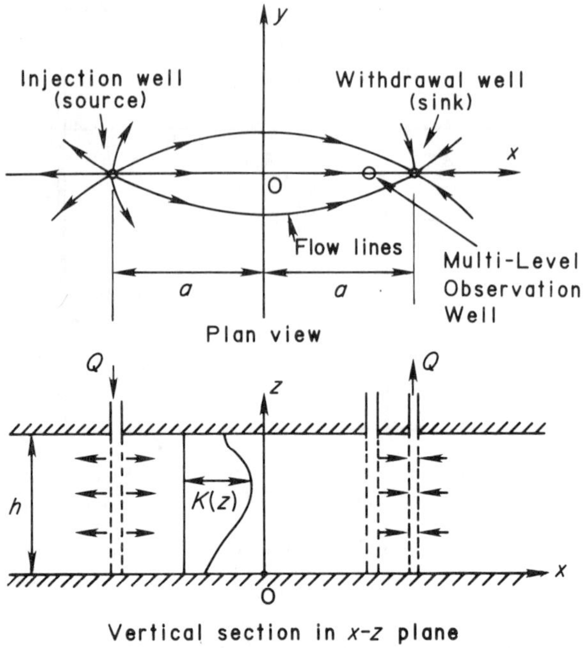

Figure 16. Plan and vertical cross-sectional diagrams illustrating the geometry and flow field associated with a two-well tracer test.

In our previous applications at the Mobile site we chose to employ the single-well tracer test as a means for inferring the hydraulic conductivity distribution over a relatively small travel distance of about 6 m. The two-well tracer test was then used to test predictions over relatively large travel distances of 38 m to 90 m (Molz *et al.*, 1988). Two single-well tests produced the I2 → E3 and I2 → E7 hydraulic conductivity distributions shown in Fig. 14. A pair of nearly identical two-well tests were performed in 1984 and 1985, respectively, that employed E1 and I2 as the tracer injection wells. In the 1984 test I2 was used as the withdrawal well, while E1 was the injection well. In 1985 I2 was used for injection, and E10 was the withdrawal well (see Fig. 1). The I2 → E3 $K/\bar{K}$ distribution of Fig. 14 was used in a three-dimensional layered model to predict the I2 breakthrough curve (Molz *et al.*, 1986a), while the I2 → E7 distribution was used to predict the E10 breakthrough curve (Molz *et al.*, 1988). In developing these predictions, no calibration of parameters was performed. In fact, the predicted breakthrough curve for well I2 was made by a subcontractor (GeoTrans, Inc.) that had no knowledge of the actual data collected during the experiment (Huyakorn *et al.*, 1986a, b).

Shown in Figs 17 and 18 are comparisons between theory and experiment for the two-well tests discussed previously. Relatively good agreement has been

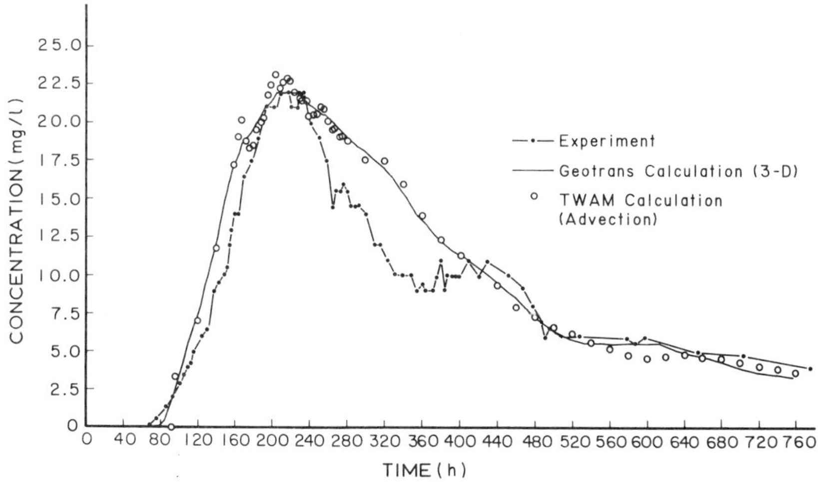

Figure 17. Comparison of measured and predicted bromide tracer concentration versus time in well I2 during the E1 → I2 two-well tracer test (Molz *et al.*, 1986b). The prediction was based on the dimensionless hydraulic conductivity distribution obtained from the I2 → E3 single-well tracer test that is shown in Fig. 14 (Molz *et al.*, 1988).

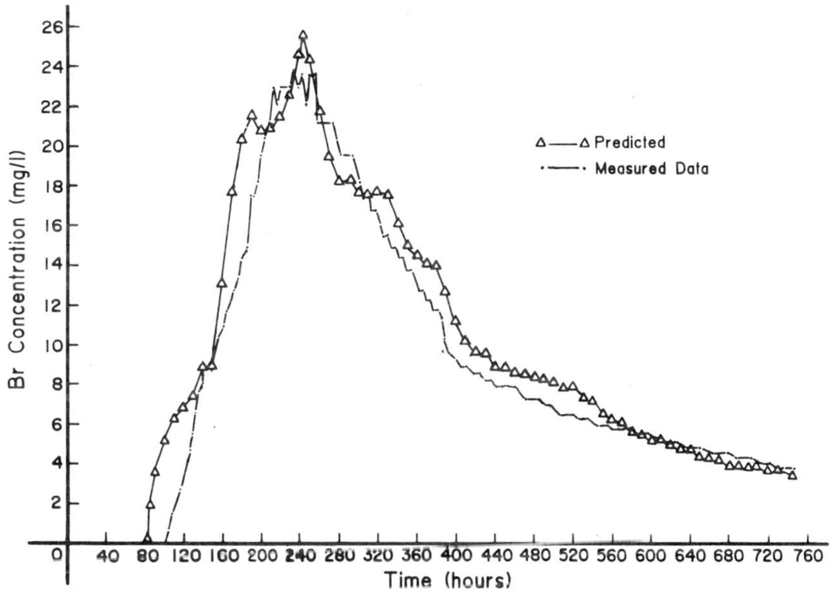

Figure 18. Comparison of measured and predicted bromide tracer concentration versus time in well E10 during the I2 → E10 two-well tracer test (Molz *et al.*, 1988). The prediction was based on the dimensionless hydraulic conductivity distribution obtained from the I2 → E7 single-well tracer test that is shown in Fig. 14 (Molz *et al.*, 1988).

achieved with two independent experiments, so the possibility of a single fortuitous result has been eliminated. A question we have asked ourselves several times, and has been asked by others, is why the predictions are as good as they are given that each one was based on a single relative hydraulic conductivity distribution? The most reasonable answer seems to be that the relative distributions used were close to the average distribution for the aquifer as a whole in the immediate vicinity of the tracer tests. We now have evidence that this is true for the I2 → E10 two-well test. Shown in Fig. 19 is a comparison of the normalized hydraulic conductivity distribution $[K(z)/K_{max}]$ obtained from the I2 → E7 single-well tracer test and the arithmetic average of the $K(z)/K_{max}$ distributions obtained from the five-foot interval impeller meter data shown in Figs 6–12. It is apparent that the two curves are quite similar. We compared the normalized distributions because in simulating a two-well test, any dimensionless $K(z)$ distribution may be used. Because a known flow is

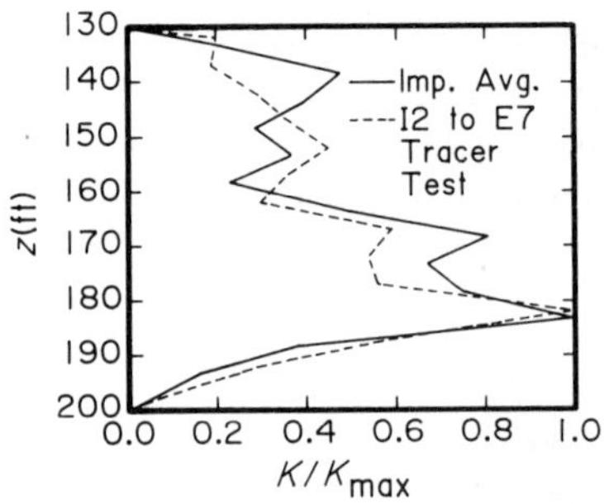

Figure 19. Comparison of the dimensionless (normalized) hydraulic conductivity distribution inferred from the I2 → E7 single-well tracer test with the arithmetic average of the seven distributions shown in Figs 6–12. The distributions based on 5-ft measurement intervals were used.

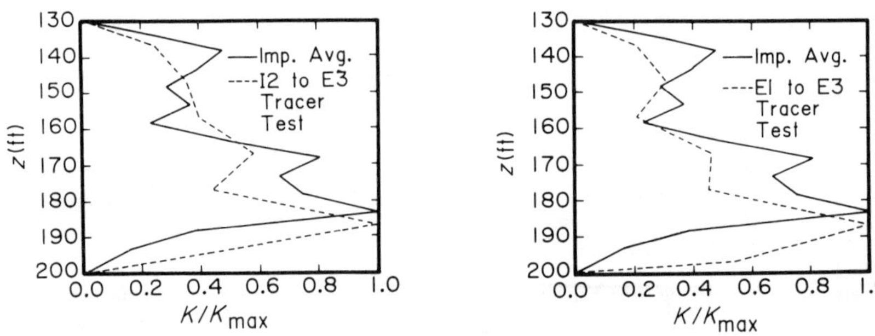

Figure 20. Comparison of dimensionless (normalized) hydraulic conductivity distributions inferred from the I2 → E3 single-well tracer test (left) and the E1 → I2 two-well tracer test (right) (Molz *et al.*, 1986a) with the arithmetic average of the seven distributions (5-ft measurement intervals) shown in Figs 6–12.

forced into the injection well and serves as a boundary condition for the calculations, only the relative distribution of K matters, not the absolute values.

It is not possible to make as direct a comparison between the average impeller meter result and the K/K_{max} distribution used to predict the I2 breakthrough curve in the E1 → I2 two-well tracer test because this test was not performed in the direct vicinity of the impeller meter tests (see Fig. 1). However, as shown in Fig. 20, two K/K_{max} distributions inferred from tracer travel times, with travel distances of approximately 6 m (I2 → E3) and 32 m (E1 → E3), have the same general nature as the arithmetic average of the seven impeller meter results. This suggests that on the average the $K(z)$ distributions between E1 and I2 and between I2 and E10 are probably rather similar.

6. Closing comments

Over the past several years at the Mobile site, a fairly large amount of hydraulic conductivity data has been developed based in part on fully penetrating pumping tests, tracer tests and impeller meter tests. As far as contaminant transport predictions are concerned, the pumping tests are of limited use because by their nature they fail to show the large amount of vertically distributed heterogeneity that is apparent to varying degrees in the tracer and impeller meter tests. This may be obvious but, nevertheless, merits emphasis because fully penetrating tests and vertically averaged properties are still the mainstay of contaminant migration studies. Obviously, this has to change. Vertically distributed information is vital to realistic simulation of contaminant transport in aquifers.

Of the three types of tests performed, the impeller meter tests had the highest degree of resolution. A different $K(z)/\bar{K}$ distribution was obtained at every vertical scale of measurement at each of seven different wells. As one would expect, the smaller the vertical scale of measurement, the larger the degree of heterogeneity that becomes apparent.

In such a system that shows increasing heterogeneity at decreasing scales of measurement, one is led to ask the question: 'What scale of measurement is appropriate for a given application?' The authors have left this question largely unanswered in a general sense. However, based on successful simulations of vertically averaged breakthrough curves measured during two-well tracer tests, it appears that a roughly constant average $K(z)/K_{max}$ distribution is obtained at a vertical measurement scale of 5–10 ft. An average $K(z)/K_{max}$ distribution measured on this scale was adequate to predict vertically averaged breakthrough curves rather well. However, one should not overemphasize the generality of the result because this type of prediction is relatively easy to make compared to predicting a tracer concentration at a particular location as a function of time. The particular location prediction was in the ballpark but did not agree with experiment nearly as well (Molz *et al.*, 1986a, b, 1988).

The reader will notice that in most situations we talk about inferring hydraulic conductivity distributions rather than actually computing them. This is because there are a lot of questions in our minds concerning what we are actually measuring. First of all, many assumptions such as constant porosity, horizontal flow, horizontal layers of homogeneous materials in the vicinity of a test well, etc. are made. Violations of these assumptions are hidden in the K distributions that result. Secondly, how do you measure 'something' that is different on every scale of measurement, perhaps right down to the scale of the REV! Various researchers are studying this question theoretically (Cushman, 1984, 1986, 1987; Baveye and Sposito, 1984), but a general solution is yet to be developed. Field measurements of various types have to be made and analysed if subsurface hydrologists are going to make progress toward solving the many problems with which they are confronted. However, until things are a little clearer perhaps it is best to say we are inferring hydraulic conductivity distributions.

Acknowledgements

This work was supported in part by the United States Environmental Protection Agency (contract CR813647010) through the Robert S. Kerr Environmental Research Laboratory. At Auburn University, the project was administered by the Water Resources Research Institute (James C. Warman, Acting Director). Although the work described in this article has been funded wholly or in part by the United States Environmental Protection Agency under assistance agreement CR813647010 to Auburn University, it has not been subjected to the agency's required peer and administrative review. Therefore, it does not necessarily reflect the views of the agency, and no official endorsement should be inferred.

References

Anderson, M.P. (1984). Movement of contaminants in groundwater: groundwater transport-advection and dispersion. *In* 'Groundwater Contamination, Studies in Geophysics', pp. 37–45. National Academy Press, Washington, D.C.

Baveye, P. and Sposito, G. (1984). The operational significance of the continuum hypothesis in the theory of water movement through soils and aquifers, *Water Resour. Res.* **20**, 521–530.

Bear, J. (1972). 'Dynamics of Fluids in Porous Media'. American Elsevier, New York.

Cooper, H.H. and Jacob, C.E. (1946). A generalized graphical method for evaluating formation constants and summarizing well-field history, *Trans. Am. Geophys. Union* **27**, 526–534.

Cushman, J.H. (1984). On unifying the concepts of scale, instrumentation, and stochastics in the development of multiphase transport theory, *Water Resour. Res.* **20**, 1668–1676.

Cushman, J.H. (1986). On measurement, scale and scaling, *Water Resour. Res.* **22**, 129–134.

Cushman, J.H. (1987). More on stochastic models, *Water Resour. Res.* **23**, 750–752.

Güven, O. and Molz, F.J. (1986). Deterministic and stochastic analyses of dispersion in an unbounded stratified porous medium, *Water Resour. Res.* **20**, 1565–1574.

Güven, O. and Molz, F.J. (1988). A field study of scale-dependent dispersion in a sandy aquifer—comment, *J. Hydrol.* **101**, 359–365.

Güven, O., Molz, F.J. and Melville, J.G. (1984). An analysis of dispersion in a stratified aquifer, *Water Resour. Res.* **20**, 1337–1354.

Güven, O., Falta, R.W., Molz, F.J. and Melville, J.G. (1985). Analysis and interpretation of single-well tracer tests in stratified aquifers, *Water Resour. Res.* **21**, 676–684.

Güven, O., Falta, R.W., Molz, F.J. and Melville, J.G. (1986). A simplified analysis of two-well tracer tests in stratified aquifers, *Ground Water* **24**, 64–82.

Huyakorn, P.S., Andersen, P.F., Güven, O. and Molz, F.J. (1986a). Curvilinear finite element model for simulating two-well tracer tests and transport in stratified aquifers, *Water Resour. Res.* **22**, 663–678.

Huyakorn, P.S., Andersen, P.F., Molz, F.J., Güven, O. and Melville, J.G. (1986b). Simulations of two-well tracer tests in stratified aquifers at the Chalk River and the Mobile sites, *Water Resour. Res.* **22**, 1016–1030.

Javandel, I. and Witherspoon, P.A. (1969). A method of analyzing transient fluid flow in multilayered aquifers, *Water Resour. Res.* **5**, 856–869.

Moltyaner, G.L. (1987). Mixing cup and through-the-wall measurements in field-scale tracer tests and their related scales of averaging, *J. Hydrol.* **89**, 281–302.

Molz, F.J., Güven, O. and Melville, J.G. (1983). An examination of scale-dependent dispersion coefficients, *Ground Water* **21**, 715–725.

Molz, F.J., Melville, J.G., Güven, O., Crocker, R.D. and Matteson, K.T. (1985). Design and performance of single-well tracer tests at the Mobile site, *Water Resour. Res.* **21**, 1497–1502.

Molz, F.J., Güven, O., Melville, J.G., Crocker, R.D. and Matteson, K.T. (1986a). Performance, analysis and simulation of a two-well tracer test at the Mobile site, *Water Resour. Res.* **22**, 1031–1037.

Molz, F.J., Güven, O., Melville, J.G. and Keely, J.F. (1986b). 'Performance and Analysis of Aquifer Tracer Tests with Implications for Contaminant Transport Modeling'. U.S. Environmental Protection Agency, Report EPA/600/2-86/062, NTIS.

Molz, F.J., Güven, O., Melville, J.G., Nohrstedt, J.S. and Overholtzer, J.K. (1988). Forced gradient tracer tests and inferred hydraulic conductivity distributions at the Mobile site, *Ground Water* **26**, 570–579.

Molz, F.J., Morin, R.H., Hess, A.E., Melville, J.G. and Güven, O. (1989). The impeller meter for measuring aquifer permeability variations: evaluation and comparison with other tests, *Water Resour. Res.* **22**, 1677–1683.

Morin, R.H., Hess, A.E. and Paillet, F.L. (1989). Determining the distribution of hydraulic conductivity in a fractured limestone aquifer by simultaneous injection and geophysical logging, *Ground Water* **26**, 587–595.

Parr, A.D., Molz, F.J. and Melville, J.G. (1983). Field determination of aquifer thermal energy storage parameters, *Ground Water* **21**, 22–35.

Pickens, J.F. and Grisak, G.E. (1981). Scale-dependent dispersion in a stratified granular aquifer, *Water Resour. Res.* **17**, 1191–1211.

Rehfeldt, K.R., Hufschmied, P., Gelhar, L.W. and Schaefer, M.E. (1989). 'The Borehole Flowmeter Technique for Measuring Hydraulic Conductivity Variability'. *Report EN-6511* Electric Power Research Institute, Palo Alto, California.

Taylor, S.R. and Howard, K.W.F. (1987). A field study of scale-dependent dispersion in a sandy aquifer, *J. Hydrol.* **90**, 11–17.

Chapter IV

Application of Central Limit Theorems to Solute Dispersion in Saturated Porous Media: From Kinetic to Field Scales

Rabi Bhattacharya
Department of Mathematics, Indiana University, Bloomington, IN 47405, USA

Vijay K. Gupta
Center for the Study of Earth from Space/CIRES and Department of Geological Sciences, The University of Colorado, Boulder, CO 80309, USA

1. Introduction

Recent theoretical developments have shown that multiple space–time scales appear naturally in the description of the transport of miscible, non-reactive solutes in saturated porous media. The problems that arise in making a transition from any one scale to the next higher scale in heterogeneous porous media are also of independent interest in the mathematics literature, and new results on these problems have been obtained within the last decade. It is therefore possible to understand the scales and scale transitions governing solute transport from a unified and rigorous mathematical point of view. It is our purpose in this chapter to explain how distinct scales arise naturally in solute transport in porous media. Specific assumptions such as the periodicity of the velocity field are made in order to provide a qualitative understanding of the phenomenon of dispersion in saturated porous media. To make the chapter accessible to a wide range of readers the mathematical exposition is kept rather informal. Specific references containing proofs of the results discussed here are cited throughout.

The principal analytical tools used in this chapter come from the theory of Markov processes. Apart from its importance in the present context, this theory has a wide variety of applications in hydrological sciences. The brief exposition of this theory given here should be helpful to the readers interested in other applications as well.

DYNAMICS OF FLUIDS IN HIERARCHICAL POROUS MEDIA
ISBN 0-12-200260-1

In order to understand the main mathematical problem in solute transport in saturated porous media, consider a small element of volume dV around a spatial point $\mathbf{x} = (x_1, x_2, x_3)$ in a porous medium of infinite extent. The spatial distances are being measured with respect to a 'unit' of space scale, e.g. cm, m, etc. Solute concentration $C(t, \mathbf{x})$ denotes the amount of solute per unit volume of porous medium at some time t. The time scale is being defined with respect to some basic unit of measurement, e.g. seconds, hours, days, etc. A unit of space scale and a unit of time scale together define a space–time scale. We will leave these units arbitrary for the present. Assume that the evolution of $C(t, \mathbf{x})$ in the space–time continuum is governed by the following parabolic differential equation, called the convective-dispersion equation (CDE),

$$\frac{\partial C}{\partial t} = \sum_{i,j=1}^{3} D_{ij} \frac{\partial^2 C}{\partial x_i \partial x_j} - \sum_{i=1}^{3} U_i(\mathbf{x}) \frac{\partial C}{\partial x_i} \tag{1.1}$$

subject to some initial condition $C(0, \mathbf{x}) = C_0(\mathbf{x})$. Here D_{ij} $[\mathrm{L}^2\ \mathrm{T}^{-1}]$ are the elements of a 3×3 positive-definite symmetric matrix and are called dispersion coefficients. Each component of the velocity vector $\mathbf{U}(\mathbf{x}) = (U_1(\mathbf{x}), U_2(\mathbf{x}), U_3(\mathbf{x}))$ denotes the volume of water passing per unit time per unit area through a small element of the porous medium around $\mathbf{x}$ in the component direction. The velocity is assumed to satisfy the following equation of continuity,

$$\sum_{i=1}^{3} \frac{\partial U_i}{\partial x_i} = 0 \tag{1.2}$$

If the velocity $\mathbf{U}$ at some scale can be assumed to be governed by another equation such as the Darcy equation then this equation and Eq. 1.2 determine the velocity field for a given specification of conductivity. The solution of the velocity field can be employed in Eq. 1.1 to study the evolution of the concentration $C(t, \mathbf{x})$. However, in the present exposition we will not consider this aspect of the problem. Rather, we will directly consider broad classes of functions $\mathbf{U}(\mathbf{x})$, e.g. periodic, almost-periodic, etc., and investigate the nature of the solution $C(t, \mathbf{x})$ for large times. The large time or asymptotic approximation of $C(t, \mathbf{x})$ enables us to obtain another partial differential equation with constant coefficients which governs solute concentration at a larger space–time scale than the scale at which Eq. 1.1 holds. This is called a *Fickian approximation* and can be expressed as

$$\frac{\partial C}{\partial t} = \sum_{i,j=1}^{3} \bar{D}_{ij} \frac{\partial^2 C}{\partial x_i \partial x_j} - \sum_{i=1}^{3} \bar{U}_i \frac{\partial C}{\partial x_i} \tag{1.3}$$

Here $\bar{\mathbf{U}}$ is the *large-scale velocity* and $((\bar{D}_{ij}))$ are the *large-scale dispersion coefficients.*

The Fickian approximation given in Eq. 1.3 is not true for arbitrary velocity fields $\mathbf{U}(\mathbf{x})$ in Eq. 1.1. For Eq. 1.3 to be even plausible, it is required that $\mathbf{U}(\mathbf{x})$ must exhibit some repetitive structure in space. Rigorous proofs of the Fickian approximation have been obtained recently in the literature for the cases of periodic $\mathbf{U}(\mathbf{x})$ (Bhattacharya, 1985), and almost-periodic $\mathbf{U}(\mathbf{x})$ (Bhattacharya and Ramasubramanian, 1988). These articles also give general formulae for $\bar{D}_{ij}$ in Eq. 1.3 in terms of the coefficients D_{ij} and $\mathbf{U}(\mathbf{x})$ in Eq. 1.1. An analysis of $\bar{D}_{ij}$ for periodic velocity fields is given here to provide qualitative insights into the nature of large-scale dispersion in the only model which has been analysed completely rigorously (Bhattacharya *et al.*, 1989).

We treat the problem of obtaining the Fickian approximation of Eq. 1.1 as a purely analytical one, in which the coefficients are deterministic and not random. A probabilistic approach is used merely as a mathematical tool for this approximation. In particular, we use a *central limit theorem* (CLT) for Markov processes. This may be contrasted with the work of Gelhar and Axness (1983) and Neuman *et al.* (1987), in which the coefficients are taken to be random. Some comments on the relationship between these two approaches are made in the last section.

The appearance of probability theory in our context is also natural physically from the point of view of kinetic theory (Chandrasekhar, 1943). Historically, the kinetic theoretic idea was introduced by Einstein in 1905 in a classic paper on molecular diffusion in aqueous solutions (Einstein, 1956).

The rest of the chapter is organized as follows. Section 2 gives an introduction to the mathematical techniques used in this chapter, namely, the theory of Markov diffusion processes, Itô's stochastic differential equations, and Kolmogorov's forward and backward equations. It also explains the mathematical connection between the CDE and the forward equation.

The statement of a CLT for stochastic processes and the large time Fickian approximation by a CLT are explained in Section 3. A general formula is given to compute the large-scale dispersion coefficients $((\bar{D}_{ij}))$ and mean velocity $\bar{\mathbf{U}}$ in Eq. 1.3 in the context of a CLT for Markov processes. From the viewpoint of measurements the large time approximation requires that space also be scaled simultaneously so that the measured solute concentration is not negligible at the large space–time scale. The precise space–time scale transformation is explained in Section 3.

The objective in Section 4 is to establish the validity of a convection–diffusion equation at an appropriate scale, referred to here as the *microscopic scale*. This scale turns out to be larger than the pore scale, but smaller than the Darcy or the laboratory scale. At the microscopic scale, the porous medium is treated as a liquid–solid continuum. Kinetic theoretic arguments are given for the derivation of this scale and of the convection–diffusion equation valid at this scale. This is contrasted with the widely held point of view that a convection–diffusion equation holds within a pore, with the solid phase acting as a smooth boundary. The results of a kinetic model calculation are described, which show that the coefficient of diffusion in a porous medium is a fraction of that in an aqueous solution. Experiments support this finding.

Section 5 derives the principal results on dispersion in porous media at different scales. It begins with a diffusion equation like Eq. 1.1 at the microscopic scale and explains how Eq. 1.3 governing dispersion at the Darcy scale is obtained using a CLT. By successive applications of this argument one can describe solute transport in aquifers which exhibit multiple length scales of spatial variability in the velocity field. These Fickian approximations are analysed by the mathematical theory developed in Sections 2 and 3. Specific examples are given to illustrate the computation of large-scale dispersion coefficients $\bar{D}_{ij}$, as functions of D_{ij} and $\mathbf{U}(\mathbf{x})$ occurring in Eq. 1.1. These examples include the classic Taylor–Aris theory of dispersion in an aqueous solution in a capillary (Taylor, 1953), and dispersion in periodic porous media. Both layered and non-layered multidimensional periodic media are analysed by three examples.

Two viewpoints on the observed growth of dispersion coefficients with distance, or time, are presented at the end of Section 5. This growth is referred to as the *scale effect*. The viewpoint of multiple scales of heterogeneity in aquifers, as proposed by Gupta and Bhattacharya (1986) and Sposito *et al.* (1986), is contrasted with the two-scale pre-asymptotic view (Dagan, 1984).

In the final section we briefly describe some related work and contrast our results with those of others.

2. Markov processes, transition probabilities and Kolmogorov's forward and backward equations

The problem of approximating Eq. 1.1 at large times by Eq. 1.3 is analytical in nature and involves no assumption of randomness. Yet this problem can be solved by appealing to probability theory purely as an analytical tool. Specifically, it makes use of the mathematical theory of Markov stochastic processes. Therefore, we begin our discussion with an informal explanation of Markov processes and Itô's stochastic differential equations. This is followed by an explanation of Kolmogorov's forward and backward equations governing the transition density of a Markov process. Finally, the mathematical connection between the forward equation and Eq. 1.1 is explained.

2.1 Markov processes and transition probabilities

A probability space is an ordered triple $(\Omega, \mathscr{C}, P)$ in which Ω is a non-empty set, $\mathscr{C}$ is a sigma field of subsets (also called 'events') of Ω, and P is a probability measure on $\mathscr{C}$. Here Ω may be viewed as the set of 'states of nature', i.e. an appropriate and adequate collection of all the different outcomes of some random phenomenon of interest. The class $\mathscr{C}$ must be large enough to include all events of interest. The probability measure P represents the chance mechanism.

A stochastic process $(\mathbf{X}(t): t \geqslant 0)$ or a random function is a map on the probability space $(\Omega, \mathscr{C}, P)$ into an appropriate class of functions. The appropriate class in our context is the class of all continuous vector-valued functions $\{\mathbf{X}(t) = [X_1(t), X_2(t), X_3(t)]: t \geqslant 0\}$ with values in the three-dimensional Euclidian space $\mathbb{R}^3$. The space $\mathbb{R}^3$ is called the state space S of the stochastic process $\{\mathbf{X}(t): t \geqslant 0\}$. Thus to each outcome $\omega \in \Omega$ corresponds a function $\mathbf{X}(\omega, t)$ of time (i.e. $t \to \mathbf{X}(\omega, t)$) called a trajectory or a *sample path.*

The most important stochastic process arising in the study of Markov processes is the *Brownian motion process.* The standard one-dimensional Brownian motion $\{B(t): t \geqslant 0\}$ is characterized by three properties: (1) it has continuous sample paths; (2) $B(t)$ is a Gaussian (or normal) random variable having mean 0 and variance t; and (3) it has independent increments, i.e. for all $0 < t_1 < \ldots < t_n$, the random variables $B(t_1), B(t_2) - B(t_1), \ldots, B(t_n) - B(t_{n-1})$ are independent.

The standard three-dimensional Brownian motion is defined as $\{\mathbf{B}(t) = [B_1(t), B_2(t), B_3(t)]: t \geqslant 0\}$, where B_1, B_2 and B_3 are three independent standard one-dimensional Brownian motions. Let $\boldsymbol{\mu}$ be a constant vector in $\mathbb{R}^3$, and $\boldsymbol{\sigma}$ a 3×3 matrix and $\boldsymbol{\sigma}'$ its transpose. Then $\mathbf{Z}(t) = t\boldsymbol{\mu} + \boldsymbol{\sigma}\mathbf{B}(t)$ is a *Brownian motion with mean $t\boldsymbol{\mu}$ and covariance matrix* $t\mathbf{M}$, where $\mathbf{M} = \boldsymbol{\sigma}\boldsymbol{\sigma}'$. Note that the process $\{\mathbf{Z}(t): t \geqslant 0\}$ has independent Gaussian increments and $\mathbf{Z}(t+s) - \mathbf{Z}(s)$ is Gaussian with mean $t\boldsymbol{\mu}$ and covariance matrix $t\mathbf{M}$.

A stochastic process $\{\mathbf{X}(t) = [X_1(t), X_2(t), X_3(t)]: t \geqslant 0\}$ is a *Markov process* if the law of future evolution of the process $\mathbf{X}$, conditionally given the state $\mathbf{x}_n$ at some present time t_n and the past states at times before t_n, does not depend on the past but depends entirely on the present time t_n and state $\mathbf{x}_n$. If $P(A \mid C)$ denotes the conditional probability of the event A given the event C, then the *Markov property* may be expressed as follows. For all $t_1 < t_2 < \ldots < t_n < t, a_i < b_i$ $(i = 1, 2, 3)$, and $\mathbf{x}_j \in \mathbb{R}^3$ $(j = 1, 2, \ldots, n)$, one has

$$
\begin{aligned}
&P(a_i < X_i(t) < b_i;\quad i = 1, 2, 3 \mid \mathbf{X}(t_j) = \mathbf{x}_j;\quad 1 \leqslant j \leqslant n) \\
&\quad = P(a_i < X_i(t) < b_i;\quad i = 1, 2, 3 \mid \mathbf{X}(t_n) = \mathbf{x}_n) \qquad (2.1)
\end{aligned}
$$

The transition probability of a Markov process, for $s < t$, is written as

$$
P(\mathbf{X}(t) \in A \mid \mathbf{X}(s) = \mathbf{x}_0) = P(s, \mathbf{x}_0;\ t, A) \qquad (2.2)
$$

where A is an arbitrary three-dimensional rectangle (parallelopiped). Thus $P(s, \mathbf{x}_0; t, A)$ denotes the probability that the particle in motion will be in region

A at time t, given that it was at $\mathbf{x}_0$ at time $s < t$. For the case in which we are interested (see the subsection on Kolmogorov's forward and backward equations) there exists a *transition probability density function* $p(s, \mathbf{x}_0;\, t, \mathbf{x})$, such that

$$\int_A p(s, \mathbf{x}_0;\, t, \mathbf{x})\, \mathrm{d}^3x = P(s, \mathbf{x}_0;\, t, A) \tag{2.3}$$

for all three-dimensional rectangles A and all $\mathbf{x}_0$, and all s, t with $s < t$. For example, the standard Brownian motion is a Markov process. Its transition probability density function is the *Gauss kernel* given by

$$p(s, \mathbf{x}_0;\, t, \mathbf{x}) = [2\pi(t-s)]^{-3/2} \exp[-\tfrac{1}{2}|\mathbf{x} - \mathbf{x}_0|^2/(t-s)] \tag{2.4}$$

The Markov property Eq. 2.1 implies the fundamental *Chapman–Kolmogorov equation*:

$$p(s, \mathbf{x}_0;\, t, \mathbf{x}) = \int_{\mathbb{R}^3} p(s, \mathbf{x}_0;\, s', \mathbf{y}) p(s', \mathbf{y};\, t, \mathbf{x})\, \mathrm{d}^3y \tag{2.5}$$

The meaning of Eq. 2.5 is simple. In order that a Markovian particle may arrive at $\mathbf{x}$ at time t, starting at $\mathbf{x}_0$ at time s, it must pass through a point $\mathbf{y}$ at an intermediate and fixed time s'. The probability of arriving in an elementary volume d^3y centred at $\mathbf{y}$ at time s' is $p(s, \mathbf{x}_0;\, s', \mathbf{y})\, \mathrm{d}^3y$. Given that at time s' it was at $\mathbf{y}$, the probability that the particle will arrive in an elementary volume d^3x centred at $\mathbf{x}$ at time t is $p(s', \mathbf{y};\, t, \mathbf{x})\, \mathrm{d}^3x$. The product of these two probabilities is integrated over $\mathbf{y}$, and d^3x cancelled from both sides, to obtain Eq. 2.5. We note here that given a function $p(s, \mathbf{x}_0;\, t, \mathbf{x})$ that satisfies Eq. 2.5, and that is a *probability density function* (pdf) in the variable $\mathbf{x}$, one can construct a unique Markov process $\mathbf{X}$ whose transition probability density function is p. This is done by computing the joint pdf of the process at any finite set of time points $t_1 < t_2 < \ldots < t_n$ as

$$p(0, \mathbf{x}_0;\, t_1, \mathbf{x}_1) p(t_1, \mathbf{x}_1;\, t_2, \mathbf{x}_2) \ldots p(t_{n-1}, \mathbf{x}_{n-1};\, t_n, \mathbf{x}_n)$$

if the process starts from the position $\mathbf{x}_0$ at time $t_0 \equiv 0$.

A Markov process is said to have homogeneous or *stationary transitions* if the transition probability density remains the same under time translation. This property can be expressed as

$$p(s, \mathbf{y};\, t, \mathbf{x}) = p(t-s;\, \mathbf{y}, \mathbf{x}) \tag{2.6}$$

where $p(t;\, \mathbf{y}, \mathbf{x})$ is a function of $t > 0$, and $\mathbf{y} \in \mathbb{R}^3$, $\mathbf{x} \in \mathbb{R}^3$. It is clear from Eq. 2.6 that if both the time points s and t are translated by an arbitrary amount $\tau > 0$, then the transition probability density does not change.

A stochastic process $\{\mathbf{X}(t): t > 0\}$ is called (*strictly*) *stationary* if the joint pdf of the process at any time points $t_1 < t_2 < \ldots < t_n$ is the same as the joint pdf at time points $t_1 + \tau < t_2 + \tau < \ldots < t_n + \tau$ for all $\tau > 0$. In general, a Markov process with stationary transitions need not be stationary. For example, Brownian motion has stationary transitions but it is not stationary. In order that a Markov process with stationary transitions may be stationary (i) there must exist a pdf $\pi(\mathbf{y})$ on the state space S, which satisfies the following identity,

$$\pi(\mathbf{x}) = \int_S \pi(\mathbf{y}) p(t;\, \mathbf{y}, \mathbf{x})\, d\mathbf{y} \tag{2.7}$$

and (ii) the distribution of $\mathbf{X}(0)$ must have this pdf. The pdf $\pi(\cdot)$ is called the *stationary* or the *steady-state density* of the Markov process. It plays a very important role, in Section 3.2, in the computation of the large-scale dispersion coefficients via a CLT for Markov processes.

2.2 Itô's stochastic differential equation

Brownian motion may be used as the building block for the construction of a large class of Markov processes known as diffusions. To understand this construction, we first note that the increment $\Delta\mathbf{B}(t) \equiv \mathbf{B}(t + \Delta t) - \mathbf{B}(t)$ in a time interval $(t, t + \Delta t)$ is Gaussian with mean $\mathbf{0} = \Delta t \mathbf{0}$ and covariance matrix

$$\Delta t \begin{pmatrix} 1 & 0 & 0 \\ 0 & 1 & 0 \\ 0 & 0 & 1 \end{pmatrix} = \Delta t \mathbf{I}$$

where $\mathbf{I}$ is the identity matrix. In general one cannot expect that for a general well-behaved Markov process $\mathbf{X}$ the increment $\Delta\mathbf{X}(t)$ will be independent of $\mathbf{X}(t)$. However, it is natural to inquire if there exists a Markov process $\mathbf{X}$ such that the conditional probability distribution of $\Delta\mathbf{X}(t)$ given $\mathbf{X}(t) = \mathbf{x}$, for sufficiently small $\Delta t > 0$, is approximately Gaussian with mean $\Delta t \mathbf{U}(\mathbf{x})$ and covariance matrix $\Delta t \mathbf{M}(\mathbf{x})$. Here $\mathbf{U}(\mathbf{x})$ is a prescribed vector-valued function of $\mathbf{x}$ and $\mathbf{M}(\mathbf{x})$ is a positive definite symmetric matrix. The answer to the above query is in the affirmative if certain mild smoothness conditions are satisfied by $\mathbf{U}$ and $\mathbf{M}$, as explained in the next section. In fact, letting $\boldsymbol{\sigma}(\mathbf{x})$ be a matrix satisfying $\boldsymbol{\sigma}(\mathbf{x})\boldsymbol{\sigma}^*(\mathbf{x}) = \mathbf{M}(\mathbf{x})$, where $\boldsymbol{\sigma}^*(\mathbf{x})$ is the transpose of the matrix $\boldsymbol{\sigma}(\mathbf{x})$

(e.g. let $\boldsymbol{\sigma}(\mathbf{x})$ be the positive definite square root of $\mathbf{M}(\mathbf{x})$), it is easy to see that

$$\mathbf{U}(\mathbf{x})\Delta t + \boldsymbol{\sigma}(\mathbf{x})\Delta\mathbf{B}(\mathbf{t})$$

has a Gaussian distribution with mean $\Delta t\mathbf{U}(\mathbf{x})$ and covariance $\Delta t\mathbf{M}(\mathbf{x})$. By 'piecing together' this local specification of the process $\mathbf{X}(t)$, one obtains the entire process. This suggests that we solve *Itô's stochastic differential equation*

$$\mathrm{d}\mathbf{X}(t) = \mathbf{U}(\mathbf{X}(t))\,\mathrm{d}t + \boldsymbol{\sigma}(\mathbf{x})\,\mathrm{d}\mathbf{B}(t) \tag{2.8}$$

in order to obtain the process $\mathbf{X}(t)$.

The mathematical difficulty with Eq. 2.8 is that it cannot be cast in the form

$$\frac{\mathrm{d}\mathbf{X}}{\mathrm{d}t} = \mathbf{U}(\mathbf{X}(t)) + \sigma(\mathbf{X}(t))\frac{\mathrm{d}\mathbf{B}(t)}{\mathrm{d}t}$$

because the sample paths of $\mathbf{B}(t)$ are not differentiable with respect to time. However, it is possible to give Eq. 2.8 a precise mathematical meaning, and this was done by Itô (1951). For the present purpose it is unnecessary to discuss further details of this rich and complex theory of stochastic differential equations, except to remark that the general mathematical theory allows both the vector $\mathbf{U}$ and matrix $\mathbf{M}$ to be functions of time t as well as space coordinates $\mathbf{x}$. For relatively simple mathematical expositions of Itô's theory, the reader may refer to Arnold (1971) and Bhattacharya and Waymire (1990, Ch. 7).

2.3 Kolmogorov's forward and backward equations

It is known, and can be simply checked by differentiation, that Eq. 2.4 is the *fundamental solution*, or the Green's function, of the diffusion equation

$$\frac{\partial}{\partial t}p(s, \mathbf{x}_0; t, \mathbf{x}) = \tfrac{1}{2}\nabla^2 p(s, \mathbf{x}_0; t, \mathbf{x}) \tag{2.9}$$

where

$$\nabla^2 = \sum_{i=1}^{3} \frac{\partial^2}{\partial x_i^2}$$

is the *Laplacian*. It was first shown by Kolmogorov (1931) that the transition probability density function of every sufficiently well behaved Markov process satisfies an equation which is analogous to the diffusion Eq. 2.9. To state precisely this fundamental result we assume that a vector-valued function $\mathbf{U}(\mathbf{x})$ and a matrix $\boldsymbol{\sigma}(\mathbf{x})$ are given which satisfy the following assumptions:

1. The elements of $\mathbf{U}$ and $\boldsymbol{\sigma}$ have bounded first-order derivatives and are thrice continuously differentiable.
2. Let $\lambda'(\mathbf{x})$ be the smallest eigenvalue of the matrix $\mathbf{M}(\mathbf{x}) = \boldsymbol{\sigma}(\mathbf{x})\boldsymbol{\sigma}^*(\mathbf{x})$. There exists a positive constant b, such that $\lambda'(\mathbf{x}) \geqslant b > 0$.

Let $p(0, \mathbf{y};\ t, \mathbf{x}) = p(t;\ \mathbf{y}, \mathbf{x})$ denote the stationary transition density of a Markov process, and let A and $\tilde{A}$ be the following 'differential operators'

$$Af(\mathbf{x}) = \frac{1}{2} \sum_{i,j=1}^{3} \frac{\partial^2}{\partial x_i \partial x_j} (M_{ij}(\mathbf{x})f) - \sum_{i=1}^{3} \frac{\partial}{\partial x_i} (U_i(\mathbf{x})f) \tag{2.10}$$

$$\tilde{A}f(\mathbf{y}) = \frac{1}{2} \sum_{i,j=1}^{3} M_{ij}(\mathbf{y}) \frac{\partial^2 f}{\partial y_i \partial y_j} + \sum_{i=1}^{3} U_i(\mathbf{y}) \frac{\partial f}{\partial y_i} \tag{2.11}$$

The main result may now be stated as follows:

Theorem 2.1. If $\mathbf{U}$ and $\boldsymbol{\sigma}$ satisfy Assumptions 1 and 2 above, then there exists a unique Markov process $\{\mathbf{X}(t);\ t \geqslant 0\}$, which satisfies Eq. 2.8. This process has a stationary transition probability density function $p(t;\ \mathbf{y}, \mathbf{x})$ which satisfies *Kolmogorov's forward equation* in the variables t and $\mathbf{x}$,

$$\frac{\partial p}{\partial t} = Ap \tag{2.12}$$

where A is given by Eq. 2.10, and *Kolmogorov's backward equation* in the variables t and $\mathbf{y}$

$$\frac{\partial p}{\partial t} = \tilde{A}p \tag{2.13}$$

where $\tilde{A}$ is given by Eq. 2.11.

The forward equation appears naturally in the context of solute dispersion because of the one-to-one correspondence between it and the CDE (Eq. 1.1). This is explained below in Section 2.4. In the physics literature, the forward equation is generally known as the *Fokker–Planck equation*. However, the backward equation is mathematically more convenient and is important for computing the asymptotic dispersion coefficients via a CLT for Markov processes. This is explained in Section 3.2, and its applications are given in Section 5.

We remark that the above theorem deals with the transition probability density function for the so-called initial value problem in the unrestricted infinite state space $\mathbb{R}^3$. In practice, of course, the space will always be restricted by boundaries. For example, the space may be $\{\mathbf{x} = (x_1, x_2, x_3); x_3 \leqslant 0\}$, for which the boundary is $\{\mathbf{x} = (x_1, x_2, x_3): x_3 = 0\}$ and boundary conditions have to be specified in order to determine the transition probability density function uniquely. These boundary conditions usually are in the form of a vanishing space derivative along a non-tangential direction, e.g. the normal direction at the boundary, or a vanishing of the solution itself, or a combination of the two. The first of these conditions corresponds to the Markovian particle being reflected to the interior on striking the boundary, while the second condition corresponds to the Markovian particle escaping from the space on striking the boundary; these are homogeneous boundary conditions. Thus one may first construct a Markovian motion that satisfies Itô's equation (Eq. 2.8) in the unrestricted space $\mathbb{R}^3$, or construct analytically a transition probability density function that satisfies Kolmogorov's forward equation (Eq. 2.12) and then impose the appropriate boundary condition in order to arrive at the corresponding $p(t; \mathbf{y}, \mathbf{x})$. This p will still satisfy Eq. 2.12, and in addition will satisfy a homogeneous boundary condition. An example of this type, discussed later on in Section 5, is the Taylor–Aris theory of solute dispersion in a capillary (Taylor, 1953; Aris, 1956; Bhattacharya and Gupta, 1984). A simple, but fairly comprehensive, treatment of the forward and backward equations for Markov processes is given in Bhattacharya and Waymire (1990, Ch. 5).

2.4 Connection between the forward equation and the CDE

Let $C_0(\mathbf{x}_0)$ denote the initial solute concentration in a porous medium at $\mathbf{x}_0$ at time 0. Assume that the total mass of solute in the medium is a constant. Any constant will suffice. For simplicity we take it to be one. Then $C_0(\mathbf{x}_0)$ can be treated as a pdf of a random variable, since $C_0(\mathbf{x}_0) \geqslant 0$ and it integrates to one over the entire space $\mathbb{R}^3$. We now define the unconditional pdf $p(t, \mathbf{x})$ of $\mathbf{X}(t)$ as

$$p(t, \mathbf{x}) = \int_{\mathbb{R}^3} p(t; \mathbf{x}_0, \mathbf{x}) C_0(\mathbf{x}_0)\, \mathrm{d}^3 x_0 \tag{2.14}$$

The reader can simply check that if both sides of Eq. 2.12 are multiplied by $C_0(\mathbf{x}_0)$ and integrated with respect to $\mathbf{x}_0$ over $\mathbb{R}^3$, then $p(t, \mathbf{x})$ also satisfies Eq. 2.12. It can be written as,

$$\frac{\partial p}{\partial t} = \frac{1}{2} \sum_{i,j=1}^{3} \frac{\partial^2}{\partial x_i \partial x_j} (M_{ij}(\mathbf{x}) p) - \sum_{i=1}^{3} \frac{\partial}{\partial x_i} (U_i(\mathbf{x}) p) \tag{2.15}$$

subject to the initial condition $p(0, \mathbf{x}) = C_0(\mathbf{x})$.

It is known from the theory of parabolic partial differential equations that the solution of Eq. 2.15 with initial condition $p(0, \mathbf{x}) = C_0(\mathbf{x})$ is unique (Friedman, 1975, pp. 139–144). Now, if we take $M_{ij}(\mathbf{x}) = 2D_{ij}$ to be constant with respect to $\mathbf{x}$, and assume that velocity $\mathbf{U}(\mathbf{x}) = (U_1(\mathbf{x}), U_2(\mathbf{x}), U_3(\mathbf{x}))$ satisfies the continuity equation (Eq. 1.2), then Eq. 2.15 and the CDE (Eq. 1.1) become identical. Therefore, by uniqueness of the solution of Eq. 2.15, it follows that

$$p(t, \mathbf{x}) = C(t, \mathbf{x}) \tag{2.16}$$

It is precisely this mathematical connection between $C(t, \mathbf{x})$ and the unconditional transition density $p(t, \mathbf{x})$ of a Markov process that enables us to use the theory of Markov processes to study the large-time behaviour of the solution of the CDE and obtain the Fickian approximation given by Eq. 1.3.

3. Asymptotic Fickian approximation and scale transition using the central limit theorem

It follows from the equality between $p(t, \mathbf{x})$ and $C(t, \mathbf{x})$ in Eq. 2.16 that the asymptotic behaviour of $C(t, \mathbf{x})$ is mathematically equivalent to that of the asymptotic behaviour of $p(t, \mathbf{x})$. Now, write the representation of the Markov process given in Eq. 2.8 in the integral form as

$$\mathbf{X}(t) = \int_0^t \mathbf{U}(\mathbf{X}(s))\, \mathrm{d}s + \boldsymbol{\sigma}\mathbf{B}(t) \tag{3.1}$$

and observe that the first term in Eq. 3.1 is the time integral of the stochastic process $\{\mathbf{U}(\mathbf{X}(s)): s \geqslant 0\}$. The second term is simply a Brownian motion process with mean $\mathbf{0}$ and covariance matrix $t\mathbf{M}$. In the applications considered in this chapter the two terms are asymptotically independent. Therefore, the asymptotic behaviour of the pdf of the first term determines the asymptotic behaviour of the pdf $p(t, \mathbf{x})$ of $\mathbf{X}(t)$. A *Central Limit Theorem* (CLT) is said to hold if the time integral of a suitably centred and scaled stochastic process can be approximated for large times by a Brownian motion with mean $\mathbf{0}$ and some covariance matrix $t\bar{\mathbf{M}}$. A great deal of probability literature has been devoted to the study of CLTs.

In this section we first give an informal description of a general CLT. This is followed by a formula for computing the covariance matrix $t\bar{\mathbf{M}}$ of the limiting Brownian motion in the context of a CLT for Markov processes. Finally, we give an explanation of how a CLT is used in making a scale transition from some initial scale to the Fickian approximation at a larger space–time scale.

The scaling of space is intimately tied to the scaling of time and is explained from the viewpoint of measurements.

3.1 On the central limit theorem and Fickian approximation

To understand a CLT in an informal manner, let $\{V(s): s \geqslant 0\}$ be a (strictly) stationary and ergodic one-dimensional stochastic process. Stationarity is the property of invariance of the joint probability distributions under time translation, and has already been explained in Section 2.1. *Ergodicity* refers to the condition under which time averages of functions of the process $V(s)$ can be approximated by their ensemble averages; see Karlin and Taylor (1975, p. 487), or Bhattacharya and Waymire (1990, Ch. 4). For example, if $\bar{V} = E\{V(s)\} = E\{V(0)\}$ (by stationarity), then by ergodicity

$$\lim_{t \to \infty} \frac{1}{t} \int_0^t V(s)\, \mathrm{d}s = \bar{V} \tag{3.2}$$

for almost all trajectories. If $\{V(s): s \geqslant 0\}$ is stationary and ergodic then a CLT holds under broad mathematical conditions. Billingsley (1968, Ch. 4) and Bhattacharya (1982) give conditions under which such a CLT holds.

To state the CLT for a stationary and ergodic stochastic process $V(s)$, let $Y(t)$ denote its time integral

$$Y(t) = \int_0^t V(s)\, \mathrm{d}s \tag{3.3}$$

Since $E\{V(s)\} = \bar{V}$, it follows simply from Eq. 3.3 that $E\{Y(t)\} = \bar{V}t$. Now, a CLT for $Y(t)$ means that

$$\lim_{t \to \infty} P\left(\frac{Y(t) - \bar{V}t}{\sqrt{t}} \leqslant y\right) = \int_{-\infty}^{y} \frac{1}{\sqrt{(2\pi\bar{\sigma}^2)}} \exp(-x^2/2\bar{\sigma}^2)\, \mathrm{d}x \tag{3.4}$$

Equation 3.4 says that the distribution function of $Y(t)$, centred at its mean $\bar{V}t$ and rescaled by $\sqrt{t}$ can be approximated for large t by a Gaussian distribution function with mean zero and variance $\bar{\sigma}^2$. The variance $\bar{\sigma}^2$ is given by

$$\bar{\sigma}^2 = 2 \int_0^{\infty} E\{(V(0) - \bar{V})(V(s) - \bar{V})\}\, \mathrm{d}s \tag{3.5}$$

We will leave it as an exercise for the reader to derive Eq. 3.5 using Eq. 3.3 and the stationarity of the process $\{V(s): s \geqslant 0\}$.

By a simple change of variables, it follows from Eq. 3.4 that for large t the pdf of $Y(t)$, say $f(t, y)$, is approximately Gaussian with mean $\bar{V}t$ and variance $\bar{\sigma}^2 t$, i.e.

$$f(t, y) \sim \frac{1}{\sqrt{(2\pi\bar{\sigma}^2 t)}} \exp\{-(y - \bar{V}t)^2/2\bar{\sigma}^2 t\} \tag{3.6}$$

and therefore it satisfies, approximately,

$$\frac{\partial f}{\partial t} = \tfrac{1}{2}\bar{\sigma}^2 \frac{\partial^2 f}{\partial y^2} - \bar{V}\frac{\partial f}{\partial y} \tag{3.7}$$

This is known as the *Fickian approximation* in the hydrology literature.

Suppose the CLT holds for the time integral $\mathbf{Y}(t) = [Y_1(t), Y_2(t), Y_3(t)]$ of a vector-valued stationary and ergodic process $\mathbf{V}(s) = [V_1(s), V_2(s), V_3(s)]$. In this case the pdf $f(t, \mathbf{y})$ of $\mathbf{Y}(t)$, for large t, can be approximated by a multivariate Gaussian pdf with mean $t\bar{\mathbf{V}}$ and covariance matrix $t\bar{\mathbf{M}} = ((t\bar{M}_{ij}))$, i.e.

$$f(t, \mathbf{y}) \sim (2\pi t)^{-3/2}(\det \bar{\mathbf{M}})^{-1/2} \exp\left\{-\frac{1}{2t}\sum_{i,j=1}^{3}(y_i - \bar{V}_i t)\bar{M}^{ij}(y_j - \bar{V}_j t)\right\} \tag{3.8}$$

where $\det \bar{\mathbf{M}}$ denotes the determinant of $\bar{\mathbf{M}}$ and $\bar{M}^{ij}$ is the (i,j) element of the inverse matrix $\bar{\mathbf{M}}^{-1}$. The elements of the matrix $\bar{\mathbf{M}}$ are given by

$$\bar{M}_{ij} = \int_0^\infty E\{(V_j(0) - \bar{V}_j)(V_i(s) - \bar{V}_i)] \, ds + \int_0^\infty E\{(V_i(0) - \bar{V}_i)(V_j(s) - \bar{V}_j)\} \, ds \quad 1 \leqslant i, j \leqslant 3 \tag{3.9}$$

The reader may check by differentiation that the Gaussian density in Eq. 3.8 satisfies

$$\frac{\partial f}{\partial t} = \sum_{i,j=1}^{3} \tfrac{1}{2}\bar{M}_{ij}\frac{\partial^2 f}{\partial y_i \partial y_j} - \sum_{i=1}^{3} \bar{V}_i \frac{\partial f}{\partial y_i} \tag{3.10}$$

The computation of the integrals in Eq. 3.9 requires specific model assumptions on the process $\mathbf{V}(s)$. In applications to dispersion in porous media, the process $\mathbf{V}(s)$ is a function of a Markov process. Therefore, we describe below a general result for computing $((\bar{M}_{ij}))$ in the context of a CLT for Markov processes.

3.2 A formula for the asymptotic covariance matrix in a central limit theorem for Markov processes

Let $\{\mathbf{X}(t): t \geqslant 0\}$ be a stationary and ergodic Markov process on the state space $S(\subset \mathbb{R}^3)$ having a stationary transition density $p(t; \mathbf{y}, \mathbf{x})$ satisfying Eqs 2.12 and 2.13 and having a stationary pdf $\pi(\cdot)$ satisfying Eq. 2.7. The density $\pi(\cdot)$ is the solution of the forward equation

$$A\pi \equiv \frac{1}{2} \sum_{i,j=1}^{3} M_{ij} \frac{\partial^2 \pi}{\partial x_i \partial x_j} - \sum_{i=1}^{3} \frac{\partial U_i \pi}{\partial x_i} = 0 \tag{3.11}$$

along with the attendant boundary conditions. Our interest is in a CLT for the time integral of a function $\mathbf{U}(\mathbf{X}(t))$ of the Markov process $\mathbf{X}$.

Let $\bar{\mathbf{U}} = (\bar{U}_1, \bar{U}_2, \bar{U}_3)$ be defined by

$$\bar{U}_i = \int_S U_i(x)\pi(x)\,\mathrm{d}x, \quad i = 1, 2, 3 \tag{3.12}$$

Then $\bar{\mathbf{U}}$ serves as the large-scale mean velocity vector about which the process $\mathbf{U}(\mathbf{X}(t))$ is to be centred in the CLT. Let $\mathbf{h}(\mathbf{x}) = (h_1(\mathbf{x}), h_2(\mathbf{x}), h_3(\mathbf{x}))$ be a vector-valued function on S defined by

$$h_i(\mathbf{x}) = U_i(\mathbf{x}) - \bar{U}_i, \quad i = 1, 2, 3 \tag{3.13}$$

Then it follows from Eqs 3.12 and 3.13 that

$$\int_S h_i(\mathbf{x})\pi(\mathbf{x})\,\mathrm{d}\mathbf{x} = 0, \quad i = 1, 2, 3 \tag{3.14}$$

Further, assume that

$$\int_S h_i^2(\mathbf{x})\pi(\mathbf{x})\,\mathrm{d}\mathbf{x} < \infty \tag{3.15}$$

Under fairly general conditions it is proved in Bhattacharya (1982) that the process $\int_0^t \mathbf{h}(\mathbf{X}(s))\,\mathrm{d}s$ can be approximated for large time t by a Brownian motion with mean $\mathbf{0}$ and covariance matrix $((t\bar{M}_{ij}))$. The elements of the matrix $\bar{\mathbf{M}}$ are given by

$$\bar{M}_{ij} = \langle h_i, g_j \rangle + \langle h_j, g_i \rangle, \quad 1 \leqslant i, j \leqslant 3 \tag{3.16a}$$

where

$$\langle h_i, g_j \rangle = \int_S h_i(\mathbf{y}) g_j(\mathbf{y}) \pi(\mathbf{y})\, d\mathbf{y} \tag{3.16b}$$

and the function $g_k(\cdot)$ is a solution of the equation

$$\tilde{A} g_k \equiv \frac{1}{2} \sum_{i,j=1}^{3} M_{ij} \frac{\partial^2 g_k}{\partial y_i \partial y_j} + \sum_{i=1}^{3} U_i(\mathbf{y}) \frac{\partial g_k}{\partial y_i} = -h_k(\mathbf{y}), \quad k = 1, 2, 3 \tag{3.17}$$

along with the attendant boundary conditions. The reader may notice that $\tilde{A}$ in Eq. 3.17 is Kolmogorov's backward operator defined in Eq. 2.11.

Equations 3.16 and 3.17 are basic for the computation of large-scale dispersion coefficients $\bar{M}_{ij}$ in the Fickian approximation. In Section 5 we will illustrate this method of computation for the Taylor–Aris theory, and for dispersion in layered and non-layered periodic porous media.

3.3 On scale change and the central limit theorem

The Fickian approximation using CLT involves a change of spatial scale as well. The scaling of time as $t = \lambda t'$, where $\lambda > 0$ is a large parameter, may be understood by expressing the result in Eq. 3.4 as follows: *For large λ, the process $\{\mathbf{Y}(\lambda t') : t' \geqslant 0\}$ at the new time scale t' may be approximated by the process* $\{\mathbf{Z}(\lambda t') : t' \geqslant 0\}$, where $\mathbf{Z}(t)$ is a Brownian motion with mean $t\bar{\mathbf{V}}$ and covariance matrix $t\bar{\mathbf{M}}$. Note that the pdf of $\mathbf{Z}(t)$ satisfies Eq. 3.10.

To understand the spatial scaling corresponding to the time scaling $t = \lambda t'$ evaluate the pdf of the Brownian motion $\mathbf{Z}(t)$, given by Eq. 3.8, at time $t = \lambda t'$ and spatial coordinate $\mathbf{y} = \lambda t' \bar{\mathbf{V}} + \sqrt{\lambda}\mathbf{x}$. One obtains

$$f(\lambda t', \lambda t' \bar{\mathbf{V}} + \sqrt{\lambda}\mathbf{x}) \simeq (\lambda)^{-3/2} (2\pi t')^{-3/2} (\det \bar{\mathbf{M}})^{-1/2} \exp\left\{ -\frac{1}{2t'} \sum \bar{M}^{ij} x_i x_j \right\} \tag{3.18}$$

Therefore the total concentration in a region of volume $\lambda^{3/2}$ centred at $\lambda t' \bar{\mathbf{V}} + \sqrt{\lambda}\mathbf{x}$ is given by

$$\lambda^{3/2} f(\lambda t', \lambda t' \bar{\mathbf{V}} + \sqrt{\lambda}\mathbf{x}) \simeq (2\pi t')^{-3/2} (\det \bar{\mathbf{M}})^{-1/2} \exp\left\{ -\frac{1}{2t'} \sum \bar{M}^{ij} x_i x_j \right\} \tag{3.19}$$

The space–time scaling leading to Eqs 3.18 and 3.19 may be best understood in terms of measurements. As time evolves, the solute concentration spreads more and more in space. For large times, most of the solute is concentrated around the large-scale mean or centre of mass $t\bar{\mathbf{V}} = \lambda t'\bar{\mathbf{V}}$. Scaling the space enables us to look at those volumes of regions of space (around $\lambda t'\bar{\mathbf{V}}$) in which the solute concentration is not negligible. Stated in another way, if one measures solute concentration over a volume of space which is much larger or much smaller than $\lambda^{3/2}$, then one would end up with either measuring almost all the solute mass or none. Hence, over such volumes the Fickian approximation will not be meaningful. In this sense the CLT provides the scaling of time and space simultaneously. In many cases the parameter λ connecting the small-scale $(t, \mathbf{y})$ and the large-scale $(t', \mathbf{x})$ can be estimated.

4. A kinetic theoretic derivation of the diffusion equation in saturated porous media at the microscopic scale

In this section, we derive a diffusion equation governing concentration of a non-reactive dilute solute in a saturated porous medium. The derivation is based on a stochastic-dynamical model of the velocity $\{\mathbf{V}(t);\ t \geqslant 0\}$ of a solute molecule at the kinetic time scale. A CLT shows that the displacements $\{\mathbf{X}(t);\ t \geqslant 0\}$, being an integral of the velocity process, is a Brownian motion at an appropriate scale. Since the solute is dilute, displacements of different molecules are approximately independent of each other. Therefore, the proportion of molecules in a small volume of space, i.e. the solute concentration, can be equated to the transition density of the Brownian motion by the law of large numbers. The diffusion equation so obtained applies at the *microscopic scale*, which is a liquid–solid continuum. This scale is larger than the pore scale and smaller than the Darcy scale.

We begin this discussion with a brief review of the kinetic theory of diffusion in aqueous solutions. This review provides the necessary background for understanding the derivation of a diffusion equation in porous media given here, and in interpreting physically the appearance of the microscopic scale at which this equation applies. The diffusion equation is also the starting point in Section 5 for the explanation of dispersion in saturated porous media.

4.1 Diffusion in aqueous solutions: from kinetic scale to fluid-mechanical scale

The concentration $C(t, \mathbf{x})$ of a non-reactive tracer in an aqueous solution under zero convection is described by the well-known Fickian equation (Landau and Lifshitz, 1959, p. 226),

$$\frac{\partial C}{\partial t} = D_0 \nabla^2 C \tag{4.1}$$

where ∇^2 is the Laplacian and D_0 is the coefficient of molecular diffusion. Albert Einstein in 1905 was the first to derive an expression for D_0 for a colloidal suspension in an aqueous solution. His derivation was based on the assumptions that colloidal molecules are approximately spherical and are large in comparison with the molecules of the aqueous solution. The successive displacements of the colloidal molecules by collisions with liquid molecules were independent giving rise to a random walk or Brownian motion. Einstein's work on Brownian movement can be found in Einstein (1956). This important work also led to a widespread acceptance of the atomic theory of matter by physicists; see Nelson (1967) for a brief historical background.

Einstein's theory was kinematic in so far as it did not explicitly consider the velocity process of colloidal molecules. A stochastic-dynamical basis of Einstein's theory of diffusion in aqueous solutions was proposed by Ornstein–Uhlenbeck some thirty years later. A great deal of physics and mathematics literature has been devoted to the study of the Ornstein–Uhlenbeck process; see Nelson (1967), Fox (1978) and the collection of articles by Chandrasekhar, Ornstein, Doob, etc., in Wax (1954).

In the Ornstein–Uhlenbeck theory, one considers a time interval $(t, t + \mathrm{d}t)$, which is large in comparison with the time scale of collisions of the molecules of the aqueous solution with the colloidal molecules. Therefore, a large number of collisions take place in $(t, t + \mathrm{d}t)$. Let us refer to this scale as the *kinetic time scale*. The net change in momentum per unit mass of a colloidal molecule in a small time interval $(t, t + \mathrm{d}t)$ due to a large number of collisions is $\sigma_1\, \mathrm{d}\mathbf{B}(t)$, where $\sigma_1 > 0$ is a scalar. One can justify this Brownian motion approximation by a CLT applied to the sum of changes in momentum due to individual collisions. In addition, the liquid exerts a frictional force on the colloidal molecule. Therefore, the total change in momentum $m\,\mathrm{d}\mathbf{V}(t)$ of a colloidal molecule of mass m is represented by the Itô's equation,

$$m\,\mathrm{d}\mathbf{V}(t) = -m\beta\mathbf{V}(t)\,\mathrm{d}t + m\sigma_1\,\mathrm{d}\mathbf{B}(t) \tag{4.2}$$

Here β is the coefficient of friction per unit mass. It is clear from Eq. 4.2 and dimensional considerations that β has the dimensions of inverse time. For spherical molecules of radius a, Stokes' law gives $m\beta = 6\pi\eta a$, where η is the coefficient of viscosity of the aqueous solution. Equation 4.2 is known as the *Langevin equation.*

The stochastic process $\mathbf{V}(t)$ obtained as a solution of Eq. 4.2 with an arbitrary initial state $\mathbf{V}(0)$ is not stationary. However, if the pdf $\pi(\mathbf{v})$ of $\mathbf{V}(0)$ is Gaussian with mean $\mathbf{0}$ and covariance matrix $(\sigma_1^2/2\beta)\mathbf{I}$, then $\mathbf{V}(t)$ becomes a stationary Markov process. The pdf π satisfies Eq. 2.7. The process $\{\mathbf{V}(t): t \geqslant 0\}$ has mean $\mathbf{0}$ and covariance (Breiman, 1968, p. 349)

$$\begin{aligned} E[V_i(s)V_j(t)] &= \frac{\sigma_1^2}{2\beta}\exp[-\beta(t-s)], \quad i = j \\ &= 0 \qquad\qquad\qquad\qquad\quad, \quad i \neq j \end{aligned} \tag{4.3}$$

The joint pdf of $\mathbf{V}(s_1), \mathbf{V}(s_2), \ldots, \mathbf{V}(s_n)$ for arbitrary times $s_1 < s_2 < \ldots < s_n$ can be shown to be Gaussian. Therefore, Eq. 4.3 completely specifies the stochastic process $\{\mathbf{V}(s): s \geqslant 0\}$.

In order to express σ_1^2 in terms of physical parameters, the law of equipartition of energy can be invoked to equate the mean kinetic energy $mE[V_i^2]/2$ with $k_B T/2$, where k_B is Boltzmann's constant and T is the absolute temperature (Nelson, 1967, p. 55). Therefore it follows from Eq. 4.3 that

$$\tfrac{1}{2} mE[V_i^2(t)] = \tfrac{1}{2} m \frac{\sigma_1^2}{2\beta} = \tfrac{1}{2} k_B T, \quad i = 1, 2, 3$$

or

$$\sigma_1^2 = 2\beta k_B T/m \tag{4.4}$$

A CLT for the time integral $\mathbf{X}(t)$ of the Ornstein–Uhlenbeck velocity process shows that at large times $\mathbf{X}(t)$ can be approximated by a Brownian motion process with mean $\mathbf{0}$ and covariance matrix $((tM_{ij}))$. To compute M_{ij} we substitute Eq. 4.3 into Eq. 3.9 to get

$$M_{ii} = 2\int_0^\infty E[V_i(0)V_i(s)]\,\mathrm{d}s = \frac{\sigma_1^2}{\beta}\int_0^\infty \exp(-\beta s)\,\mathrm{d}s = \frac{\sigma_1^2}{\beta^2}, \quad i = 1, 2, 3$$

$$M_{ij} = 0, \quad i \neq j \tag{4.5}$$

Substituting Eq. 4.4 into Eq. 4.5, Einstein's celebrated expression for the coefficient of molecular diffusion D_0 is obtained

$$D_0 = \tfrac{1}{2} M_{ii} = \frac{k_B T}{m\beta}, \quad i = 1, 2, 3 \tag{4.6}$$

A detailed calculation shows that the time at which the Brownian motion approximation holds is large compared to β^{-1} (Nelson, 1967, p. 57). For colloidal molecules β^{-1} is of the order of 10^{-8} s.

It is simple to show that the colloidal concentration $C(t, \mathbf{x})$ is the same as the transition density $p(t, \mathbf{x})$ of the coordinate process $\mathbf{X}(t)$. For this, let $I(\mathbf{y})$ be a counting function: $I(\mathbf{y}) = 1$ if $\mathbf{y}$ belongs to a small volume $\mathrm{d}V$ centred on $\mathbf{x}$, and 0 otherwise. Since the colloidal suspension is dilute, the positions $\mathbf{X}_i(t)$, $1 \leqslant i \leqslant n$, of the n colloidal molecules are approximately independent. Now, $C(t, \mathbf{x})\,\mathrm{d}V$ is equal to the proportion of molecules in $\mathrm{d}V$ at time t. This proportion is approximately equal to $p(t, \mathbf{x})\,\mathrm{d}V$ for large n by the law of large numbers (Feller, 1971, p. 238). This can be displayed as

$$C(t, \mathbf{x})\,\mathrm{d}V = \frac{1}{n}\sum_{i=1}^{n} I(\mathbf{X}_i(t)) \simeq p(t, \mathbf{x})\,\mathrm{d}V \tag{4.7}$$

Since at the large time scale $\mathbf{X}(t)$ is a Brownian motion with mean $\mathbf{0}$ and covariance matrix $D_0\mathbf{I}$ its transition density $p(t, \mathbf{x})$ satisfies Eq. 4.1. Therefore, it follows from Eq. 4.7 that $C(t, \mathbf{x})$ obeys Eq. 4.1. The space–time scale at which this equation applies can be called the *fluid-mechanical scale.*

4.2 Diffusion in saturated porous media: from kinetic scale to microscopic scale

It has been known experimentally that diffusion in a water-saturated porous medium can be described by a Fickian equation, such as displayed in Eq. 4.1. However, the coefficient of molecular diffusion D_0' in a porous medium is always observed to be a fraction of D_0 in aqueous solutions. Typically, the ratio D_0'/D_0 varies in the range 0.4–0.8 (Fried and Combarnous, 1971). The appearance of the Fickian approximation can be justified by the fact that a CLT holds under great generality for the time integral of a stationary, ergodic velocity process $\{\mathbf{V}(s);\ s \geqslant 0\}$ of a solute molecule. However, a precise theoretical demonstration of the experimental observation that $0 < D_0'/D_0 < 1$ requires detailed model calculations. This was carried out in two articles by Bhattacharya and Gupta (1979) and Gupta, Bhattacharya and Sposito (1981), as briefly explained below.

As before, let $\mathbf{V}(s)$ denote the velocity of a solute molecule in a saturated porous medium. Consider the kinetic time scale, and assume that the molecule undergoes contact collisions with the solid grains of the porous medium at successive random times $\tau_1 < \tau_2 < \ldots < \tau_n < \ldots$. Within any time interval $(\tau_i, \tau_{i+1}]$, the velocity is governed by the Ornstein–Uhlenbeck velocity process. To complete the specification of stochastic dynamics, it is necessary to describe the dynamics governing the collisions of a solute molecule with the solid grains. It is assumed that the velocity $\mathbf{V}(\tau_i +)$, soon after a collision with a solid grain at time τ_i, is given by

$$\mathbf{V}(\tau_i +) = \rho|\mathbf{V}(\tau_i)|\boldsymbol{\varepsilon}_i \tag{4.8}$$

where $|\cdot|$ is the magnitude of a vector, and $\{\boldsymbol{\varepsilon}_i\}$ is a sequence of independent random vectors of length one. The vector $|\mathbf{V}(\tau_i)|\boldsymbol{\varepsilon}_i$ gives a random direction to the molecule without any change in its velocity. The parameter ρ accounts for a possible decrease in velocity magnitude due to dissipation of kinetic energy during a collision. It can take any value between 0 and 1. The highly irregular nature of the porous medium geometry and an assumption of local isotropy of the porous medium imply that every random direction is equally likely. This means that ε_i, $i \geqslant 1$ have a uniform probability distribution on the surface of the unit sphere.

Let $f(s)$ be the common pdf of $(\tau_{i+1} - \tau_i)/\mu$, where $\mu = E[\tau_{i+1} - \tau_i]$ is the mean collision time. Bhattacharya and Gupta (1979) proved a CLT and showed that at a time scale which is large compared to μ the time integral $\mathbf{X}(t)$ of the velocity process $\mathbf{V}(s)$ is a Brownian motion with mean $\mathbf{0}$ and covariance matrix $t\mathbf{M}'$, where

$$M'_{ii} = g(\beta\mu)\frac{\sigma_1^2}{\beta^2}, \quad i = 1, 2, 3$$

$$M'_{ij} = 0, \quad i \neq j \tag{4.9}$$

The function $g(\cdot)$ is given by,

$$g(z) = 1 - \frac{1}{z}\int_0^\infty \{2(1 - \mathrm{e}^{-zs}) - \tfrac{1}{2}(1 - \mathrm{e}^{-2zs})\} f(s)\,\mathrm{d}s \tag{4.10}$$

It follows from Eqs 4.4, 4.6 and 4.9, that the coefficient of molecular diffusion in a porous medium D'_0 is related to D_0 in aqueous solutions as,

$$\tfrac{1}{2}M'_{ii} = D'_0 = g(\beta\mu)D_0, \quad i = 1, 2, 3 \tag{4.11}$$

The reader should check using Eq. 4.10 that $0 < g(z) < 1$. It follows from the fact that, for $z > 0$ and all $s > 0$, the integrand in Eq. 4.10 is strictly positive. This finding gives a rigorous theoretical explanation of the experimental observation that D'_0 is a fraction of D_0 (Fried and Combarnous, 1971).

The above results provide an insight into the structure of scales in saturated porous media at the 'pore level'. A typical model calculation of $g(\beta\mu)$, using Eq. 4.10 and a model assumption of $f(\cdot)$ (e.g. exponential) shows that $\beta\mu$ lies in the range 2.0–8.0, corresponding to $0.4 < D'_0/D_0 < 0.8$ (Gupta *et al.*, 1981, p. 364). This means that μ and β^{-1} are of the same order of magnitude. Thus the presence of the solid phase prevents a molecule from travelling unimpeded in the liquid phase for times which are large compared to β^{-1}. This leads to two important conclusions. First, a Fickian diffusion approximation does not hold within a 'pore' of a porous medium containing the liquid phase. We discussed below Eq. 4.6 that according to the Ornstein–Uhlenbeck theory a diffusion approximation in the liquid phase is only true at time scales which are large compared to β^{-1}. However, this condition is violated because of the disordered nature of the solid phase geometry. Secondly, at time scales which are large compared to $\mu \simeq \beta^{-1}$ the porous medium should be viewed as a liquid–solid continuum. In this continuum the solute concentration is governed by the following diffusion equation

$$\frac{\partial C}{\partial t} = D'_0 \nabla^2 C \tag{4.12}$$

The appearance of this equation follows from the fact that, at times large compared to μ, $\mathbf{X}(t)$ is a Brownian motion, and that the displacements of different solute molecules are approximately independent so that Eq. 4.7 holds. The presence of both the liquid and solid phases in Eq. 4.12 is reflected in the expression for D'_0 through the physical parameters μ, T, η, etc. This scale is termed the *microscopic scale* by Bhattacharya and Gupta (1983). It should be clearly distinguished from the notion of a *pore scale* which implies a separation of the liquid and solid phases in the diffusion approximation.

It should be emphasized that the Langevin Eq. 4.2 holds for large colloidal molecules in dilute concentration. Consequently the theory developed above applies to this case. For small solute molecules the velocity within a pore may be taken to be governed by a *generalized Langevin equation* (Fox, 1978; Cushman, 1987). Consequently the velocity process $\{\mathbf{V}(s): s \geqslant 0\}$ at the pore scale, including collisions with the solid grains, may be shown to be stationary and ergodic, and a CLT may be applied to its time integral $\mathbf{X}(t)$. Thus the qualitative conclusions concerning the appearance of a Fickian approximation at the microscopic scale would still hold.

In the presence of a non-zero constant convective velocity the Langevin Eq. 4.2 is modified to include a drift velocity. But as is proved by Bhattacharya and Gupta (1979) the above conclusions remain valid. In this case, one arrives at the following modification of the diffusion Eq. 4.12 at the microscopic scale

$$\frac{\partial C}{\partial t} = D'_0 \sum_{i=1}^{3} \frac{\partial^2 C}{\partial x_i^2} - \sum_{i=1}^{3} U_i \frac{\partial C}{\partial x_i} \tag{4.13}$$

where $\mathbf{U} = (U_1, U_2, U_3)$ represents the velocity at the microscopic scale.

It has been widely assumed in the literature (with no *a priori* justification) that a diffusion approximation holds within a pore of a porous medium while the highly irregular solid phase serves as mathematical boundaries. This viewpoint has led to the introduction of many *ad hoc* models of porous medium geometry. It has also posed numerous difficulties in a rigorous understanding of diffusion and dispersion in porous media; see the review article by Sposito, Gupta and Bhattacharya (1979) and discussion by Brenner (1980a, p. 127) regarding the role of diffusion in explaining dispersion in porous media. These difficulties are circumvented by the theoretical demonstration of the presence of a microscopic scale larger than the pore scale and the validity of a Fickian approximation given by Eq. 4.13 at the microscopic scale. This equation constitutes the starting point for the explanation of dispersion in saturated porous media at the Darcy and field scales.

5. Dispersion in saturated porous media

The difference between 'dispersion' and 'diffusion' of solutes in aqueous solutions and in saturated porous media can be best explained from two observations. First, *dispersion* can only arise if the liquid is in motion with a spatially variable velocity $\mathbf{U}(\mathbf{x})$. Second, the space–time scale at which dispersion manifests is always larger than the scale at which diffusion appears. The transition from diffusion to dispersion is made by application of an appropriate CLT. Let us first explain these two observations within the context of the general mathematical theory described in Sections 2 and 3.

Let $\mathbf{U}(\mathbf{x})$ denote a spatially variable convective liquid velocity at the microscopic scale in a saturated porous medium. Recall from our discussion at the end of Section 4 that $\mathbf{U}$ is taken to be a constant in the derivation of Eq. 4.13 from the kinetic scale to the microscopic scale. However, due to the presence of heterogeneities at the microscopic scale, spatial variability in $\mathbf{U}(\mathbf{x})$ appears at this scale. Therefore, Eq. 4.13 may be expressed as

$$\frac{\partial C}{\partial t} = \sum_{i=1}^{3} D'_0 \frac{\partial^2 C}{\partial x_i^2} - \sum_{i=1}^{3} U_i(\mathbf{x}) \frac{\partial C}{\partial x_i} \tag{5.1}$$

subject to some initial concentration $C(0, \mathbf{x}) = C_0(\mathbf{x})$. The velocity $\mathbf{U}(\mathbf{x})$ is assumed to satisfy the continuity condition

$$\sum_{i=1}^{3} \frac{\partial U_i}{\partial x_i} = 0 \tag{5.2}$$

From the two equations above and Eq. 2.16 it follows that the Markov process $\mathbf{X}(t)$ at the microscopic scale is governed by the following Itô's equation

$$\mathbf{X}(t) = \int_0^t \mathbf{U}(\mathbf{X}(s))\, \mathrm{d}s + \sqrt{(2D'_0)}\mathbf{B}(t) \tag{5.3}$$

It was explained in Section 3 that, under certain conditions on $\mathbf{U}(\mathbf{x})$, the first term on the right-hand side of Eq. 5.3 can be approximated for large t by a Brownian motion process. Let $\mathbf{W}(t)$ denote this limiting Brownian motion with mean $t\bar{\mathbf{U}}$ and the covariance matrix $((t\bar{M}_{ij}))$. In the special case when $\mathbf{B}(t)$ is asymptotically (as $t \to \infty$) independent of the time integral in Eq. 5.3, then the sum $\mathbf{X}(t)$ is also a Brownian motion in the limit with mean $t\bar{\mathbf{U}}$ and covariance matrix $t\bar{\mathbf{D}}$. Now the sum of two independent Brownian motions with mean vectors $t\mathbf{V}_1$ and $t\mathbf{V}_2$ and covariance matrices $t\mathbf{\Sigma}_1$ and $t\mathbf{\Sigma}_2$ is also a Brownian motion with mean vector $t(\mathbf{V}_1 + \mathbf{V}_2)$ and covariance matrix $t(\mathbf{\Sigma}_1 + \mathbf{\Sigma}_2)$. Therefore

$$\bar{D}_{ij} = \tfrac{1}{2}\bar{M}_{ij} + D'_0\delta_{ij} \tag{5.4}$$

where $\delta_{ij} = 1$ if $i = j$ and 0 otherwise.

The parameters $\bar{D}_{ij}$ are called the *dispersion coefficients.* Since $\bar{M}_{ii} \geqslant 0$, it follows that $\bar{D}_{ii} \geqslant D'_0$. Therefore, the diagonal dispersion coefficients are larger in magnitude than diffusion coefficients. Since $\mathbf{X}(t)$ is a Brownian motion at large times, it follows from Eq. 2.16 that the solute concentration $C(t, \mathbf{x})$ satisfies the Fickian equation

$$\frac{\partial C}{\partial t} = \sum_{i,j=1}^{3} \bar{D}_{ij} \frac{\partial^2 C}{\partial x_i \partial x_j} - \sum_{i=1}^{3} \bar{U}_i \frac{\partial C}{\partial x_i} \tag{5.5}$$

Based on our discussion in Section 3.3, it should be noted that the space–time scale at which the dispersion Eq. 5.5 applies is much larger than the scale at which the diffusion Eq. 5.1 applies. We call this the *laboratory*, or the *Darcy Scale.* Indeed, dispersion as a phenomenon different from diffusion can be observed experimentally in saturated porous media in laboratory columns (Fried and Combarnous, 1971). The experimentally measured breakthrough curves for solutes can be fitted well by solutions of Eq. 5.5, and the fitted diagonal dispersion coefficients $\bar{D}_{ii}$ are observed to increase in magnitude as the convective velocity is increased. Experiments have shown that $\bar{D}_{ii}$ are generally much larger in magnitude than the coefficient of diffusion D'_0, and this feature is exhibited theoretically by Eq. 5.4. This explains how dispersion is different from diffusion and also shows how a dispersion equation at the Darcy scale derives from a diffusion equation at the microscopic scale by a CLT.

Now suppose that in an aquifer the convective velocity $\bar{\mathbf{U}}$ of Eq. 5.5 is not uniform at the Darcy scale but varies spatially due to spatial variability in aquifer properties such as hydraulic conductivity. One then writes $\bar{\mathbf{U}} \equiv \bar{\mathbf{U}}(\mathbf{x})$ and assumes that $\bar{\mathbf{U}}(\mathbf{x})$ satisfies the continuity condition. In this case, the solute concentration in an aquifer at the Darcy scale is governed by the CDE

$$\frac{\partial C}{\partial t} = \sum_{i,j=1}^{3} \bar{D}_{ij} \frac{\partial^2 C}{\partial x_i \partial x_j} - \sum_{i=1}^{3} \bar{U}_i(\mathbf{x}) \frac{\partial C}{\partial x_i} \tag{5.6}$$

Notice that Eqs 5.6 and 5.1 are formally similar. Therefore, under appropriate conditions on $\bar{\mathbf{U}}(\mathbf{x})$, it follows from a similar argument as above that $C(t, \mathbf{x})$ at large times can be approximated by another Fickian equation with mean velocity say $\bar{\bar{\mathbf{U}}}$, and dispersion coefficients $((\bar{\bar{D}}_{ij}))$. The scale at which this Fickian approximation applies is clearly larger than the Darcy scale. Let us call this a *field scale*. Moreover, it also follows from an equation similar to Eq. 5.4 that $\bar{\bar{D}}_{ii} \geqslant \bar{D}_{ii}$, i.e. the field-scale dispersion coefficients are larger in magnitude than the dispersion coefficients at the laboratory scale. This feature is observed experimentally; see Pickens and Grisak (1981). The field-scale dispersion is sometimes called *macrodispersion* to distinguish it from the laboratory-scale dispersion.

Historically, from the mid-fifties to mid-seventies an understanding of laboratory-scale dispersion was considered to be an important problem; see Sposito *et al.* (1979) for a critical review. Since then the focus of attention in the hydrology literature has shifted to the field-scale dispersion. However, the above discussion shows clearly that understanding macrodispersion at some field scale on the basis of the CDE (Eq. 5.6) at the Darcy scale is mathematically equivalent to understanding dispersion at the Darcy scale based on the CDE (Eq. 5.1) at the microscopic scale.

One can now further generalize the basic idea explained above to solute transport in aquifers which exhibit multiple length scales of spatial variability in the convective velocity. In this case, the solute transport over the entire medium can be approximated by 'piecing together' the Fickian approximations over successive pairs of scales. Although some of the qualitative aspects have been described in Gupta and Bhattacharya (1986), much work needs to be done on this multiple-scales theory. These issues are explained at the end of Section 5.

In summary, the basic theory developed in Sections 2 and 3 applies to dispersion at different scales. However, as we shall show, different geometries of the velocity field $\mathbf{U}(\mathbf{x})$ may lead to different qualitative behaviour in the growth of dispersion coefficients. Therefore, our emphasis in this section is on precise computation of dispersion coefficients for different geometries of the velocity $\mathbf{U}(\mathbf{x})$. These consumptions are based on the general formula given in Section 3.2.

The remainder of this section is organized as follows. Section 5.1 explains a general result on the functional dependence of the asymptotic dispersion coefficients on *velocity* and *scale parameters* U_0 and a, respectively. It says that if a CLT holds then dispersion coefficients depend on a and U_0 only through their product aU_0. Section 5.2 gives the computation of dispersion coefficient for the classic Taylor–Aris theory of solute dispersion in an aqueous solution in a capillary (Taylor, 1953; Aris, 1956; Bhattacharya and Gupta, 1984). Section 5.3 analyses solute dispersion in *periodic* porous media, i.e. when $\mathbf{U}(\mathbf{x})$ is periodic in $\mathbf{x}$. Dispersion coefficients are computed in several examples. In Section 5.4 we briefly discuss some experimental and theoretical results on laboratory-scale dispersion. Section 5.5 is devoted to an explanation of *scale effect* in dispersion in aquifers.

In Eq. 5.5 $\bar{D}_{ij}$s are taken to be constants for simplicity. Even if one takes spatially dependent $\bar{D}_{ij}$ the Fickian approximation at large times remains valid and the qualitative nature of the growth of dispersion coefficients with respect to velocity and scale remains unchanged.

5.1 Interchangeability of velocity and spatial scale parameters in the large-scale dispersion coefficients

Consider a CDE governing solute concentration at some space–time scale in an infinite porous medium,

$$\frac{\partial C}{\partial t} = \sum_{i,j=1}^{3} D_{ij} \frac{\partial^2 C}{\partial x_i \partial x_j} - \sum_{i=1}^{3} U_i(\mathbf{x}) \frac{\partial C}{\partial x_i} \tag{5.7}$$

Physically, as explained above, the choice of the coefficients D_{ij} and the convective velocity $\mathbf{U}(\mathbf{x})$ depends on the scale at which Eq. 5.7 applies. For example, if the above CDE applies at the microscopic scale and the porous medium is isotropic, then $D_{ij} = D'_0$ for $i = j$ and 0 for $i \neq j$, and $\mathbf{U}(\mathbf{x})$ is the microscopic convective velocity. If, on the other hand, this CDE applies at the Darcy scale, then D_{ij}s represent dispersion coefficients at the Darcy scale and $\mathbf{U}(\mathbf{x})$ is the convective velocity at this scale. Since the choice of scale does not affect the mathematical treatment, in our subsequent discussion the scale at which Eq. 5.7 applies is left arbitrary. However, we assume that $\mathbf{U}(\mathbf{x})$ satisfies the continuity condition (Eq. 5.2) at every scale.

The dependence of dispersion coefficients on spatial scale and velocity can be investigated explicitly by introducing a velocity scale parameter U_0 and a spatial scale parameter a as follows:

$$\mathbf{U}(\mathbf{x}) = U_0 \mathbf{b}(\mathbf{x}/a) \tag{5.8}$$

The positive scalars U_0 and a are dimensionless parameters, whereas $\mathbf{b}$ has dimensions of velocity and $\mathbf{x}$ has dimensions of length, and $\mathbf{b}$ does not depend on U_0.

In view of Eq. 2.16, the Markov process $\mathbf{X}(t)$ corresponding to Eqs 5.7 and 5.8 can be represented as,

$$\mathbf{X}(t) = U_0 \int_0^t \mathbf{b}(\mathbf{X}(s)/a)\, \mathrm{d}s + \boldsymbol{\sigma}\mathbf{B}(t) \tag{5.9}$$

where

$$\tfrac{1}{2}\boldsymbol{\sigma}\boldsymbol{\sigma}^* - \mathbf{D}$$

Let the large-scale dispersion matrix of $\mathbf{X}(t)$ be denoted by $\bar{\mathbf{D}}(U_0, a)$. The following result is proved in Bhattacharya *et al.* (1989) regarding the functional dependence of $\bar{\mathbf{D}}$ on a and U_0.

Proposition 5.1. If the Central Limit Theorem holds for the stochastic process $\mathbf{X}(t)$ given by Eq. 5.9, then the dispersion matrix $\bar{\mathbf{D}}$ depends on a and U_0 only through their product aU_0. In particular,

$$\bar{\mathbf{D}}(a, U_0) = \bar{\mathbf{D}}(U_0, a) = \bar{\mathbf{D}}(aU_0, 1) \tag{5.10}$$

In view of Eq. 5.10 we will take the spatial scale parameter $a = 1$ in our subsequent discussion. Whenever we need to bring a back into the discussion regarding the behaviour of $\bar{\mathbf{D}}$ as a function of a and U_0 we will simply replace U_0 by aU_0 in the theoretical expressions for $\bar{\mathbf{D}}$. Moreover $\bar{\mathbf{D}}$ will be called the *large-scale dispersion matrix* and $\mathbf{D}$ the *small-scale dispersion matrix* from here on.

5.2 Taylor–Aris theory of dispersion in a capillary

Consider a long capillary of radius a in which water is moving under laminar flow conditions. Since a scales the velocity we take it to be unity. The Navier–Stokes equation and the continuity equation show that the convective velocity is given by $\mathbf{U}(\mathbf{x}) = U_0(b_1(x_2, x_3), 0, 0)$, where $\mathbf{x} = (x_1, x_2, x_3)$, and

$$b_1(x_2, x_3) = 1 - (x_2^2 + x_3^2) \tag{5.11}$$

Diffusion under convection at the fluid-mechanical scale in a capillary is described by Eq. 5.7 with

$$\begin{aligned} D_{ij} &= D_0, \quad \text{if } i = j \\ &= 0, \quad\ \ \text{if } i \neq j \end{aligned} \tag{5.12}$$

In addition, a no-flux boundary condition is imposed on Eq. 5.7, i.e. $\partial C/\partial n = 0$ at the capillary boundary. Here $\partial/\partial n$ denotes differentiation normal to the boundary.

The large time Fickian approximation of the solution of Eq. 5.7 in a capillary was analysed in a classic paper by G. I. Taylor (1953). A more systematic mathematical analysis of Taylor's theory was given later by Aris (1956). We will now illustrate the computation of the large-scale dispersion coefficient in the Taylor–Aris theory using the formula given in Section 3.2.

In view of Eqs 5.11 and 5.12, the coordinate $X_1(t)$ of the stochastic process

$\mathbf{X}(t) = (X_1(t), X_2(t), X_3(t))$ may be represented as

$$X_1(t) = U_0 \int_0^t [1 - (X_2^2(s) + X_3^2(s))]\, ds + \sigma_0 B_1(t) \tag{5.13}$$

Here $\sigma_0^2 = 2D_0$, B_1 is a standard Brownian motion, and (X_2, X_3) is a two-dimensional Brownian motion on the unit disc $E = \{x_2^2 + x_3^2 \leqslant 1\}$ reflected at the boundary having the diffusion coefficient D_0. Also (X_2, X_3) is independent of B_1. The main problem is to compute the large scale dispersion coefficient of the coordinate process $X_1(t)$.

Since the integral in Eq. 5.13 is a function of $(X_2(t), X_3(t))$, we first need to compute the stationary probability density $\pi(x_2, x_3)$ of this process. It follows from Eq. 3.11 that $\pi(x_2, x_3)$ is the solution of $\nabla^2 \pi = 0$ on the unit disc E, subject to the boundary condition $\partial\pi/\partial n = 0$ on the boundary ∂E of E. It is simple to check that the solution of this equation is a constant on E. Since $\pi(x_2, x_3)$ integrates to 1 on E and π is the area of the unit disc, this constant is simply $1/\pi$.

Now define the radial Markov process $R(t)$ as

$$R(t) = (X_2^2(t) + X_3^2(t))^{1/2} \tag{5.14}$$

and note that the drift velocity $U_0 b_1$ (see Eq. 5.11) is a function of $R(t)$. The stationary density $\pi(r)$ of the process $R(t)$ can be obtained by changing Cartesian coordinates (x_2, x_3) into polar coordinates (r, θ). Since $dx_2\, dx_3 = r\, dr\, d\theta$, integrating the uniform density with respect to θ from 0 to 2π gives,

$$\pi(r) = 2r, \quad 0 < r < 1 \tag{5.15}$$

The large scale mean drift velocity $\bar{U}_1$ can be computed using Eqs 3.12, 5.11 and 5.15, as

$$\bar{U}_1 = U_0 \int_0^1 (1 - r^2) 2r\, dr = \frac{U_0}{2} \tag{5.16}$$

Therefore, the function $\mathbf{h}(\mathbf{x})$ given by Eq. 3.13 becomes

$$h_1(r) = U_0(1 - r^2) - \frac{U_0}{2} = U_0(\tfrac{1}{2} - r^2) \tag{5.17}$$

Write $X_1(t)$ in Eq. 5.13 as

$$X_1(t) - \tfrac{1}{2}U_0 t = \int_0^t h_1(R(s))\,\mathrm{d}s + \sigma_0 B_1(t)$$

$$= W_1(t) + \sigma_0 B_1(t) \tag{5.18}$$

where $W_1(t)$ is the integral of a function of the radial Markov process $R(t)$, which is both stationary and ergodic with stationary density $2r$. It obeys a CLT. In view of Eq. 3.16 the asymptotic variance parameter σ_1^2 of $W_1(t)$ is given by,

$$\sigma_1^2 = 2\int_0^1 g(r)h_1(r)2r\,\mathrm{d}r \tag{5.19}$$

where $g(r)$ is the solution of Kolmogorov's backward equation for $R(t)$ (see Eq. 3.17)

$$\tilde{A}g(r) \equiv D_0\left(\frac{\partial^2 g}{\partial r^2} + \frac{1}{r}\frac{\partial g}{\partial r}\right) = -h_1(r), \quad \left.\frac{\partial g}{\partial r}\right|_{r=1} = 0 \tag{5.20}$$

The reader simply can check that the solution of Eq. 5.20 is

$$g(r) = \frac{U_0}{4D_0}\left(\frac{r^2}{2} - \frac{r^4}{4}\right) + \text{const.} \tag{5.21}$$

It now follows from Eq. 5.19 and Proposition 5.1 that for a capillary of radius a,

$$\bar{M}_1 = \tfrac{1}{2}\sigma_1^2 = \frac{a^2U_0^2}{192D_0} \tag{5.22}$$

Since $W_1(t)$ and $B_1(t)$ are statistically independent, the variance of the asymptotic Brownian motion approximation of $X_1(t)$ is $t\bar{D}_1 = t(D_0 + \bar{M}_1)$. This is the well-known expression for the Taylor dispersion coefficient. Notice from Eq. 5.22 that as U_0 increases the term $\bar{M}_1$ due to convection in the large-scale dispersion coefficient completely dominates the diffusion coefficient D_0. Taylor conducted experiments and verified the expression for the large-scale dispersion coefficient $\bar{D}_1$. For further mathematical details and proofs the reader can refer to Gupta and Bhattacharya (1983) and Bhattacharya and Gupta (1984).

5.3 Dispersion in periodic porous media

Next consider an infinite periodic porous medium in which the solute concentration $C(t, \mathbf{x})$ is governed by Eq. 5.7. The assumption of periodicity means that $\mathbf{b}(\mathbf{x}/a)$ in Eq. 5.8 is periodic in $\mathbf{x}$ with period a units in each coordinate. A CLT for the solution $\mathbf{X}(t)$ of Eq. 5.9 for periodic $\mathbf{b}$ has been proved by Bhattacharya (1985). However, for periodic media it is generally difficult to obtain closed-form analytical expressions for $((\bar{D}_{ij}))$ using Eq. 3.16. Therefore, an analytical theory has been developed recently by Bhattacharya *et al.* (1989) which investigates the behaviour of $\bar{D}_{ij}$s as functions of aU_0 for large values of aU_0. This theory shows that, for most periodic **b**s, $\bar{D}_{ij}$s either grow quadratically with aU_0 or become asymptotically constant as $aU_0 \to \infty$. Bhattacharya *et al.* (1989) give precise analytical conditions on the geometry of **b**s which determine how each $\bar{D}_{ij}$ behaves as a function of aU_0 for large aU_0. In this section, we will give two examples which provide closed-form solutions of $\bar{D}_{ij}$ and exhibit their quadratic growth and asymptotic constancy as functions of aU_0 (Gupta and Bhattacharya, 1986). Both of these examples assume that the porous medium is layered. In a third example for a non-layered porous medium the results for $\bar{D}_{ij}$ are taken from Bhattacharya *et al.* (1989) and explained.

In view of Proposition 5.1 we will assume throughout that period $a = 1$. The steady-state probability density $\pi(\mathbf{x})$ of the solution $\mathbf{X}(t)$ of Eq. 5.9 is obtained by solving Eq. 3.11, subject to the periodicity condition that $\pi(\mathbf{x})$ is the same on opposite surfaces of the unit cube $[0, 1]^3$. It is simple to check that the solution $\pi(\mathbf{x})$ of Eq. 3.11 is a constant. Since $\pi(\mathbf{x})$ integrates to 1 on the unit cube, $\pi(\mathbf{x}) = 1$ is the desired steady-state density.

Example 1 (layered periodic media)

The layering of the media is reflected in the condition that $b_1(\mathbf{x}) \equiv b_1(x_3)$, $b_2(\mathbf{x}) \equiv b_2(x_3)$, because the spatial variability in b_1 and b_2 manifests itself only in the vertical direction x_3 due to layering. Suppose also that $b_3(\mathbf{x}) = 0$. The state space S in Eqs. 3.12 and 3.16 is the unit cube $[0, 1]^3$ due to periodicity with opposite faces identified. As a concrete example let us take the dispersion matrix $\mathbf{D}$ to be diagonal and $b_1 = 1 + \sin(2\pi x_3)$, $b_2 = \sin(2\pi x_3)$ and $b_3 = 0$. Then it is simple to check from Eq. 3.12 that $\bar{b}_1 = 1, \bar{b}_2 = 0, \bar{b}_3 = 0$.

The components of $\mathbf{h}(\mathbf{x})$ in Eq. 3.13 are $h_3(\mathbf{x}) = 0$, and

$$h_1(\mathbf{x}) = U_0 \sin(2\pi x_3) = h_2(\mathbf{x}) \tag{5.23}$$

For the stochastic process $\mathbf{X}(t)$ in Eq. 5.9 one has

$$\mathbf{X}(t) - U_0\bar{\mathbf{b}}t = \int_0^t \mathbf{h}(X_3(s))\,\mathrm{d}s + \boldsymbol{\sigma}\mathbf{B}(t)$$

$$= \mathbf{W}(t) + \boldsymbol{\sigma}\mathbf{B}(t) \tag{5.24}$$

In order to compute the asymptotic dispersion matrix of $\mathbf{W}(t)$ first notice that $\mathbf{h}(\cdot)$ depends only on the vertical coordinate process $X_3(t)$. Therefore, $g_i(\cdot)$s in Eq. 3.16 are functions of x_3 and are obtained as solutions of Kolmogorov's backward equation

$$D_{33}\frac{\mathrm{d}^2 g_i}{\mathrm{d}x_3^2} = -U_0 \sin(2\pi x_3), \quad i = 1, 2 \tag{5.25}$$

subject to the periodicity condition, $g_i(0) = g_i(1)$. The reader can check, using Eqs 5.25, 3.16a and Proposition 5.1, that

$$\left.\begin{aligned} \bar{M}_{11} = \bar{M}_{22} = \bar{M}_{12} = \bar{M}_{21} &= \frac{U_0^2 a^2}{4\pi^2 D_{33}} \\ \bar{M}_{31} = \bar{M}_{13} = \bar{M}_{32} = \bar{M}_{23} &= 0 \end{aligned}\right\} \tag{5.26}$$

Since $((D_{ij}))$ are constants and the continuity condition (Eq. 5.2) holds, it follows from the CLT by Bhattacharya (1985) that $\mathbf{W}(t)$ and $\mathbf{B}(t)$ in Eq. 5.9 are asymptotically independent. Therefore, the large-scale dispersion coefficients of $\mathbf{X}(t)$ are given by $D_{ij} + \frac{1}{2}\bar{M}_{ij}$, where $\bar{M}_{ij}$ are displayed in Eq. 5.26. This example shows that $\bar{D}_{ij}$s grow quadratically with aU_0.

Example 2 (layered porous media)

This example is identical to Example 1 except that there is a non-zero uniform velocity $\bar{U}_0 b_3$ in the vertical direction. It follows from Example 1 that the large-scale mean is $\bar{\mathbf{U}} = U_0(1, 0, \bar{b}_3)$. The coordinate processes $X_1(t)$ and $X_2(t)$ in Eq. 5.24 remain unchanged but Eq. 5.25 changes. Specifically, the functions $g_i(\mathbf{x})$ are now obtained as solutions of

$$D_{33}\frac{\mathrm{d}^2 g_i}{\mathrm{d}x_3^2} + U_0\bar{b}_3\frac{\mathrm{d}g_i}{\mathrm{d}x_3} = -U_0 \sin(2\pi x_3), \quad i = 1, 2 \tag{5.27}$$

subject to the periodicity condition, $g_i(0) = g_i(1)$. The reader can check that the covariance matrix $((t\bar{M}_{ij}))$ is given by

$$\left.\begin{aligned}\bar{M}_{11} = \bar{M}_{22} = \bar{M}_{12} = \bar{M}_{21} &= \frac{D_{33}U_0^2a^2}{2\pi D_{33}^2 + (U_0a\bar{b}_3)^2}\\ \bar{M}_{31} = \bar{M}_{13} = \bar{M}_{23} = \bar{M}_{32} &= 0\end{aligned}\right\} \tag{5.28}$$

From the asymptotic independence of $\mathbf{W}(t)$ and $\mathbf{B}(t)$ in Eq. 5.24 it follows that $\bar{D}_{ij}$ are given by $D_{ij} + \frac{1}{2}\bar{M}_{ij}$, where $\bar{M}_{ij}$s are as displayed above. It is simple to see that for small values of aU_0, $\bar{D}_{ij}$s depend quadratically on aU_0. However, as $aU_0 \to \infty$, $\bar{D}_{ij}$s converge to constants. Specifically, it follows from Eq. 5.28 that,

$$\left.\begin{aligned}\lim_{aU_0\to\infty} \bar{D}_{ii} &= D_{ii} + \frac{D_{33}}{2}\left(\frac{1}{\bar{b}_3}\right)^2, \quad i = 1, 2\\ \lim_{aU_0\to\infty} \bar{D}_{21} &= \bar{D}_{12} = \frac{D_{33}}{2}\left(\frac{1}{\bar{b}_3}\right)^2\end{aligned}\right\} \tag{5.29}$$

Example 3 (non-layered media)

Consider a two-dimensional non-layered periodic porous medium with $\mathbf{b}(\mathbf{x})$ given by

$$\left.\begin{aligned}b_1(\mathbf{x}) &= 2 + \sin(2\pi(\sin(2\pi x_1) - x_2))\\ b_2(\mathbf{x}) &= b_1(\mathbf{x})2\pi\cos(2\pi x_1)\end{aligned}\right\} \tag{5.30}$$

The reader can check that $\mathbf{b}$ satisfies the continuity condition and that $\bar{b}_1 = 2$, $\bar{b}_2 = 0$. For this example closed form solutions of $g_i(\mathbf{x})$ using Eq. 3.17 cannot be obtained. However, the analytical theory developed by Bhattacharya *et al.* (1989, p. 96) shows that as $aU_0 \to \infty$,

$$\left.\begin{aligned}\bar{M}_{11} &= O(a^2U_0^2), \quad \bar{M}_{22} = O(1)\\ \bar{M}_{12} &= \bar{M}_{21} = O(aU_0)\end{aligned}\right\} \tag{5.31}$$

Moreover, by asymptotic independence between $\mathbf{W}(t)$ and $\mathbf{B}(t)$ in Eq. 5.9, the large-scale dispersion coefficients $\bar{D}_{ij}$ of $\mathbf{X}(t)$ are given by $D_{ij} + \frac{1}{2}\bar{M}_{ij}$

5.4 Linearity of the dispersion coefficients with velocity at the laboratory scale

Suppose that the x_1-axis is aligned along the direction of the mean velocity, i.e. $\mathbf{U} = (U_0\bar{b}_1, 0, 0)$. Experiments in laboratory columns show that the dispersion coefficients $\bar{D}_{ii}$ grow linearly as functions of aU_0 for large *Peclet number* aU_0/D'_0, i.e.

$$\bar{D}_{11} \sim c_{11}aU_0, \quad \bar{D}_{22} = \bar{D}_{33} \sim c_{22}aU_0 \tag{5.32}$$

Typically, the ratio c_{11}/c_{22} is between 4 and 10. The scale parameter a may be identified with the so-called *dispersivity* in the hydrology literature. It should be noted that although Example 3 above shows that $\bar{D}_{11} > \bar{D}_{22}$, the observed linear growth of $\bar{D}_{ii}$ at the Darcy scale is not predicted by the periodic model $\mathbf{U}(x)$ at the microscopic scale. Another calculation carried out in Bhattacharya and Gupta (1983) shows that for large Peclet numbers, $\bar{D}_{ii} \sim O(aU_0)$. Even though this method is independent of any specific model assumptions on $\mathbf{U}(\mathbf{x})$ it is somewhat heuristic and has not been justified rigorously. In this sense a precise theoretical explanation of the experimental finding regarding linearity of the dispersion coefficients with velocity and scale remains an open problem. Perhaps a more general class of model assumptions on $\mathbf{U}(\mathbf{x})$ than periodic, such as almost-periodic, should be analysed to prove linearity of $\bar{D}_{ii}$ as functions of aU_0. A CLT for *almost-periodic* velocity functions $\mathbf{U}(\mathbf{x})$ has been recently proved by Bhattacharya and Ramasubramanian (1988), where an expression for large-scale dispersion coefficients similar to Eq. 3.16 is derived. However, the growth of dispersion coefficients as functions of aU_0 has not been analysed in this case.

5.5 Explanation of scale effect in dispersion in natural formations: two scales versus multiple scales of heterogeneity

Many experimental studies show that when Gaussian densities are fitted to solute concentrations at successively larger distances (or times), the dispersion coefficients increase with distance. This is referred to as the *scale effect* in dispersion in aquifers; see for example Sudicky, Cherry and Frind (1983), and other references on this topic in Neuman *et al.* (1987, p. 454). Two contrasting theoretical viewpoints have been proposed to explain the scale effect. These two approaches are explained below.

Many authors assume the presence of only *two distinct scales* to describe dispersion in aquifers. These are the Darcy scale and a field scale. Due to spatial variability in the velocity at the Darcy scale, the diagonal dispersion coefficients at the field scale are larger than those at the Darcy scale. However, the dispersion coefficients grow with time until they reach their limiting large-scale values. After the Fickian approximation takes hold with these larger dispersion coefficients, there is no further growth in dispersion coefficients with distance and time. This may be called the *pre-asymptotic explanation* of the scale effect; see Gelhar and Axness (1983), Dagan (1984), and Neuman *et al.* (1987).

The second approach is based on the notion of *multiple scales.* Consider the experimentally observed growth in dispersion from the microscopic to the Darcy scale (Fried and Combarnous, 1971), and again from the Darcy to a larger field scale (Pickens and Grisak, 1981). The theoretical explanation for this growth is that newer heterogeneities in the medium properties arise from one scale to the next larger scale. It is then natural to envision that multiple scales of heterogeneity govern spatial variability in the velocity field in a typical aquifer. Such media have been labelled *evolving heterogeneous media* by Sposito *et al.* (1986) and have been further discussed by Wheatcraft and Tyler (1988).

To explain the growth in dispersion from the point of view of Gaussian approximations it is mathematically convenient to introduce a discrete sequence of appropriately separated scales, $a_1 < a_2 < \ldots < a_n < \ldots$. Under certain conditions on $\mathbf{U}(\mathbf{x})$, a CLT holds approximately over each pair of scales. So different Fickian approximations govern solute concentration over successively larger scales. In view of Proposition 5.1 let us take $U_0 = 1$, and let the diagonal dispersion coefficients at the nth scale be denoted by $\bar{D}_{ii}(a_n)$. It follows from an equation similar to Eq. 5.4 that $\bar{D}_{ii}(a_1) < \bar{D}_{ii}(a_2) < \ldots < \bar{D}_{ii}(a_n) < \ldots$, etc. Since time and spatial scales are related these Fickian approximations may be pieced together in a single equation of the form Eq. 5.5 in which $\bar{D}_{ii}(t)$ grow as functions of time. This explains the scale effect in dispersion. However, much work remains to be done on obtaining the precise nature of functional dependence on $\bar{D}_{ii}(t)$ on t.

An important feature distinguishing the pre-asymptotic and the multi-scale views of dispersion is that in general while the former predicts an improvement of the Fickian approximation with time, the latter does not.

6. Brief comments on the relationship between this work and other approaches

A widely used technique in the dispersion literature for deriving the asymptotic Gaussian, or Fickian, approximation is the *method of moments*; see for example Aris (1956), Brenner (1980a, b), Güven and Molz (1986). This is a classical general method in probability theory, in which one proves convergence of moments of all orders to prove convergence in distribution (Uspensky, 1937, p. 384). A rigorous proof based on the method of moments is often cumbersome. For example, in the derivation of a CLT by this method, it is not enough to prove convergence of only the first two moments to those of the limiting Gaussian distribution. However, this method has been applied successfully in the study of dispersion in a capillary and in periodic porous media by the above-mentioned authors and others. In contrast, the method adopted in this chapter stems from a general CLT which applies to all ergodic Markov processes. For example, this CLT and the computation of the variance of the limiting Gaussian is applicable also to almost-periodic coefficients in Eq. 1.1 (Bhattacharya and Ramasubramanian, 1988).

Another common method for the study of large time behaviour of solutions of parabolic partial differential equations, such as Eq. 1.1, is *homogenization.* A standard reference for periodic coefficients is Bensoussan, Lions and Papanicolaou (1978). The case $\bar{\mathbf{U}} = 0$ (see Eq. 3.12 for $\bar{\mathbf{U}}$) analysed by these authors is not applicable to the study of dispersion. As indicated by Bensoussan *et al.* (1978, p. 529), their analysis does not cover the case $\bar{\mathbf{U}} \neq 0$. This case is dealt with by Bhattacharya (1985) and is considered in Section 5.

We would like to mention here the work of Brenner (1980a) in periodic porous media. In this work a convection-diffusion equation with periodic coefficients is assumed at the pore scale with the surface of the solid grains acting as reflecting mathematical boundaries. From this a limiting Gaussian distribution of concentration is derived which applies at the Darcy scale. The expression for the dispersion coefficients is the same as that given by Eq. 3.16 when reflecting boundary conditions are imposed. It is our view, as pointed out in Section 4 and in Gupta *et al.* (1981), that a convection-diffusion equation is not appropriate at the pore scale. Kinetic model calculations show that such an equation applies at a larger scale which we have referred to as the microscopic scale in this chapter. Further, in contrast to Brenner (1980a), a principal focus in the present chapter is on the nature of functional dependence of dispersion coefficients on the velocity and scale parameters.

Gelhar and Axness (1983) and Neuman *et al.* (1987) consider a different class of velocity models in their studies of field-scale dispersion. They assume that the velocity $\mathbf{U}(\mathbf{x})$ in the CDE (Eq. 1.1) is of the form $\boldsymbol{\mu} + \varepsilon\mathbf{V}(\mathbf{x})$, where $\mathbf{V}(\mathbf{x})$ is a stationary, isotropic, ergodic, mean zero random field, $\boldsymbol{\mu}$ is the constant mean velocity and ε is a dimensionless parameter. Some arguments are given by the authors to justify a Fickian approximation for the mean concentration $\bar{C}(t, \mathbf{x})$ at large times. For any given sample realization of the velocity field $\mathbf{U}(\mathbf{x})$, the solution $C(t, \mathbf{x})$ is uniquely determined. The mean $\bar{C}(t, \mathbf{x})$ is the average of these solutions over the ensemble of all such realizations. In this Fickian approximation, and for small values of ε, the dispersion coefficients are expanded in powers of ε and the principal terms of the form $d_0 + \varepsilon^2 d_2$ are analysed. Here d_0 is the dispersion coefficient at the Darcy scale. It is shown by Neuman *et al.* (1987) that the term $\varepsilon^2 d_2$ for the longitudinal dispersion coefficient (parallel to the mean velocity) is asymptotically linear in $|\boldsymbol{\mu}|L$ for large Peclet numbers, where L is the characteristic correlation length of the random field $\mathbf{V}(\mathbf{x})$. The term $\varepsilon^2 d_2$ for the transverse dispersion coefficient (orthogonal to the mean velocity) is asymptotically constant for large Peclet numbers. Similar results for anisotropic random fields are also given in the above articles.

In this random field model a rigorous derivation of the asymptotic Fickian approximation for $\bar{C}(t, \mathbf{x})$ is still an open problem for the case $\boldsymbol{\mu} \neq \mathbf{0}$. For the case $\boldsymbol{\mu} = \mathbf{0}$, such a derivation is given by Papanicolaou and Varadhan (1979). Of course the case $\boldsymbol{\mu} = \mathbf{0}$ is not applicable to solute dispersion in saturated porous media. Also, a Fickian approximation for the mean concentration $\bar{C}(t, \mathbf{x})$ over an ensemble of aquifers does not directly apply to individual aquifers; see Sposito *et al.* (1986) for additional discussion. For the latter application one needs to prove that for each sample realization of the random field $\mathbf{U}(\mathbf{x})$ a Fickian approximation for $C(t, \mathbf{x})$ holds. Equivalently one needs to prove a CLT. This is also a difficult open problem. In contrast, the CLTs referred to in the present chapter have been proved for two classes of non-random velocity fields—periodic and almost-periodic. The Fickian approximation in these cases therefore applies to individual aquifers. Also, the growth in dispersion coefficients for periodic velocity fields has been shown to be either asymptotically quadratic in aU_0 or constant for large Peclet numbers.

Acknowledgements

This research was supported by grants ECE-8513980, ECE-8513638, and DMS-8503358 from the National Science Foundation.

References

Aris, R. (1956). On the dispersion of a solute in a fluid flowing through a tube, *Proc. Roy. Soc., Ser. A* **235**, 67–78.

Arnold, L. (1971). 'Stochastic Differential Equations: Theory and Applications'. John Wiley, New York.

Bensoussan, A., Lions, J.L. and Papanicolaou, G. (1978). 'Asymptotic Analysis of Periodic Structures'. North-Holland, New York.

Bhattacharya, R.N. (1982). On the functional central limit theorem and the law of iterated logarithm for Markov processes, *Z. Wahr. v. Geb.* **60**, 185–201.

Bhattacharya, R.N. (1985). A central limit theorem for diffusions with periodic coefficients, *Ann. Prob.* **13**, 385–396.

Bhattacharya, R.N. and Gupta, V.K. (1979). On a statistical theory of Solute transport in porous media, *SIAM J. Appl. Math.* **37**, 485–498.

Bhattacharya, R.N. and Gupta, V.K. (1983). A theoretical explanation of solute dispersion in saturated porous media at the Darcy scale, *Water Resour. Res.* **19**, 938–944.

Bhattacharya, R.N. and Gupta, V.K. (1984). On the Taylor–Aris theory of solute transport in a capillary, *SIAM J. Appl. Math.* **44**, 33–39.

Bhattacharya, R.N. and Ramasubramanian, S. (1988). On the central limit theorem for diffusions with almost periodic coefficients, *Sankhya Ser. A* **50**, 9–25.

Bhattacharya, R.N. and Waymire, E. (1990). 'Stochastic Processes with Applications'. John Wiley, New York (in press).

Bhattacharya, R.N., Gupta, V.K. and Walker, H. (1989). Asymptotics of solute dispersion in periodic porous media, *SIAM J. Appl. Math.* **49**, 86–98.

Billingsley, P. (1968). 'Convergence of Probability Measures'. John Wiley, New York.

Breiman, L. (1968). 'Probability'. Addison-Wesley, Reading, MA.

Brenner, H. (1980a). Dispersion resulting from flow through spatially periodic porous media, *Phil. Trans. Roy. Soc. Lond. A* **297**, 81–133.

Brenner, H. (1980b). A general theory of Taylor dispersion phenomena, *Physico Chem. Hydrodyn.* **1**, 91–123.
Chandrasekhar, S. (1943). Stochastic problems in physics and astronomy, *Rev. Mod. Phys.* **15**, 1–89.
Cushman, J.F. (1987). Development of stochastic partial differential equations for subsurface hydrology, *Stochas. Hydrol. Hydraul.* **1**, 241–262.
Dagan, G. (1984). Solute transport in heterogeneous porous formations, *J. Fluid Mech.* **145**, 151–177.
Einstein, A. (1956). 'Investigations on the Theory of Brownian Movement'. Dover, New York.
Feller, W. (1971). 'An Introduction to Probability Theory and its Applications', Vol. 2. John Wiley, New York.
Fox, R.F. (1978). Gaussian stochastic processes in physics, *Phys. Rep.* (A review section of *Phys. Lett.*) **48**, 179–283.
Fried, J.J. and Combarnous, M.A. (1971). Dispersion in porous media, *Adv. Hydrosci.* **7**, 169–282.
Friedman, A. (1975). 'Stochastic Differential Equations and Applications', Vol. 1. Academic Press, New York.
Gelhar, L.W. and Axness, C.L. (1983). Three-dimensional stochastic analysis of macrodispersion in aquifers. *Water Resourc. Res.* **19**, 161–180.
Gupta, V.K. and Bhattacharya, R.N. (1983). A new derivation of Taylor–Aris theory of solute dispersion in a capillary, *Water Resour. Res.* **19**, 945–951.
Gupta, V.K. and Bhattacharya, R.N. (1986). Solute dispersion in multidimensional periodic saturated porous media, *Water Resour. Res.* **22**, 156–164.
Gupta, V.K., Bhattacharya, R.N. and Sposito, G. (1981). A molecular approach to the foundations of the theory of solute transport in porous media: I. Conservative solutes in homogeneous isotropic saturated media, *J. Hydrol.* **50**, 355–370.
Güven, O. and Molz, F.J. (1986). Deterministic and stochastic analyses of dispersion in an unbounded stratified porous medium, *Water Resour. Res.* **22**, 1565–1574.
Itô, K. (1951). On stochastic differential equations, *Mem. Am. Math. Soc.* **4**, 1–51.
Karlin, S. and Taylor, H.M. (1975). 'A First Course in Stochastic Processes'. Academic Press, New York.
Kolmogorov, A. (1931). On analytical methods in probability theory, *Math. Ann.* **104**, 415–458 (in German).
Landau, L.D. and Lifshitz, E.M. (1959). 'Fluid Mechanics'. Pergamon, New York.
Nelson, E. (1967). 'Dynamical Theories of Brownian Motion'. Princeton University Press, Princeton, NJ.
Neuman, S.P., Winter, C.L. and Newman, C.M. (1987). Stochastic theory of field scale Fickian dispersion in anisotropic porous media, *Water Resour. Res.* **23**, 453–466.
Papanicolaou, G. and Varadhan, S.R.S. (1979). Boundary value problems with rapidly oscillating random coefficients, *Colloq. Math. Soc. Janos Bolyai* **27**, 835–873.
Pickens, J.F. and Grisak, G.E. (1981). Scale-dependent dispersion in a stratified granular aquifer, *Water Resour. Res.* **17**, 1191–1211.
Sposito, G., Gupta, V.K. and Bhattacharya, R.N. (1979). Foundational theories of solute transport in porous media: A critical review, *Adv. Water Resour.* **2**, 59–68.
Sposito, G., Jury, W.A. and Gupta, V.K. (1986). Fundamental problems in the stochastic convection-dispersion model of solute transport in aquifers and field soils, *Water Resour. Res.* **22**, 77–88.
Sudicky, E.A., Cherry, J.A. and Frind, E.O. (1983). Migration of contaminants in groundwater at a landfill. 4. A natural-gradient dispersion test, *J. Hydrol.* **63**, 81–108.
Taylor, G.I. (1953). Dispersion of soluble matter in a solvent flowing slowly through a tube, *Proc. Roy Soc., Ser. A* **219**, 186–203.
Uspensky, J.V. (1937). 'Introduction to Mathematical Probability'. McGraw-Hill, New York.
Wax, N. (ed.) (1954). 'Selected Papers on Noise and Stochastic Processes'. Dover, New York.
Wheatcraft, S.W. and Tyler, S.W. (1988). An explanation of scale-dependent dispersivity in heterogeneous aquifers using concepts of fractal geometry, *Water Resour. Res.* **24**, 566–578.

Chapter V

Diffusion, Adsorption and Dispersion in Porous Media: Small-Scale Averaging and Local Volume Averaging*

O.A. Plumb
Department of Mechanical and Materials Engineering, Washington State University, Pullman, WA 99164, USA

S. Whitaker
Department of Chemical Engineering, University of California, Davis, CA 95616, USA

1. Introduction

The problem of transport of chemical species in porous media is a generic problem for soil scientists (Angelakis *et al.*, 1987), chemical engineers (Jackson, 1977), and petroleum reservoir engineers (Pfannkuch, 1963; Heller, 1972). The process is often complicated by the presence of two flowing fluids (Nielsen *et al.*, 1986), and recent studies of two-phase flow in heterogeneous porous media (Quintard and Whitaker, 1988) clearly establish that solute transport in the vadose zone and the capillary fringe is a very, very complicated process. We will eventually attack the problem of solute transport in two-phase flow; however, our development at this time is restricted to single-phase flow phenomena. The system under consideration is illustrated in Fig. 1 where we have shown a two-region model of a heterogeneous porous medium. The development that we will present is not restricted to distinct heterogeneities such as those shown in Fig. 1, and in an analysis of single-phase flow Quintard and Whitaker (1987) have carefully outlined the treatment of both discrete and continuous variations of the properties of a heterogeneous porous medium.

We wish to develop spatially smoothed equations for the transport of a chemical species in the heterogeneous porous medium illustrated in Fig. 1. To accomplish this we will need to use the large-scale averaging volume, $\mathscr{V}_\infty$, shown

* The unabridged version of this article is available from Professor Whitaker.

DYNAMICS OF FLUIDS IN HIERARCHICAL POROUS MEDIA
ISBN 0-12-200260-1

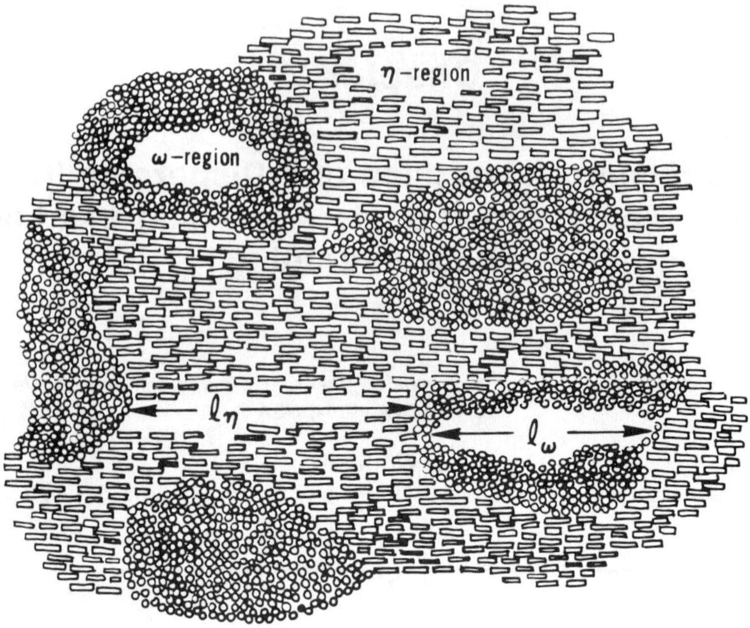

Figure 1. Heterogeneous porous medium.

in Fig. 2. Spatial smoothing over $\mathscr{V}_\infty$ requires a knowledge of the transport equations within the ω and η-regions, and this requires the use of the local averaging volume, $\mathscr{V}$, that contains the β-phase and σ-region illustrated in Fig. 2. In order to understand the transport process taking place within the averaging volume $\mathscr{V}$, we need to develop the transport equation for the σ-region and this leads to the use of the small-scale averaging volume, $\mathscr{V}_\sigma$. The γ-phase contained within the σ-region is identical to the β-phase; however, it is convenient to identify this phase differently when it exists within the porous particles. In order to develop an equation for the transport of a chemical species in the heterogeneous porous medium depicted in Fig. 1, we must spatially smooth various point equations over the volumes $\mathscr{V}_\sigma$, $\mathscr{V}$, and $\mathscr{V}_\infty$. This represents the three-scale variation of Cushman's (1984) N-scale problem. In this chapter we consider only the first two scales of this three-scale problem.

2. Diffusion and adsorption in porous particles

The process of diffusion and heterogeneous reaction in porous catalysts has been discussed in detail elsewhere (Ryan *et al.*, 1980; Whitaker, 1983, 1986c, d),

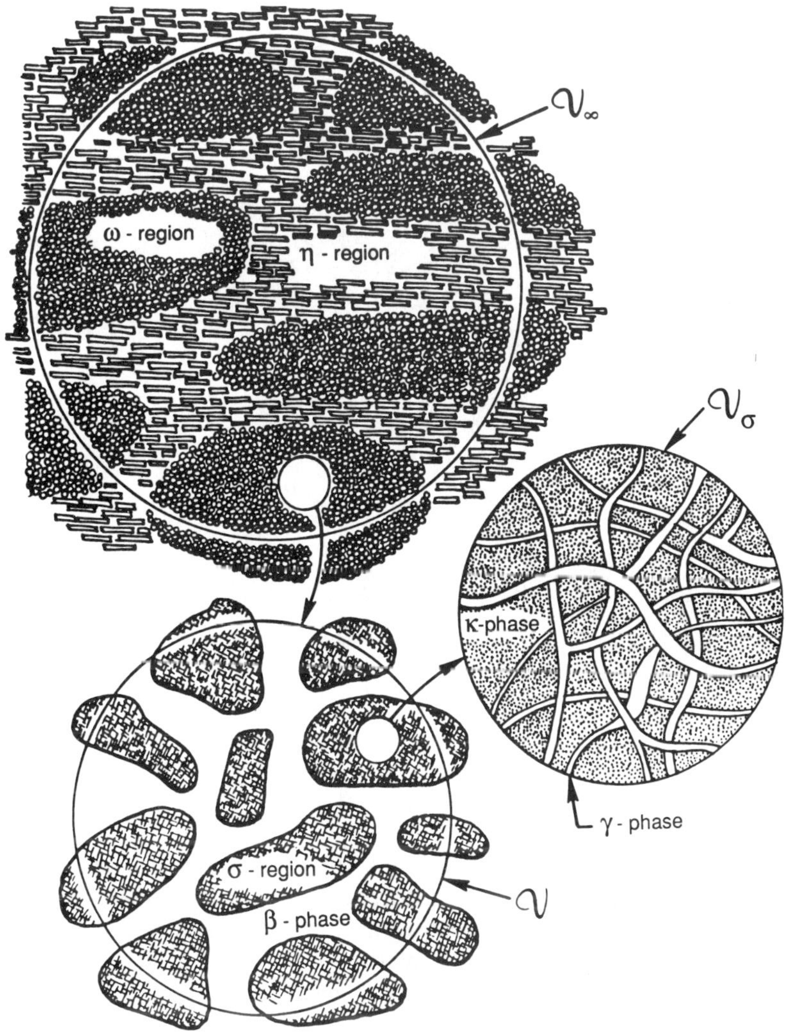

Figure 2. Averaging volumes for diffusion, adsorption, and dispersion in heterogeneous porous media.

and we will follow these previous studies in our analysis of diffusion and adsorption. The boundary value problem for the γ–κ system illustrated in Fig. 3 can be expressed as

$$\frac{\partial c_\gamma}{\partial t} = \nabla \cdot (\mathscr{D}_\gamma \nabla c_\gamma), \quad \text{in the } \gamma\text{-phase} \tag{2.1}$$

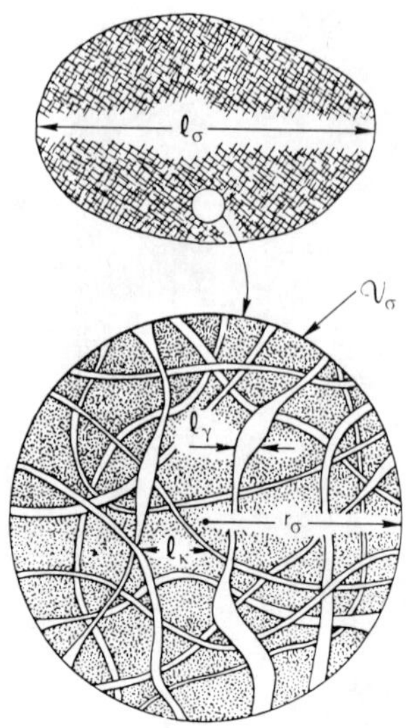

Figure 3. The porous particle.

B.C.1
$$-\mathbf{n}_{\gamma\kappa}\cdot\mathscr{D}_\gamma\nabla c_\gamma = \frac{\partial c_{\mathrm{ad}}}{\partial t}, \quad \text{at } \mathscr{A}_{\gamma\kappa} \tag{2.2}$$

B.C.2
$$c_\gamma = f(\mathbf{r}, t), \quad \text{at } \mathscr{A}_{\gamma\mathrm{e}} \tag{2.3}$$

$$c_{\mathrm{ad}} = K_{\mathrm{eq}} c_\gamma, \quad \text{at } \mathscr{A}_{\gamma\kappa} \tag{2.4}$$

Here c_γ represents the molar concentration of the chemical species under consideration and c_{ad} represents the adsorbed or surface concentration (mol/m^2) of this species. Equation 2.4 indicates that our analysis is restricted to linear adsorption isotherms and a state of local mass equilibrium at the γ–κ interface.

The volume-averaged form of Eq. 2.1 is given by

$$\frac{1}{\mathscr{V}_\sigma}\int_{V_\gamma}\left(\frac{\partial c_\gamma}{\partial t}\right)\mathrm{d}V = \frac{1}{\mathscr{V}_\sigma}\int_{V_\gamma}\nabla\cdot(\mathscr{D}_\gamma\nabla c_\gamma)\,\mathrm{d}V \tag{2.5}$$

where V_γ is the volume of the γ-phase contained in the averaging volume, $\mathscr{V}_\sigma$, illustrated in Fig. 3. When the κ-phase is rigid, Eq. 2.5 can be expressed as

$$\frac{\partial}{\partial t}\left(\frac{1}{\mathscr{V}_\sigma}\int_{V_\gamma} c_\gamma\,\mathrm{d}V\right) = \frac{1}{\mathscr{V}_\sigma}\int_{V_\gamma}\nabla\cdot(\mathscr{D}_\gamma\nabla c_\gamma)\,\mathrm{d}V \tag{2.6}$$

and this naturally leads to the definition of the *phase-average concentration*,

$\langle c_\gamma \rangle$, according to

$$\frac{1}{\mathscr{V}_\sigma}\int_{\mathscr{V}_\gamma} c_\gamma \, dV = \langle c_\gamma \rangle \tag{2.7}$$

With this nomenclature we can express Eq. 2.6 as

$$\frac{\partial \langle c_\gamma \rangle}{\partial t} = \langle \nabla \cdot (\mathscr{D}_\gamma \nabla c_\gamma) \rangle \tag{2.8}$$

The phase-average concentration defined by Eq. 2.7 is not the preferred dependent variable, and we seek instead the governing equation for the *intrinsic phase-average concentration* defined by

$$\frac{1}{V_\gamma}\int_{V_\gamma} c_\gamma \, dV = \langle c_\gamma \rangle^\gamma \tag{2.9}$$

These two averaged concentrations are related by

$$\langle c_\gamma \rangle = \varepsilon_\gamma \langle c_\gamma \rangle^\gamma \tag{2.10}$$

where ε_γ is the volume fraction of the γ-phase given explicitly by $\varepsilon_\gamma = V_\gamma / \mathscr{V}_\sigma$. Use of this relation in Eq. 2.8 leads to

$$\varepsilon_\gamma \frac{\partial \langle c_\gamma \rangle^\gamma}{\partial t} = \langle \nabla \cdot (\mathscr{D}_\gamma \nabla c_\gamma) \rangle \tag{2.11}$$

provided ε_γ is independent of time. The spatial averaging theorem (Howes and Whitaker, 1985) can be used to interchange integration and differentiation on the right-hand side of Eq. 2.11. This allows us to write the volume-averaged diffusion equation as

$$\varepsilon_\gamma \frac{\partial \langle c_\gamma \rangle^\gamma}{\partial t} = \nabla \cdot \langle \mathscr{D}_\gamma \nabla c_\gamma \rangle + \frac{1}{\mathscr{V}_\sigma}\int_{A_{\gamma\kappa}} \mathbf{n}_{\gamma\kappa} \cdot \mathscr{D}_\gamma \nabla c_\gamma \, dA \tag{2.12}$$

where $A_{\gamma\kappa}$ represents the interfacial area contained within the averaging volume illustrated in Fig. 3. Use of Eq. 2.4 in the boundary condition given by Eq. 2.2 allows us to express that result as

B.C.1 $$-\mathbf{n}_{\gamma\kappa} \cdot \mathscr{D}_\gamma \nabla c_\gamma = K_{\text{eq}} \frac{\partial c_\gamma}{\partial t}, \quad \text{at } \mathscr{A}_{\gamma\kappa} \tag{2.13}$$

We can now substitute this form of the boundary condition into Eq. 2.12 to obtain

$$\varepsilon_\gamma \frac{\partial \langle c_\gamma \rangle^\gamma}{\partial t} = \nabla \cdot \langle \mathscr{D}_\gamma \nabla c_\gamma \rangle - K_{\text{eq}} a_{\gamma\kappa} \frac{\partial \langle c_\gamma \rangle_{\gamma\kappa}}{\partial t} \tag{2.14}$$

Here we have neglected variations of K_{eq} within the averaging volume, and we have used the following two definitions

$$a_{\gamma\kappa} = A_{\gamma\kappa} / \mathscr{V}_\sigma \tag{2.15}$$

$$\langle c_\gamma \rangle_{\gamma\kappa} = \frac{1}{A_{\gamma\kappa}} \int_{A_{\gamma\kappa}} c_\gamma \, \mathrm{d}A \tag{2.16}$$

In general, it is quite acceptable to neglect variations in parameters such as $\mathscr{D}_\gamma$ and K_{eq} within the averaging volume, and this allows us to use the spatial averaging theorem a second time to obtain

$$\langle \mathscr{D}_\gamma \nabla c_\gamma \rangle = \mathscr{D}_\gamma \langle \nabla c_\gamma \rangle = \mathscr{D}_\gamma \left(\nabla \langle c_\gamma \rangle + \frac{1}{\mathscr{V}_\sigma} \int_{A_{\gamma\kappa}} \mathbf{n}_{\gamma\kappa} c_\gamma \, \mathrm{d}A \right) \tag{2.17}$$

Use of this result in Eq. 2.14 leads to a volume-averaged diffusion equation containing four different concentrations.

$$\varepsilon_\gamma \frac{\partial \langle c_\gamma \rangle^\gamma}{\partial t} = \nabla \cdot \left[\mathscr{D}_\gamma \left(\nabla \langle c_\gamma \rangle + \frac{1}{\mathscr{V}_\sigma} \int_{A_{\gamma\kappa}} \mathbf{n}_{\gamma\kappa} c_\gamma \, \mathrm{d}A \right) \right] - K_{\text{eq}} a_{\gamma\kappa} \frac{\partial \langle c_\gamma \rangle_{\gamma\kappa}}{\partial t} \tag{2.18}$$

We can eliminate $\langle c_\gamma \rangle$ by the use of Eq. 2.10 leading to:

$$\varepsilon_\gamma \frac{\partial \langle c_\gamma \rangle^\gamma}{\partial t} = \nabla \cdot \left[\mathscr{D}_\gamma \left(\varepsilon_\gamma \nabla \langle c_\gamma \rangle^\gamma + \langle c_\gamma \rangle^\gamma \nabla \varepsilon_\gamma + \frac{1}{\mathscr{V}_\sigma} \int_{A_{\gamma\kappa}} \mathbf{n}_{\gamma\kappa} c_\gamma \, \mathrm{d}A \right) \right] - K_{\text{eq}} a_{\gamma\kappa} \frac{\partial \langle c_\gamma \rangle_{\gamma\kappa}}{\partial t} \tag{2.19}$$

However, we are still confronted with the preferred average concentration $\langle c_\gamma \rangle^\gamma$, the point concentration c_γ, and the area-averaged concentration $\langle c_\gamma \rangle_{\gamma\kappa}$.

In one of the early studies of dispersion in porous media (Whitaker, 1967), it was found to be convenient to decompose the point concentration in terms of an average concentration and a *spatial deviation concentration.* Gray (1975) explored this point of view in more detail and we adopt Gray's decomposition by expressing c_γ as

$$c_\gamma = \langle c_\gamma \rangle^\gamma + \tilde{c}_\gamma \tag{2.20}$$

It is important to recognize that this decomposition represents a *decomposition of length scales.* The characteristic length associated with $\langle c_\gamma \rangle^\gamma$ is designated by ℓ_σ, while the pore diameter ℓ_γ represents the characteristic length associated with $\tilde{c}_\gamma$. Both these lengths are illustrated in Fig. 3.

When Eq. 2.20 is used in Eq. 2.19, we obtain a rather unwieldy form of the volume-averaged diffusion equation given by

$$\varepsilon_\gamma \frac{\partial \langle c_\gamma \rangle^\gamma}{\partial t} = \nabla \cdot \left[\mathscr{D}_\gamma \left(\varepsilon_\gamma \nabla \langle c_\gamma \rangle^\gamma + \langle c_\gamma \rangle^\gamma \nabla \varepsilon_\gamma + \frac{1}{\mathscr{V}_\sigma} \int_{A_{\gamma\kappa}} \mathbf{n}_{\gamma\kappa} \langle c_\gamma \rangle^\gamma \, \mathrm{d}A \right.\right.$$
$$\left.\left. + \frac{1}{\mathscr{V}_\sigma} \int_{A_{\gamma\kappa}} \mathbf{n}_{\gamma\kappa} \tilde{c}_\gamma \, \mathrm{d}A \right) \right] - K_{\mathrm{eq}} a_{\gamma\kappa} \frac{\partial \langle \langle c_\gamma \rangle^\gamma \rangle_{\gamma\kappa}}{\partial t} - K_{\mathrm{eq}} a_{\gamma\kappa} \frac{\partial \langle \tilde{c}_\gamma \rangle_{\gamma\kappa}}{\partial t} \tag{2.21}$$

It is intuitively appealing to neglect variations of $\langle c_\gamma \rangle^\gamma$ within the averaging volume, and Carbonell and Whitaker (1984) have shown that this is acceptable when

$$\left(\frac{r_\sigma}{\ell_\sigma} \right)^2 \ll 1 \tag{2.22}$$

The characteristic lengths ℓ_σ and r_σ are identified in Fig. 3, and when the length-scale constraint given by Eq. 2.22 is valid we can use the approximations

$$\frac{1}{\mathscr{V}_\sigma} \int_{A_{\gamma\kappa}} \mathbf{n}_{\gamma\kappa} \langle c_\gamma \rangle^\gamma \, \mathrm{d}A = \left(\frac{1}{\mathscr{V}_\sigma} \int_{A_{\gamma\kappa}} \mathbf{n}_{\gamma\kappa} \, \mathrm{d}A \right) \langle c_\gamma \rangle^\gamma = -(\nabla \varepsilon_\gamma) \langle c_\gamma \rangle^\gamma \tag{2.23a}$$

$$\langle \tilde{c}_\gamma \rangle^\gamma = 0 \tag{2.23b}$$

While a rigorous proof is still unavailable for the area averages in Eq. 2.21, the results given by Eqs 2.23 strongly suggest that the following approximations are also acceptable when Eq. 2.22 is in force.

$$\langle \langle c_\gamma \rangle^\gamma \rangle_{\gamma\kappa} = \langle c_\gamma \rangle^\gamma \tag{2.24a}$$

$$\langle \tilde{c}_\gamma \rangle_{\gamma\kappa} = 0 \tag{2.24b}$$

Use of Eqs 2.23a and 2.24 in Eq. 2.21 provides a simplified version of the volume-averaged diffusion equation given by

$$\varepsilon_\gamma\left(1+\frac{a_{\gamma\kappa}K_{\mathrm{eq}}}{\varepsilon_\gamma}\right)\frac{\partial\langle c_\gamma\rangle^\gamma}{\partial t}=\nabla\cdot\left[\varepsilon_\gamma\mathscr{D}_\gamma\left(\nabla\langle c_\gamma\rangle^\gamma+\frac{1}{V_\gamma}\int_{A_{\gamma\kappa}}\mathbf{n}_{\gamma\kappa}\tilde{c}_\gamma\,\mathrm{d}A\right)\right] \tag{2.25}$$

Here we are faced with a single problem—the presence of $\tilde{c}_\gamma$ in the area integral over $A_{\gamma\kappa}$. To obtain a closed form of Eq. 2.25 we need to represent $\tilde{c}_\gamma$ in terms of $\langle c_\gamma\rangle^\gamma$ and the route to this representation is described in the next section.

3. The closure problem for diffusion and adsorption

The development of the closure problem follows the method outlined by Crapiste *et al.* (1986). In this approach, one begins with the governing equation and boundary condition given by

$$\frac{\partial c_\gamma}{\partial t}=\nabla\cdot(\mathscr{D}_\gamma\nabla c_\gamma),\quad \text{in the } \gamma\text{-phase} \tag{3.1}$$

B.C.1
$$-\mathbf{n}_{\beta\gamma}\cdot\mathscr{D}_\gamma\nabla c_\gamma=K_{\mathrm{eq}}\frac{\partial c_\gamma}{\partial t},\quad \text{at } \mathscr{A}_{\gamma\kappa} \tag{3.2}$$

and one makes use of the decomposition represented by Eq. 2.20 to obtain

$$\frac{\partial\tilde{c}_\gamma}{\partial t}-\nabla\cdot(\mathscr{D}_\gamma\nabla\tilde{c}_\gamma)=-\left[\frac{\partial\langle c_\gamma\rangle^\gamma}{\partial t}-\nabla\cdot(\mathscr{D}_\gamma\nabla\langle c_\gamma\rangle^\gamma)\right] \tag{3.3}$$

B.C.1
$$-\mathbf{n}_{\gamma\kappa}\cdot\mathscr{D}_\gamma\nabla\tilde{c}_\gamma-K_{\mathrm{eq}}\frac{\partial\tilde{c}_\gamma}{\partial t}=\mathbf{n}_{\gamma\kappa}\cdot\mathscr{D}_\gamma\nabla\langle c_\gamma\rangle^\gamma+K_{\mathrm{eq}}\frac{\partial\langle c_\gamma\rangle^\gamma}{\partial t},\quad \text{at } \mathscr{A}_{\gamma\kappa} \tag{3.4}$$

B.C.2
$$\tilde{c}_\gamma=g(\mathbf{r},t),\quad \text{at } \mathscr{A}_{\gamma\mathrm{e}} \tag{3.5}$$

I.C.
$$\tilde{c}_\gamma=h(\mathbf{r}),\quad t=0 \tag{3.6}$$

Here we have listed a boundary condition for $\tilde{c}_\gamma$ at $\mathscr{A}_{\gamma\mathrm{e}}$ and an initial condition for $\tilde{c}_\gamma$ at $t=0$ primarily to remind the reader that the functions $g(\mathbf{r},t)$ and $h(\mathbf{r})$ are *generally not known.* If the functions $g(\mathbf{r},t)$ and $h(\mathbf{r})$ have an important influence on $\tilde{c}_\gamma$, it is unlikely that we will be able to develop a successful closure problem for $\tilde{c}_\gamma$.

The form of the boundary value problem for $\tilde{c}_\gamma$ given by Eqs 3.3–3.6 clearly identifies the *sources* of the $\tilde{c}_\gamma$-field. We list the three dominant sources as:

1. $\left[\dfrac{\partial \langle c_\gamma \rangle^\gamma}{\partial t} - \nabla \cdot (\mathscr{D}_\gamma \nabla \langle c_\gamma \rangle^\gamma)\right]$ a *volumetric source* that depends on the spatial and temporal gradients of the average concentration;
2. $\mathbf{n}_{\gamma\kappa} \cdot \mathscr{D}_\gamma \nabla \langle c_\gamma \rangle^\gamma$, a *surface source* that depends on the geometry (through $\mathbf{n}_{\gamma\kappa}$) and the spatial gradient of the average concentration;
3. $K_{\text{eq}} \dfrac{\partial \langle c_\gamma \rangle^\gamma}{\partial t}$, a *surface source* that depends on the adsorption isotherm (through K_{eq}) and the temporal gradient of the average concentration.

In addition, the functions $g(\mathbf{r}, t)$ and $h(\mathbf{r})$ represent sources of the $\tilde{c}_\gamma$-field; however, we are assured that the source $g(\mathbf{r}, t)$ contributes to the $\tilde{c}_\gamma$-field only in a region close to the σ–β interface and that the source $h(\mathbf{r})$ can be ignored at sufficiently long times.

The three dominant sources can be reduced to two sources by following the method of Crapiste *et al.* (1986). This requires that we form the intrinsic phase average of Eq. 3.3

$$\frac{1}{V_\gamma}\int_{V_\gamma}\left[\frac{\partial \tilde{c}_\gamma}{\partial t} - \nabla \cdot (\mathscr{D}_\gamma \nabla \tilde{c}_\gamma)\right] \mathrm{d}V = -\frac{1}{V_\gamma}\int_{V_\gamma}\left[\frac{\partial \langle c_\gamma \rangle^\gamma}{\partial t} - \nabla \cdot (\mathscr{D}_\gamma \nabla \langle c_\gamma \rangle^\gamma)\right] \mathrm{d}V \tag{3.7}$$

and then make use of Eq. 2.22 to neglect variations of averaged quantities within the averaging volume. This allows us to write Eq. 3.7 as

$$\frac{1}{V_\gamma}\int_{V_\gamma}\left[\frac{\partial \tilde{c}_\gamma}{\partial t} - \nabla \cdot (\mathscr{D}_\gamma \nabla \tilde{c}_\gamma)\right] \mathrm{d}V = \left[\frac{\partial \langle c_\gamma \rangle^\gamma}{\partial t} - \nabla \cdot (\mathscr{D}_\gamma \nabla \langle c_\gamma \rangle^\gamma\right] \tag{3.8}$$

and when this result is used with Eq. 3.3 we obtain the following integro–differential equation for $\tilde{c}_\gamma$.

$$\frac{\partial \tilde{c}_\gamma}{\partial t} - \nabla \cdot (\mathscr{D}_\gamma \nabla \tilde{c}_\gamma) = \frac{1}{V_\gamma}\int_{V_\gamma}\left[\frac{\partial \tilde{c}_\gamma}{\partial t} - \nabla \cdot (\mathscr{D}_\gamma \nabla \tilde{c}_\gamma)\right] \mathrm{d}V \tag{3.9}$$

Since the average of $\tilde{c}_\gamma$ is zero, as indicated by Eq. 2.23b, we can simplify Eq. 3.9 to obtain

$$\frac{\partial \tilde{c}_\gamma}{\partial t} = \nabla \cdot (\mathscr{D}_\gamma \nabla \tilde{c}_\gamma) - \frac{1}{V_\gamma}\int_{V_\gamma} \nabla \cdot (\mathscr{D}_\gamma \nabla \tilde{c}_\gamma)\, \mathrm{d}V \tag{3.10}$$

Use of the divergence theorem along with the boundary condition given by Eq. 3.4 leads to the following form of our governing equation for $\tilde{c}_\gamma$.

$$\frac{\partial \tilde{c}_\gamma}{\partial t} = \nabla \cdot (\mathscr{D}_\gamma \nabla \tilde{c}_\gamma) - \frac{1}{V_\gamma} \int_{A_{\gamma e}} \mathbf{n}_{\gamma e} \cdot \mathscr{D}_\gamma \nabla \tilde{c}_\gamma \, dA + \frac{1}{V_\gamma} \int_{A_{\gamma\kappa}} \mathbf{n}_{\gamma\kappa} \cdot \mathscr{D}_\gamma \nabla \langle c_\gamma \rangle^\gamma \, dA$$
$$+ \frac{1}{V_\gamma} \int_{A_{\gamma\kappa}} K_{\text{eq}} \frac{\partial \langle c_\gamma \rangle^\gamma}{\partial t} \, dA + \frac{1}{V_\gamma} \int_{A_{\gamma\kappa}} K_{\text{eq}} \frac{\partial \tilde{c}_\gamma}{\partial t} \, dA \tag{3.11}$$

Here $A_{\gamma e}$ represents the area of entrances and exits for the γ-phase contained in the averaging volume illustrated in Fig. 3. By removing averaged quantities and the parameters $\mathscr{D}_\gamma$ and K_{eq} from the area integrals, and by using Eq. 2.24b, we can obtain the following boundary value problem for $\tilde{c}_\gamma$.

$$\frac{\partial \tilde{c}_\gamma}{\partial t} = \nabla \cdot (\mathscr{D}_\gamma \nabla \tilde{c}_\gamma) - \frac{1}{V_\gamma} \int_{A_{\gamma e}} \mathbf{n}_{\gamma e} \cdot \mathscr{D}_\gamma \nabla \tilde{c}_\gamma \, dA - \nabla \varepsilon_\gamma \cdot \mathscr{D}_\gamma \nabla \langle c_\gamma \rangle^\gamma + a_{\gamma\kappa} K_{\text{eq}} \frac{\partial \langle c_\gamma \rangle^\gamma}{\partial t} \tag{3.12}$$

B.C.1 $$-\mathbf{n}_{\gamma\kappa} \cdot \mathscr{D}_\gamma \nabla \tilde{c}_\gamma - K_{\text{eq}} \frac{\partial \tilde{c}_\gamma}{\partial t} = \mathbf{n}_{\gamma\kappa} \cdot \mathscr{D}_\gamma \nabla \langle c_\gamma \rangle^\gamma + K_{\text{eq}} \frac{\partial \langle c_\gamma \rangle^\gamma}{\partial t}, \quad \text{at } a_{\gamma\kappa} \tag{3.13}$$

B.C.2 $$\tilde{c}_\gamma = g(\mathbf{r}, t), \quad \text{at } \mathscr{A}_{\gamma e} \tag{3.14}$$

I.C. $$\tilde{c}_\gamma = h(\mathbf{r}), \quad t = 0 \tag{3.15}$$

We can use Eq. 3.13 and the type of analysis presented by Ryan *et al.* (1981) to estimate $\tilde{c}_\gamma$ as

$$\tilde{c}_\gamma = O\left\{ \frac{\left(\frac{\ell_\gamma}{\ell_\sigma}\right) + O\left(\frac{K_{\text{eq}} \ell_\gamma}{\mathscr{D}_\gamma t}\right)}{1 + O\left(\frac{K_{\text{eq}} \ell_\gamma}{\mathscr{D}_\gamma t}\right)} \right\} \langle c_\gamma \rangle^\gamma \tag{3.16}$$

Here we have assumed that it is impossible to generate an initial condition, given by Eq. 3.14, that would not be consistent with our estimate of $\tilde{c}_\gamma$ from Eq. 3.13.

Since ℓ_γ is small compared to ℓ_σ (see Fig. 3), we see that $\tilde{c}_\gamma$ is constrained by

$$\tilde{c}_\gamma \ll \langle c_\gamma \rangle^\gamma \tag{3.17}$$

when the following constraint on the time is satisfied

$$\frac{\mathscr{D}_\gamma t}{\ell_\gamma^2} \gg K_{\text{eq}}/\ell_\gamma \tag{3.18}$$

Under these circumstances, Eq. 3.13 can be simplified by neglecting $\partial\tilde{c}_\gamma/\partial t$ with respect to $\partial\langle c_\gamma\rangle^\gamma/\partial t$. In addition, the estimate given by Eq. 3.16 takes the form

$$\tilde{c}_\gamma = O\left[\left(\frac{\ell_\gamma}{\ell_\sigma}\right)\langle c_\gamma\rangle^\gamma\right] \tag{3.19}$$

and this means that the *volumetric source*, $\nabla\varepsilon_\gamma \cdot \mathscr{D}_\gamma \nabla\langle c_\gamma\rangle^\gamma$, makes a small contribution relative to the *surface source*, $\mathbf{n}_{\gamma\kappa} \cdot \mathscr{D}_\gamma \nabla\langle c_\gamma\rangle^\gamma$. Because of this, the boundary value problem for $\tilde{c}_\gamma$ simplifies to

$$\frac{\partial\tilde{c}_\gamma}{\partial t} = \nabla\cdot(\mathscr{D}_\gamma\nabla\tilde{c}_\gamma) - \frac{1}{V_\gamma}\int_{A_{\gamma e}} \mathbf{n}_{\gamma e}\cdot\mathscr{D}_\gamma\nabla\tilde{c}_\gamma \, \mathrm{d}A + a_{\gamma\kappa}K_{\text{eq}}\frac{\partial\langle c_\gamma\rangle^\gamma}{\partial t} \tag{3.20}$$

B.C.1 $$-\mathbf{n}_{\gamma\kappa}\cdot\mathscr{D}_\gamma\nabla\tilde{c}_\gamma = \mathbf{n}_{\gamma\kappa}\cdot\mathscr{D}_\gamma\nabla\langle c_\gamma\rangle^\gamma + K_{\text{eq}}\frac{\partial\langle c_\gamma\rangle^\gamma}{\partial t}, \quad \text{at } \mathscr{A}_{\gamma\kappa} \tag{3.21}$$

B.C.2 $$\tilde{c}_\gamma = g(\mathbf{r}, t), \quad \text{at } \mathscr{A}_{\gamma e} \tag{3.22}$$

I.C. $$\tilde{c}_\gamma = h(\mathbf{r}), \quad t = 0 \tag{3.23}$$

About the value of $\tilde{c}_\gamma$ at $\mathscr{A}_{\gamma e}$, we know only the order of magnitude as indicated by Eq. 3.19. In addition, we know that the characteristic length scale associated with $\tilde{c}_\gamma$ is the length ℓ_γ indicated in Fig. 3. This means that the source $g(\mathbf{r}, t)$ will influence the $\tilde{c}_\gamma$-field within a distance of the β–σ interface of the order of ℓ_γ. Under these circumstances, it is acceptable to ignore the influence of the boundary condition at $\mathscr{A}_{\gamma e}$ and use the model of a spatially periodic porous medium (Brenner, 1980) to develop a *local* boundary value problem for $\tilde{c}_\gamma$ given by

$$\frac{\partial\tilde{c}_\gamma}{\partial t} = \mathscr{D}_\gamma\nabla^2\tilde{c}_\gamma + a_{\gamma\kappa}K_{\text{eq}}\frac{\partial\langle c_\gamma\rangle^\gamma}{\partial t} \tag{3.24}$$

B.C.1 $$-\mathbf{n}_{\gamma\kappa}\cdot\mathscr{D}_\gamma\nabla\tilde{c}_\gamma = \mathbf{n}_{\gamma\kappa}\cdot\mathscr{D}_\gamma\nabla\langle c_\gamma\rangle^\gamma + K_{\text{eq}}\frac{\partial\langle c_\gamma\rangle^\gamma}{\partial t}, \quad \text{at } A_{\gamma\kappa} \tag{3.25}$$

$$\tilde{c}_\gamma(\mathbf{r} + \boldsymbol{\ell}_i) = \tilde{c}_\gamma(\mathbf{r}), \quad i = 1, 2, 3 \tag{3.26}$$

I.C. $$\tilde{c}_\gamma = h(\mathbf{r}), \quad t = 0 \tag{3.27}$$

Here ℓ_i represents the three non-unique lattice vectors that are needed to describe a spatially periodic porous medium, and in Fig. 4 we have shown an example of a unit cell of a spatially periodic porous media. Since $\tilde{c}_\gamma$ is spatially periodic, we know that $\nabla\tilde{c}_\gamma$ is also spatially periodic and this provides the simplification

$$\frac{1}{V_\gamma}\int_{A_{\gamma e}} \mathbf{n}_{\gamma e}\cdot\mathscr{D}_\gamma\nabla\tilde{c}_\gamma\,\mathrm{d}A = 0 \tag{3.28}$$

that was used in going from Eq. 3.20 to 3.24. In addition, we have simplified Eq. 3.20 by neglecting the variation of $\mathscr{D}_\gamma$ in the formulation of the local problem given by Eqs 3.24–3.27.

We are now faced with a boundary value problem for $\tilde{c}_\gamma$ that contains two well-defined source terms and this motivates a representation of the form

$$\tilde{c}_\gamma = \mathbf{f}\cdot\nabla\langle c_\gamma\rangle^\gamma + s\frac{\partial\langle c_\gamma\rangle^\gamma}{\partial t} + \psi \tag{3.29}$$

where $\mathbf{f}$ and s are referred to as the *closure variables*. Here we think of ψ as an

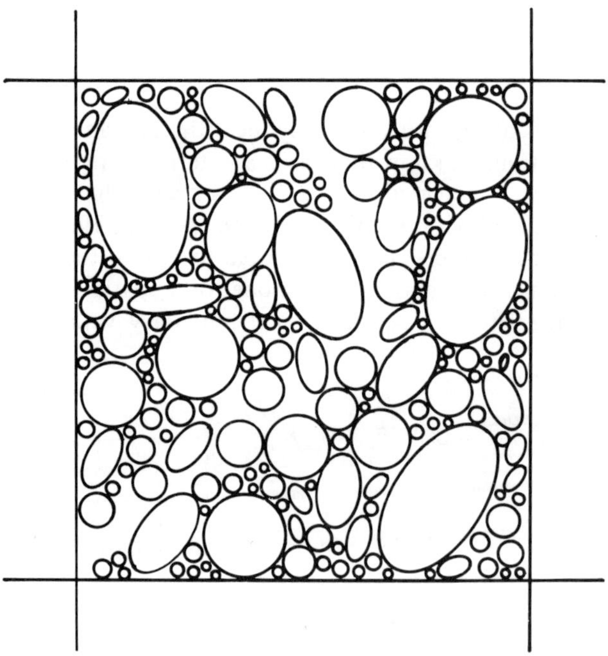

Figure 4. Complex unit cell.

arbitrary function, thus we are free to specify $\mathbf{f}$ and s according to the following two boundary value problems

Problem I

$$\frac{\partial \mathbf{f}}{\partial t} = \mathscr{D}_\gamma \nabla^2 \mathbf{f} \tag{3.30}$$

B.C.1 $$-\mathbf{n}_{\gamma\kappa} \cdot \nabla \mathbf{f} = \mathbf{n}_{\gamma\kappa}, \quad \text{at } A_{\gamma\kappa} \tag{3.31}$$

$$\mathbf{f}(\mathbf{r} + \boldsymbol{\ell}_i) = \mathbf{f}(\mathbf{r}) \tag{3.32}$$

I.C. $$\mathbf{f} = 0, \quad t = 0 \tag{3.33}$$

Problem II

$$\frac{\partial s}{\partial t} = \mathscr{D}_\gamma \nabla^2 s + a_{\gamma\kappa} K_{\text{eq}} \tag{3.34}$$

B.C.1 $$-\mathbf{n}_{\gamma\kappa} \cdot \nabla s = K_{\text{eq}}/\mathscr{D}_\gamma, \quad \text{at } A_{\gamma\kappa} \tag{3.35}$$

$$s(\mathbf{r} + \boldsymbol{\ell}_i) = s(\mathbf{r}), \quad i = 1, 2, 3 \tag{3.36}$$

I.C. $$s = 0, \quad t = 0 \tag{3.37}$$

When Eq. 3.29 is used in Eqs 3.24–3.27, and the two boundary value problems for $\mathbf{f}$ and s are imposed, we find the following problem for ψ.

Problem III

$$\frac{\partial \psi}{\partial t} = \mathscr{D}_\gamma \nabla^2 \psi \tag{3.38}$$

B.C.1 $$-\mathbf{n}_{\gamma\kappa} \cdot \mathscr{D}_\gamma \nabla \psi = 0, \quad \text{at } A_{\gamma\kappa} \tag{3.39}$$

$$\psi(\mathbf{r} + \boldsymbol{\ell}_i) = \psi(\mathbf{r}) \tag{3.40}$$

I.C. $$\psi = h(\mathbf{r}), \quad t = 0 \tag{3.41}$$

Since the average value of $h(\mathbf{r})$ must be zero in order to satisfy Eq. 2.23b, and the characteristic length for $\tilde{c}_\gamma$ is ℓ_γ, we can estimate the magnitude of ψ as

$$\psi = O[h(\mathbf{r}) \exp(-\mathscr{D}_\gamma t/\ell_\gamma^2)] \tag{3.42}$$

Clearly the contribution of ψ to the representation of $\tilde{c}_\gamma$ given by Eq. 3.29 can be neglected when

$$\frac{\mathscr{D}_\gamma t}{\ell_\gamma^2} > 1 \tag{3.43}$$

Under these circumstances the boundary value problems for $\mathbf{f}$ and s are simplified to:

Problem I

$$0 = \nabla^2 \mathbf{f} \tag{3.44}$$

B.C.1

$$-\mathbf{n}_{\gamma\kappa} \cdot \nabla \mathbf{f} = \mathbf{n}_{\gamma\kappa}, \quad \text{at } A_{\gamma\kappa} \tag{3.45}$$

$$\mathbf{f}(\mathbf{r} + \boldsymbol{\ell}_i) = \mathbf{f}(\mathbf{r}) \tag{3.46}$$

Problem II

$$0 = \nabla^2 s + a_{\gamma\kappa} K_{\text{eq}} / \mathscr{D}_\gamma \tag{3.47}$$

B.C.1

$$-\mathbf{n}_{\gamma\kappa} \cdot \nabla s = K_{\text{eq}} / \mathscr{D}_\gamma \tag{3.48}$$

$$s(\mathbf{r} + \boldsymbol{\ell}_i) = s(\mathbf{r}) \tag{3.49}$$

Based on the constraint given by Eq. 3.43 we can simplify Eq. 3.29 to

$$\tilde{c}_\gamma = \mathbf{f} \cdot \nabla \langle c_\gamma \rangle^\gamma + s \frac{\partial \langle c_\gamma \rangle^\gamma}{\partial t} \tag{3.50}$$

and use of this result in Eq. 2.25 provides the *closed form* of the volume-averaged diffusion equation.

$$\varepsilon_\gamma (1 + K) \frac{\partial \langle c_\gamma \rangle^\gamma}{\partial t} = \nabla \cdot \left[\varepsilon_\gamma \mathscr{D}_\gamma \left(\mathbf{I} + \frac{1}{V_\gamma} \int_{A_{\gamma\kappa}} \mathbf{n}_{\gamma\kappa} \mathbf{f} \, dA \right) \cdot \nabla \langle c_\gamma \rangle^\gamma \right] + \nabla \cdot \left[\varepsilon_\gamma \mathscr{D}_\gamma \left(\frac{1}{V_\gamma} \int_{A_{\gamma\kappa}} \mathbf{n}_{\gamma\kappa} s \, dA \right) \frac{\partial \langle c_\gamma \rangle^\gamma}{\partial t} \right] \tag{3.51}$$

Here we have used K to represent the dimensionless adsorption isotherm which is defined by

$$K = a_{\gamma\kappa} K_{\text{eq}} / \varepsilon_\gamma \tag{3.52}$$

There are two important points to be made concerning this closed form. The effective diffusivity tensor is defined by

$$\varepsilon_\gamma \mathbf{D}^\gamma_{\text{eff}} = \varepsilon_\gamma \mathscr{D}_\gamma \left(\mathbf{I} + \frac{1}{V_\gamma} \int_{A_{\gamma\kappa}} \mathbf{n}_{\gamma\kappa} \mathbf{f} \, dA \right) \tag{3.53}$$

and Eqs 3.44–3.46 clearly indicate that $\mathbf{D}^\gamma_{\text{eff}}$ depends only upon the geometry of the porous medium and it is independent of time. It must be remembered that this latter comment relies on both Eqs 3.18 and 3.43 and a conservative combination of these constraints could be expressed as

$$\frac{\mathscr{D}_\gamma t}{\ell_\gamma^2} \gg (1 + K_{\text{eq}} / \ell_\gamma) \tag{3.54}$$

The second point to be made concerning Eq. 3.51 is that the adsorption process does not influence the effective diffusivity tensor but instead gives rise to the mixed derivative term involving the area integral of $\mathbf{n}_{\gamma\kappa} s$.

From the boundary value problem for s, we can extract the estimate

$$s = O(K_{\text{eq}} \ell_\gamma / \mathscr{D}_\gamma) \tag{3.55}$$

and this leads to

$$\begin{aligned} \nabla \cdot \left[\varepsilon_\gamma \mathscr{D}_\gamma \left(\frac{1}{V_\gamma} \int_{A_{\gamma\kappa}} \mathbf{n}_{\gamma\kappa} s \, dA \right) \frac{\partial \langle c_\gamma \rangle^\gamma}{\partial t} \right] &= O\left[\varepsilon_\gamma K_{\text{eq}} \nabla \left(\frac{\partial \langle c_\gamma \rangle^\gamma}{\partial t} \right) \right] \\ &= O\left[\frac{\varepsilon_\gamma K_{\text{eq}}}{\ell_\sigma} \left(\frac{\partial \langle c_\gamma \rangle^\gamma}{\partial t} \right) \right] \end{aligned} \tag{3.56}$$

From Eq. 3.52 we can see that $\varepsilon_\gamma K_{\text{eq}} / \ell_\sigma$ will always be much less than $\varepsilon_\gamma K$, thus we can drop the last term from Eq. 3.51 and express the volume-averaged diffusion equation as

$$\varepsilon_\gamma (1 + K) \frac{\partial \langle c_\gamma \rangle^\gamma}{\partial t} = \nabla \cdot (\varepsilon_\gamma \mathbf{D}^\gamma_{\text{eff}} \nabla \langle c_\gamma \rangle^\gamma) \tag{3.57}$$

Since the effective diffusivity tensor is independent of the adsorption process when Eq. 3.54 is valid, a valid comparison between theory and experiment can be carried out in passive systems.

4. Effective diffusivities: comparison between theory and experiment

Rather than use a complex unit cell such as the one illustrated in Fig. 4, we have carried out two-dimensional calculations using unit cells such as those illustrated in Figs 5 and 6 (Ryan *et al.*, 1981; Kim *et al.*, 1987). These are systems for which the diffusional process is isotropic (even though the porous media are not) and it is sufficient to calculate the single distinct value of $\mathbf{D}_{\text{eff}}^{\gamma}$. The comparison between theory and experiment is shown in Fig. 7 where the curve identified by *Ryan* represents the solution of Eqs 3.44–3.46 and the effective diffusivity given by Eq. 3.53. The experimental results of Kim *et al.* (1987) were

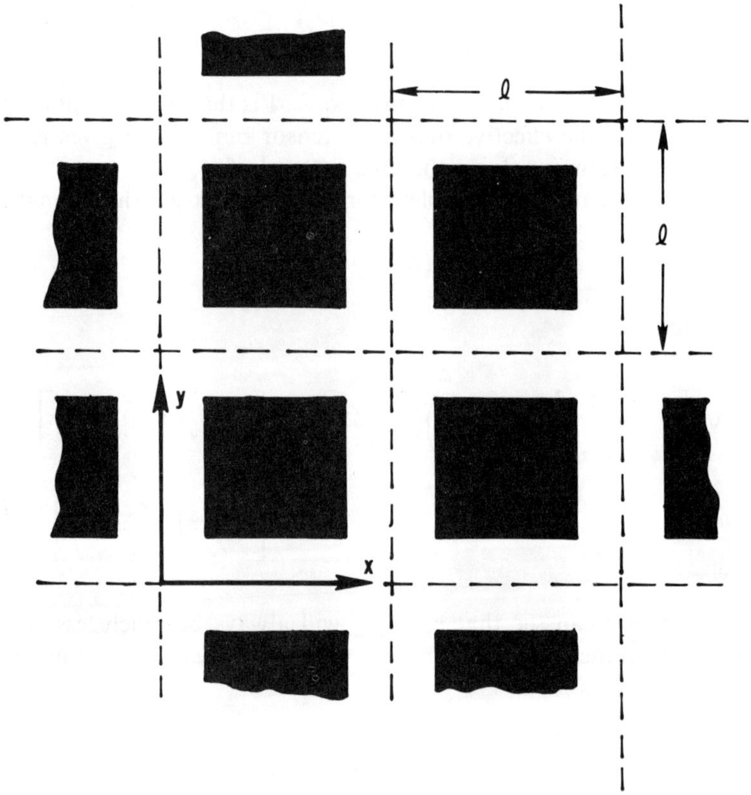

Figure 5. Two-dimensional square array as a model of a spatially periodic porous medium.

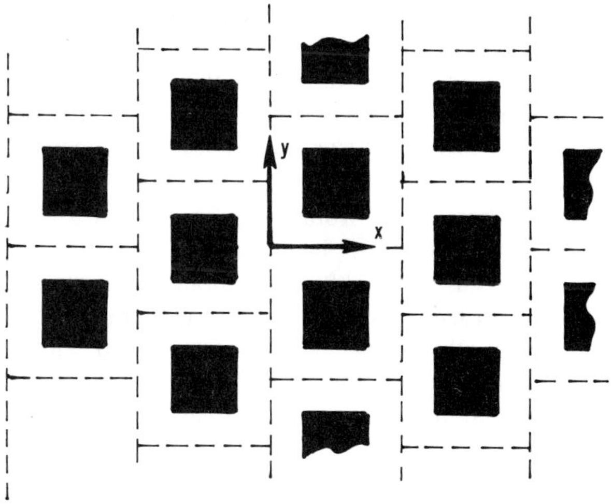

Figure 6. Two-dimensional body-centred array as a model of a spatially periodic porous medium.

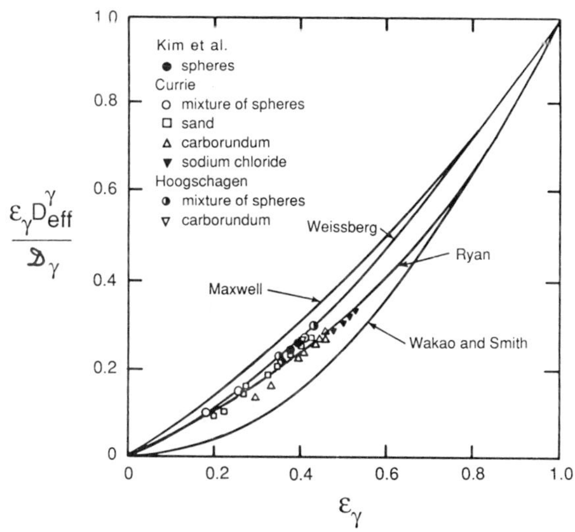

Figure 7. Effective diffusivities for isotropic systems.

obtained with glass spheres having a diameter of 3.95 mm, while the work of Currie (1960) and Hoogschagen (1955) represents a wide range of unconsolidated systems. Neither the analysis of Maxwell (1881) nor the model of Wakao and Smith (1962) fit the experimental data very well; however, the analysis of Weissberg (1963) does well at the lower void fractions. From the comparison between theory and experiment it would appear that simple geometrical models, such as those illustrated in Figs 5 and 6, provide good theoretical results for *unconsolidated porous media.* We doubt that this is the case for consolidated porous media, and the work of Kim *et al.* (1987) clearly indicates that simplistic models *cannot* be used to predict successfully the components of the effective diffusivity tensor for anisotropic systems.

In thinking about the closure problems given by Eqs 3.44–3.46, one must remark that it represents a great simplification over the original boundary value problem for $\tilde{c}_\gamma$ given by Eqs 3.3–3.6. The simplification results from the fact that there are *two length scales* associated with the point concentration c_γ. One of these is the long length scale ℓ_σ, which is associated with $\langle c_\gamma \rangle^\gamma$, and the other is the short length scale ℓ_γ, associated with $\tilde{c}_\gamma$. Since the ratio of these length scales can be thought of as a small parameter, perturbation methods can be used to study problems of this type. The classic exposition of this approach is given by Bensoussan *et al.* (1978), and a direct application to the problem of diffusion in porous media is available in the work of Chang (1982). A discussion of the homogenization method is provided by Ene in Chapter VIII and Douglas and Arbogast in Chapter VII, this volume. That the method of volume averaging and the method of spatial homogenization are essentially equivalent has been noted by Bourgeat *et al.* (1988) and by Pironti and Whitaker (1986).

5. Diffusion, adsorption and convection in porous media composed of porous particles

In this section we wish to develop the spatially smoothed equations associated with the averaging volume $\mathscr{V}$ shown in Fig. 3. The length scales associated with this averaging volume are identified in Fig. 8, and the boundary value problem associated with *local volume averaging* is given by

$$\varepsilon_\gamma(1+K)\frac{\partial c_\sigma}{\partial t} = \nabla\cdot(\mathbf{D}_\sigma\cdot\nabla c_\sigma), \quad \text{in the } \sigma\text{-phase} \tag{5.1}$$

B.C.1
$$c_\sigma = c_\beta, \quad \text{at } \mathscr{A}_{\beta\sigma} \tag{5.2}$$

B.C.2
$$-\mathbf{n}_{\sigma\beta}\cdot(\mathbf{D}_\sigma\cdot\nabla c_\sigma) = -\mathbf{n}_{\sigma\beta}\cdot(\mathscr{D}_\beta\nabla c_\beta), \quad \text{at } \mathscr{A}_{\beta\sigma} \tag{5.3}$$

$$\frac{\partial c_\beta}{\partial t} + \nabla\cdot(c_\beta\mathbf{v}_\beta) = \nabla\cdot(\mathscr{D}_\beta\nabla c_\beta), \quad \text{in the } \beta\text{-phase} \tag{5.4}$$

One should think of these equations as describing the transport process in both the ω- and η-regions. In this representation we have simplified the nomenclature

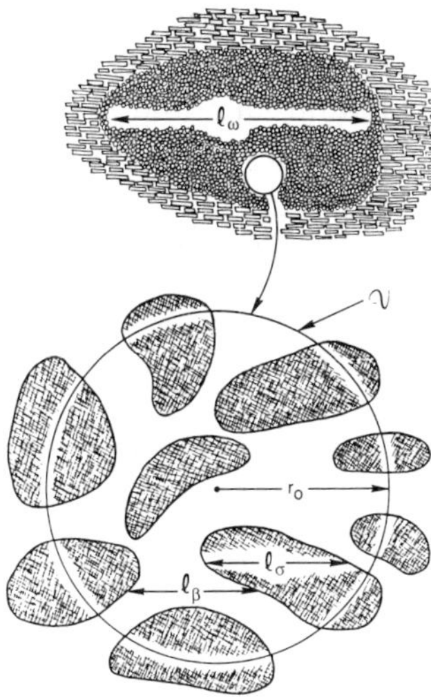

Figure 8. Averaging volume for the β–σ system.

for the σ-phase by using

$$\langle c_\gamma \rangle^\gamma = c_\sigma, \quad \varepsilon_\gamma \mathbf{D}^\gamma_{\text{eff}} = \mathbf{D}_\sigma \tag{5.5}$$

and the boundary conditions given by Eqs 5.2 and 5.3 have been adopted from an earlier study of diffusion and reaction in a micropore–macropore model of a porous medium (Whitaker, 1983).

We begin the process of spatial smoothing with Eq. 5.1 and express the phase average form as

$$\frac{1}{\mathscr{V}} \int_{V_\sigma} \varepsilon_\gamma (1 + K) \left(\frac{\partial c_\sigma}{\partial t} \right) \mathrm{d}V = \frac{1}{\mathscr{V}} \int_{V_\sigma} \nabla \cdot (\mathbf{D}_\sigma \cdot \nabla c_\sigma) \, \mathrm{d}V \tag{5.6}$$

Here $\mathscr{V}$ represents the averaging volume illustrated in Figs 2 and 8, and V_σ represents the volume of the σ-phase contained within $\mathscr{V}$.

In the analysis leading to Eq. 5.1 we assumed that the variations of ε_γ and K were negligible in the averaging volume $\mathscr{V}_\sigma$ illustrated in Fig. 2. Now we are faced with an integral of ε_γ and K over a much larger region, i.e. the averaging volume $\mathscr{V}$ illustrated in Fig. 2. In this part of our analysis we are going to impose the severe restriction that ε_γ and K can be treated as constants with respect to the integration indicated in Eq. 5.6, and this allows us to write the phase average form of the σ-phase transport equation as

$$\varepsilon_\gamma (1 + K) \frac{\partial \langle c_\sigma \rangle}{\partial t} = \langle \nabla \cdot (\mathbf{D}_\sigma \cdot \nabla c_\sigma) \rangle \tag{5.7}$$

Here we have again imposed the restriction that the porous medium under consideration is rigid. It is important to keep in mind that the derivation of Eq. 5.1 required negligible variations of ε_γ and K over the distance r_σ illustrated in Fig. 3, while Eq. 5.7 requires negligible variations of ε_γ and K over the distance r_0 illustrated in Fig. 8.

The analysis of Eq. 5.7 follows that given previously by Eqs 2.8–2.19 and we express the result as

$$\varepsilon_\sigma\varepsilon_\gamma(1+K)\frac{\partial\langle c_\sigma\rangle^\sigma}{\partial t} = \nabla\cdot\left[\mathbf{D}_\sigma\cdot\left(\varepsilon_\sigma\nabla\langle c_\sigma\rangle^\sigma + \langle c_\sigma\rangle^\sigma\nabla\varepsilon_\sigma + \frac{1}{\mathscr{V}}\int_{A_{\sigma\beta}}\mathbf{n}_{\sigma\beta}c_\sigma\,\mathrm{d}A\right)\right]$$

$$+\frac{1}{\mathscr{V}}\int_{A_{\beta\sigma}}\mathbf{n}_{\sigma\beta}\cdot\mathbf{D}_\sigma\cdot\nabla c_\sigma\,\mathrm{d}A \tag{5.8}$$

Once again we decompose the concentration according to Gray (1975) and express c_σ as

$$c_\sigma = \langle c_\sigma\rangle^\sigma + \tilde{c}_\sigma \tag{5.9}$$

Here one should think of $\langle c_\sigma\rangle^\sigma$ as having a characteristic length ℓ_λ which represents either ℓ_ω or ℓ_η, while the length scale associated with $\tilde{c}_\sigma$ is represented by ℓ_σ in Fig. 8. We neglect variations of $\langle c_\sigma\rangle^\sigma$ with the averaging volume $\mathscr{V}$ on the basis of the constraint (Carbonell and Whitaker, Section 2, 1984)

$$\left(\frac{r_0}{\ell_\lambda}\right)^2 \ll 1 \tag{5.10}$$

and this allows us to use Eq. 5.9 in Eq. 5.8 to obtain

$$\varepsilon_\sigma\varepsilon_\gamma(1+K)\frac{\partial\langle c_\sigma\rangle^\sigma}{\partial t} = \nabla\cdot\left[\varepsilon_\sigma\mathbf{D}_\sigma\cdot\left(\nabla\langle c_\sigma\rangle^\sigma + \frac{1}{V_\sigma}\int_{A_{\sigma\beta}}\mathbf{n}_{\sigma\beta}\tilde{c}_\sigma\,\mathrm{d}A\right)\right]$$

$$+\frac{1}{\mathscr{V}}\int_{A_{\beta\sigma}}\mathbf{n}_{\sigma\beta}\cdot\mathbf{D}_\sigma\cdot\nabla c_\sigma\,\mathrm{d}A \tag{5.11}$$

Turning our attention to the β-phase, we note that the phase average form of Eq. 5.4 is given by

$$\frac{1}{\mathscr{V}}\int_{V_\beta}\left(\frac{\partial c_\beta}{\partial t}\right)\mathrm{d}V + \frac{1}{\mathscr{V}}\int_{V_\beta}\nabla\cdot(c_\beta\mathbf{v}_\beta)\,\mathrm{d}V = \frac{1}{\mathscr{V}}\int_{V_\beta}\nabla\cdot(\mathscr{D}_\beta\nabla c_\beta)\,\mathrm{d}V \tag{5.12}$$

where V_β represents the volume for the β-phase contained within the averaging volume $\mathscr{V}$. The first term in this result can be expressed as

$$\frac{1}{\mathscr{V}}\int_{V_\beta}\left(\frac{\partial c_\beta}{\partial t}\right)\mathrm{d}V = \frac{\partial\langle c_\beta\rangle}{\partial t} = \varepsilon_\beta\frac{\partial\langle c_\beta\rangle^\beta}{\partial t} \tag{5.13}$$

and the spatial averaging theorem can be used with the convective and diffusive fluxes to obtain

$$\varepsilon_\beta\frac{\partial\langle c_\beta\rangle^\beta}{\partial t} + \nabla\cdot\langle c_\beta\mathbf{v}_\beta\rangle + \frac{1}{\mathscr{V}}\int_{A_{\beta\sigma}}\mathbf{n}_{\beta\sigma}\cdot(c_\beta\mathbf{v}_\beta)\,\mathrm{d}A$$

$$= \nabla\cdot\langle\mathscr{D}_\beta\nabla c_\beta\rangle + \frac{1}{\mathscr{V}}\int_{A_{\beta\sigma}}\mathbf{n}_{\beta\sigma}\cdot\mathscr{D}_\beta\nabla c_\beta\,\mathrm{d}A \tag{5.14}$$

If one can use Darcy's law (Whitaker, 1986a) to describe the flow in the β- and γ-phases, the following estimate for the Peclet number in the two phases can be constructed.

$$\left\{\begin{matrix}\text{Peclet number}\\ \text{for the }\gamma\text{-phase}\end{matrix}\right\} = O\left(\frac{\ell_\gamma^3\mathscr{D}_\beta}{\ell_\beta^3\mathscr{D}_\gamma}\right)\left\{\begin{matrix}\text{Peclet number}\\ \text{for the }\beta\text{-phase}\end{matrix}\right\} \tag{5.15}$$

Our formulation of the diffusion problem in the γ-phase suggests that bulk diffusion occurs within the pores of the σ-phase and this means that $\mathscr{D}_\beta$ is equal to $\mathscr{D}_\gamma$. Taking this into account and remembering that ℓ_γ is small compared to ℓ_β, we see that the γ-phase Peclet number will be very small compared to the β-phase Peclet number. Under normal circumstances this means that diffusion dominates in the γ-phase as indicated by Eqs 2.1–2.4. This is consistent with the inequality

$$\mathbf{n}_{\beta\sigma}\cdot(c_\beta\mathbf{v}_\beta) \ll \mathbf{n}_{\beta\sigma}\cdot\mathscr{D}_\beta\nabla c_\beta, \quad \text{at } A_{\beta\sigma} \tag{5.16}$$

and this allows us to simplify Eq. 5.14 to

$$\varepsilon_\beta\frac{\partial\langle c_\beta\rangle^\beta}{\partial t} + \nabla\cdot\langle c_\beta\mathbf{v}_\beta\rangle = \nabla\cdot\langle\mathscr{D}_\beta\nabla c_\beta\rangle + \frac{1}{\mathscr{V}}\int_{A_{\beta\sigma}}\mathbf{n}_{\beta\sigma}\cdot\mathscr{D}_\beta\nabla c_\beta\,\mathrm{d}A \tag{5.17}$$

To analyse the diffusive term on the right-hand side of this result, we need only follow the development from Eq. 2.8 to Eq. 2.25 in order to express Eq. 5.17 as

$$\varepsilon_\beta \frac{\partial \langle c_\beta \rangle^\beta}{\partial t} + \nabla \cdot \langle c_\beta \mathbf{v}_\beta \rangle = \nabla \cdot \left[\varepsilon_\beta \mathscr{D}_\beta \left(\nabla \langle c_\beta \rangle^\beta + \frac{1}{V_\beta} \int_{A_{\beta\sigma}} \mathbf{n}_{\beta\sigma} \tilde{c}_\beta \, \mathrm{d}A \right) \right]$$

$$+ \frac{1}{\mathscr{V}} \int_{A_{\beta\sigma}} \mathbf{n}_{\beta\sigma} \cdot \mathscr{D}_\beta \nabla c_\beta \, \mathrm{d}A \tag{5.18}$$

In the development of this result we have decomposed the concentration according to

$$c_\beta = \langle c_\beta \rangle^\beta + \tilde{c}_\beta \tag{5.19}$$

In order to analyse the convective term in Eq. 5.18 we express the velocity as

$$\mathbf{v}_\beta = \langle \mathbf{v}_\beta \rangle^\beta + \tilde{\mathbf{v}}_\beta \tag{5.20}$$

and consider the average of the product, $\langle c_\beta \mathbf{v}_\beta \rangle$. This term has been considered in detail by Carbonell and Whitaker (Section 2, 1984) and their analysis leads to

$$\langle c_\beta \mathbf{v}_\beta \rangle = \varepsilon_\beta \langle c_\beta \rangle^\beta \langle \mathbf{v}_\beta \rangle^\beta + \varepsilon_\beta \langle \tilde{c}_\beta \tilde{\mathbf{v}}_\beta \rangle^\beta \tag{5.21}$$

provided the following constraint is satisfied

$$\left(\frac{\ell_\beta}{r_0} \right) \gg \left(\frac{r_0}{\ell_\lambda} \right) \qquad \lambda = \omega \text{ or } \eta \tag{5.22}$$

Equation 5.22 places very severe constraints on ℓ_β, r_0, and ℓ_λ; however, it must be noted that this result is based on order of magnitude estimates and detailed numerical results of the type described by Prat (1989) are needed to identify the precise nature of this constraint.

Use of Eq. 5.21 in Eq. 5.18 leads to the β-phase analogy to Eq. 5.11 which we express as

$$\varepsilon_\beta \frac{\partial \langle c_\beta \rangle^\beta}{\partial t} + \varepsilon_\beta \langle \mathbf{v}_\beta \rangle^\beta \cdot \nabla \langle c_\beta \rangle^\beta = \nabla \cdot \left[\varepsilon_\beta \mathscr{D}_\beta \left(\nabla \langle c_\beta \rangle^\beta + \frac{1}{V_\beta} \int_{A_{\beta\sigma}} \mathbf{n}_{\beta\sigma} \tilde{c}_\beta \, \mathrm{d}A \right) \right]$$

$$- \nabla \cdot (\varepsilon_\beta \langle \tilde{c}_\beta \tilde{\mathbf{v}}_\beta \rangle^\beta) + \frac{1}{\mathscr{V}} \int_{a_{\beta\sigma}} \mathbf{n}_{\beta\sigma} \cdot \mathscr{D}_\beta \nabla c_\beta \, \mathrm{d}A \tag{5.23}$$

In obtaining this result we have treated the flow as incompressible and used the point continuity equation

$$\nabla \cdot \mathbf{v}_\beta = 0 \tag{5.24}$$

to develop the local volume averaged form

$$\nabla \cdot \langle \mathbf{v}_\beta \rangle + \frac{1}{\mathscr{V}} \int_{A_{\beta\sigma}} \mathbf{n}_{\beta\sigma} \cdot \mathbf{v}_\beta \, \mathrm{d}A = 0 \tag{5.25}$$

It is consistent with Eq. 5.16 to neglect the convective flux at the β–σ interface and simplify Eq. 5.25 to

$$\nabla \cdot \langle \mathbf{v}_\beta \rangle = 0 \tag{5.26}$$

This, in turn, can be expressed as

$$\nabla \cdot (\varepsilon_\beta \langle \mathbf{v}_\beta \rangle^\beta) = 0 \tag{5.27}$$

and it is this form of the local volume-averaged continuity equation that has been used to simplify the convective transport term in Eq. 5.23.

In Eq. 5.23 we refer to $\varepsilon_\beta \langle \tilde{c}_\beta \tilde{\mathbf{v}}_\beta \rangle^\beta$ as the hydrodynamic dispersion vector and evaluation of this term, along with the area integral of $\mathbf{n}_{\beta\sigma} \tilde{c}_\beta$, requires the solution of a closure problem for $\tilde{c}_\beta$. If it were not for the interfacial flux terms in Eqs 5.11 and 5.23, we would be ready to develop the closure problem for $\tilde{c}_\sigma$ and $\tilde{c}_\beta$ and then go on to the comparison between theory and experiment. However, these flux terms can be eliminated by the use of a one-equation model and that model is considered in the following paragraphs.

On a purely intuitive basis, one might guess that conditions exist for which the concentration in the σ-phase,

$$\langle c_\sigma \rangle^\sigma = \frac{1}{V_\sigma} \int_{V_\sigma} \langle c_\gamma \rangle^\gamma \, \mathrm{d}V \tag{5.28}$$

is essentially equal to the concentration in the β-phase,

$$\langle c_\beta \rangle^\beta = \frac{1}{V_\beta} \int_{V_\beta} c_\beta \, \mathrm{d}V \tag{5.29}$$

For example, it the size of the porous particles were very *small*, or the effective diffusivity were very *large*, one might be tempted to use the approximation $\langle c_\sigma \rangle^\sigma = \langle c_\beta \rangle^\beta$. We can formalize this approximation by decomposing $\langle c_\sigma \rangle^\sigma$

and $\langle c_\beta \rangle^\beta$ in terms of the *intrinsic spatial average concentration* and a spatial deviation. These decompositions are given by

$$\langle c_\sigma \rangle^\sigma = \langle c \rangle + \hat{c}_\sigma \tag{5.30a}$$

$$\langle c_\beta \rangle^\beta = \langle c \rangle + \hat{c}_\beta \tag{5.30b}$$

where the intrinsic spatial average concentration is defined by

$$\langle c \rangle = \varepsilon_\sigma \langle c_\sigma \rangle^\sigma + \varepsilon_\beta \langle c_\beta \rangle^\beta \quad , \tag{5.31}$$

When Eqs 5.30a and b are used in Eqs 5.11 and 5.23, we can add the results in order to cancel the interfacial flux terms and arrive at a single transport equation which takes the form

$$\begin{aligned}
&\varepsilon_\beta \left[1 + \frac{\varepsilon_\sigma \varepsilon_\gamma}{\varepsilon_\beta} (1 + K) \right] \frac{\partial \langle c \rangle}{\partial t} + \varepsilon_\beta \langle \mathbf{v}_\beta \rangle^\beta \cdot \nabla \langle c \rangle = \nabla \cdot \left[\varepsilon_\beta \mathscr{D}_\beta \left(\nabla \langle c \rangle \right.\right. \\
&\left. + \frac{1}{V_\beta} \int_{A_{\beta\sigma}} \mathbf{n}_{\beta\sigma} \tilde{c}_\beta \, dA \right) + \varepsilon_\sigma \mathbf{D}_\sigma \cdot \left. \left(\nabla \langle c \rangle + \frac{1}{V_\sigma} \int_{A_{\sigma\beta}} \mathbf{n}_{\sigma\beta} \tilde{c}_\sigma \right) \right] - \nabla \cdot (\varepsilon_\beta \langle \tilde{c}_\beta \tilde{\mathbf{v}}_\beta \rangle^\beta \\
&- \varepsilon_\beta \frac{\partial \hat{c}_\beta}{\partial t} - \varepsilon_\sigma \varepsilon_\gamma (1 + K) \frac{\partial \hat{c}_\sigma}{\partial t} - \varepsilon_\beta \langle \mathbf{v}_\beta \rangle^\beta \cdot \nabla \hat{c}_\beta + \nabla \cdot (\varepsilon_\beta \mathscr{D}_\beta \nabla \hat{c}_\beta) + \nabla \cdot (\varepsilon_\sigma \mathbf{D}_\sigma \nabla \hat{c}_\sigma)
\end{aligned} \tag{5.32}$$

If the last five terms in this result are negligible, we can express the one-equation model as

$$\begin{aligned}
&\varepsilon_\beta (1 + \mathscr{K}) \frac{\partial \langle c \rangle}{\partial t} + \langle \mathbf{v}_\beta \rangle \cdot \langle c \rangle = \nabla \cdot \left[\varepsilon_\beta \mathscr{D}_\beta \left(\nabla \langle c \rangle + \frac{1}{V_\beta} \int_{A_{\beta\sigma}} \mathbf{n}_{\beta\sigma} \tilde{c}_\beta \, dA \right) \right. \\
&\qquad \left. + \varepsilon_\sigma \mathbf{D}_\sigma \cdot \left(\nabla \langle c \rangle + \frac{1}{V_\sigma} \int_{A_{\sigma\beta}} \mathbf{n}_{\sigma\beta} \tilde{c}_\sigma \, dA \right) \right] + \nabla \cdot (\varepsilon_\beta \langle \tilde{c}_\beta \tilde{\mathbf{v}}_\beta \rangle^\beta)
\end{aligned} \tag{5.33}$$

Here we have replaced $\varepsilon_\beta \langle \mathbf{v}_\beta \rangle^\beta$ with the phase average velocity $\langle \mathbf{v}_\beta \rangle$ which is available from Darcy's law, and we have used a new adsorption isotherm defined by

$$\mathscr{K} = \frac{\varepsilon_\sigma \varepsilon_\gamma}{\varepsilon_\beta} (1 + K) \tag{5.34}$$

In order to obtain a closed form for Eq. 5.33, we need to develop representations for $\tilde{c}_\beta$ and $\tilde{c}_\sigma$; however, we first need to justify the simplifications that led us from Eq. 5.32 to Eq. 5.33.

6. Local mass equilibrium

The simplification provided by Eq. 5.33 as a replacement for Eqs 5.11 and 5.23 is enormous, and this will become apparent in the next chapter (Chapter VI) when we examine the two-equation model for transport in the ω–η system illustrated in Fig. 1. However, when Eq. 5.33 is *not valid* it becomes a dangerous instrument if the diffusive and dispersive terms on the right-hand side are represented in terms of an adjustable parameter to be determined by experiment. The complexities of two-equation models have been explored by Whitaker (1983, Section 6) for diffusion and reaction in micropore–macropore systems, by Zanotti and Carbonell (1984) and Levec and Carbonell (1985) for heat transfer in porous media, and by Whitaker (1989) for heat transfer in packed bed catalytic reactors. However, at this time no detailed solutions for the several published closure problems exist, thus we can only estimate the magnitude of the coefficients that appear in the two-equation model.

In order to justify the neglect of the last five terms in Eq. 5.32, we must be able to estimate the spatial deviation concentrations $\hat{c}_\sigma$ and $\tilde{c}_\beta$. In considering these two concentrations we must remember that they represent the difference between two average concentrations, thus they are associated with a characteristic length ℓ_λ which is much larger than either ℓ_β or ℓ_σ. To obtain estimates of $\hat{c}_\sigma$ and $\tilde{c}_\beta$, we follow an earlier study of *local thermal equilibrium* (Whitaker, 1980, Section 3.3) and note that Eqs 5.30 and 5.31 can be used to obtain

$$\hat{c}_\sigma = \varepsilon_\beta(\langle c_\sigma\rangle^\sigma - \langle c_\beta\rangle^\beta) \tag{6.1a}$$

$$\hat{c}_\beta = \varepsilon_\sigma(\langle c_\beta\rangle^\beta - \langle c_\sigma\rangle^\sigma) \tag{6.1b}$$

Here it becomes clear that we need an estimate of $\langle c_\sigma\rangle^\sigma - \langle c_\beta\rangle^\beta$ and to obtain this estimate we turn our attention to the interfacial flux terms in Eqs 5.11 and 5.23. We estimate these terms as

$$-\frac{1}{\mathscr{V}}\int_{A_{\sigma\beta}} \mathbf{n}_{\sigma\beta}\cdot\mathbf{D}_\sigma\cdot\nabla c_\sigma\,\mathrm{d}A = O\left[\frac{a_{\beta\sigma}D_\sigma(\langle c_\sigma\rangle^\sigma - \langle c_\sigma\rangle_{\beta\sigma})}{\delta_\sigma}\right] \tag{6.2a}$$

$$-\frac{1}{\mathscr{V}}\int_{A_{\beta\sigma}} \mathbf{n}_{\beta\sigma}\cdot\mathscr{D}_\beta\nabla c_\beta\,\mathrm{d}A = O\left[\frac{a_{\beta\sigma}\mathscr{D}_\beta(\langle c_\beta\rangle^\beta - \langle c_\beta\rangle_{\beta\sigma})}{\delta_\beta}\right] \tag{6.2b}$$

where we have used $a_{\beta\sigma}$ to represent the interfacial area per unit volume, $A_{\beta\sigma}/\mathscr{V}$, and we have represented the area average concentrations by $\langle c_\sigma \rangle_{\beta\sigma}$ and $\langle c_\beta \rangle_{\beta\sigma}$. The terms δ_σ and δ_β represent characteristic lengths associated with the transport processes in the σ- and β-phases, respectively, and these lengths are difficult to estimate even though we have some knowledge of the details of the transport processes through the solution of special boundary value problems.

On the basis of the boundary condition given by Eq. 5.2 we know that

$$\langle c_\sigma \rangle_{\beta\sigma} = \langle c_\beta \rangle_{\beta\sigma} \tag{6.3}$$

and we can use the flux boundary condition given by Eq. 5.3 to eliminate $\langle c_\sigma \rangle_{\beta\sigma}$ and $\langle c_\beta \rangle_{\beta\sigma}$ from Eqs 6.2. This leads to the estimates

$$-\frac{1}{\mathscr{V}} \int_{A_{\sigma\beta}} \mathbf{n}_{\sigma\beta} \cdot \mathbf{D}_\sigma \cdot \nabla c_\sigma \, \mathrm{d}A = O\left[\frac{a_{\beta\sigma}(\langle c_\sigma \rangle^\sigma - \langle c_\beta \rangle^\beta)}{\left(\dfrac{\delta_\sigma}{D_\sigma} + \dfrac{\delta_\beta}{\mathscr{D}_\beta}\right)}\right] \tag{6.4a}$$

$$-\frac{1}{\mathscr{V}} \int_{A_{\beta\sigma}} \mathbf{n}_{\beta\sigma} \cdot \mathscr{D}_\beta \nabla c_\beta \, \mathrm{d}A = O\left[\frac{a_{\beta\sigma}(\langle c_\beta \rangle^\beta - \langle c_\sigma \rangle^\sigma)}{\left(\dfrac{\delta_\sigma}{D_\sigma} + \dfrac{\delta_\beta}{\mathscr{D}_\beta}\right)}\right] \tag{6.4b}$$

At this point we have connected the values of $\hat{c}_\sigma$ and $\hat{c}_\beta$, given by Eqs 6.1, to the σ- and β-phase transport equations given by Eqs 5.11 and 5.23.

We now return to Eq. 5.32 and note that the one-equation model will be valid when the following constraints are satisfied

$$\left.\begin{aligned} &\varepsilon_\beta \frac{\partial \hat{c}_\beta}{\partial t} \ll \varepsilon_\beta \frac{\partial \langle c \rangle}{\partial t}, \quad \varepsilon_\sigma \varepsilon_\gamma (1 + K) \frac{\partial \hat{c}_\sigma}{\partial t} \ll \varepsilon_\sigma \varepsilon_\gamma (1 + K) \frac{\partial \langle c \rangle}{\partial t}, \\ &\varepsilon_\beta \langle \mathbf{v}_\beta \rangle^\beta \cdot \nabla \hat{c}_\beta \ll \varepsilon_\beta \langle \mathbf{v}_\beta \rangle^\beta \cdot \nabla \langle c \rangle, \quad \nabla \hat{c}_\beta \ll \nabla \langle c \rangle, \\ &\nabla \hat{c}_\sigma \ll \nabla \langle c \rangle \end{aligned}\right\} \tag{6.5}$$

Considering the first of these, we use the estimates

$$\frac{\partial \hat{c}_\beta}{\partial t} = O\left(\frac{\hat{c}_\beta}{t}\right), \quad \frac{\partial \langle c \rangle}{\partial t} = O\left(\frac{\nabla \langle c \rangle}{t}\right) \tag{6.6}$$

since the change of $\hat{c}_\beta$ with respect to time and space will be of the order of $\hat{c}_\beta$. The characteristic times for $\hat{c}_\beta$ and $\langle c \rangle$ will be the same, thus the first constraint given by Eq. 6.5 will be satisfied when

$$\hat{c}_\beta \ll \Delta \langle c \rangle \tag{6.7}$$

A more severe form of this can be extracted from Eq. 5.31 which suggests that Eq. 6.7 will always be true when the following constraint is satisfied

$$\hat{c}_\beta \ll \varepsilon_\beta \Delta \langle c_\beta \rangle^\beta \tag{6.8}$$

Since the length scales associated with $\hat{c}_\beta$ and $\langle c \rangle$ will be the same order of magnitude, the fourth constraint in Eq. 6.5

$$\nabla \hat{c}_\beta \ll \nabla \langle c \rangle \tag{6.9}$$

can be replaced by Eq. 6.7 and subsequently by Eq. 6.8. A little thought will indicate that all the constraints in Eq. 6.5 can be satisfied by Eq. 6.8 and the analogous version for $\hat{c}_\sigma$ which is given by

$$\hat{c}_\sigma \ll \varepsilon_\sigma \Delta \langle c_\sigma \rangle^\sigma \tag{6.10}$$

When these results are combined with Eqs 6.1 we find that the condition of local mass equilibrium is valid when the following constraints are satisfied

$$\varepsilon_\sigma \Delta \langle c_\sigma \rangle^\sigma \gg \varepsilon_\beta (\langle c_\sigma \rangle^\sigma - \langle c_\beta \rangle^\beta) \tag{6.11a}$$

$$\varepsilon_\beta \Delta \langle c_\beta \rangle^\beta \gg \varepsilon_\sigma (\langle c_\beta \rangle^\beta - \langle c_\sigma \rangle^\sigma) \tag{6.11b}$$

In order to put these constraints in useable form we need estimates of the right-hand side in terms of the left-hand side. The flux estimates given by Eqs 6.4 and the transport equations given by Eqs 5.11 and 5.23 will provide the information that we seek.

We begin our analysis with Eq. 5.11 and express that result as

$$\frac{1}{\mathscr{V}} \int_{A_{\beta\sigma}} \mathbf{n}_{\sigma\beta} \cdot \mathbf{D}_\sigma \cdot \nabla c_\sigma \, dA = O \begin{cases} \varepsilon_\sigma \varepsilon_\gamma (1 + K) \dfrac{\partial \langle c_\sigma \rangle^\sigma}{\partial t} \\ \nabla \cdot \left[\varepsilon_\sigma \mathbf{D}_\sigma \cdot \left(\nabla \langle c_\sigma \rangle^\sigma + \dfrac{1}{V_\sigma} \displaystyle\int_{A_{\sigma\beta}} \mathbf{n}_{\sigma\beta} \tilde{c}_\sigma \, dA \right) \right] \end{cases} \tag{6.12}$$

The idea here is that the interfacial flux term in Eq. 5.11 might be as large as either the accumulation term or the diffusive term, but it certainly cannot be much larger than either of these terms. In addition, it is possible that the interfacial flux term is *small* compared to either the accumulation term or the diffusive term, but it is difficult to see how this could occur. The terms on the right-hand side of Eq. 6.12 can be estimated to obtain

$$\frac{1}{\mathscr{V}}\int_{A_{\sigma\beta}} \mathbf{n}_{\sigma\beta}\cdot\mathbf{D}_\sigma\cdot\nabla c_\sigma\,dA = O\begin{cases} \varepsilon_\sigma\varepsilon_\gamma(1+K)\dfrac{\Delta\langle c_\sigma\rangle^\sigma}{t} \\ \dfrac{\varepsilon_\sigma D_\sigma\Delta\langle c_\sigma\rangle^\sigma}{\ell_\lambda^2} \end{cases} \tag{6.13}$$

Here we have assumed that the diffusive flux in the σ-phase could be represented in terms of an effective diffusivity tensor $\mathbf{D}_{\sigma,\mathrm{eff}}$, and D_σ represents an estimate of the magnitude of the σ-phase effective diffusivity.

At this point we are in a position to combine the estimate given by Eq. 6.4a with the estimate given by Eq. 6.13 in order to express the inequality given by Eq. 6.11a as

$$\frac{a_{\beta\sigma}t}{\left(\dfrac{\delta_\sigma}{D_\sigma}+\dfrac{\delta_\beta}{\mathscr{D}_\beta}\right)} \gg \varepsilon_\beta\varepsilon_\gamma(1+K) \tag{6.14}$$

It is convenient to express this result in the form

$$\varepsilon_\beta\varepsilon_\gamma(1+K)\left(\frac{\delta_\sigma}{a_{\beta\sigma}D_\sigma t}\right)\left(1+\frac{D_\sigma}{\mathscr{D}_\beta}\frac{\delta_\beta}{\delta_\sigma}\right) \ll 1 \tag{6.15}$$

so that the restriction takes the same form as that given by Whitaker (1986b, Eq. 54) for the case of local thermal equilibrium in a packed bed catalytic reactor. In this case the constraint is modified by the factor $\varepsilon_\gamma(1+K)$ and for large values of K it may be difficult to satisfy Eq. 6.15 for times of interest. The operating characteristics of packed bed reactors make it easy to say something about δ_σ and δ_β; however, these two length scales are difficult to estimate for mass transport in geological formations. It would seem reasonable to replace δ_σ with ℓ_σ and express the constraint given by Eq. 6.15 as

$$\varepsilon_\beta\varepsilon_\gamma(1+K)\left(\frac{\ell_\sigma^2}{D_\sigma t}\right)\left(1+\frac{D_\sigma}{\mathscr{D}_\beta}\frac{\delta_\beta}{\ell_\sigma}\right) \ll 1 \tag{6.16}$$

If convective transport in the β-phase is important, one might think of δ_β as a boundary layer thickness. Under these circumstances our estimate of δ_β takes the form

$$\delta_\beta = O(\ell_\sigma \mathrm{Re}^{-1/2}), \quad \text{if } \ell_\sigma \mathrm{Re}^{-1/2} \leqslant \ell_\beta \tag{6.17a}$$

$$\delta_\beta = O(\ell_\beta), \quad \text{if } \ell_\sigma \mathrm{Re}^{-1/2} \geqslant \ell_\beta \tag{6.17b}$$

and the constraint given by Eq. 6.16 can be expressed as

$$\varepsilon_\beta \varepsilon_\gamma (1 + K)\left(\frac{\ell_\sigma^2}{D_\sigma t}\right)\left(1 + \frac{D_\sigma \ell_\beta}{\mathscr{D}_\beta \ell_\sigma}\right) \ll 1, \quad \text{if } \ell_\sigma \mathrm{Re}^{-1/2} \geqslant \ell_\beta \tag{6.18a}$$

$$\varepsilon_\beta \varepsilon_\gamma (1 + K)\left(\frac{\ell_\sigma}{D_\sigma t}\right)\left(1 + \frac{D_\sigma}{\mathscr{D}_\beta} \mathrm{Re}^{-1/2}\right) \ll 1, \quad \text{if } \ell_\sigma \mathrm{Re}^{-1/2} \leqslant \ell_\beta \tag{6.18b}$$

At this point we have considered the possible constraints that result from the basic inequality, Eq. 6.11a, the estimate given by Eq. 6.4a, and the estimate given by the first term on the right-hand side of Eq. 6.13. Moving on to the second term on the right-hand side, we find a constraint given by

$$\varepsilon_\beta \left(\frac{\ell_\sigma}{\ell_\lambda}\right)^2 \left(1 + \frac{D_\sigma}{\mathscr{D}_\beta} \frac{\delta_\beta}{\ell_\sigma}\right) \gg 1 \tag{6.19}$$

Here we have used ℓ_σ as an estimate of δ_σ and $a_{\beta\sigma}^{-1}$ and Eqs 6.17 can be used to produce estimates of δ_β.

In order to develop useable constraints associated with Eq. 6.11b, we need an estimate of $\Delta\langle c_\beta\rangle^\beta$ that is connected to $(\langle c_\beta\rangle^\beta - \langle c_\sigma\rangle^\sigma)$. To obtain this we return to Eq. 5.23 and write the β-phase analogue of Eq. 6.12 as

$$\frac{1}{\mathscr{V}} \int_{A_{\beta\sigma}} \mathbf{n}_{\beta\sigma} \cdot \mathscr{D}_\beta \nabla c_\beta \, \mathrm{d}A = O \begin{cases} \varepsilon_\beta \dfrac{\partial \langle c_\beta\rangle^\beta}{\partial t} \\ \varepsilon_\beta \langle \mathbf{v}_\beta\rangle^\beta \cdot \nabla \langle c_\beta\rangle^\beta \\ \nabla \cdot \left[\varepsilon_\beta \mathscr{D}_\beta \left(\nabla \langle c_\beta\rangle^\beta + \dfrac{1}{V_\beta} \displaystyle\int_{A_{\beta\sigma}} \mathbf{n}_{\beta\sigma} \tilde{c}_\beta \, \mathrm{d}A\right)\right] \\ \nabla \cdot (\varepsilon_\beta \langle \tilde{c}_\beta \tilde{\mathbf{v}}_\beta\rangle^\beta) \end{cases} \tag{6.20}$$

This result can be expressed in terms of order of magnitude estimates according to

$$\frac{1}{\mathscr{V}} \int_{A_{\beta\sigma}} \mathbf{n}_{\beta\sigma} \cdot \mathscr{D}_\beta \nabla c_\beta \, \mathrm{d}A = O \begin{cases} \varepsilon_\beta \dfrac{\Delta \langle c_\beta\rangle^\beta}{t} \\ \varepsilon_\beta \langle v_\beta\rangle^\beta \Delta \langle c_\beta\rangle^\beta / L_\mathrm{c} \\ \varepsilon_\beta D_\beta \Delta \langle c_\beta\rangle^\beta / L_\mathrm{D}^2 \end{cases} \tag{6.21}$$

Here D_β represents an estimate of the β-phase dispersion tensor, L_c represents a *convective length*, and L_D represents a *dispersive length*. The convective and dispersive lengths are discussed in more detail elsewhere (Whitaker, 1982, Sections 1.1.2.5 and 1.1.3.3). We can now combine Eq. 6.21 with the estimate given by Eq. 6.4b in order to express the constraint given by Eq. 6.11b according to

$$\varepsilon_\sigma\left(\frac{\delta_\beta}{a_{\beta\sigma}\mathscr{D}_\beta t}\right)\left(1+\frac{\mathscr{D}_\beta}{D_\sigma}\frac{\ell_\sigma}{\delta_\beta}\right) \ll 1 \tag{6.22a}$$

$$\varepsilon_\sigma\left(\frac{\langle v_\beta\rangle^\beta \ell_\beta}{\mathscr{D}_\beta}\right)\left(\frac{\delta_\beta}{\ell_\beta}\right)\left(1+\frac{\mathscr{D}_\beta}{D_\sigma}\frac{\ell_\sigma}{\delta_\beta}\right)\left(\frac{\ell_\sigma}{L_\mathrm{c}}\right) \ll 1 \tag{6.22b}$$

$$\varepsilon_\sigma\left(\frac{D_\beta}{\mathscr{D}_\beta}\right)\left(1+\frac{\mathscr{D}_\beta}{D_\sigma}\frac{\ell_\sigma}{\delta_\beta}\right)\left(\frac{\delta_\beta \ell_\sigma}{L_\mathrm{D}^2}\right) \ll 1 \tag{6.22c}$$

In general, these constraints are analogous to these given for the condition of local thermal equilibrium (Whitaker, 1986b), and they are all satisfied when the time is sufficiently long and the length scales associated with $\langle c_\beta\rangle^\beta$ are large compared to ℓ_σ or ℓ_β. In packed bed reactors, the Peclet number

$$\mathrm{Pe} = \langle v_\beta\rangle^\beta \ell_\beta/\mathscr{D}_\beta \tag{6.23}$$

is often large compared to one and the failure of Eq. 6.22b necessitates the use of a two-equation model (Whitaker, 1989). For groundwater flows the Peclet number is generally not large compared to one and it is more likely that the constraints associated with the σ-phase will fail. This is especially true when the value of K is large compared to one.

When the constraints presented in this section are satisfied, one can use the principle of local mass equilibrium to justify the one-equation model given by Eq. 5.33. Under these circumstances we need only develop the closure problems for $\tilde{c}_\beta$ and $\tilde{c}_\sigma$ so that the theory can be compared with experiment.

7. Closure problem associated with the one-equation model

Our development of the closure problem for $\tilde{c}_\beta$ and $\tilde{c}_\sigma$ begins with Eqs 5.1–5.4 and makes use of the decompositions given by Eqs 5.9 and 5.19. We follow the analysis given by Eqs 3.1–3.9 to arrive at

$$\varepsilon_\gamma(1+K)\frac{\partial\tilde{c}_\sigma}{\partial t} - \nabla\cdot(\mathbf{D}_\sigma\nabla\tilde{c}_\sigma) = \frac{1}{V_\sigma}\int_{V_\sigma}\left[\varepsilon_\gamma(1+K)\frac{\partial\tilde{c}_\sigma}{\partial t} - \nabla\cdot(\mathbf{D}_\sigma\tilde{c}_\sigma)\right]dV \tag{7.1}$$

B.C.1 $$\tilde{c}_\sigma = \tilde{c}_\beta + \langle c_\beta\rangle^\beta - \langle c_\sigma\rangle^\sigma, \quad \text{at } \mathscr{A}_{\beta\sigma} \tag{7.2}$$

B.C.2
$$-\mathbf{n}_{\sigma\beta}\cdot(D_\sigma\cdot\nabla\tilde{c}_\sigma) = -\mathbf{n}_{\sigma\beta}\cdot(\mathscr{D}_\beta\nabla\tilde{c}_\beta) - \mathbf{n}_{\sigma\beta}\cdot\mathscr{D}_\beta\nabla\langle c_\beta\rangle^\beta + \mathbf{n}_{\sigma\beta}\cdot\mathbf{D}_\sigma\nabla\langle c_\sigma\rangle^\sigma, \quad \text{at } \mathscr{A}_{\beta\sigma} \tag{7.3}$$

$$\frac{\partial\tilde{c}_\beta}{\partial t} + \nabla\cdot(\mathbf{v}_\beta\tilde{c}_\beta) + \nabla\cdot(\tilde{\mathbf{v}}_\beta\langle c_\beta\rangle^\beta) - \nabla\cdot(\mathscr{D}_\beta\nabla\tilde{c}_\beta)$$
$$= \frac{1}{V_\beta}\int_{V_\beta}\left[\frac{\partial\tilde{c}_\beta}{\partial t} + \nabla\cdot(\mathbf{v}_\beta\tilde{c}_\beta) + \nabla\cdot(\tilde{\mathbf{v}}_\beta\langle c_\beta\rangle^\beta) - \nabla\cdot(\mathscr{D}_\beta\nabla\tilde{c}_\beta)\right]\mathrm{d}V \tag{7.4}$$

Here we have taken the approach recommended by Crapiste *et al.* (1986) in the manipulation of Eqs 5.1 and 5.4, and we have imposed the length-scale constraint

$$\left(\frac{r_0}{\ell_\lambda}\right)^2 \ll 1 \tag{7.5}$$

so that averaged quantities could be treated as constants with respect to integration over V_σ or V_β. The constraint given by Eq. 7.5 also permits us to use

$$\langle\tilde{c}_\sigma\rangle^\sigma = 0, \quad \langle\tilde{c}_\beta\rangle^\beta = 0 \tag{7.6}$$

and the first of these allows us to write Eq. 7.1 as

$$\varepsilon_\gamma(1+K)\frac{\partial\tilde{c}_\sigma}{\partial t} = \nabla\cdot(\mathbf{D}_\sigma\cdot\nabla\tilde{c}_\sigma) - \frac{1}{V_\sigma}\int_{V_\sigma}\nabla\cdot(\mathbf{D}_\sigma\cdot\nabla\tilde{c}_\sigma)\,\mathrm{d}V \tag{7.7}$$

A similar simplification can be carried out on the right-hand side of Eq. 7.4, but before doing so we need to reconsider the continuity equation for the β-phase in order to learn something about $\tilde{\mathbf{v}}_\beta$.

7.1 The continuity equation

We assume that the flow of the β-phase can be treated as incompressible so that the continuity equation takes the form

$$\nabla\cdot\mathbf{v}_\beta = 0 \tag{7.8}$$

The volume average of Eq. 7.8 can be expressed as

$$\nabla\cdot(\varepsilon_\beta\langle\mathbf{v}_\beta\rangle^\beta) + \frac{1}{\mathscr{V}}\int_{A_{\beta\sigma}}\mathbf{n}_{\beta\sigma}\cdot\mathbf{v}_\beta\,\mathrm{d}A - 0 \tag{7.9}$$

and when the interfacial flux is neglected we arrive at the form given earlier by Eq. 5.27. One can use that result to express the divergence of $\langle \mathbf{v}_\beta \rangle^\beta$ as

$$\nabla \cdot \langle \mathbf{v}_\beta \rangle^\beta = -\varepsilon_\beta^{-1} \langle \mathbf{v}_\beta \rangle^\beta \cdot \nabla \varepsilon_\beta \tag{7.10}$$

When the decomposition given by Eq. 5.20 is used in Eq. 7.8 we obtain

$$\nabla \cdot \tilde{\mathbf{v}}_\beta = -\nabla \cdot \langle \mathbf{v}_\beta \rangle^\beta \tag{7.11}$$

and this leads to the continuity equation for the spatial deviation velocity

$$\nabla \cdot \tilde{\mathbf{v}}_\beta = \varepsilon_\beta^{-1} \langle \mathbf{v}_\beta \rangle^\beta \cdot \nabla \varepsilon_\beta \tag{7.12}$$

One can think of $\varepsilon_\beta^{-1} \langle \mathbf{v}_\beta \rangle^\beta \cdot \nabla \varepsilon_\beta$ as a *source* for the $\tilde{\mathbf{v}}_\beta$-field. The three terms on the left-hand side of Eq. 7.12 are all of the order of v_β / ℓ_β while the single term on the right-hand side is of the order of v_β / ℓ_λ. Given that $\ell_\beta \ll \ell_\lambda$, we express Eq. 7.12 as

$$\nabla \cdot \tilde{\mathbf{v}}_\beta = 0 \tag{7.13}$$

and when this simplification is imposed on Eq. 7.4 we obtain the following form of the transport equation for $\tilde{c}_\beta$.

$$\frac{\partial \tilde{c}_\beta}{\partial t} + \mathbf{v}_\beta \cdot \nabla \tilde{c}_\beta + \tilde{\mathbf{v}}_\beta \cdot \nabla \langle c_\beta \rangle^\beta = \nabla \cdot (\mathscr{D}_\beta \nabla \tilde{c}_\beta) + \frac{1}{V_\beta} \int_{V_\beta} [\nabla \cdot (\mathbf{v}_\beta \tilde{c}_\beta) - \nabla \cdot (\mathscr{D}_\beta \nabla \tilde{c}_\beta)] \, \mathrm{d}V \tag{7.14}$$

Here we have used

$$\frac{1}{V_\beta} \int_{V_\beta} \frac{\partial \tilde{c}_\beta}{\partial t} \, \mathrm{d}V = \frac{\partial}{\partial t} \left(\frac{1}{V_\beta} \int_{V_\beta} \tilde{c}_\beta \, \mathrm{d}V \right) = 0 \tag{7.15}$$

$$\frac{1}{V_\beta} \int_{V_\beta} \nabla \cdot (\tilde{\mathbf{v}}_\beta \langle c_\beta \rangle^\beta) \, \mathrm{d}V = \frac{1}{V_\beta} \int_{V_\beta} \tilde{\mathbf{v}}_\beta \cdot \nabla \langle c_\beta \rangle^\beta \, \mathrm{d}V = \left(\frac{1}{V_\beta} \int_{V_\beta} \tilde{\mathbf{v}}_\beta \, \mathrm{d}V \right) \cdot \nabla \langle c_\beta \rangle^\beta = 0 \tag{7.16}$$

on the basis that the averages of $\tilde{c}_\beta$ and $\tilde{\mathbf{v}}_\beta$ can be set equal to zero.

The right-hand side of Eq. 7.14 can be transformed by the divergence theorem to obtain

$$\frac{1}{V_\beta}\int_{V_\beta}[\nabla\cdot(\mathbf{v}_\beta\tilde{c}_\beta)-\nabla\cdot(\mathscr{D}_\beta\nabla\tilde{c}_\beta)]\,\mathrm{d}V=\frac{1}{V_\beta}\int_{A_{\beta\sigma}}\mathbf{n}_{\beta\sigma}\cdot(\mathbf{v}_\beta\tilde{c}_\beta-\mathscr{D}_\beta\nabla\tilde{c}_\beta)\,\mathrm{d}A$$
$$+\frac{1}{V_\beta}\int_{A_{\beta e}}\mathbf{n}_{\beta e}\cdot(\mathbf{v}_\beta\tilde{c}_\beta-\mathscr{D}_\beta\nabla\tilde{c}_\beta)\,\mathrm{d}A \tag{7.17}$$

Here $A_{\beta e}$ represents the area of entrances and exits for the β-phase contained in the averaging volume illustrated in Fig. 8, and $\mathbf{n}_{\beta e}$ represents the outwardly directed unit normal vector at the surface, $A_{\beta e}$. We are going to solve the closure problem for $\tilde{c}_\sigma$ and $\tilde{c}_\beta$ in a representative unit cell, such as the one illustrated in Fig. 4, and we are going to treat $\tilde{c}_\sigma$ and $\tilde{c}_\beta$ as spatially periodic functions since the unit cell represents an element of a spatially periodic porous medium. This means that the velocity field will also be spatially periodic, thus the terms $\mathbf{v}_\beta\tilde{c}_\beta$ and $\mathscr{D}_\beta\nabla\tilde{c}_\beta$ will be spatially periodic and the area integral over $A_{\beta e}$ will be zero for any unit cell. This allows us to simplify Eq. 7.17 to

$$\frac{1}{V_\beta}\int_{V_\beta}[\nabla\cdot(\mathbf{v}_\beta\tilde{c}_\beta)-\nabla\cdot(\mathscr{D}_\beta\nabla\tilde{c}_\beta)]\,\mathrm{d}V=\frac{1}{V_\beta}\int_{A_{\beta\sigma}}\mathbf{n}_{\beta\sigma}\cdot(\mathbf{v}_\beta\tilde{c}_\beta-\mathscr{D}_\beta\nabla\tilde{c}_\beta)\,\mathrm{d}A \tag{7.18}$$

In order to simplify the right-hand side of this result, we return to Eq. 5.16 and use the decomposition for c_β to obtain

$$\mathbf{n}_{\beta\sigma}\cdot(\tilde{c}_\beta\mathbf{v}_\beta)+\mathbf{n}_{\beta\sigma}\cdot(\langle c_\beta\rangle^\beta\mathbf{v}_\beta)\ll\mathbf{n}_{\beta\sigma}\cdot(\mathscr{D}_\beta\nabla\tilde{c}_\beta)+\mathbf{n}_{\beta\sigma}\cdot(\mathscr{D}_\beta\nabla\langle c_\beta\rangle^\beta) \tag{7.19}$$

When $\mathbf{D}_\sigma$ is small compared to $\mathscr{D}_\beta$, one can put forth arguments (Carbonell and Whitaker, 1983) that suggest

$$\tilde{c}_\beta\ll\langle c_\beta\rangle^\beta,\quad\nabla\tilde{c}_\beta=\mathbf{0}(\nabla\langle c_\beta\rangle^\beta) \tag{7.20}$$

Under these conditions, an acceptable extension of the inequality given by Eq. 5.16 is the following

$$\mathbf{n}_{\beta\sigma}\cdot(\tilde{c}_\beta\mathbf{v}_\beta)\ll\mathbf{n}_{\beta\sigma}\cdot(\mathscr{D}_\beta\nabla\tilde{c}_\beta),\quad\text{at }A_{\beta\sigma} \tag{7.21}$$

This allows us to write Eq. 7.18 as

$$\frac{1}{V_\beta}\int_{V_\beta}[\nabla\cdot(\mathbf{v}_\beta\tilde{c}_\beta)-\nabla\cdot(\mathscr{D}_\beta\nabla\tilde{c}_\beta)]\,\mathrm{d}V=-\frac{1}{V_\beta}\int_{A_{\beta\sigma}}\mathbf{n}_{\beta\sigma}\cdot\mathscr{D}_\beta\nabla\tilde{c}_\beta\,\mathrm{d}A \tag{7.22}$$

and this result can be used in Eq. 7.14 to produce the final form of the transport equation for $\tilde{c}_\beta$. A similar simplification can be applied to Eq. 7.7 so that the closure problem takes the form shown below.

7.2 The closure problem

$$\varepsilon_\gamma(1+K)\frac{\partial \tilde{c}_\sigma}{\partial t} = \nabla\cdot(\mathbf{D}_\sigma \nabla \tilde{c}_\sigma) - \frac{1}{V_\sigma}\int_{A_{\sigma\beta}} \mathbf{n}_{\sigma\beta}\cdot \mathbf{D}_\sigma \nabla \tilde{c}_\sigma \, \mathrm{d}A \tag{7.23}$$

B.C.1
$$\tilde{c}_\sigma = \tilde{c}_\beta + \langle c_\beta\rangle^\beta - \langle c_\sigma\rangle^\sigma, \quad \text{at } A_{\beta\sigma} \tag{7.24}$$

B.C.2
$$-\mathbf{n}_{\sigma\beta}\cdot(\mathbf{D}_\sigma\cdot\nabla\tilde{c}_\sigma) = -\mathbf{n}_{\sigma\beta}\cdot(\mathscr{D}_\beta\cdot\nabla\tilde{c}_\beta) - \mathbf{n}_{\sigma\beta}\cdot\mathscr{D}_\beta\nabla\langle\tilde{c}_\beta\rangle^\beta + \mathbf{n}_{\sigma\beta}\cdot\mathbf{D}_\sigma\cdot\nabla\tilde{c}_\sigma, \quad \text{at } A_{\beta\sigma} \tag{7.25}$$

$$\frac{\partial \tilde{c}_\beta}{\partial t} + \mathbf{v}_\beta\cdot\nabla\tilde{c}_\beta + \tilde{\mathbf{v}}_\beta\cdot\nabla\langle c_\beta\rangle^\beta = \nabla\cdot(\mathscr{D}_\beta\nabla\tilde{c}_\beta) - \frac{1}{V_\beta}\int_{A_{\beta\sigma}} \mathbf{n}_{\beta\sigma}\cdot\mathscr{D}_\beta\nabla\tilde{c}_\beta \, \mathrm{d}A \tag{7.26}$$

At this point we need to remember that we seek a closure to the one-equation model represented by Eq. 5.33. This means that $\tilde{c}_\sigma$ and $\tilde{c}_\beta$ must be represented in terms of $\langle c\rangle$, thus we employ the decompositions given by Eqs 5.30 so that the boundary conditions given by Eqs 7.24 and 7.25 take the form

B.C.1
$$\tilde{c}_\sigma = \tilde{c}_\beta + \hat{c}_\beta - \hat{c}_\sigma, \quad \text{at } A_{\beta\sigma} \tag{7.27}$$

B.C.2
$$-\mathbf{n}_{\sigma\beta}\cdot(\mathbf{D}_\sigma\cdot\nabla\tilde{c}_\sigma) = -\mathbf{n}_{\sigma\beta}\cdot(\mathscr{D}_\beta\nabla\tilde{c}_\beta) - \mathbf{n}_{\sigma\beta}\cdot(\mathscr{D}_\beta\mathbf{I} - \mathbf{D}_\sigma)\cdot\nabla\langle c\rangle - \mathbf{n}_{\sigma\beta}\cdot(\mathscr{D}_\beta\nabla\hat{c}_\beta) + \mathbf{n}_{\sigma\beta}\cdot(\mathbf{D}_\sigma\cdot\nabla\hat{c}_\sigma), \quad \text{at } A_{\beta\sigma} \tag{7.28}$$

On the basis of the last two inequalities listed under Eq. 6.5, the second of these two boundary conditions can be simplified to

B.C.2
$$-\mathbf{n}_{\sigma\beta}\cdot(\mathbf{D}_\sigma\cdot\nabla\tilde{c}_\sigma) = -\mathbf{n}_{\sigma\beta}\cdot(\mathscr{D}_\beta\nabla\tilde{c}_\beta) - \mathbf{n}_{\sigma\beta}\cdot(\mathscr{D}_\beta\mathbf{I} - \mathbf{D}_\sigma)\cdot\nabla\langle c\rangle, \quad \text{at } A_{\beta\sigma} \tag{7.29}$$

This simplification is based on the restrictions associated with *local mass equilibrium*, and these are listed in Section 6. In order to simplify Eq. 7.27 we return to Eqs 6.1b, 6.2b and 6.21 to estimate $\hat{c}_\beta$ as

$$\hat{c}_\beta = \mathbf{0}\begin{Bmatrix} \varepsilon_\sigma\varepsilon_\beta\left(\dfrac{\delta_\beta}{a_{\beta\sigma}\mathscr{D}_\beta t}\right) \\ \varepsilon_\sigma\varepsilon_\beta\left(\dfrac{\langle v_\beta\rangle^\beta \ell_\beta}{\mathscr{D}_\beta}\right)\left(\dfrac{\delta_\beta}{a_{\beta\sigma}L_c\ell_\beta}\right) \\ \varepsilon_\sigma\varepsilon_\beta\left(\dfrac{D_\beta}{\mathscr{D}_\beta}\right)\left(\dfrac{\delta_\beta}{a_{\beta\sigma}L_D^2}\right) \end{Bmatrix}\Delta\langle c_\beta\rangle^\beta \tag{7.30}$$

From Eq. 7.29 we can estimate $\tilde{c}_\beta$ as

$$\tilde{c}_\beta = \mathbf{0}\left(\frac{\ell_\beta}{\ell_\lambda}\Delta\langle c_\beta\rangle^\beta\right) \tag{7.31}$$

where ℓ_λ should be thought of as the smallest of either L_c or L_D and $\Delta\langle c\rangle$ has been estimated by $\Delta\langle c_\beta\rangle^\beta$. This result is just a variation of Eq. 3.19, and estimates of this type for diffusion in cellular systems and heat conduction in porous media are available from Ochoa *et al.* (1986) and Nozad *et al.* (1985). The requirement that $\hat{c}_\beta \ll \tilde{c}_\beta$ leads to

$$\varepsilon_\sigma\left(\frac{\delta_\beta}{a_{\beta\sigma}\mathscr{D}_\beta t}\right)\left(\frac{\varepsilon_\beta\ell_\lambda}{\ell_\beta}\right) \ll 1 \tag{7.32a}$$

$$\varepsilon_\sigma\left(\frac{\langle v_\beta\rangle^\beta\ell_\beta}{\mathscr{D}_\beta}\right)\left(\frac{\delta_\beta}{\ell_\beta}\right)\left(\frac{\ell_\beta}{L_c}\right)\left(\frac{\varepsilon_\beta\ell_\lambda}{\ell_\beta}\right) \ll 1 \tag{7.32b}$$

$$\varepsilon_\sigma\left(\frac{D_\beta}{\mathscr{D}_\beta}\right)\left(\frac{\delta_\beta\ell_\sigma}{L_D^2}\right)\left(\frac{\varepsilon_\beta\ell_\lambda}{\ell_\beta}\right) \ll 1 \tag{7.32c}$$

Here δ_β is given by the estimates listed earlier by Eqs 6.17.

We can repeat the procedure leading to Eqs 7.32 for the σ-phase, and we find that the constraints associated with $\hat{c}_\sigma \ll \tilde{c}_\sigma$ are given by

$$\varepsilon_\beta\varepsilon_\gamma(1+K)\left(\frac{\ell_\sigma^2}{D_\sigma t}\right)\left(\frac{\varepsilon_\sigma\ell_\lambda}{\ell_\sigma}\right) \ll 1 \tag{7.33a}$$

$$\varepsilon_\beta\left(\frac{\ell_\sigma}{\ell_\lambda}\right)^2\left(\frac{\varepsilon_\sigma\ell_\lambda}{\ell_\sigma}\right) \ll 1 \tag{7.33b}$$

It is important to note that the constraints given by Eqs 7.32 and 7.33 are associated with the simplification of Eq. 7.27 to the form given by

B.C.1
$$\tilde{c}_\sigma = \tilde{c}_\beta, \quad \text{at } A_{\beta\sigma} \tag{7.34}$$

It is also important to note that the constraints given by Eqs 7.32 and 7.33 are generally *more severe* than the analogous constraints associated with *local mass equilibrium* (see Eqs 6.15, 6.18 and 6.22). This means that the *closure problem* associated with the *one-equation model* given by Eq. 5.33 is subject to more severe constraints than those that are necessary for the derivation of Eq. 5.33.

In actual fact there is very little reliable information concerning the use of one-equation models, since the only solution of the two-equation closure problem is restricted to uniform flow in a bundle of capillary tubes (Zanotti and Carbonell, 1984). Some computational studies have been performed by Pereira Duarte *et al.* (1984) for heat transfer in packed beds; however, these studies were not performed with the idea of testing constraints of the type given by Eqs 7.32 and 7.33. In a study of transient heat conduction in porous media, Nozad *et al.* (1985) found good agreement between theory and experiment when a one-equation model was used and their experiments *clearly violated* constraints of the type given by Eqs 7.33. Since these constraints are all based on order of magnitude estimates they can be used to suggest the *form* of the appropriate constraints, but we suspect that the actual constraints given by Eqs 7.32 and 7.33 are *overly severe.*

When the simplifications illustrated by Eqs 7.29 and 7.34 are valid, the closure problem expressed by Eqs 7.23–7.26 takes the form

$$\varepsilon_\gamma(1+K)\frac{\partial \tilde{c}_\sigma}{\partial t} = \mathbf{D}_\sigma : \nabla\nabla\tilde{c}_\sigma - \frac{1}{V_\sigma}\int_{A_{\sigma\beta}} \mathbf{n}_{\sigma\beta}\cdot\mathbf{D}_\sigma\cdot\nabla\tilde{c}_\sigma \, dA \tag{7.35}$$

B.C.1
$$\tilde{c}_\sigma = \tilde{c}_\beta, \quad \text{at } A_{\beta\sigma} \tag{7.36}$$

B.C.2
$$-\mathbf{n}_{\sigma\beta}\cdot(\mathbf{D}_\sigma\cdot\nabla\tilde{c}_\sigma) = -\mathbf{n}_{\sigma\beta}\cdot(\mathscr{D}_\beta\nabla\tilde{c}_\beta) - \mathbf{n}_{\sigma\beta}\cdot(\mathscr{D}_\beta\mathbf{I} - \mathbf{D}_\sigma)\cdot\nabla\langle c\rangle, \quad \text{at } A_{\beta\sigma} \tag{7.37}$$

$$\frac{\partial \tilde{c}_\beta}{\partial t} + \mathbf{v}_\beta\cdot\nabla\tilde{c}_\beta + \tilde{\mathbf{v}}_\beta\cdot\nabla\langle c\rangle = \mathscr{D}_\beta\nabla^2\tilde{c}_\beta - \frac{1}{V_\beta}\int_{A_{\beta\sigma}} \mathbf{n}_{\beta\sigma}\cdot\mathscr{D}_\beta\nabla\tilde{c}_\beta \, dA \tag{7.38}$$

$$\tilde{c}_\sigma(\mathbf{r}+\boldsymbol{\ell}_i) = \tilde{c}_\sigma(\mathbf{r}), \quad \tilde{c}_\beta(\mathbf{r}+\boldsymbol{\ell}_i) = \tilde{c}_\beta(\mathbf{r}), \quad i = 1, 2, 3 \tag{7.39}$$

in which $\nabla\langle c_\beta\rangle^\beta$ has been replaced by $\nabla\langle c\rangle$ on the basis of Eqs 5.30b and 6.5. In Eq. 7.39 we have used $\boldsymbol{\ell}_i$ to represent the three non-unique lattice vectors that are needed to describe a unit cell such as the one illustrated in Fig. 4. In writing Eqs 7.35 and 7.38 we have neglected the variations of $\mathbf{D}_\sigma$ and $\mathscr{D}_\beta$ within the unit cell, and this is entirely consistent with the length-scale constraints that have been imposed within this development.

In addition to the length-scale constraints that have been imposed in the development of Eqs 7.35–7.39, we have also imposed time-scale constraints that are entirely consistent with

$$\left(\frac{D_\sigma t}{\ell_\sigma^2}\right) \gg \varepsilon_\gamma(1+K), \quad \frac{\mathscr{D}_\beta t}{\ell_\beta^2} \gg 1 \tag{7.40}$$

This means that the closure problem must be quasi-steady in order to be consistent with the condition of local mass equilibrium, and we are forced to express Eqs 7.35–7.39 as

$$\mathbf{D}_\sigma : \nabla\nabla\tilde{c}_\sigma = \frac{1}{V_\sigma}\int_{A_{\sigma\beta}} \mathbf{n}_{\sigma\beta}\cdot\mathbf{D}_\sigma\cdot\nabla\tilde{c}_\sigma \,\mathrm{d}A \tag{7.41}$$

B.C.1
$$\tilde{c}_\sigma = \tilde{c}_\beta, \quad \text{at } A_{\sigma\beta} \tag{7.42}$$

B.C.2
$$-\mathbf{n}_{\beta\sigma}\cdot(\mathbf{D}_\sigma\cdot\nabla\tilde{c}_\sigma) = -\mathbf{n}_{\sigma\beta}\cdot(\mathscr{D}_\beta\nabla\tilde{c}_\beta) - \mathbf{n}_{\sigma\beta}\cdot(\mathscr{D}_\beta\mathbf{I} - \mathbf{D}_\sigma)\cdot\nabla\langle c\rangle, \quad \text{at } A_{\beta\sigma} \tag{7.43}$$

$$\mathbf{v}_\beta\cdot\nabla\tilde{c}_\beta + \tilde{\mathbf{v}}_\beta\cdot\nabla\langle c\rangle = \mathscr{D}_\beta\nabla^2\tilde{c}_\beta - \frac{1}{V_\beta}\int_{A_{\beta\sigma}} \mathbf{n}_{\beta\sigma}\cdot\mathscr{D}_\beta\nabla\tilde{c}_\beta \,\mathrm{d}A \tag{7.44}$$

$$\tilde{c}_\sigma(\mathbf{r}+\boldsymbol{\ell}_i) = \tilde{c}_\sigma(\mathbf{r}), \quad \tilde{c}_\beta(\mathbf{r}+\boldsymbol{\ell}_i) = \tilde{c}_\beta(\mathbf{r}), \quad i = 1, 2, 3 \tag{7.45}$$

Obviously this problem represents a great simplification over the original form given by Eqs 7.1–7.4. Further simplification results when we recognize that both the *surface source*, $\mathbf{n}_{\sigma\beta}\cdot(\mathscr{D}_\beta\mathbf{I} - \mathbf{D}_\sigma)\cdot\nabla\langle c\rangle$, and the *volume source*, $\tilde{\mathbf{v}}_\beta\cdot\nabla\langle c\rangle$, are directly proportional to $\nabla\langle c\rangle$. This leads to a *single closure variable* as opposed to the two closure variables encountered in Section 3.

7.3 Closure variables

Since Eqs 7.41–7.45 contain only a single source parameter, $\nabla\langle c\rangle$, we can express $\tilde{c}_\sigma$ and $\tilde{c}_\beta$ as

$$\tilde{c}_\sigma = \mathbf{g}\cdot\nabla\langle c\rangle + \psi \tag{7.46}$$

$$\tilde{c}_\beta = \mathbf{f}\cdot\nabla\langle c\rangle + \xi \tag{7.47}$$

where ψ and ξ are arbitrary scalar functions and $\mathbf{g}$ and $\mathbf{f}$ are determined by the following boundary value problem

$$\mathbf{D}_\sigma : \nabla\nabla \mathbf{g} = \frac{1}{V_\sigma} \int_{A_{\sigma\beta}} \mathbf{n}_{\sigma\beta} \cdot \mathbf{D}_\sigma \cdot \nabla \mathbf{g} \, dA \tag{7.48}$$

B.C.1 $$\mathbf{g} = \mathbf{f}, \quad \text{at } A_{\beta\sigma} \tag{7.49}$$

B.C.2 $$-\mathbf{n}_{\sigma\beta} \cdot (\mathbf{D}_\sigma \cdot \nabla \mathbf{g}) = -\mathbf{n}_{\sigma\beta} \cdot (\mathscr{D}_\beta \nabla \mathbf{f}) - \mathbf{n}_{\sigma\beta} \cdot (\mathscr{D}_\beta \mathbf{I} - \mathbf{D}_\sigma), \quad \text{at } A_{\beta\sigma} \tag{7.50}$$

$$\tilde{\mathbf{v}}_\beta + \mathbf{v}_\beta \cdot \nabla \mathbf{f} = \mathscr{D}_\beta \nabla^2 \mathbf{f} - \frac{1}{V_\beta} \int_{A_{\beta\sigma}} \mathbf{n}_{\beta\sigma} \cdot \mathscr{D}_\beta \nabla \mathbf{f} \, dA \tag{7.51}$$

$$\mathbf{g}(\mathbf{r} + \boldsymbol{\ell}_i) = \mathbf{g}(\mathbf{r}), \quad \mathbf{f}(\mathbf{r} + \boldsymbol{\ell}_i) = \mathbf{f}(\mathbf{r}), \quad i = 1, 2, 3 \tag{7.52}$$

The proof that ψ and ξ are zero follows previous developments given by Carbonell and Whitaker (1983), Nozad *et al.* (1985), and Ochoa *et al.* (1986).

When Eqs 7.46 and 7.47 (with $\psi = \xi = 0$) are used in Eq. 5.33 we obtain the closed form of the one-equation model which is given by

$$\varepsilon_\beta (1 + \mathscr{K}) \frac{\partial \langle c \rangle}{\partial t} + \langle \mathbf{v}_\beta \rangle \cdot \nabla \langle c \rangle = \nabla \cdot [(\varepsilon_\beta \mathbf{D}_{\text{eff}} + \varepsilon_\beta \mathbf{D}_\beta) \cdot \nabla \langle c \rangle] \tag{7.53}$$

Here we think of $\mathbf{D}_{\text{eff}}$ as the effective diffusivity tensor defined by

$$\varepsilon_\beta \mathbf{D}_{\text{eff}} = \varepsilon_\beta \mathscr{D}_\beta \mathbf{I} + \varepsilon_\sigma \mathbf{D}_\sigma + \frac{(\mathscr{D}_\beta \mathbf{I} - \mathbf{D}_\sigma)}{\mathscr{V}} \cdot \int_{A_\beta} \mathbf{n}_{\beta\sigma} \mathbf{f} \, dA \tag{7.54}$$

while $\mathbf{D}_\beta$ is the hydrodynamic dispersion tensor which is represented by

$$\mathbf{D}_\beta = -\langle \tilde{\mathbf{v}}_\beta \mathbf{f} \rangle^\beta \tag{7.55}$$

In general it is convenient to express Eq. 7.53 as

$$\varepsilon_\beta (1 + \mathscr{K}) \frac{\partial \langle c \rangle}{\partial t} + \langle \mathbf{v}_\beta \rangle \cdot \nabla \langle c \rangle = \nabla \cdot (\varepsilon_\beta \mathbf{D}^* \cdot \nabla \langle c \rangle) \tag{7.56}$$

where the total dispersion tensor, $\mathbf{D}^*$, is defined by

$$\varepsilon_\beta \mathbf{D}^* = \varepsilon_\beta \mathscr{D}_\beta \mathbf{I} + \varepsilon_\sigma \mathbf{D}_\sigma + \frac{(\mathscr{D}_\beta \mathbf{I} - \mathbf{D}_\sigma)}{\mathscr{V}} \cdot \int_{A_{\beta\sigma}} \mathbf{n}_{\beta\sigma} \mathbf{f} \, dA - \varepsilon_\beta \langle \tilde{\mathbf{v}}_\beta \mathbf{f} \rangle^\beta \tag{7.57}$$

To our knowledge, the closure problem given by Eqs 7.48–7.52 has never been solved, thus *predicted values* of $\mathbf{D}^*$ are not available for the case of dispersion with adsorption in porous particles. What is available is that the closure problem tells us that $\mathbf{D}^*$ depends only on:

(1) the geometry of the β–σ system;
(2) the dimensionless tensor, $\mathbf{D}_\sigma/\mathscr{D}_\beta$;
(3) the Peclet number, $\langle v_\beta \rangle^\beta \ell_\beta/\mathscr{D}_\beta$.

That $\mathbf{D}^*$ is independent of $\mathscr{K}$ has important consequences since it means that the theory can be *tested by experiment* in the absence of adsorption. This represents a significant simplification in the experimental program; however, one must remember that this observation is based on the one-equation model represented by Eq. 7.56 and the constraints imposed by local mass equilibrium in addition to those associated with the closure problem. Two-equation models, or non-equilibrium models, have been proposed (Valocchi, 1986; Nielsen *et al.*, 1986); however, the length scales associated with these models were not defined. Since the constraints associated with the concept of local mass equilibrium presented in Section 6 are highly length-scale dependent, it is not possible to comment on the validity of the one-equation model without identifying the appropriate length scales.

8. Passive dispersion in homogeneous porous media: comparison between theory and experiment

By *passive* we mean the transport of a chemical species in the absence of either chemical reaction or adsorption. Most real transport processes of interest are *not passive*; however, the analysis presented in the previous section indicated that the total dispersion tensor was independent of the adsorption process when a series of length and time-scale constraints were satisfied. This situation certainly motivates the theoretical and experimental study of passive transport in a porous medium made of porous particles. A complete comparison between theory and experiment is extremely difficult to carry out because of the problem of measuring $\mathbf{D}_\sigma$ for porous particles (Whitaker, 1988) or predicting this parameter by theoretical analysis (Kim *et al.*, 1987). In addition, there are currently no general solutions available of the closure problem defined by Eqs 7.41–7.45. While there appear to be no well-defined experiments for systems similar to the one described in Section 5, there are numerous experimental results for systems in which $\mathbf{D}_\sigma = 0$. Under these circumstances, Eqs 5.1–5.4 reduce to

$$\frac{\partial c_\beta}{\partial t} + \nabla \cdot (c_\beta \mathbf{v}_\beta) = \nabla \cdot (\mathscr{D}_\beta \nabla c_\beta) \tag{8.1}$$

B.C.1
$$\mathbf{n}_{\beta\sigma} \cdot (\mathscr{D}_\beta \nabla c_\beta) = 0, \quad \text{at } A_{\beta\sigma} \tag{8.2}$$

From these equations one obtains the obvious simplification of Eq. 5.23 given by

$$\varepsilon_\beta \frac{\partial \langle c_\beta \rangle^\beta}{\partial t} + \varepsilon_\beta \langle \mathbf{v}_\beta \rangle^\beta \cdot \nabla \langle c_\beta \rangle^\beta = \nabla \cdot \left[\varepsilon_\beta \mathscr{D}_\beta \left(\nabla \langle c_\beta \rangle^\beta + \frac{1}{V_\beta} \int_{A_{\beta\sigma}} \mathbf{n}_{\beta\sigma} \tilde{c}_\beta \, \mathrm{d}A \right) \right]$$

$$- \nabla \cdot (\varepsilon_\beta \langle \tilde{c}_\beta \tilde{\mathbf{v}}_\beta \rangle^\beta) \tag{8.3}$$

subject only to the length-scale constraint

$$\left(\frac{\ell_\beta}{r_0} \right) \gg \left(\frac{r_0}{\ell_\lambda} \right) \tag{8.4}$$

All of the constraints associated with *local mass equilibrium* are irrelevant for this case, and the closure problem for $\tilde{c}_\beta$

$$\frac{\partial \tilde{c}_\beta}{\partial t} + \mathbf{v}_\beta \cdot \nabla \tilde{c}_\beta + \tilde{\mathbf{v}}_\beta \cdot \nabla \langle c_\beta \rangle^\beta = \mathscr{D}_\beta \nabla^2 \tilde{c}_\beta \tag{8.5a}$$

B.C.1

$$\mathbf{n}_{\beta\sigma} \cdot (\mathscr{D}_\beta \nabla \tilde{c}_\beta) = -\mathbf{n}_{\beta\sigma} \cdot (\mathscr{D}_\beta \nabla \langle c_\beta \rangle^\beta), \quad \text{at } A_{\beta\sigma} \tag{8.5b}$$

$$\tilde{c}_\beta(\mathbf{r} + \boldsymbol{\ell}_i) = \tilde{c}_\beta(\mathbf{r}), \quad i = 1, 2, 3 \tag{8.5c}$$

can be obtained by imposing only the constraint

$$\left(\frac{r_0}{\ell_\lambda} \right)^2 \ll 1 \tag{8.6}$$

One can follow the procedure discussed in Section 7 in order to show that

$$\tilde{c}_\beta = \mathbf{f} \cdot \nabla \langle c_\beta \rangle^\beta \tag{8.7}$$

and this leads to a convective-dispersion equation of the form

$$\varepsilon_\beta \frac{\partial \langle c_\beta \rangle^\beta}{\partial t} + \varepsilon_\beta \langle \mathbf{v}_\beta \rangle^\beta \cdot \nabla \langle c_\beta \rangle^\beta = \nabla \cdot (\varepsilon_\beta \mathbf{D}^* \cdot \nabla \langle c_\beta \rangle^\beta) \tag{8.8}$$

In this case $\mathbf{D}^*$ is defined by

$$\varepsilon_\beta \mathbf{D}^* = \varepsilon_\beta \mathscr{D}_\beta \mathbf{I} + \frac{\mathscr{D}_\beta}{\mathscr{V}} \int_{A_{\beta\sigma}} \mathbf{n}_{\beta\sigma} \mathbf{f} \, \mathrm{d}A - \varepsilon_\beta \langle \tilde{\mathbf{v}}_\beta \mathbf{f} \rangle^\beta \tag{8.9}$$

and the closure problem for $\mathbf{f}$ can be expressed as

$$\frac{\partial \mathbf{f}}{\partial t} + \mathbf{v}_\beta \cdot \nabla \mathbf{f} + \tilde{\mathbf{v}}_\beta = \mathscr{D}_\beta \nabla^2 \mathbf{f} \tag{8.10a}$$

B.C.1 $$\mathbf{n}_{\beta\sigma} \cdot \nabla \mathbf{f} = -\mathbf{n}_{\beta\sigma}, \quad \text{at } A_{\beta\sigma} \tag{8.10b}$$

I.C. $$\mathbf{f} = 0, \quad t = 0 \tag{8.10c}$$

$$\mathbf{f}(\mathbf{r} + \boldsymbol{\ell}_i) = \mathbf{f}(\mathbf{r}), \quad i = 1, 2,3 \tag{8.10d}$$

Here we must think of the initial condition for $\mathbf{f}$ as an uncertainty since our knowledge of $\tilde{c}_\beta$ at $t = 0$ is rather vague. For pulse-injection experiments, we know that $\tilde{c}_\beta$ will be zero over most of the region of interest while non-zero values of $\tilde{c}_\beta$ will exist in the neighbourhood of the pulse. This means that the correct initial condition for $\mathbf{f}$ will depend on whether the unit cell under consideration is in the neighbourhood of the pulse-injection point or whether it is far removed from this point. In terms of traditional experimental studies, the measuring station is far removed (in terms of the magnitude of $\boldsymbol{\ell}_i$) from the injection point. Under these circumstances the initial condition given by Eq. 8.10c is plausible.

The steady-state form of Eqs 8.10 is achieved when the time is constrained by

$$\frac{\mathscr{D}_\beta t}{\ell_\beta^2} \gg 1 \tag{8.11}$$

and this leads to *quasi-steady* values of $\tilde{c}_\beta$. Even though the accumulation term, $\partial \tilde{c}_\beta / \partial t$, can be discarded from Eq. 8.5a on the basis of Eq. 8.11, the concentration $\tilde{c}_\beta$ will still depend on time through the presence of $\nabla \langle c_\beta \rangle^\beta$ in the governing equation and the boundary condition given by Eq. 8.5b.

The steady-state form of the boundary value problem given by Eqs 8.10 can be expressed as

$$\mathbf{v}_\beta \cdot \nabla \mathbf{f} + \tilde{\mathbf{v}}_\beta = \mathscr{D}_\beta \nabla^2 \mathbf{f} \tag{8.12a}$$

$$\mathbf{n}_{\beta\sigma} \cdot \nabla \mathbf{f} = -\mathbf{n}_{\beta\sigma}, \quad \text{at } A_{\beta\sigma} \tag{8.12b}$$

$$\mathbf{f}(\mathbf{r} + \boldsymbol{\ell}_i) = \mathbf{f}(\mathbf{r}), \quad i = 1, 2, 3 \tag{8.12c}$$

and the two-dimensional version of this problem has been solved by Eidsath *et al.* (1983) for several simple geometrical models, one of which is illustrated in Fig. 9. In order to determine $\mathbf{D}^*$, as defined by Eq. 8.9, one must first solve the

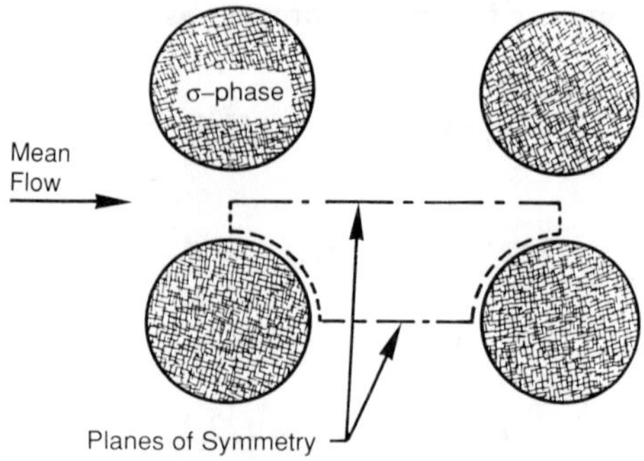

Figure 9. Geometric model of a unit cell in a spatially periodic porous medium.

Navier–Stokes equations to produce $\mathbf{v}_\beta$ and $\tilde{\mathbf{v}}_\beta$. These are used in Eqs 8.10 which are solved to produce the vector field $\mathbf{f}$. Both these solution procedures made use of a finite element code (Eidsath, 1981). A comparison between the theory and the experimental results of Gunn and Pryce (1969) is shown in Fig. 10. Gunn and Pryce performed axial dispersion experiments for *cubic arrays* of spheres ($\varepsilon_\beta = 0.48$) and for *randomly packed* spheres ($\varepsilon_\beta = 0.37$), and the theoretical calculations for $\varepsilon_\beta = 0.37$ are generally in good agreement with the data. However, the dependence of D^*_{xx} on the particle Peclet number

$$\mathrm{Pe_p} = \varepsilon_\beta \langle v_\beta \rangle^\beta d_\mathrm{p}/(1 - \varepsilon_\beta)\mathscr{D}_\beta \tag{8.13}$$

is quite different for the two sets of experimental data and the theory when the Peclet number is larger than 50. We list this functional dependence as

$$\left(\frac{D^*_{xx}}{\mathscr{D}_\beta}\right)_{\text{cubic array}} \sim \mathrm{Pe_p^{2.0}} \tag{8.14a}$$

$$\left(\frac{D^*_{xx}}{\mathscr{D}_\beta}\right)_{\text{theory}} \sim \mathrm{Pe_p^{1.7}} \tag{8.14b}$$

$$\left(\frac{D^*_{xx}}{\mathscr{D}_\beta}\right)_{\text{random packing}} \sim \mathrm{Pe_p^{1.2}} \tag{8.14c}$$

and in this form the disparity between theory and experiment becomes clear. One must keep in mind that the theoretical calculations were carried out for two-dimensional arrays of cylinders, while the cubic array studied by Gunn and Pryce represents a three-dimensional system. This may be the reason for the difference between the functional dependence represented by Eqs 8.14a and 8.14b. For random packing, the difference between the theory and

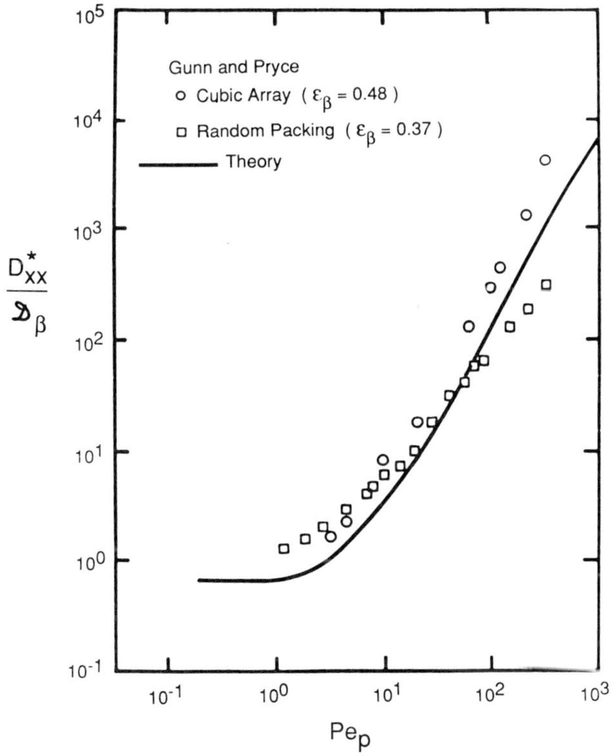

Figure 10. Comparison between theory and experiment for random and ordered arrays of spheres.

experimental results appears to be quite significant when we examine the broader range of data shown in Fig. 11. The departure between theory and experiment is especially obvious for values of $\mathrm{Pe_p}$ greater than 10^3; however, one must remember that the steady-state solution of Eqs 8.10 is only valid when Eq. 8.11 is in force. In fact the detailed theoretical study of Paine *et al.* (1983) and the careful experimental study of Han *et al.* (1985) both indicate that Eq. 8.11 should be replaced by

$$\frac{\mathscr{D}_\beta t}{d_p^2} \geqslant 0.3 \tag{8.15}$$

where d_p is the effective particle diameter defined by

$$d_p = 6V_p/A_p \tag{8.16}$$

If we let L be the distance at which a pulse is measured, the time in Eq. 8.15

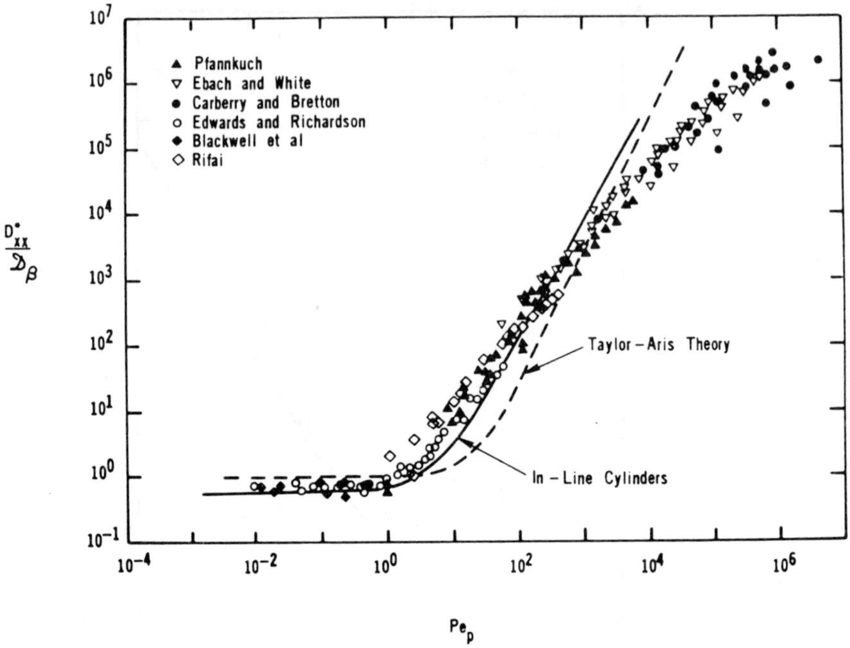

Figure 11. Comparison between theory and experiment for randomly packed particles.

is given by $L/\langle v_\beta\rangle^\beta$ and the constraint given by Eq. 8.15 can be cast in the form

$$\left(\frac{L}{d_p}\right) \geqslant 0.3\left(\frac{1-\varepsilon_\beta}{\varepsilon_\beta}\right)\mathrm{Pe}_p \tag{8.17}$$

For Peclet numbers greater than 10^3 this constraint becomes very difficult to satisfy in the typical laboratory experiment, and some evidence of this difficulty appears to be visible in the experimental data shown in Fig. 11. At particle Peclet numbers greater than 10^3, the dependence of $D^*_{xx}/\mathscr{D}_\beta$ on Pe_p diminishes and this is an indication that the steady form of the closure problem is no longer applicable.

The transient closure problem for dispersion in capillary tubes has been solved by Paine *et al.* (1983) in order to produce values of $D^*_{xx}/\mathscr{D}_\beta$ as a function of the dimensionless time, $\mathscr{D}_\beta t/r_0^2$, where r_0 is the radius of the capillary tube. The dimensionless time can be expressed as

$$\frac{\mathscr{D}_\beta t}{r_0^2} = \frac{\mathscr{D}_\beta L}{\langle v_\beta\rangle^\beta r_0^2} \tag{8.18}$$

where L is the distance from the pulse-injection point to the measuring station where $\langle c_\beta \rangle^\beta$ is recorded as a function of time. With a little algebra one can express Eq. 8.18 as

$$\frac{\mathscr{D}_\beta t}{r_0^2} = \frac{9}{\mathrm{Pe_p}} \left(\frac{1 - \varepsilon_\beta}{\varepsilon_\beta} \right) \frac{L}{d_\mathrm{p}} \tag{8.19}$$

where d_p for a bundle of capillary tubes of radius r_0 is given by

$$d_\mathrm{p} = 3r_0 \left(\frac{1 - \varepsilon_\beta}{\varepsilon_\beta} \right) \tag{8.20}$$

The representation given by Eq. 8.19 allows us to express the values of $D^*_{xx}/\mathscr{D}_\beta$ determined by Paine *et al.* (1983) as

$$\frac{D^*_{xx}}{\mathscr{D}_\beta} = \mathscr{F}\left[\mathrm{Pe_p}, \left(\frac{1 - \varepsilon_\beta}{\varepsilon_\beta} \right) \frac{L}{d_\mathrm{p}} \right] \tag{8.21}$$

These values are shown in Fig. 12 for two values of $[(1 - \varepsilon_\beta)/\varepsilon_\beta]L/d_\mathrm{p}$, and when this parameter is equal to ten we find a striking correspondence between the theory for dispersion in a capillary tube and the experimental data for packed beds. Almost all of the high Peclet number data were obtained by Ebach and White (1958) and Carberry and Bretton (1958). In the experimental studies

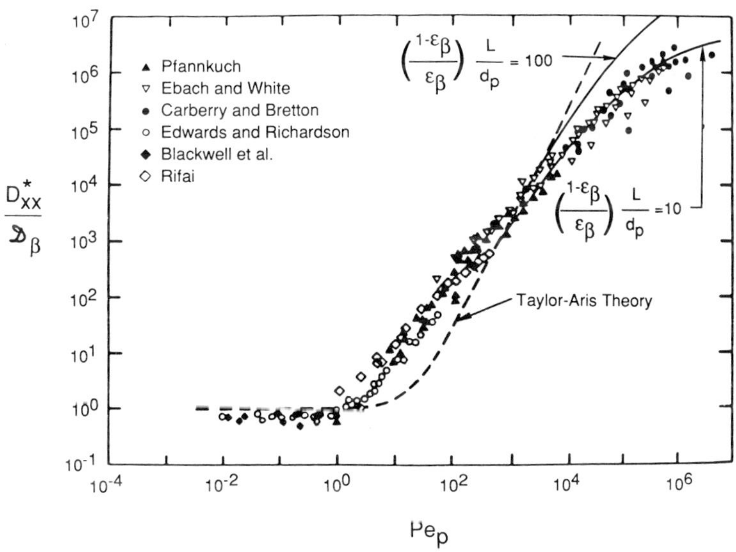

Figure 12. Influence of the transient closure problem.

of Ebach and White, the *minimum value* of $[(1-\varepsilon_\beta)/\varepsilon_\beta]L/d_p$ was about 240 while the minimum value used by Carberry and Bretton was about 60. Both these values are much larger than the value of 10 that provides such a good fit between theory and experiment; however, the general behaviour of both theory and experiment strongly suggests that the transient closure problem given by Eqs 8.10 needs to be solved.

Since lateral dispersion experiments are typically performed under steady-state conditions, one can use solutions of Eqs 8.12 with confidence to predict values of D^*_{yy}. The comparison between the experimental results of several investigators and the theoretical calculations of Eidsath *et al.* (1983) is shown in Fig. 13. The shocking disparity between theory and experiment would appear to be entirely associated with the use of overly simplistic two-dimensional models such as the one shown in Fig. 9. For simple unit cells the *lateral convective transport* is limited to a distance of the order of one particle diameter, however, the use of more realistic, complex unit cells such as the one illustrated in Fig. 4 would obviously lead to greater lateral convective transport. This would, in turn, create a larger lateral dispersion coefficient. The studies of Debbas and Rumpf (1966) have made it clear that local heterogeneities are to be expected in packed beds of spheres, and these effects are clearly evident in the experimental studies of Gunn and Pryce (1969). Such effects are not captured by models such as the one illustrated in Fig. 9, and the importance of *local heterogeneities* or *non-uniformities* has been explored by Koch and Brady (1985).

9. Conclusions

In this work we have considered two levels of averaging: small-scale averaging associated with the process of diffusion and adsorption in porous particles, and

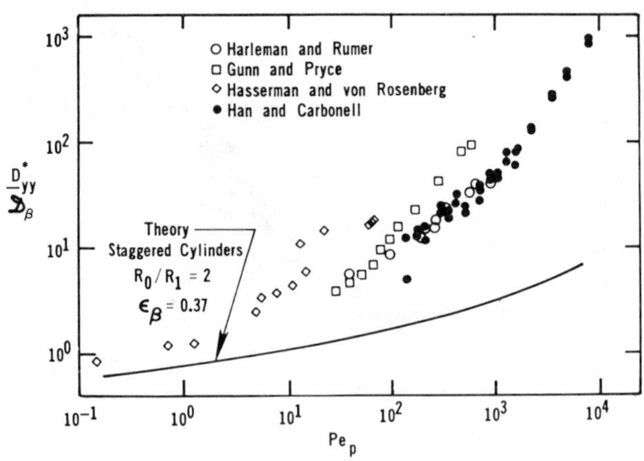

Figure 13. Theoretical and experimental values for the lateral dispersion coefficient in randomly packed beds.

local volume averaging associated with convection and diffusion in a porous medium composed of porous particles. This type of analysis indicates how physical phenomena that occur at one length scale are manifested in the transport equations at a higher length scale. The comparison between theory and experiment clearly indicates that more complex models of the microstructure need to be studied and that the transient form of the closure problem should always be considered.

10. Nomenclature

The equation in which each symbol first appears is given in parentheses. When appropriate, the figure number is given instead of the equation number.

$\mathscr{A}_{\gamma\kappa}$	interfacial area of the γ–κ system (2.2), m^2.
$\mathscr{A}_{\gamma e}$	area of entrances and exits for the γ-phase (2.3), m^2.
$A_{\gamma\kappa}$	the γ–κ interfacial area contained within the averaging volume $\mathscr{V}_\sigma$ (2.12), m^2.
$a_{\gamma\kappa}$	the γ–κ interfacial area per unit volume (2.15), m^{-1}
$A_{\gamma e}$	area of entrances and exits for the γ-phase contained within the averaging volume $\mathscr{V}_\sigma$ (3.11), m^2.
$A_{\sigma\beta}$	$A_{\beta\sigma}$, the β–σ interfacial area contained within the averaging volume $\mathscr{V}$ (5.8), m^2.
$a_{\beta\sigma}$	the β–σ interfacial area per unit volume (6.2), m^{-1}.
c_γ	point concentration in the γ-phase (2.1), kg mol/m^3.
c_{ad}	point surface concentration at the γ–κ interface (2.2), kg mol/m^2.
$\langle c_\gamma \rangle$	phase average concentration in the γ-phase (2.7), kg mol/m^3.
$\langle c_\gamma \rangle^\gamma$	intrinsic phase concentration in the γ-phase (2.9), kg mol/m^3.
$\tilde{c}_\gamma$	$c_\gamma - \langle c_\gamma \rangle^\gamma$, spatial deviation concentration in the γ-phase (2.20), kg mol/m^3.
$\langle c_\gamma \rangle_{\gamma\kappa}$	area-averaged concentration at the γ–κ interface (2.14), kg mol/m^3.
c_σ	$\langle c_\gamma \rangle^\gamma$, point concentration in the σ-phase (5.5), kg mol/m^3.
c_β	point concentration in the β-phase (5.2), kg mol/m^3.
$\langle c_\sigma \rangle$	phase average concentration in the σ-phase (5.7), kg mol/m^3.
$\langle c_\sigma \rangle^\sigma$	intrinsic phase average concentration in the σ-phase (5.9), kg mol/m^3.
$\tilde{c}_\sigma$	$c_\sigma - \langle c_\sigma \rangle^\sigma$, spatial deviation concentration in the σ-phase (5.9), kg mol/m^3.
$\langle c_\beta \rangle$	phase average concentration in the β-phase (5.13), kg mol/m^3.
$\langle c_\beta \rangle^\beta$	intrinsic phase average concentration in the β-phase (5.13), kg mol/m^3.
$\tilde{c}_\beta$	$c_\beta - \langle c_\beta \rangle^\beta$, spatial deviation concentration in the β-phase (5.19), kg mol/m^3.
$\langle c \rangle$	intrinsic spatial average concentration for the β–σ system (5.31), kg mol/m^3.
$\hat{c}_\sigma$	$\langle c_\sigma \rangle^\sigma - \langle c \rangle$, σ-phase spatial deviation concentration for the β–σ system (5.30a), kg mol/m^3.

$\hat{c}_\beta$ — $\langle c_\beta \rangle^\beta - \langle c \rangle$, β-phase spatial deviation concentration for the β–σ system (5.30b), kg mol/m^3.

$\langle c_\sigma \rangle_{\beta\sigma}$ — area-averaged σ-phase concentration at the β–σ interface (6.2a), kg mol/m^3.

$\langle c_\beta \rangle_{\beta\sigma}$ — area-averaged β-phase concentration at the β–σ interface (6.2b), kg mol/m^3.

$\mathscr{D}_\gamma$ — the γ-phase molecular diffusivity (2.1), m^2/s.

$\mathbf{D}^\gamma_{\text{eff}}$ — the γ-phase effective diffusivity tensor (3.53), m^2/s.

D^γ_{eff} — the effective diffusivity for isotropic systems (Fig. 7), m^2/s.

$\mathbf{D}_\sigma$ — $\varepsilon_\gamma \mathbf{D}^\gamma_{\text{eff}}$, the σ-phase diffusivity tensor (5.5), m^2/s.

D_σ — order of magnitude estimate for $\mathbf{D}_\sigma$ (6.2), m^2/s.

$\mathscr{D}_\beta$ — the β-phase molecular diffusivity (5.4), m^2/s.

$\mathbf{D}_\beta$ — hydrodynamic dispersion tensor for the β-phase (7.55), m^2/s.

$\mathbf{D}_{\text{eff}}$ — effective diffusivity tensor for the β–σ system (7.54), m^2/s.

$\mathbf{D}^*$ — total dispersivity tensor for the β–σ system (7.57), m^2/s.

d_{p} — effective particle diameter (8.16), m.

$\mathbf{f}$ — the vector field that maps $\nabla \langle c_\gamma \rangle^\gamma$ onto $\tilde{c}_\gamma$ (3.29), and the vector field that maps $\nabla \langle c \rangle$ onto $\tilde{c}_\beta$ (7.47), m.

$\mathbf{g}$ — the vector field that maps $\nabla \langle c \rangle$ onto $\tilde{c}_\sigma$ (7.46), m.

$\mathbf{I}$ — unit tensor.

K_{eq} — equilibrium coefficient for the linear isotherm (2.4), m.

K — $a_{\gamma\kappa} K_{\text{eq}} / \varepsilon_\gamma$, dimensionless equilibrium coefficient for the σ-phase (3.52).

$\mathscr{K}$ — $\varepsilon_\sigma \varepsilon_\gamma (1 + K)/\varepsilon_\beta$, dimensionless equilibrium coefficient for the β–σ system (5.34).

ℓ_ω — characteristic length for the ω-region (Fig. 1), m.

ℓ_η — characteristic length for the η-region (Fig. 1), m.

ℓ_σ — characteristic length for the σ-phase (Fig. 3), m.

ℓ_γ — characteristic length for the γ-phase (Fig. 3), m.

ℓ_κ — characteristic length for the κ-phase (Fig. 3), m.

ℓ_β — characteristic length for the β-phase (Fig. 8), m.

$\boldsymbol{\ell}_i$ — lattice vectors describing a spatially periodic porous medium (3.26), m.

ℓ_λ — characteristic length for volume-averaged quantities associated with the β–σ system (5.22), m.

L_{c} — convective length for volume-averaged quantities associated with the β–σ system (6.27), m.

L_{D} — dispersive length for volume-averaged quantities associated with the β–σ system (6.27), m.

L — length of a packed bed (8.17), m.

$\mathbf{n}_{\gamma\kappa}$ — outwardly directed unit normal vector pointing from the γ-phase toward the κ-phase (2.2).

$\mathbf{n}_{\gamma e}$ — outwardly directed unit normal vector at the entrances and exits of the γ-phase (3.11).

$\mathbf{n}_{\sigma\beta}$	$-\mathbf{n}_{\beta\sigma}$, outwardly directed unit normal vector pointing from the σ-phase toward the β-phase (5.3).
$\mathbf{n}_{\beta e}$	outwardly directed unit normal vector at the entrances and exits of the β-phase (7.17).
Re	Reynolds number (6.23).
r_σ	radius of the σ-phase averaging volume $\mathscr{V}_\sigma$ (Fig. 3), m.
$\mathbf{r}$	position vector (2.3), m.
r_0	radius of the β–σ system averaging volume $\mathscr{V}$ (Fig. 8), m.
R_0	radius of cylinders used in the solution of the closure problem (Fig. 13), m.
R_1	radius of cylinders used in the solution of the closure problem (Fig. 13), m.
s	the scalar field that maps $\partial\langle c_\gamma\rangle^\gamma/\partial t$ onto $\tilde{c}_\gamma$ (3.29), s.
t	time (2.1), s.
$\mathscr{V}_\infty$	large-scale averaging volume (Fig. 2), m^3.
$\mathscr{V}$	local averaging volume (Fig. 2), m^3.
$\mathscr{V}_\sigma$	small-scale averaging volume (Fig. 2), m^3.
$\mathbf{v}_\beta$	the β-phase fluid velocity vector (5.4), m/s.
V_σ	volume of the σ-phase contained within $\mathscr{V}_\sigma$ (5.6), m^3.
V_β	volume of the β-phase contained within $\mathscr{V}$ (5.12), m/s.
$\langle\mathbf{v}_\beta\rangle^\beta$	the β-phase intrinsic phase average velocity (5.20), m/s.
$\langle\mathbf{v}_\beta\rangle$	the β-phase phase average velocity (5.25), m/s.
$\tilde{\mathbf{v}}_\beta$	$\mathbf{v}_\beta - \langle\mathbf{v}_\beta\rangle^\beta$, the β-phase spatial deviation velocity (5.20), m/s.
δ_σ	a characteristic length for the interphase flux in the σ-phase (6.2a), m.
δ_β	a characteristic length for the interphase flux in the β-phase (6.2b), m.
ε_γ	the γ-phase volume fraction in the γ–κ system (2.10).
ε_σ	the σ-phase volume fraction in the β–σ system (5.8).
ε_β	the β-phase volume fraction in the β–σ system (5.13).

Acknowledgements

This work was supported in part (S.W.) by NSF Grant CBT 8812870. Thanks are due to Ayse G. Batigün for the preparation of Fig. 12.

References

Angelakis, A.N., Kadir, T.N. and Rolston, D.E. (1987). Solutions for transport of two sorbed solutes with differing dispersion coefficients in soil, *Soil Sci. Soc. Am. J.* **51**, 1428–1434.

Bensoussan, A., Lions, J.-L. and Papanicolaou, G. (1978). 'Asymptotic Analysis for Periodic Structures'. North-Holland, New York.

Blackwell, R.J., Rayne, J.R. and Terry, W.M. (1959). Factors influencing the efficiency of miscible displacement, *Trans. AIME* **217**, 1–8.

Bourgeat, A., Quintard, M. and Whitaker, S. (1988). Eléments de comparaison entre la méthode d'homogénéisation et la méthode de prise de moyenne avec fermature, *C.R. Acad. Sci. Paris, t.306, Série II*, 463–466.

Brenner, H. (1980). Dispersion resulting from flow through spatially periodic porous media, *Trans. Roy. Soc. London* **297**, 81–133.

Carberry, J.J. and Bretton, R.H. (1958). Axial dispersion of mass in flow through fixed beds, *AIChE J.*, **4**, 367–375.

Carbonell, R.G. and Whitaker, S. (1983). Dispersion in pulsed systems—II. Theoretical developments for passive dispersion in porous media, *Chem. Engng Sci.* **38**, 1795–1802.

Carbonell, R.G. and Whitaker, S. (1984). Heat and mass transfer in porous media. *In* 'Fundamentals of Transport Phenomena in Porous Media' (J. Bear and M.Y. Corapcioglu, eds), pp. 123–198. Martinus Nijhoff Publishers, Dordrecht, The Netherlands.

Chang, H.-C. (1982). Multiscale analysis of effective transport in periodic heterogeneous media, *Chem. Engng Commun.* **15**, 83–91.

Crapiste, G.H., Rotstein, E. and Whitaker, S. (1986). A general closure scheme for the method of volume averaging, *Chem. Engng Sci.* **41**, 227–235.

Currie, J.A. (1960). Gaseous diffusion in porous media I: A non-steady state method, *Br. J. Appl. Phys.* **II**, 314–324.

Cushman, J.H. (1984). On unifying concepts of scale, instrumentation and stochastics in the development of multiphase transport theory, *Water Resour. Res.* **20**, 1668–1672.

Debbas, S. and Rumpf, H.J. (1966). On the randomness of beds packed with spheres of irregular shaped particles, *Chem. Engng Sci.* **21**, 583–607.

Ebach, E.A. and White, R.R. (1958). Mixing of fluids flowing through beds of packed solids, *AIChEJ.* **4**, 161–164, 1958.

Edwards, M.F. and Richardson, J.F. (1968). Gas dispersion in packed beds, *Chem. Engng Sci.* **23**, 109–123.

Eidsath, A.B. (1981). Flow and dispersion in spatially periodic porous media: A finite element study. M.Sc. Thesis, Department of Chemical Engineering, University of California, Davis, CA.

Eidsath, A., Carbonell, R.G., Whitaker, S. and Hermann, L.R. (1983). Dispersion in pulsed systems —III. Comparison between theory and experiments for packed beds. *Chem. Engng Sci.* **38**, 1803–1816.

Gray, W.G. (1975). A derivation of the equations for multiphase transport, *Chem. Engng Sci.* **30**, 229–233.

Gunn, D.J. and Pryce, C. (1969). Dispersion in packed beds, *Trans. Inst. Chem. Eng.* **47**, 341–350.

Han, N.-W., Bhakta, J. and Carbonell, R.G. (1985). Longitudinal and lateral dispersion in packed beds: Effect of column length and particle size distribution, *AIChE J.* **31**, 277–288.

Harleman, D.R.E. and Rumer, R.R. (1963). Longitudinal and lateral dispersion in isotropic porous media. *J. Fluid Mech.* **16**, 385–394.

Hassinger, R.C. and von Rosenberg, R.V. (1968). A mathematical and experimental examination of transverse dispersion coefficients. *Trans. Am. Inst. Min. Metall. Pet Eng.* **243**, 195–210.

Heller, J.P. (1972). Observations of mixing and diffusion in porous media. *In* 'Proceedings of the Second Symposium on Fundamentals of Transport Phenomena in Porous Media', University of Guelph, IAHR-1SSS, Vol. 1, pp. 1–26.

Hoogschagen, J. (1955). Diffusion in porous catalysts and absorbents, *Ind. Eng. Chem.* **47**, 906–916.

Howes, F.A. and Whitaker, S. (1985). The spatial averaging theorem revisited, *Chem. Engng Sci.* **40**, 1387–1392.

Jackson, R. (1977). 'Transport in Porous Catalysts'. Elsevier, New York.

Kim, J.-H., Ochoa, J.A. and Whitaker, S. (1987). Diffusion in anisotropic porous media, *Transport in Porous Media* **2**, 327–356.

Koch, D.L. and Brady, J.F. (1985). Dispersion in packed beds, *J. Fluid Mech.* **154**, 399–427.

Levec, J. and Carbonell, R.G. (1985). Longitudinal and lateral thermal dispersions in packed beds —Part I: Theory, *AIChE J.* **31**, 581–590.

Maxwell, J.C. (1881). 'Treatise on Electricity and Magnetism', Vol. I. Clarendon Press, Oxford.

Nielsen, D.R., van Genuchten, M.Th. and Biggar, J.W. (1986). Water flow and solute transport processes in the unsaturated zone, *Water Resour. Res.* **22**, 895–1085.

Nozad, I., Carbonell, R.G. and Whitaker, S. (1985). Heat conduction in multiphase systems—I. Theory and experiment for two-phase systems, *Chem. Engng Sci.* **40**, 843–855.

Ochoa, J.A., Stroeve, P. and Whitaker, S. (1986). Diffusion and reaction in cellular media, *Chem. Engng Sci.* **41**, 2999–3013.

Paine, M.A., Carbonell, R.G. and Whitaker, S. (1983). Dispersion in pulsed systems—I. Heterogeneous reaction and reversible adsorption in capillary tubes, *Chem. Engng Sci.* **38**, 1781–1793.

Pereira Duarte, S.I., Ferretti, O.A. and Lemcoff, N.O. (1984). A heterogeneous one-dimensional model for nonadibatic fixed bed catalytic reactors, *Chem. Engng Sci.* **39**, 1017–1024.

Pfannkuch, H.O. (1963). Contributions à l'étude des déplacements de fluides miscibles dans un milieu poreux, *Rev. Inst. Fr. Pétrol.* **18**, 215–227.

Pironti, F. and Whitaker, S. (1986). Aplication de Metodo de Escalas Multiples al Proceso de Diffusion con Reaccion en Catalizadores Bimodales, *Acta Cientifica Venezolana* **37**, 394–405.

Prat, M. (1989). On the boundary conditions at the macroscopic level, *Transport in Porous Media* **4**, 259–280.

Quintard, M. and Whitaker, S. (1987). Ecoulement monophasique en milieu poreux: effet des Hétérogénéités locales. *J. Mécanique théorique et appliquée* **6**, 691–726.

Quintard, M. and Whitaker, S. (1988). Two-phase flow in heterogeneous porous media: The method of large-scale averaging, *Transport in Porous Media* **3**, 357–413.

Quintard, M. and Whitaker, S. (1990). Two-phase flow in heterogeneous porous media I: The influence of large spatial and temporal gradients, *Transport in Porous Media* (in press).

Rifai, M.N.E., Kaufman, W.J. and Todd, D.K. (1956). Dispersion phenomena in laminar flow through porous media, *Sanit. Eng. Rep.* **3**, University of California, Berkeley, CA.

Ryan, D., Carbonell, R.G. and Whitaker, S. (1980). Effective diffusivities for catalyst pellets under reactive conditions, *Chem. Engng Sci.* **35**, 10–16.

Ryan, D., Carbonell, R.G. and Whitaker, S. (1981). 'A Theory of Diffusion and Reaction in Porous Media'. American Institute of Chemical Engineers Symposium Series, No. 202, Vol. 71, pp. 46–62.

Valocchi, A.J. (1986). Effect of radial flow on deviations from local equilibrium during sorbing solute transport through homogeneous soils, *Water Resour. Res.* **22**, 1693–1701.

Wakao, N. and Smith, J.M. (1962). Diffusion in catalyst pellets, *Chem. Engng Sci.* **17**, 825–834.

Weissberg, H.L. (1963). Effective diffusion coefficients in porous media, *J. Appl. Phys.* **34**, 2636–2639.

Whitaker, S. (1967). Diffusion and dispersion in porous media. *Amer. Inst. Chem. Eng. J.* **13**, 420–427.

Whitaker, S. (1980). Heat and mass transfer in granular porous media. *In* 'Advances in Drying', Vol. 1, pp. 23–61. Hemisphere Publishing Corporation, New York.

Whitaker, S. (1982). Laws of continuum physics for single-phase, single-component systems. *In* 'Handbook of Multiphase Systems' (G. Hetsroni, ed.), pp. 1-5 to 1-35. Hemisphere Publishing Corporation, New York.

Whitaker, S. (1983). Diffusion and reaction in a micropore-macropore model of a porous medium. *Latin Amer. J. Chem. Engng Appl. Chem.* **13**, 143–183.

Whitaker, S. (1986a). Flow in porous media I: A theoretical derivation of Darcy's Law, *Transport in Porous Media* **1**, 3–25.

Whitaker, S. (1986b). Local thermal equilibrium: An application to packed bed catalytic reactor design, *Chem. Engng Sci.* **41**, 2029–2039.

Whitaker, S. (1986c). Transient diffusion, adsorption and reaction in porous catalysts: the reaction controlled, quasi-steady catalytic surface, *Chem. Engng Sci.* **41**, 3015–3022.

Whitaker, S. (1986d). Transport processes with heterogeneous reaction. *In* 'Concepts and Design of Chemical Reactors' (S. Whitaker and A.E. Cassano, eds), pp. 1–94. Gordon and Breach, New York.

Whitaker, S. (1988). Diffusion in packed beds of porous particles, *AIChE J.* **34**, 679–683.

Whitaker, S. (1989). Heat transfer in catalytic packed bed reactors. In 'Handbook of Heat and Mass Transfer' (N.P. Cheremisinoff, ed.), Vol. 3, Ch. 10. Gulf Publishers, NJ.

Zanotti, F. and Carbonell, R.G. (1984). Development of transport equations for multiphase systems III. Applications to heat transfer in packed beds, *Chem. Engng Sci.* **39**, 299–311.

Chapter VI

Diffusion, Adsorption, and Dispersion in Heterogeneous Porous Media: The Method of Large-scale Averaging*

O.A. Plumb
Department of Mechanical and Materials Engineering, Washington State University, Pullman, WA 99164, USA

S. Whitaker
Department of Chemical Engineering, University of California, Davis, CA 95616, USA

1. Introduction

In the previous chapter we referred to *heterogeneous porous media* and a two-region model of a heterogeneous porous medium was illustrated. While multi-region models of heterogeneous porous media have a definite intuitive appeal, it would be attractive to have a mathematical definition for heterogeneous media. Elsewhere (Quintard and Whitaker, 1987, 1988; Plumb and Whitaker, 1988a) we have considered this problem and the best we have to offer at this time is the following definition by exclusion:

> A porous medium is homogeneous with respect to a given process and a given averaging volume when the effective transport coefficients (such as the Darcy's law permeability tensor) in the volume averaged transport equations are independent of position: If a porous medium is not homogeneous, it is heterogeneous.

It is particularly important to keep in mind that the definition of homogeneous depends on the process, thus a particular porous medium may appear to be homogeneous with respect to diffusion while it is obviously heterogeneous with respect to dispersion. Previous studies of diffusion (Kim *et al.*, 1987) and

* The unabridged version of this article is available from Professor Whitaker.

DYNAMICS OF FLUIDS IN HIERARCHICAL POROUS MEDIA
ISBN 0-12-200260-1

dispersion (Eidsath *et al.*, 1983) in packed beds of uniform spheres tend to confirm this point of view. The problem of adsorption in porous media is a particularly delicate one since a heterogeneity can be developed by a variation in the *chemical characteristics* of the fluid–solid interface when no heterogeneity exists in the geometry of the fluid–solid interface. Under these circumstances the porous medium would appear to be homogeneous with respect to single-phase flow and heterogeneous with respect to mass transport and adsorption. For the latter process the method of large-scale averaging is required in order to develop the appropriate mass transport equations.

2. Large-scale averaging

In this section we are finally ready to analyse the process of diffusion, adsorption and dispersion in the heterogeneous medium illustrated in Fig. 1 of the previous chapter. The averaging volume to be used is illustrated in Fig. 1 where we have suggested that yet another domain may require spatial smoothing. The length

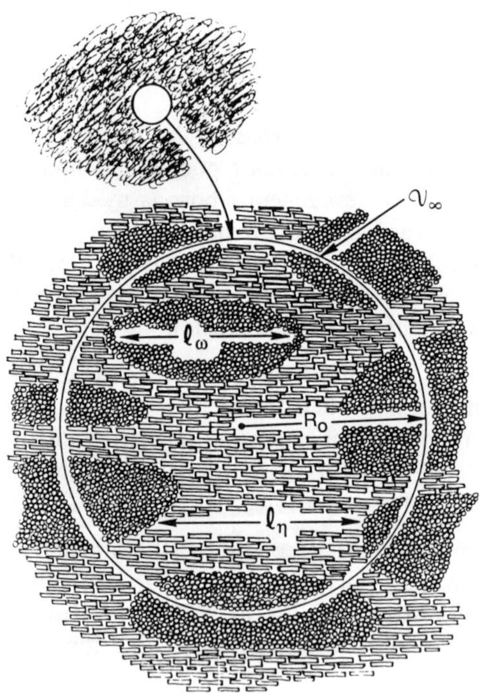

Figure 1. Large-scale averaging.

scales associated with the ω- and η-regions are designated by ℓ_ω and ℓ_η, and the volume fractions of these regions are identified by ϕ_ω and ϕ_η, respectively.

The boundary value problem that forms the basis for large-scale averaging is given by

$$\varepsilon_{\beta\eta}(1 + \mathscr{K}_\eta)\frac{\partial\langle c\rangle_\eta}{\partial t} + \langle \mathbf{v}_\beta\rangle_\eta\cdot\nabla\langle c\rangle_\eta = \nabla\cdot(\varepsilon_{\beta\eta}\mathbf{D}_\eta^*\cdot\nabla\langle c\rangle_\eta) \tag{2.1}$$

B.C.1
$$\langle c\rangle_\eta = \langle c\rangle_\omega, \quad \text{at } A_{\eta\omega} \tag{2.2}$$

B.C.2
$$(\langle \mathbf{v}_\beta\rangle_\eta\langle c\rangle_\eta - \varepsilon_{\beta\eta}\mathbf{D}_\eta^*\cdot\nabla\langle c\rangle_\eta)\cdot\mathbf{n}_{\eta\omega}$$
$$= (\langle \mathbf{v}_\beta\rangle_\omega\langle c\rangle_\omega - \varepsilon_{\beta\omega}\mathbf{D}_\omega^*\cdot\nabla\langle c\rangle)\cdot\mathbf{n}_{\eta\omega}, \quad \text{at } A_{\eta\omega} \tag{2.3}$$

$$\varepsilon_{\beta\omega}(1 + \mathscr{K}_\omega)\frac{\partial\langle c\rangle_\omega}{\partial t} + \langle \mathbf{v}_\beta\rangle_\omega\cdot\nabla\langle c\rangle_\omega = \nabla\cdot(\varepsilon_{\beta\omega}\mathbf{D}_\omega^*\cdot\nabla\langle c\rangle_\omega) \tag{2.4}$$

Equations 2.1 and 2.4 are derived directly from Eq. 7.56 of Chapter V where $\varepsilon_{\beta\eta}$ and $\varepsilon_{\beta\omega}$ have been used to indicate the porosity of the η- and ω-regions, and a similar notation has been used to identify $\mathscr{K}_\eta$, $\mathscr{K}_\omega$, $\langle c\rangle_\eta$, etc. The boundary conditions given by Eqs 2.2 and 2.3 are based on the idea that the local volume averaged concentration and flux will be continuous at the η–ω interface. The matter of boundary conditions for single- and two-phase flow have been discussed at some length by Quintard and Whitaker (1987, 1988), and those studies support the intuitive representations given by Eqs 2.2 and 2.3.

This approach to dispersion in heterogeneous porous media is slightly different from our original study (Plumb and Whitaker, 1988a, b) in which it was assumed, *a priori*, that a one-equation model would be satisfactory. When adsorption in a heterogeneous porous medium is considered, it seems entirely plausible that large differences may exist between $\mathscr{K}_\eta$ and $\mathscr{K}_\omega$ and this will require the use of separate equations for the η- and ω-regions.

We begin by developing the average form of Eq. 2.1, and from this we can extract the average form of Eq. 2.4 by analogy. In our analysis we will encounter the following large-scale averages

$$\{\langle c\rangle_\eta\} = \frac{1}{\mathscr{V}_\infty}\int_{V_\eta}\langle c\rangle_\eta\,\mathrm{d}V, \quad \text{large-scale phase average} \tag{2.5a}$$

$$\{\langle c\rangle_\eta\}^\eta = \frac{1}{V_\eta}\int_{V_\eta}\langle c\rangle_\eta\,\mathrm{d}V, \quad \text{large-scale intrinsic phase average} \tag{2.5b}$$

Here V_η represents the volume of the η-region contained within $\mathscr{V}_\infty$, and in terms of V_η we express the volume fraction ϕ_η as

$$\phi_\eta = V_\eta/\mathscr{V}_\infty \tag{2.6}$$

One can use Eq. 5.26 of the previous chapter in the form

$$\nabla \cdot \langle v_\beta \rangle_\eta = 0 \tag{2.7}$$

to express Eq. 2.1 as

$$\varepsilon_{\beta\eta}(1 + \mathscr{K}_\eta) \frac{\partial \langle c \rangle_\eta}{\partial t} + \nabla \cdot (\langle \mathbf{v}_\beta \rangle_\eta \langle c \rangle_\eta) = \nabla \cdot (\varepsilon_{\beta\eta} \mathbf{D}_\eta^* \cdot \nabla \langle c \rangle_\eta) \tag{2.8}$$

and the average of this result is given by

$$\frac{1}{\mathscr{V}_\infty} \int_{V_\eta} \varepsilon_{\beta\eta}(1 + \mathscr{K}_\eta) \frac{\partial \langle c \rangle_\eta}{\partial t} \, \mathrm{d}V + \frac{1}{\mathscr{V}_\infty} \int_{V_\eta} \nabla \cdot (\langle \mathbf{v}_\beta \rangle_\eta \langle c \rangle_\eta) \, \mathrm{d}V$$

$$= \frac{1}{\mathscr{V}_\infty} \int_{V_\eta} \nabla \cdot (\varepsilon_{\beta\eta} \mathbf{D}_\eta^* \cdot \nabla \langle c \rangle_\eta) \, \mathrm{d}V \tag{2.9}$$

At this point we will impose the severe restriction that $\varepsilon_{\beta\eta}$ and $\mathscr{K}_\eta$ undergo negligible variation in the η-region, thus the first term in Eq. 2.9 can be expressed as

$$\frac{1}{\mathscr{V}_\infty} \int_{V_\eta} \varepsilon_{\beta\eta}(1 + \mathscr{K}_\eta) \frac{\partial \langle c \rangle_\eta}{\partial t} \, \mathrm{d}V = \varepsilon_{\beta\eta}(1 + \mathscr{K}_\eta) \frac{\partial}{\partial t} \{\langle c \rangle_\eta\} \tag{2.10}$$

The interchanging of differentiation and integration is based on the restriction that V_η is independent of time. This allows us to use

$$\{\langle c \rangle_\eta\} = \phi_\eta \{\langle c \rangle_\eta\}^\eta \tag{2.11}$$

and express Eq. 2.10 as

$$\frac{1}{\mathscr{V}_\infty} \int_{V_\eta} \varepsilon_{\beta\eta}(1 + \mathscr{K}_\eta) \frac{\partial \langle c \rangle_\eta}{\partial t} \, \mathrm{d}V = \varepsilon_{\beta\eta}(1 + \mathscr{K}_\eta) \phi_\eta \frac{\partial}{\partial t} \{\langle c \rangle_\eta\}^\eta \tag{2.12}$$

This result can be used in Eq. 2.9, along with the averaging theorem, to obtain

$$\varepsilon_{\beta\eta}(1 + \mathscr{K}_\eta)\phi_\eta \frac{\partial}{\partial t}\{\langle c\rangle_\eta\}^\eta + \nabla\cdot\{\langle \mathbf{v}_\beta\rangle_\eta \langle c\rangle_\eta\} + \frac{1}{\mathscr{V}_\infty}\int_{A_{\eta\omega}} \mathbf{n}_{\eta\omega}\cdot(\langle \mathbf{v}_\beta\rangle_\eta \langle c\rangle_\eta)\,\mathrm{d}A$$

$$= \nabla\cdot\{\varepsilon_{\beta\eta}\mathbf{D}^*_\eta\cdot\nabla\langle c\rangle_\eta\} + \frac{1}{\mathscr{V}_\infty}\int_{A_{\eta\omega}} \mathbf{n}_{\eta\omega}\cdot(\varepsilon_{\beta\eta}\mathbf{D}^*_\eta\cdot\nabla\langle c\rangle_\eta)\,\mathrm{d}A \tag{2.13}$$

where $\mathbf{n}_{\eta\omega}$ is the unit normal vector directed from the η-region toward the ω-region. Since we have already assumed that $\varepsilon_{\beta\eta}$ can be treated as a constant in the η-region, it is plausible to extend this simplification to $\mathbf{D}^*_\eta$ and this leads to

$$\nabla\cdot\{\varepsilon_{\beta\eta}\mathbf{D}^*_\eta\cdot\nabla\langle c\rangle_\eta\} = \nabla\cdot[\varepsilon_{\beta\eta}\mathbf{D}^*_\eta\cdot\{\nabla\langle c\rangle_\eta\}]$$

$$= \nabla\cdot\left[\varepsilon_{\beta\eta}\mathbf{D}^*_\eta\cdot\left(\nabla\{\langle c\rangle_\eta\} + \frac{1}{\mathscr{V}_\infty}\int_{A_{\eta\omega}} \mathbf{n}_{\eta\omega}\langle c\rangle_\eta\,\mathrm{d}A\right)\right] \tag{2.14}$$

Substitution of this result into Eq. 2.13 yields

$$\varepsilon_{\beta\eta}(1 + \mathscr{K}_\eta)\phi_\eta \frac{\partial}{\partial t}\{\langle c\rangle_\eta\}^\eta + \nabla\cdot\{\langle \mathbf{v}_\beta\rangle_\eta \langle c\rangle_\eta\} + \frac{1}{\mathscr{V}_\infty}\int_{A_{\eta\omega}} \mathbf{n}_{\eta\omega}\cdot(\langle \mathbf{v}_\beta\rangle_\eta \langle c\rangle_\eta)\,\mathrm{d}A$$

$$= \nabla\cdot\left[\varepsilon_{\beta\eta}\mathbf{D}^*_\eta\cdot\left(\nabla\{\langle c\rangle_\eta\} + \frac{1}{\mathscr{V}_\infty}\int_{A_{\eta\omega}} \mathbf{n}_{\eta\omega}\langle c\rangle_\eta\,\mathrm{d}A\right)\right]$$

$$+ \frac{1}{\mathscr{V}_\infty}\int_{A_{\eta\omega}} \mathbf{n}_{\eta\omega}\cdot(\varepsilon_{\beta\eta}\mathbf{D}^*_\eta\cdot\nabla\langle c\rangle_\eta)\,\mathrm{d}A \tag{2.15}$$

This form of the η-region average equation is similar to Eq. 5.18 of Chapter V, and it presents us with a similar set of problems. To begin with, we want an equation for $\{\langle c\rangle_\eta\}^\eta$ and Eq. 2.11 will be a necessary relation to use in order to eliminate $\{\langle c\rangle_\eta\}$ from Eq. 2.15. Next we must deal with the product $\{\langle \mathbf{v}_\beta\rangle_\eta\langle c\rangle_\eta\}$, and here we follow the development in Section 5 of Chapter V and use the analogues of Eqs 5.19 and 5.20. These are given by

$$\langle c\rangle_\eta = \{\langle c\rangle_\eta\}^\eta + \hat{c}_\eta \tag{2.16a}$$

$$\langle \mathbf{v}_\beta\rangle_\eta = \{\langle \mathbf{v}_\beta\rangle_\eta\}^\eta + \hat{\mathbf{v}}_{\beta\eta} \tag{2.16b}$$

and when the convective transport is represented in terms of these large-scale decompositions we obtain

$$\{\langle \mathbf{v}_\beta \rangle_\eta \langle c \rangle_\eta\} = \phi_\eta \{\langle \mathbf{v}_\beta \rangle_\eta\}^\eta \{\langle c \rangle_\eta\}^\eta + \phi_\eta \{\hat{\mathbf{v}}_{\beta\eta} \hat{c}_\eta\}^\eta \tag{2.17}$$

This representation of the convective transport is analogous to that given by Eq. 5.21 in Chapter V, and the validity of Eq. 2.17 requires the large-scale version of Eq. 5.22. This can be expressed as

$$\frac{\ell_\eta}{R_0} \gg \frac{R_0}{\mathscr{L}_\eta} \tag{2.18}$$

in which $\mathscr{L}_\eta$ represents the characteristic length associated with quantities such as ϕ_η, $\{\langle \mathbf{v}_\beta \rangle_\eta\}^\eta$, and $\{\langle c \rangle_\eta\}^\eta$. Use of Eq. 2.17 in Eq. 2.15 allows us to write the η-region transport equation as

$$\begin{aligned} &\varepsilon_{\beta\eta}(1 + \mathscr{K}_\eta)\phi_\eta \frac{\partial}{\partial t}\{\langle c \rangle_\eta\}^\eta + \nabla \cdot (\phi_\eta \{\langle \mathbf{v}_\beta \rangle_\eta\}^\eta \{\langle c \rangle_\eta\}^\eta) \\ &= \nabla \cdot \left[\varepsilon_{\beta\eta} \mathbf{D}_\eta^* \left(\nabla \{\langle c \rangle_\eta\} + \frac{1}{\mathscr{V}_\infty} \int_{A_{\eta\omega}} \mathbf{n}_{\eta\omega} \langle c \rangle_\eta \, \mathrm{d}A \right) \right] - \nabla \cdot (\phi_\eta \{\hat{\mathbf{v}}_{\beta\eta} \hat{c}_\eta\}^\eta) \\ &\quad - \frac{1}{\mathscr{V}_\infty} \int_{A_{\eta\omega}} \mathbf{n}_{\eta\omega} \cdot [\langle \mathbf{v}_\beta \rangle_\eta \langle c \rangle_\eta - \varepsilon_{\beta\eta} \mathbf{D}_\eta^* \cdot \nabla \langle c \rangle_\eta] \, \mathrm{d}A \end{aligned} \tag{2.19}$$

Here the first term represents the accumulation modified by the adsorption process that is taking place in the σ-phase, and the second term represents the large-scale convective transport in the η-region. The first two terms on the right-hand side of Eq. 2.19 represent the large-scale 'diffusive' and dispersive transport, while the last term represents the inter-region flux that is constrained by Eq. 2.3. Our analysis of the 'diffusive' transport term follows the development from Eq. 2.17 to 2.25 in the previous chapter and we express the final result as

$$\begin{aligned} &\varepsilon_{\beta\eta}\phi_\eta(1 + \mathscr{K}_\eta) \frac{\partial}{\partial t}\{\langle c \rangle_\eta\}^\eta + \nabla \cdot (\phi_\eta \{\langle \mathbf{v}_\beta \rangle_\eta\}^\eta \{\langle c \rangle_\eta\}^\eta) \\ &= \nabla \cdot \left[\varepsilon_{\beta\eta} \phi_\eta \mathbf{D}_\eta^* \cdot \left(\nabla \{\langle c \rangle_\eta\}^\eta + \frac{1}{\mathscr{V}_\infty} \int_{A_{\eta\omega}} \mathbf{n}_{\eta\omega} \hat{c}_\eta \, \mathrm{d}A \right) \right] - \nabla \cdot (\phi_\eta \{\hat{\mathbf{v}}_{\beta\eta} \hat{c}_\eta\}^\eta) \\ &\quad - \frac{1}{\mathscr{V}_\infty} \int_{A_{\eta\omega}} \mathbf{n}_{\eta\omega} \cdot [\langle \mathbf{v}_\beta \rangle_\eta \langle c \rangle_\eta - \varepsilon_{\beta\eta} \mathbf{D}_\eta^* \cdot \nabla \langle c \rangle_\eta] \, \mathrm{d}A \end{aligned} \tag{2.20}$$

This result requires the large-scale version of Eqs 2.22 and 5.10 given in the previous chapter, and we list this result explicitly as

$$\left(\frac{R_0}{\mathscr{L}_\eta}\right)^2 \ll 1 \tag{2.21}$$

The development leading from Eq. 2.1 to 2.20 can be repeated for Eq. 2.4 which describes the transport in the ω-region. For completeness we list this result

$$\varepsilon_{\beta\omega}\phi_\omega(1 + \mathscr{K}_\omega)\frac{\partial}{\partial t}\{\langle c\rangle_\omega\}^\omega + \nabla\cdot(\phi_\omega\{\langle \mathbf{v}_\beta\rangle_\omega\}^\omega\{\langle c\rangle_\omega\}^\omega)$$

$$= \nabla\cdot\left[\varepsilon_{\beta\omega}\phi_\omega \mathbf{D}^*_\omega\cdot\left(\nabla\{\langle c\rangle_\omega\}^\omega + \frac{1}{V_\omega}\int_{A_{\omega\eta}} \mathbf{n}_{\omega\eta}\hat{c}_\omega \, dA\right)\right] - \nabla\cdot(\phi_\omega\{\hat{\mathbf{v}}_{\beta\omega}\hat{c}_\omega\}^\omega)$$

$$-\frac{1}{\mathscr{V}_\infty}\int_{A_{\omega\eta}} \mathbf{n}_{\omega\eta}\cdot[\langle \mathbf{v}_\beta\rangle_\omega\langle c\rangle_\omega - \varepsilon_{\beta\omega}\mathbf{D}^*_\omega\cdot\nabla\langle c\rangle_\omega]\, dA \tag{2.22}$$

If a one-equation model were our objective, we would add Eqs 2.20 and 2.22 in order to eliminate the inter-region flux terms and then go on to develop the appropriate closure problems for $\hat{c}_\eta$ and $\hat{c}_\omega$. However, in this case we want to illustrate the structure of the two-equation model and the associated closure problem.

It is important to recognize that our analysis has been set up with the specific intention of developing a two-equation model for the ω–η system shown in Fig. 1. This means that the velocity decomposition given by Eq. 2.16b has made use of the large-scale *intrinsic phase average velocity* $\{\langle \mathbf{v}_\beta\rangle_\eta\}^\eta$, with the thought that this velocity would be determined using some large-scale form of Darcy's law. At this time there are only two theoretical studies of flow in heterogeneous porous media (Quintard and Whitaker, 1987, 1988), and both of these lead to governing equations for the large-scale averaged velocity defined by

$$\{\langle \mathbf{v}_\beta\rangle\} = \frac{1}{\mathscr{V}_\infty}\int_{V_\eta} \langle \mathbf{v}_\beta\rangle_\eta \, dV + \frac{1}{\mathscr{V}_\infty}\int_{V_\omega} \langle \mathbf{v}_\beta\rangle_\omega \, dV \tag{2.23}$$

For single- and two-phase flows in heterogeneous porous media this velocity is usually the variable of interest and is related to $\{\langle \mathbf{v}_\beta\rangle_\eta\}^\eta$ and $\{\langle \mathbf{v}_\beta\rangle_\omega\}^\omega$ by

$$\{\langle \mathbf{v}_\beta\rangle\} = \phi_\eta\{\langle \mathbf{v}_\beta\rangle_\eta\}^\eta + \phi_\omega\{\langle \mathbf{v}_\beta\rangle_\omega\}^\omega \tag{2.24}$$

Knowledge of $\{\langle \mathbf{v}_\beta\rangle_\eta\}^\eta$ and $\{\langle \mathbf{v}_\beta\rangle_\omega\}^\omega$ can be extracted from $\{\langle \mathbf{v}_\beta\rangle\}$ only through the Darcy's law closure problem, and this is not a convenient route to follow in order to solve Eqs 2.20 and 2.22. It seems clear that if we wish to solve two-equation models of dispersion and adsorption in heterogeneous porous media, we will need to extend previous studies of flow in heterogeneous porous media in order to provide the average velocities and velocity deviations required for the solution of Eqs 2.20 and 2.22. This extension will lead to governing equations for $\{\langle \mathbf{v}_\beta \rangle_\eta\}^\eta$ and $\{\langle \mathbf{v}_\beta \rangle_\omega\}^\omega$ in terms of the appropriate pressure gradients and large-scale permeabilities, $\mathbf{K}_\eta^*$ and $\mathbf{K}_\omega^*$. Knowledge of these two permeabilities would allow one to comment on the extreme cases that could be associated with Eqs 2.20 and 2.22. If the permeability tensor for the η-region is much, much less than that for the ω-region, it is possible that there will be negligible convective transport in the η-region and the structure of the η–ω region transport problem would be comparable to the σ–β phase transport problem as defined by Eqs 5.11 and 5.23 in Chapter V. Under these circumstances we would be confronted with a problem of diffusion and adsorption in the η-region, and a problem of dispersion, convection and adsorption in the ω-region. Because ℓ_η is so much larger than ℓ_σ, it is unlikely that the constraint analogous to Eq. 6.18 of Chapter V would be satisfied, and a one-equation model would not be acceptable. If the permeability tensor for the η-region is comparable to that for the ω-region *and* convective transport is large compared to diffusive transport, it is possible that $\{\langle c \rangle_\eta\}^\eta$ and $\{\langle c \rangle_\omega\}^\omega$ are nearly equal and a one-equation model would be sufficient. This is the point of view taken by Plumb and Whitaker (1988a, b); however, that work did not take into account the process of adsorption where significant differences between $\mathscr{K}_\eta$ and $\mathscr{K}_\omega$ could invalidate the use of one-equation models.

One-equation models have the advantage that the inter-region flux terms are eliminated; however, if we wish to develop a two-equation model we must deal directly with these flux terms. We begin with the continuity equation

$$\nabla \cdot \langle \mathbf{v}_\beta \rangle_\eta = 0 \tag{2.25}$$

and form the large-scale average to obtain

$$\nabla \cdot \{\langle \mathbf{v}_\beta \rangle_\eta\} + \frac{1}{\mathscr{V}_\infty} \int_{A_{\eta\omega}} \mathbf{n}_{\eta\omega} \cdot \langle \mathbf{v}_\beta \rangle_\eta \, \mathrm{d}A = 0 \tag{2.26}$$

We can now use the analogue of Eq. 2.11 for the velocity to arrive at

$$\nabla \cdot (\phi_\eta \{\langle \mathbf{v}_\beta \rangle_\eta\}^\eta) + \frac{1}{\mathscr{V}_\infty} \int_{A_{\eta\omega}} \mathbf{n}_{\eta\omega} \cdot \langle \mathbf{v}_\beta \rangle_\eta \, \mathrm{d}A = 0 \tag{2.27}$$

It is important to note here that one must be very careful about the various forms of the continuity equation. Equation 2.25 is a special form of Eq. 5.25 in Chapter V which is analogous to Eq. 2.26. The simplification of Eq. 5.25 to

Eq. 2.25 is achieved because the convective flux at the β–σ interface is negligible; however, this same type of simplification may or may not be permissible for Eq. 2.26 since the flux at the η–ω interface may or may not be negligible.

We are now in a position to use Eq. 2.27 to simplify the flux terms in Eq. 2.20. Since we are willing to treat $\{\langle c\rangle_\eta\}^\eta$ as a constant with respect to integration over V_η and $A_{\eta\omega}$, we can multiply Eq. 2.27 by $\{\langle c\rangle_\eta\}^\eta$ and use the result in Eq. 2.20 to obtain

$$\varepsilon_{\beta\eta}\phi_\eta(1+\mathscr{K}_\eta)\frac{\partial}{\partial t}\{\langle c\rangle_\eta\}^\eta + \phi_\eta\{\langle \mathbf{v}_\beta\rangle_\eta\}^\eta\cdot\nabla\{\langle c\rangle_\eta\}^\eta$$
$$= \nabla\cdot\left[\varepsilon_{\beta\eta}\phi_\eta\mathbf{D}^*_\eta\cdot\left(\nabla\{\langle c\rangle_\eta\}^\eta + \frac{1}{V_\eta}\int_{A_{\eta\omega}}\mathbf{n}_{\eta\omega}\hat{c}_\eta\,dA\right)\right] - \nabla\cdot(\phi_\eta\{\hat{\mathbf{v}}_{\beta\eta}\hat{c}_\eta\}^\eta)$$
$$-\frac{1}{\mathscr{V}_\infty}\int_{A_{\eta\omega}}\mathbf{n}_{\eta\omega}\cdot[\langle \mathbf{v}_\beta\rangle_\eta\hat{c}_\eta - \varepsilon_{\beta\eta}\mathbf{D}^*_\eta\cdot\nabla\langle c\rangle_\eta]\,dA \tag{2.28}$$

We now make use of the decomposition given by Eq. 2.16a to write

$$\frac{1}{\mathscr{V}_\infty}\int_{A_{\eta\omega}}\mathbf{n}_{\eta\omega}\cdot\varepsilon_{\beta\eta}\mathbf{D}^*_\eta\cdot\nabla\langle c\rangle_\eta\,dA = \left(\frac{1}{\mathscr{V}_\infty}\int_{A_{\eta\omega}}\mathbf{n}_{\eta\omega}\,dA\right)\cdot\varepsilon_{\beta\eta}\mathbf{D}^*_\eta\cdot\nabla\{\langle c\rangle_\eta\}^\eta$$
$$+\frac{1}{\mathscr{V}_\infty}\int_{A_{\eta\omega}}\mathbf{n}_{\eta\omega}\cdot\varepsilon_{\beta\eta}\mathbf{D}^*_\eta\cdot\nabla\hat{c}_\eta\,dA \tag{2.29}$$

since $\varepsilon_{\beta\eta}\mathbf{D}^*_\eta\cdot\nabla\{\langle c\rangle_\eta\}^\eta$ undergoes negligible variations over the surface $A_{\omega\eta}$. We can use this result, along with a result from the averaging theorem

$$\frac{1}{\mathscr{V}_\infty}\int_{A_{\eta\omega}}\mathbf{n}_{\eta\omega}\,dA = -\nabla\phi_\eta \tag{2.30}$$

to express Eq. 2.28 as

$$\varepsilon_{\beta\eta}\phi_\eta(1+\mathscr{K}_\eta)\frac{\partial}{\partial t}\{\langle c\rangle_\eta\}^\eta + \phi_\eta\{\langle \mathbf{v}_\beta\rangle_\eta\}^\eta\cdot\nabla\{\langle c\rangle_\eta\}^\eta$$
$$= \nabla\cdot\left[\varepsilon_{\beta\eta}\phi_\eta\mathbf{D}^*_\eta\cdot\left(\nabla\{\langle c\rangle_\eta\}^\eta + \frac{1}{V_\eta}\int_{A_{\eta\omega}}\mathbf{n}_{\eta\omega}\hat{c}_\eta\,dA\right)\right] - \nabla\cdot(\phi_\eta\{\hat{\mathbf{v}}_{\beta\eta}\hat{c}_\eta\}^\eta)$$
$$-\nabla\phi_\eta\cdot(\varepsilon_{\beta\eta}\mathbf{D}^*_\eta\cdot\nabla\{\langle c\rangle_\eta\}^\eta) - \frac{1}{\mathscr{V}_\infty}\int_{A_{\eta\omega}}\mathbf{n}_{\eta\omega}\cdot[\langle \mathbf{v}_\beta\rangle_\eta\hat{c}_\eta - \varepsilon_{\beta\eta}\mathbf{D}^*_\eta\cdot\nabla\hat{c}_\eta]\,dA \tag{2.31}$$

The result for the ω-region can be written by analogy; however, we will ignore the ω-region transport equation for the present and move on to the closure problem which will involve both the η-region and the ω-region.

3. Closure problem for the large-scale, two-equation model

Our development of the closure problem for $\hat{c}_\eta$ and $\hat{c}_\omega$ parallels that given in Sections 3 and 7 of Chapter V. We begin with the boundary value problem stated by Eqs 2.1–2.4 and utilize the decompositions

$$\langle c\rangle_\eta = \{\langle c\rangle_\eta\}^\eta + \hat{c}_\eta, \qquad \langle c\rangle_\omega = \{\langle c\rangle_\omega\}^\omega + \hat{c}_\omega \tag{3.1a}$$

$$\langle \mathbf{v}_\beta\rangle_\eta = \{\langle \mathbf{v}_\beta\rangle_\eta\}^\eta + \hat{\mathbf{v}}_{\beta\eta}, \quad \langle \mathbf{v}_\beta\rangle_\omega = \{\langle \mathbf{v}_\beta\rangle_\omega\}^\omega + \hat{\mathbf{v}}_{\beta\omega} \tag{3.1b}$$

to arrive at

$$\varepsilon_{\beta\eta}(1+\mathscr{K}_\eta)\frac{\partial \hat{c}_\eta}{\partial t} + \langle \mathbf{v}_\beta\rangle_\eta\cdot\nabla\hat{c}_\eta + \hat{\mathbf{v}}_{\beta\eta}\cdot\nabla\{\langle c\rangle_\eta\}^\eta - \nabla\cdot(\varepsilon_{\beta\eta}\mathbf{D}^*_\eta\cdot\nabla\hat{c}_\eta)$$

$$= -\left[\varepsilon_{\beta\eta}(1+\mathscr{K}_\eta)\frac{\partial}{\partial t}\{\langle c\rangle_\eta\}^\eta + \{\langle \mathbf{v}_\beta\rangle_\eta\}^\eta\cdot\nabla\{\langle c\rangle_\eta\}^\eta - \nabla\cdot(\varepsilon_{\beta\eta}\mathbf{D}^*_\eta\cdot\nabla\{\langle c\rangle_\eta\}^\eta)\right] \tag{3.2}$$

B.C.1
$$\hat{c}_\eta = \hat{c}_\omega + \{\langle c\rangle_\omega\}^\omega - \{\langle c\rangle_\eta\}^\eta, \quad \text{at } A_{\eta\omega} \tag{3.3}$$

B.C.2
$$-\mathbf{n}_{\eta\omega}\cdot(\varepsilon_{\beta\eta}\mathbf{D}^*_\eta\cdot\nabla\hat{c}_\eta) = -\mathbf{n}_{\eta\omega}\cdot(\varepsilon_{\beta\omega}\mathbf{D}^*_\omega\cdot\nabla\hat{c}_\omega)$$
$$-\mathbf{n}_{\eta\omega}\cdot\left(\varepsilon_{\beta\omega}\mathbf{D}^*_\omega\cdot\nabla\{\langle c\rangle_\omega\}^\omega - \varepsilon_{\beta\eta}\mathbf{D}^*_\eta\cdot\nabla\{\langle c\rangle_\eta\}^\eta\right), \quad \text{at } A_{\eta\omega} \tag{3.4}$$

$$\varepsilon_{\beta\omega}(1+\mathscr{K}_\omega)\frac{\partial \hat{c}_\omega}{\partial t} + \langle \mathbf{v}_\beta\rangle_\omega\cdot\nabla\hat{c}_\omega + \hat{\mathbf{v}}_{\beta\omega}\cdot\nabla\{\langle c\rangle_\omega\}^\omega - \nabla\cdot(\varepsilon_{\beta\omega}\mathbf{D}^*_\omega\cdot\nabla\hat{c}_\omega)$$

$$= -\left[\varepsilon_{\beta\omega}(1+\mathscr{K}_\omega)\frac{\partial}{\partial t}\{\langle c\rangle_\omega\}^\omega + \{\langle \mathbf{v}_\beta\rangle_\omega\}^\omega\cdot\nabla\{\langle c\rangle_\omega\} - \nabla\cdot(\varepsilon_{\beta\omega}\mathbf{D}^*_\omega\cdot\nabla\{\langle c\rangle_\omega\}^\omega)\right] \tag{3.5}$$

In our treatment of the boundary condition given by Eq. 2.3, we have used Eq. 2.2 along with the result

$$\mathbf{n}_{\eta\omega}\cdot\langle \mathbf{v}_\beta\rangle_\eta = \mathbf{n}_{\eta\omega}\cdot\langle \mathbf{v}_\beta\rangle_\omega, \quad \text{at } A_{\eta\omega} \tag{3.6}$$

(Quintard and Whitaker, 1987, Appendix A) in order to arrive at Eq. 3.4. The treatment of Eqs 3.2 and 3.5 follows the development of Crapiste *et al.* (1986), and the result for Eq. 3.2 takes the form

$$\varepsilon_{\beta\eta}(1 + \mathscr{K}_\eta)\frac{\partial \hat{c}_\eta}{\partial t} + \langle \mathbf{v}_\beta \rangle_\eta \cdot \nabla \hat{c}_\eta + \hat{\mathbf{v}}_{\beta\eta} \cdot \nabla \{\langle c \rangle_\eta\}^\eta - \nabla \cdot (\varepsilon_{\beta\eta} \mathbf{D}_\eta^* \cdot \nabla \hat{c}_\eta)$$

$$= -\frac{1}{V_\eta}\int_{V_\eta}\left[\varepsilon_{\beta\eta}(1 + \mathscr{K}_\eta)\frac{\partial \hat{c}_\eta}{\partial t} + \langle \mathbf{v}_\beta \rangle_\eta \cdot \nabla \hat{c}_\eta + \hat{\mathbf{v}}_{\beta\eta} \cdot \nabla \{\langle c \rangle_\eta\}^\eta - \nabla \cdot (\varepsilon_{\beta\eta} \mathbf{D}_\eta^* \cdot \nabla \hat{c}_\eta)\right] \mathrm{d}V \tag{3.7}$$

This result is analogous to Eq. 7.4 in Chapter V and we can use the constraints on $\hat{c}_\eta$ and $\hat{\mathbf{v}}_{\beta\eta}$

$$\frac{1}{V_\eta}\int_{V_\eta} \hat{c}_\eta \,\mathrm{d}V = 0, \quad \frac{1}{V_\eta}\int_{V_\eta} \hat{\mathbf{v}}_{\beta\eta} \,\mathrm{d}V = 0 \tag{3.8}$$

to simplify Eq. 3.7 to

$$\varepsilon_{\beta\eta}(1 + \mathscr{K}_\eta)\frac{\partial \hat{c}_\eta}{\partial t} + \langle \mathbf{v}_\beta \rangle_\eta \cdot \nabla \hat{c}_\eta + \hat{\mathbf{v}}_{\beta\eta} \cdot \nabla \{\langle c \rangle_\eta\}^\eta - \nabla \cdot (\varepsilon_{\beta\eta} \mathbf{D}_\eta^* \cdot \nabla \hat{c}_\eta)$$

$$= \frac{1}{V_\eta}\int_{V_\eta} \nabla \cdot [\langle \mathbf{v}_\beta \rangle_\eta \hat{c}_\eta - \varepsilon_{\beta\eta} \mathbf{D}_\eta^* \cdot \nabla \hat{c}_\eta] \,\mathrm{d}V \tag{3.9}$$

Since the closure problem will be solved in some representative unit cell, such as the one illustrated in Fig. 2, the large-scale spatial deviation concentrations and the velocity will be spatially periodic. This can be expressed explicitly as

$$\hat{c}_\eta(\mathbf{r} + \boldsymbol{\ell}_i) = \hat{c}_\eta(\mathbf{r}), \quad \langle \mathbf{v}_\beta \rangle_\eta(\mathbf{r} + \boldsymbol{\ell}_i) = \langle \mathbf{v}_\beta \rangle_\eta(\mathbf{r}), \quad i = 1, 2, 3 \tag{3.10a}$$

$$\hat{c}_\omega(\mathbf{r} + \boldsymbol{\ell}_i) = \hat{c}_\omega(\mathbf{r}), \quad \langle \mathbf{v}_\beta \rangle_\omega(\mathbf{r} + \boldsymbol{\ell}_i) = \langle \mathbf{v}_\beta \rangle_\omega(\mathbf{r}), \quad i = 1, 2, 3 \tag{3.10b}$$

and these periodicity conditions allow us to express the last term in Eq. 3.9 as

$$\frac{1}{V_\eta}\int_{V_\eta} \nabla \cdot [\langle \mathbf{v}_\beta \rangle_\eta \hat{c}_\eta - \varepsilon_{\beta\eta} \mathbf{D}_\eta^* \cdot \nabla \hat{c}_\eta] \,\mathrm{d}V = \frac{1}{V_\eta}\int_{A_{\eta\omega}} \mathbf{n}_{\eta\omega} \cdot (\langle \mathbf{v}_\beta \rangle_\eta \hat{c}_\eta - \varepsilon_{\beta\eta} \mathbf{D}_\eta^* \cdot \nabla \hat{c}_\eta) \,\mathrm{d}A \tag{3.11}$$

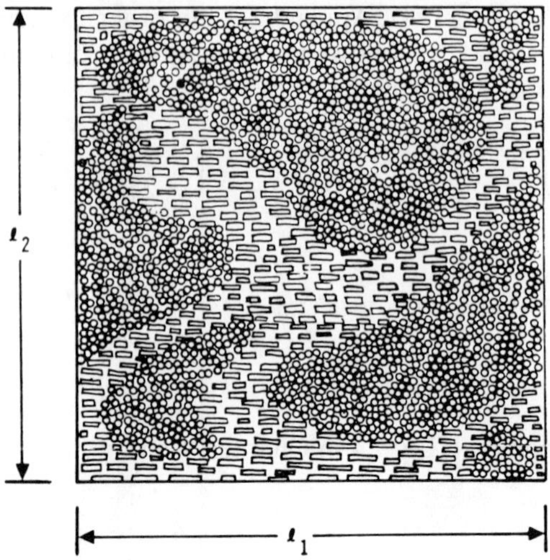

Figure 2. Unit cell for a spatially periodic, heterogeneous porous medium.

We are now in position to re-write the closure problem given by Eqs 3.2–3.5 in the following simplified form

$$\varepsilon_{\beta\eta}(1 + \mathscr{K}_\eta)\frac{\partial \hat{c}_\eta}{\partial t} + \langle \mathbf{v}_\beta \rangle_\eta \cdot \nabla \hat{c}_\eta + \hat{\mathbf{v}}_{\beta\eta} \cdot \nabla \{\langle c \rangle_\eta\}^\eta$$

$$= \nabla \cdot (\varepsilon_{\beta\eta} \mathbf{D}_\eta^* \cdot \nabla \hat{c}_\eta) + \frac{1}{V_\eta}\int_{A_{\eta\omega}} \mathbf{n}_{\eta\omega} \cdot (\langle \mathbf{v}_\beta \rangle_\eta \hat{c}_\eta - \varepsilon_{\beta\eta} \mathbf{D}_\eta^* \cdot \nabla \hat{c}_\eta)\, \mathrm{d}A \tag{3.12}$$

B.C.1
$$\hat{c}_\eta = \hat{c}_\omega + \{\langle c \rangle_\omega\}^\omega - \{\langle c \rangle_\eta\}^\eta, \quad \text{at } A_{\eta\omega} \tag{3.13}$$

B.C.2
$$-\mathbf{n}_{\eta\omega} \cdot (\varepsilon_{\beta\eta} \mathbf{D}_\eta^* \cdot \nabla \hat{c}_\eta) = -\mathbf{n}_{\eta\omega} \cdot (\varepsilon_{\beta\omega} \mathbf{D}_\omega^* \cdot \nabla \hat{c}_\omega)$$

$$-\mathbf{n}_{\eta\omega} \cdot \left(\varepsilon_{\beta\omega} \mathbf{D}_\omega^* \cdot \nabla \{\langle c \rangle_\omega\}^\omega - \varepsilon_{\beta\eta} \mathbf{D}_\eta^* \cdot \nabla \{\langle c \rangle_\eta\}^\eta\right), \quad \text{at } A_{\eta\omega} \tag{3.14}$$

$$\varepsilon_{\beta\omega}(1 + \mathscr{K}_\omega)\frac{\partial \hat{c}_\omega}{\partial t} + \langle \mathbf{v}_\beta \rangle_\omega \cdot \nabla \hat{c}_\omega + \hat{\mathbf{v}}_{\beta\omega} \cdot \nabla \{\langle c \rangle_\omega\}^\omega$$

$$= \nabla \cdot (\varepsilon_{\beta\omega} \mathbf{D}_\omega^* \cdot \nabla \hat{c}_\omega) + \frac{1}{V_\omega}\int_{A_{\omega\eta}} \mathbf{n}_{\omega\eta} \cdot (\langle \mathbf{v}_\beta \rangle_\omega \hat{c}_\omega - \varepsilon_{\beta\omega} \mathbf{D}_\omega^* \cdot \nabla \hat{c}_\omega)\, \mathrm{d}A \tag{3.15}$$

I.C.
$$\hat{c}_\eta = 0, \quad \hat{c}_\omega = 0, \quad t = 0 \tag{3.16}$$

$$\hat{c}_\eta(\mathbf{r} + \boldsymbol{\ell}_i) = \hat{c}_\eta(\mathbf{r}), \quad \hat{c}_\omega(\mathbf{r} + \boldsymbol{\ell}_i) = \hat{c}_\omega(\mathbf{r}), \quad i = 1, 2, 3 \tag{3.17}$$

$$\{\hat{c}_\eta\}^\eta = 0, \quad \{\hat{c}_\omega\}^\omega = 0 \tag{3.18}$$

Here we have added an initial condition to the closure problem, and our hope is that Eq. 3.16 is an acceptable approximation to the actual initial condition which is generally unknown. In writing Eq. 3.16 we have in mind that the *sources* in Eqs 3.12 and 3.15, and in the boundary conditions given by Eqs 3.13 and 3.14, will dominate the evolution of the $\hat{c}_\eta$ and $\hat{c}_\omega$ fields and that the *source* present in the actual initial condition will be unimportant.

Closure variables

Our solution of the closure problem for $\hat{c}_\eta$ and $\hat{c}_\omega$ follows the developments given in Sections 3 and 7 of Chapter V. We express $\hat{c}_\eta$ and $\hat{c}_\omega$ in terms of the *sources* that appear in the boundary value problem given by Eqs 3.12–3.18. This leads to the representations

$$\hat{c}_\eta = \mathbf{F}_\eta \cdot \nabla\{\langle c\rangle_\eta\}^\eta + \mathbf{G}_\eta \cdot \nabla\{\langle c\rangle_\omega\}^\omega + R_\eta(\{\langle c\rangle_\omega\}^\omega - \{\langle c\rangle_\eta\}^\eta) \tag{3.19}$$

$$\hat{c}_\omega = \mathbf{F}_\omega \cdot \nabla\{\langle c\rangle_\eta\}^\eta + \mathbf{G}_\omega \cdot \nabla\{\langle c\rangle_\omega\}^\omega + R_\omega(\{\langle c\rangle_\omega\}^\omega - \{\langle c\rangle_\eta\}^\eta) \tag{3.20}$$

which are identical in form to those encountered in a study of heat transfer in catalytic packed bed reactors (Whitaker, 1989, Eqs 288). There are three boundary value problems for the closure variables, and these three problems take the form

Problem I

$$\varepsilon_{\beta\eta}(1 + \mathscr{K}_\eta)\frac{\partial \mathbf{F}_\eta}{\partial t} + \langle \mathbf{v}_\beta\rangle_\eta \cdot \nabla \mathbf{F}_\eta + \hat{\mathbf{v}}_{\beta\eta} = \varepsilon_{\beta\eta}\mathbf{D}^*_\eta : \nabla\nabla \mathbf{F}_\eta$$

$$+ \frac{1}{V_\eta}\int_{A_{\eta\omega}} \mathbf{n}_{\eta\omega} \cdot (\langle \mathbf{v}_\beta\rangle_\eta \mathbf{F}_\eta - \varepsilon_{\beta\eta}\mathbf{D}^*_\eta \cdot \nabla \mathbf{F}_\eta)\, \mathrm{d}A \tag{3.21a}$$

B.C.1
$$\mathbf{F}_\eta = \mathbf{F}_\omega, \quad \text{at } A_{\eta\omega} \tag{3.21b}$$

B.C.2
$$-\mathbf{n}_{\eta\omega} \cdot \varepsilon_{\beta\eta}\mathbf{D}^*_\eta \cdot \nabla \mathbf{F}_\eta = -\mathbf{n}_{\eta\omega} \cdot \varepsilon_{\beta\omega}\mathbf{D}^*_\omega \cdot \nabla \mathbf{F}_\omega + \mathbf{n}_{\eta\omega} \cdot \varepsilon_{\beta\eta}\mathbf{D}^*_\eta, \quad \text{at } A_{\eta\omega} \tag{3.21c}$$

$$\varepsilon_{\beta\omega}(1 + \mathscr{K}_\omega)\frac{\partial \mathbf{F}_\omega}{\partial t} + \langle \mathbf{v}_\beta\rangle_\omega \cdot \nabla \mathbf{F}_\omega = \varepsilon_{\beta\omega}\mathbf{D}^*_\omega : \nabla\nabla \mathbf{F}_\omega$$

$$+ \frac{1}{V_\omega}\int_{A_{\omega\eta}} \mathbf{n}_{\omega\eta} \cdot (\langle \mathbf{v}_\beta\rangle_\omega \mathbf{F}_\omega - \varepsilon_{\beta\omega}\mathbf{D}^*_\omega \cdot \nabla \mathbf{F}_\omega)\, \mathrm{d}A \tag{3.21d}$$

I.C. $$\mathbf{F}_\eta = 0, \quad \mathbf{F}_\omega = 0, \quad t = 0 \tag{3.21e}$$

$$\mathbf{F}_\eta(\mathbf{r} + \boldsymbol{\ell}_i) = \mathbf{F}_\eta(\mathbf{r}), \quad \mathbf{F}_\omega(\mathbf{r} + \boldsymbol{\ell}_i) = \mathbf{F}_\omega(\mathbf{r}), \quad i = 1, 2, 3 \tag{3.21f}$$

$$\{\mathbf{F}_\eta\}^\eta = 0, \quad \{\mathbf{F}_\omega\}^\omega = 0 \tag{3.21g}$$

Problem II

$$\varepsilon_{\beta\eta}(1 + \mathscr{K}_\eta)\frac{\partial \mathbf{G}_\eta}{\partial t} + \langle \mathbf{v}_\beta \rangle_\eta \cdot \nabla \mathbf{G}_\eta = \varepsilon_{\beta\eta}\mathbf{D}^*_\eta : \nabla\nabla \mathbf{G}_\eta$$

$$+ \frac{1}{V_\eta}\int_{A_{\eta\omega}} \mathbf{n}_{\eta\omega}\cdot(\langle \mathbf{v}_\beta \rangle_\eta \mathbf{G}_\eta - \varepsilon_{\beta\eta}\mathbf{D}^*_\eta \cdot \nabla \mathbf{G}_\eta)\, \mathrm{d}A \tag{3.22a}$$

B.C.1 $$\mathbf{G}_\eta = \mathbf{G}_\omega, \quad \text{at } A_{\eta\omega} \tag{3.22b}$$

B.C.2 $$-\mathbf{n}_{\eta\omega}\cdot\varepsilon_{\beta\eta}\mathbf{D}^*_\eta \cdot \nabla \mathbf{G}_\eta = -\mathbf{n}_{\eta\omega}\cdot\varepsilon_{\beta\omega}\mathbf{D}^*_\omega \cdot \nabla \mathbf{G}_\omega - \mathbf{n}_{\eta\omega}\cdot\varepsilon_{\beta\omega}\mathbf{D}^*_\omega, \quad \text{at } A_{\eta\omega} \tag{3.22c}$$

$$\varepsilon_{\beta\omega}(1 + \mathscr{K}_\omega)\frac{\partial \mathbf{G}_\omega}{\partial t} + \langle \mathbf{v}_\beta \rangle_\omega \cdot \nabla \mathbf{G}_\omega + \hat{\mathbf{v}}_{\beta\omega} = \varepsilon_{\beta\omega}\mathbf{D}^*_\omega : \nabla\nabla \mathbf{G}_\omega$$

$$+ \frac{1}{V_\omega}\int_{A_{\omega\eta}} \mathbf{n}_{\omega\eta}\cdot(\langle \mathbf{v}_\beta \rangle_\omega \mathbf{G}_\omega - \varepsilon_{\beta\omega}\mathbf{D}^*_\omega \cdot \nabla \mathbf{G}_\omega)\, \mathrm{d}A \tag{3.22d}$$

I.C. $$\mathbf{G}_\eta = 0, \quad \mathbf{G}_\omega = 0, \quad t = 0 \tag{3.22e}$$

$$\mathbf{G}_\eta(\mathbf{r} + \boldsymbol{\ell}_i) = \mathbf{G}_\eta(\mathbf{r}), \quad \mathbf{G}_\omega(\mathbf{r} + \boldsymbol{\ell}_i) = \mathbf{G}_\omega(\mathbf{r}), \quad i = 1, 2, 3 \tag{3.22f}$$

$$\{\mathbf{G}_\eta\}^\eta = 0, \quad \{\mathbf{G}_\omega\}^\omega = 0 \tag{3.22g}$$

Problem III

$$\varepsilon_{\beta\eta}(1 + \mathscr{K}_\eta)\frac{\partial \mathbf{R}_\eta}{\partial t} + \langle \mathbf{v}_\beta \rangle_\eta \cdot \nabla \mathbf{R}_\eta = \varepsilon_{\beta\eta}\mathbf{D}^*_\eta : \nabla\nabla \mathbf{R}_\eta$$

$$+ \frac{1}{V_\eta}\int_{A_{\eta\omega}} \mathbf{n}_{\eta\omega}\cdot(\langle \mathbf{v}_\beta \rangle_\eta \mathbf{R}_\eta - \varepsilon_{\beta\eta}\mathbf{D}^*_\eta \cdot \nabla \mathbf{R}_\eta)\, \mathrm{d}A \tag{3.23a}$$

B.C.1 $$\mathbf{R}_\eta = \mathbf{R}_\omega + 1, \quad \text{at } A_{\eta\omega} \tag{3.23b}$$

B.C.2
$$-\mathbf{n}_{\eta\omega}\cdot\varepsilon_{\beta\eta}\mathbf{D}_{\eta}^{*}\cdot\nabla\mathbf{R}_{\eta}=-\mathbf{n}_{\eta\omega}\cdot\varepsilon_{\beta\omega}\mathbf{D}_{\omega}^{*}\cdot\nabla\mathbf{R}_{\omega},\quad \text{at } A_{\eta\omega} \tag{3.23c}$$

$$\varepsilon_{\beta\omega}(1+\mathscr{K}_{\omega})\frac{\partial\mathbf{R}_{\omega}}{\partial t}+\langle\mathbf{v}_{\beta}\rangle_{\omega}\cdot\nabla\mathbf{R}_{\omega}=\varepsilon_{\beta\omega}\mathbf{D}_{\omega}^{*}:\nabla\nabla\mathbf{R}_{\omega}$$

$$+\frac{1}{V_{\omega}}\int_{A_{\omega\eta}}\mathbf{n}_{\omega\eta}\cdot(\langle\mathbf{v}_{\beta}\rangle_{\omega}\mathbf{R}_{\omega}-\varepsilon_{\beta\omega}\mathbf{D}_{\omega}^{*}\cdot\nabla\mathbf{R}_{\omega})\,\mathrm{d}A \tag{3.23d}$$

I.C.
$$\mathbf{R}_{\eta}=0,\quad \mathbf{R}_{\omega}=0,\quad t=0 \tag{3.23e}$$

$$\mathbf{R}_{\eta}(\mathbf{r}+\boldsymbol{\ell}_{i})=\mathbf{R}_{\eta}(\mathbf{r}),\quad \mathbf{R}_{\omega}(\mathbf{r}+\boldsymbol{\ell}_{i})=\mathbf{R}_{\omega}(\mathbf{r}),\quad i=1,2,3 \tag{3.23f}$$

$$\{\mathbf{R}_{\eta}\}^{\eta}=0,\quad \{\mathbf{R}_{\omega}\}^{\omega}=0 \tag{3.23g}$$

In the development of these three closure problems, the *source terms*, $\nabla\{\langle c\rangle_{\eta}\}^{\eta}$, $\nabla\{\langle c\rangle_{\omega}\}^{\omega}$, and $(\{\langle c\rangle_{\omega}\}^{\omega}-\{\langle c\rangle_{\eta}\}^{\eta})$ are treated as constants with respect to temporal and spatial differentiation. This is based on the idea that the characteristic time and length associated with the averaged quantities are much larger than the characteristic time and length associated with $\hat{c}_{\eta}$ and $\hat{c}_{\omega}$. When this is not the case, the entire analysis becomes much more complex, and one must use the type of analysis presented by Quintard and Whitaker (1990) for two-phase flow in heterogeneous porous media.

While the closure problem presented by Eqs 3.21–3.23 appears to be enormously complex, one must keep in mind that the *form* of each separate problem is the same, thus minor modifications to a single computer code will suffice to produce calculated results for all six closure variables. If the time is constrained by

$$t\gg\frac{(1+\mathscr{K}_{\eta})\ell_{\eta}^{2}}{D_{\eta}^{*}},\quad t\gg\frac{(1+\mathscr{K}_{\omega})\ell_{\omega}^{2}}{D_{\omega}^{*}} \tag{3.24}$$

the closure problems become quasi-steady; however, these constraints may be difficult to satisfy.

4. Closed form of the two-equation model

In order to develop the closed form of the ω- and η-region transport equations, we begin with Eq. 2.20 which describes the transport in the η-region

$$\varepsilon_{\beta\eta}\phi_{\eta}(1+\mathscr{K}_{\eta})\frac{\partial}{\partial t}\{\langle c\rangle_{\eta}\}^{\eta}+\nabla\cdot(\phi_{\eta}\{\langle\mathbf{v}_{\beta}\rangle_{\eta}\}^{\eta}\{\langle c\rangle_{\eta}\}^{\eta})$$

$$
= \nabla \cdot \left[\varepsilon_{\beta\eta}\phi_\eta \mathbf{D}^*_\eta \cdot \left(\nabla\{\langle c\rangle_\eta\}^\eta + \frac{1}{V_\eta}\int_{A_{\eta\omega}} \mathbf{n}_{\eta\omega}\hat{c}_\eta \, \mathrm{d}A \right)\right] - \nabla\cdot(\phi_\eta\{\hat{\mathbf{v}}_{\beta\eta}\hat{c}_\eta\}^\eta)
$$

$$
- \frac{1}{\mathscr{V}_\infty}\int_{A_{\eta\omega}} \mathbf{n}_{\eta\omega}\cdot[\langle \mathbf{v}_\beta\rangle_\eta \langle c\rangle_\eta - \varepsilon_{\beta\eta}\mathbf{D}^*_\eta\cdot\nabla\langle c\rangle_\eta]\,\mathrm{d}A \tag{4.1}
$$

The first term in this equation represents the accumulation of the chemical species under consideration with the coefficient $\mathscr{K}_\eta$ accounting for the adsorption in the σ-phase. The second term represents the convective transport, while the third term represents the 'diffusive' transport. In reality, the third term represents both *diffusive and dispersive* transport at the second level of averaging; however, we reserve the word *dispersive* to describe the transport represented by the fourth term in Eq. 4.1. The final term in Eq. 4.1 represents the net flux of the chemical species leaving the η-region and entering the ω-region.

In the development presented in Section 2 we showed that the inter-region flux could be expressed as

$$
\frac{1}{V_\eta}\int_{A_{\eta\omega}} \mathbf{n}_{\eta\omega}\cdot[\langle \mathbf{v}_\beta\rangle_\eta\langle c\rangle_\eta - \varepsilon_{\beta\eta}\mathbf{D}^*_\eta\cdot\nabla\langle c\rangle_\eta]\,\mathrm{d}A = \nabla\phi_\eta\cdot(\varepsilon_{\beta\eta}\mathbf{D}^*_\eta\cdot\nabla\{\langle c\rangle_\eta\}^\eta)
$$

$$
- \{\langle c\rangle_\eta\}^\eta[\nabla\cdot(\phi_\eta\{\langle \mathbf{v}_\beta\rangle_\eta\}^\eta)] + \frac{1}{V_\eta}\int_{A_{\eta\omega}} \mathbf{n}_{\eta\omega}\cdot[\langle \mathbf{v}_\beta\rangle_\eta\hat{c}_\eta - \varepsilon_{\beta\eta}\mathbf{D}^*_\eta\cdot\nabla\hat{c}_\eta]\,\mathrm{d}A \tag{4.2}
$$

and the area integral on the right-hand side of this result will be evaluated on the basis of the closure problem presented in Section 3. Since the closure problem is based on the spatially periodic model of a heterogeneous porous medium, and for such a model we require that $\nabla\phi_\eta = 0$, one could question the validity of retaining the first term on the right-hand side of Eq. 4.2 since that term would be set equal to zero in the closure problem. From our point of view it seems best to use Eq. 4.2 in Eq. 4.1 to obtain

$$
\varepsilon_{\beta\eta}\phi_\eta(1 + \mathscr{K}_\eta)\frac{\partial}{\partial t}\{\langle c\rangle_\eta\}^\eta + \phi_\eta\{\langle \mathbf{v}_\beta\rangle_\eta\}^\eta\cdot\nabla\{\langle c\rangle_\eta\}^\eta
$$

$$
= \nabla \cdot \left[\varepsilon_{\beta\eta}\phi_\eta \mathbf{D}^*_\eta \cdot \left(\nabla\{\langle c\rangle_\eta\}^\eta + \frac{1}{V_\eta}\int_{A_{\eta\omega}} \mathbf{n}_{\eta\omega}\hat{c}_\eta \, \mathrm{d}A \right)\right] - \nabla\phi_\eta\cdot(\varepsilon_{\beta\eta}\mathbf{D}^*_\eta\cdot\nabla\{\langle c\rangle_\eta\}^\eta)
$$

$$
- \nabla\cdot(\phi_\eta\{\hat{\mathbf{v}}_{\beta\eta}\hat{c}_\eta\}^\eta) - \frac{1}{\mathscr{V}_\infty}\int_{A_{\eta\omega}} \mathbf{n}_{\eta\omega}\cdot[\langle \mathbf{v}_\beta\rangle_\eta\hat{c}_\eta - \varepsilon_{\beta\eta}\mathbf{D}^*_\eta\cdot\nabla\hat{c}_\eta]\,\mathrm{d}A \tag{4.3}
$$

with the idea that this result is valid subject to a series of length scale constraints that began with Eq. 2.22 in the previous chapter and ended with Eq. 2.21 in this chapter. All that remains at this point is to develop a mathematically consistent representation for $\hat{c}_\eta$ and to develop a method of accurately calculating $\hat{c}_\eta$. This led us to the closure problem described in Section 3 and the use of a spatially periodic model for the solution of the closure problem.

We are now in a position to use Eq. 3.19 in Eq. 4.3 to obtain the closed form for the η-region transport equation.

$$\varepsilon_{\beta\eta}\phi_\eta(1+\mathscr{K}_\eta)\frac{\partial}{\partial t}\{\langle c\rangle_\eta\}^\eta + \phi_\eta\{\langle \mathbf{v}_\beta\rangle_\eta\}^\eta\cdot\nabla\{\langle c\rangle_\eta\}^\eta + \mathbf{u}_{\eta\eta}\cdot\nabla\{\langle c\rangle_\eta\}^\eta$$
$$+ \mathbf{u}_{\eta\omega}\cdot\nabla\{\langle c\rangle_\omega\}^\omega + \nabla\cdot(\mathbf{v}_{\eta\eta}\{\langle c\rangle_\eta\}^\eta) + \nabla\cdot(\mathbf{v}_{\eta\omega}\{\langle c\rangle_\omega\}^\omega)$$
$$= \nabla\cdot(\phi_\eta\mathbf{D}^{**}_{\eta\eta}\cdot\nabla\{\langle c\rangle_\eta\}^\eta) + \nabla\cdot(\phi_\eta\mathbf{D}^{**}_{\eta\omega}\cdot\nabla\{\langle c\rangle_\omega\}^\omega)$$
$$+\left(\frac{A_{\eta\omega}}{\mathscr{V}_\infty}\right)k_{\eta\omega}(\{\langle c\rangle_\omega\}^\omega - \{\langle c\rangle_\eta\}^\eta) \tag{4.4}$$

The transport coefficients can all be determined theoretically on the basis of solutions of the three closure problems given by Eqs 3.21–3.23. The various vector and tensor transport coefficients are defined by

$$\mathbf{u}_{\eta\eta} = \nabla\phi_\eta\cdot\varepsilon_{\beta\eta}\mathbf{D}^*_\eta + \frac{1}{\mathscr{V}_\infty}\int_{A_{\eta\omega}}\mathbf{n}_{\eta\omega}\cdot(\langle\mathbf{v}_\beta\rangle_\eta\mathbf{F}_\eta - \varepsilon_{\beta\eta}\mathbf{D}^*_\eta\cdot\nabla\mathbf{F}_\eta)\,\mathrm{d}A \tag{4.5}$$

$$\mathbf{u}_{\eta\omega} = \frac{1}{\mathscr{V}_\infty}\int_{A_{\eta\omega}}\mathbf{n}_{\eta\omega}\cdot(\langle\mathbf{v}_\beta\rangle_\eta\mathbf{G}_\eta - \varepsilon_{\beta\eta}\mathbf{D}^*_\eta\cdot\nabla\mathbf{G}_\eta)\,\mathrm{d}A \tag{4.6}$$

$$\mathbf{v}_{\eta\eta} = -\phi_\eta\{\hat{\mathbf{v}}_{\beta\eta}R_\eta\}^\eta + \varepsilon_{\beta\eta}\phi_\eta\mathbf{D}^*_\eta\cdot\left(\frac{1}{V_\eta}\int_{A_{\eta\omega}}\mathbf{n}_{\eta\omega}R_\eta\,\mathrm{d}A\right) \tag{4.7}$$

$$\mathbf{v}_{\eta\omega} = \phi_\eta\{\hat{\mathbf{v}}_{\beta\eta}R_\eta\}^\eta - \varepsilon_{\beta\eta}\phi_\eta\mathbf{D}^*_\eta\cdot\left(\frac{1}{V_\eta}\int_{A_{\eta\omega}}\mathbf{n}_{\eta\omega}R_\eta\,\mathrm{d}A\right) \tag{4.8}$$

$$\phi_\eta\mathbf{D}^{**}_{\eta\eta} = \varepsilon_{\beta\eta}\phi_\eta\mathbf{D}^*_\eta\cdot\left(\mathbf{I} + \frac{1}{V_\eta}\int_{A_{\eta\omega}}\mathbf{n}_{\eta\omega}\mathbf{F}_\eta\,\mathrm{d}A\right) - \phi_\eta\{\hat{\mathbf{v}}_{\beta\eta}\mathbf{F}_\eta\}^\eta \tag{4.9}$$

$$\phi_\eta\mathbf{D}^{**}_{\eta\omega} = \varepsilon_{\beta\eta}\phi_\eta\mathbf{D}^*_\eta\cdot\left(\frac{1}{V_\eta}\int_{A_{\eta\omega}}\mathbf{n}_{\eta\omega}\mathbf{G}_\eta\,\mathrm{d}A\right) - \phi_\eta\{\hat{\mathbf{v}}_{\beta\eta}\mathbf{G}_\eta\}^\eta \tag{4.10}$$

$$\left(\frac{A_{\eta\omega}}{\mathscr{V}_\infty}\right)k_{\eta\omega} = -\frac{1}{\mathscr{V}_\infty}\int_{A_{\eta\omega}} \mathbf{n}_{\eta\omega}\cdot(\langle \mathbf{v}_\beta\rangle_\eta R_\eta - \varepsilon_{\beta\eta}\mathbf{D}^*_\eta\cdot\nabla R_\eta)\,\mathrm{d}A \tag{4.11}$$

In Eq. 4.4, the first and second terms represent accumulation and convection and these terms are found in any model of transport of a chemical species in porous media. The first term on the right-hand side of Eq. 4.4 represents the traditional dispersive transport and the last term represents the exchange term that is always part of any two-region model. The non-traditional terms in Eq. 4.4 are generally ignored since little is known about the magnitude of the associated vector and tensor coefficients, and the experimental determination of these coefficients is nearly impossible at this time. Certainly the value of the approach put forth in this chapter is that the transport coefficients can be predicted theoretically, thus we are not in the position of extracting coefficients from experimental studies. Instead we are in a position to compare theory and experiment in order to *test the theory.*

An examination of the exchange coefficient, $k_{\eta\omega}$, defined by Eq. 4.11 indicates that it could be quite sensitive to the flow field. This means that we must be able to calculate $\langle \mathbf{v}_\beta\rangle_\eta$ for some representative unit cell, such as the one shown in Fig. 2, and we must be able to calculate $\{\langle \mathbf{v}_\beta\rangle_\eta\}^\eta$ in order to evaluate the convective transport term in Eq. 4.4. These two velocities are related by Eq. 3.1b which we repeat here as

$$\langle \mathbf{v}_\beta\rangle_\eta = \{\langle \mathbf{v}_\beta\rangle_\eta\}^\eta + \hat{\mathbf{v}}_{\beta\eta} \tag{4.12}$$

Clearly the problem under consideration requires the knowledge of $\{\langle \mathbf{v}_\beta\rangle_\eta\}^\eta$ and $\hat{\mathbf{v}}_{\beta\eta}$. This means that the momentum transport problem, *and the associated closure problem*, must be solved simultaneously with Eq. 4.4 and the closure problem given by Eqs 3.21–3.23.

The ω-region transport equation can be extracted starting with Eq. 2.22 and following the steps that led from Eq. 2.20 to Eq. 2.31. This yields

$$\varepsilon_{\beta\omega}\phi_\omega(1+\mathscr{K}_\omega)\frac{\partial}{\partial t}\{\langle c\rangle_\omega\}^\omega + \phi_\omega\{\langle \mathbf{v}_\beta\rangle_\omega\}^\omega\cdot\nabla\{\langle c\rangle_\omega\}^\omega$$

$$= \nabla\cdot\left[\varepsilon_{\beta\omega}\phi_\omega\mathbf{D}^*_\omega\cdot\left(\nabla\{\langle c\rangle_\omega\}^\omega + \frac{1}{V_\omega}\int_{A_{\eta\omega}}\mathbf{n}_{\eta\omega}\hat{c}_\omega\,\mathrm{d}A\right)\right] - \nabla\cdot(\phi_\omega\{\hat{\mathbf{v}}_{\beta\omega}\hat{c}_\omega\}^\omega)$$

$$- \nabla\phi_\omega\cdot(\varepsilon_{\beta\omega}\mathbf{D}^*_\omega\cdot\nabla\{\langle c_\omega\rangle\}^\omega)\frac{1}{\mathscr{V}_\infty}\int_{A_{\omega\eta}}\mathbf{n}_{\omega\eta}\cdot[\langle \mathbf{v}_\beta\rangle_\omega\hat{c}_\omega - \varepsilon_{\beta\omega}\mathbf{D}^*_\omega\cdot\nabla\hat{c}_\omega]\,\mathrm{d}A \tag{4.13}$$

and when the representation for $\hat{c}_\omega$ given by Eq. 3.20 is utilized we find the following closed form for the ω-region transport equation.

$$
\begin{aligned}
&\varepsilon_{\beta\omega}\phi_\omega(1+\mathscr{K}_\omega)\frac{\partial}{\partial t}\{\langle c\rangle_\omega\}^\omega + \phi_\omega\{\langle \mathbf{v}_\beta\rangle_\omega\}^\omega\cdot\nabla\{\langle c\rangle_\omega\}^\omega + \mathbf{u}_{\omega\omega}\cdot\nabla\{\langle c\rangle_\omega\}^\omega \\
&\qquad + \mathbf{u}_{\omega\eta}\cdot\nabla\{\langle c\rangle_\eta\}^\eta + \nabla\cdot(\mathbf{v}_{\omega\omega}\{\langle c\rangle_\omega\}^\omega) + \nabla\cdot(\mathbf{v}_{\omega\eta}\{\langle c\rangle_\eta\}^\eta) \\
&\qquad = \nabla\cdot\left(\phi_\omega\mathbf{D}^{**}_{\omega\omega}\cdot\nabla\{\langle c\rangle_\omega\}^\omega\right) + \nabla\cdot\left(\phi_\omega\mathbf{D}^{**}_{\omega\eta}\cdot\nabla\{\langle c\rangle_\eta\}^\eta\right) \\
&\qquad + \left(\frac{A_{\omega\eta}}{\mathscr{V}_\infty}\right)k_{\omega\eta}\left(\{\langle c\rangle_\eta\}^\eta - \{\langle c\rangle_\omega\}^\omega\right)
\end{aligned}
\tag{4.14}
$$

The vector and tensor transport coefficients can be determined by solving the closure problems given by Eqs 3.21–3.23, and these coefficients are expressed in terms of the closure variables according to

$$\mathbf{u}_{\omega\omega} = \nabla\phi_\omega\cdot\varepsilon_{\beta\omega}\mathbf{D}^*_\omega + \frac{1}{\mathscr{V}_\infty}\int_{A_{\omega\eta}}\mathbf{n}_{\omega\eta}\cdot(\langle\mathbf{v}_\beta\rangle_\omega\mathbf{G}_\omega - \varepsilon_{\beta\omega}\mathbf{D}^*_\omega\cdot\nabla\mathbf{G}_\omega)\,\mathrm{d}A \tag{4.15}$$

$$\mathbf{u}_{\omega\eta} = \frac{1}{\mathscr{V}_\infty}\int_{A_{\omega\eta}}\mathbf{n}_{\omega\eta}\cdot(\langle\mathbf{v}_\beta\rangle_\omega\mathbf{F}_\omega - \varepsilon_{\beta\omega}\mathbf{D}^*_\omega\cdot\nabla\mathbf{F}_\omega)\,\mathrm{d}A \tag{4.16}$$

$$\mathbf{v}_{\omega\omega} = \phi_\omega\{\hat{\mathbf{v}}_{\beta\omega}R_\omega\}^\omega - \varepsilon_{\beta\omega}\phi_\omega\mathbf{D}^*_\omega\cdot\left(\frac{1}{V_\omega}\int_{A_{\omega\eta}}\mathbf{n}_{\omega\eta}R_\omega\,\mathrm{d}A\right) \tag{4.17}$$

$$\mathbf{v}_{\omega\eta} = -\phi_\omega\{\hat{\mathbf{v}}_{\beta\omega}R_\omega\}^\omega + \varepsilon_{\beta\omega}\phi_\omega\mathbf{D}^*_\omega\cdot\left(\frac{1}{V_\omega}\int_{A_{\omega\eta}}\mathbf{n}_{\omega\eta}R_\omega\,\mathrm{d}A\right) \tag{4.18}$$

$$\phi_\omega\mathbf{D}^{**}_{\omega\omega} = \varepsilon_{\beta\omega}\phi_\omega\mathbf{D}^*_\omega\cdot\left(\mathbf{I} + \frac{1}{V_\omega}\int_{A_{\omega\eta}}\mathbf{n}_{\omega\eta}\mathbf{G}_\omega\,\mathrm{d}A\right) - \phi_\omega\{\hat{\mathbf{v}}_{\beta\omega}\mathbf{G}_\omega\}^\omega \tag{4.19}$$

$$\phi_\omega\mathbf{D}^{**}_{\omega\eta} = \varepsilon_{\beta\omega}\phi_\omega\mathbf{D}^*_\omega\cdot\left(\frac{1}{V_\omega}\int_{A_{\omega\eta}}\mathbf{n}_{\omega\eta}\mathbf{F}_\omega\,\mathrm{d}A\right) - \phi_\omega\{\hat{\mathbf{v}}_{\beta\omega}\mathbf{F}_\omega\}^\omega \tag{4.20}$$

$$\left(\frac{A_{\eta\omega}}{\mathscr{V}_\infty}\right)k_{\omega\eta} - \frac{1}{\mathscr{V}_\infty}\int_{A_{\omega\eta}}\mathbf{n}_{\omega\eta}\cdot(\langle\mathbf{v}_\beta\rangle_\omega R_\omega - \varepsilon_{\beta\omega}\mathbf{D}^*_\omega\cdot\nabla R_\omega)\,\mathrm{d}A \tag{4.21}$$

At this point we should remark that the inter-region flux terms in Eqs 2.20 and 2.22 are *equal and opposite* since these terms are expressed in terms of the local volume-averaged velocities and concentrations. This is *not the* case for the last two terms in Eqs 2.31 and 4.13 since these terms contain only a portion of the inter-region flux. This is made clear by Eq. 4.2 in which the inter-region flux is decomposed into three parts, and it is only the third part of that decomposition that has the *appearance* of an inter-region flux. One must keep in mind that *only a portion* of that third part, which appears as the last term in Eqs 4.13, is proportional to the concentration difference $\{\langle c\rangle_\omega\}^\omega - \{\langle c\rangle_\eta\}^\eta$. Because of this, we should not expect $k_{\eta\omega}$ to be equal to $k_{\omega\eta}$.

5. Passive dispersion in heterogeneous porous media: comparison between theory and experiment for the one-equation model

While neither definitive experimental results nor theoretical results exist for the two-equation model of diffusion, adsorption and dispersion in heterogeneous porous media, some results of this type are available for *passive* transport. In our original study of dispersion in heterogeneous porous media (Plumb and Whitaker, 1988a, b), we assumed that a one-equation model would be sufficient for the analysis of the passive case ($\mathbf{D}_\sigma = 0$ in Eq. 7.57 of Chapter V). Under these circumstances, the starting point for large-scale averaging given by Eqs 2.1–2.4 is replaced by

$$\varepsilon_{\beta\eta}\frac{\partial\langle c_\beta\rangle_\eta^\beta}{\partial t} + \langle \mathbf{v}_\beta\rangle_\eta\cdot\nabla\langle c_\beta\rangle_\eta^\beta = \nabla\cdot(\varepsilon_{\beta\eta}\mathbf{D}_\eta^*\cdot\nabla\langle c_\beta\rangle_\eta^\beta) \tag{5.1}$$

B.C.1
$$\langle c_\beta\rangle_\eta^\beta = \langle c_\beta\rangle_\omega^\beta, \quad \text{at } A_{\eta\omega} \tag{5.2}$$

B.C.2
$$\mathbf{n}_{\eta\omega}\cdot(\langle \mathbf{v}_\beta\rangle_\eta\langle c_\beta\rangle_\eta^\beta - \varepsilon_{\beta\eta}\mathbf{D}_\eta^*\cdot\nabla\langle c_\beta\rangle_\eta^\beta)$$

$$= \mathbf{n}_{\eta\omega}\cdot\big(\langle \mathbf{v}_\beta\rangle_\omega\langle c_\beta\rangle_\omega^\beta - \varepsilon_{\beta\omega}\mathbf{D}_\omega^*\cdot\nabla\langle c_\beta\rangle_\omega^\beta\big), \quad \text{at } A_{\eta\omega} \tag{5.3}$$

$$\varepsilon_{\beta\omega}\frac{\partial\langle c_\beta\rangle_\omega^\beta}{\partial t} + \langle \mathbf{v}_\beta\rangle_\omega\cdot\nabla\langle c_\beta\rangle_\omega^\beta = \nabla\cdot(\varepsilon_{\beta\omega}\mathbf{D}_\omega^*\cdot\nabla\langle c_\beta\rangle_\omega^\beta) \tag{5.4}$$

In our original theoretical development (Plumb and Whitaker, 1988a), Eqs 5.1–5.4 are represented by Eq. 31 with the continuity of flux and concentration at the inter-region boundaries being understood. In fact, the original large-scale spatial smoothing was not limited to a two-region model; however, all of our theoretical calculations were restricted to the η–ω system shown in Fig. 1. Thus, in terms of available theoretical results, Eqs 5.1–5.4 are representative of our original study of passive transport.

To develop the large-scale averaged one-equation model, one begins with the continuity equation

$$\nabla \cdot \langle \mathbf{v}_\beta \rangle = 0 \tag{5.5}$$

so that Eqs 5.1 and 5.4 can be expressed as

$$\varepsilon_\beta \frac{\partial \langle c_\beta \rangle^\beta}{\partial t} + \nabla \cdot (\langle \mathbf{v}_\beta \rangle \langle c_\beta \rangle^\beta) = \nabla \cdot (\varepsilon_\beta \mathbf{D}^* \cdot \nabla \langle c_\beta \rangle^\beta) \tag{5.6}$$

The large-scale average is defined by

$$\{\psi\} = \frac{1}{\mathscr{V}_\infty} \int_{\mathscr{V}_\infty} \psi \, \mathrm{d}V, \quad \text{large-scale average} \tag{5.7}$$

and the averaged forms of Eqs 5.5 and 5.6 are written as

$$\{\nabla \cdot \langle \mathbf{v}_\beta \rangle\} = 0 \tag{5.8}$$

$$\left\{ \varepsilon_\beta \frac{\partial \langle c_\beta \rangle^\beta}{\partial t} \right\} + \{\nabla \cdot (\langle \mathbf{v}_\beta \rangle \langle c_\beta \rangle^\beta)\} = \{\nabla \cdot (\varepsilon_\beta \mathbf{D}^* \cdot \nabla \langle c_\beta \rangle^\beta)\} \tag{5.9}$$

In addition to assuming that the normal components of the flux and concentration are continuous, as indicated by Eqs 5.2 and 5.3, we also make use of Eq. 3.6 for the velocity field. Under these circumstances the averaging theorem can be used in order to express Eqs 5.8 and 5.9 as

$$\nabla \cdot \{\langle \mathbf{v}_\beta \rangle\} = 0 \tag{5.10}$$

$$\left\{ \varepsilon_\beta \frac{\partial \langle c_\beta \rangle^\beta}{\partial t} \right\} + \nabla \cdot \{\langle \mathbf{v}_\beta \rangle \langle c_\beta \rangle^\beta\} = \nabla \cdot \{\varepsilon_\beta \mathbf{D}^* \cdot \nabla \langle c_\beta \rangle^\beta\} \tag{5.11}$$

Obviously the first of these can be used to express the second as

$$\left\{ \varepsilon_\beta \frac{\partial \langle c_\beta \rangle^\beta}{\partial t} \right\} + \{\langle \mathbf{v}_\beta \rangle\} \cdot \nabla \{\langle c_\beta \rangle^\beta\} = \nabla \cdot \{\varepsilon_\beta \mathbf{D}^* \cdot \nabla \langle c_\beta \rangle^\beta\} \tag{5.12}$$

To deal with the various products that appear in this result, we made use of the following decompositions (Plumb and Whitaker, 1988a, Section 3)

$$\varepsilon_\beta = \{\varepsilon_\beta\} + \hat{\varepsilon}_\beta \tag{5.13}$$

$$\langle c_\beta \rangle^\beta = \{\langle c_\beta \rangle^\beta\} + \hat{c}_\beta \tag{5.14}$$

$$\mathbf{D}^* = \{\mathbf{D}^*\} + \hat{\mathbf{D}} \tag{5.15}$$

and on the basis of the length-scale constraint

$$\left(\frac{R_0}{\mathscr{L}}\right)^2 \ll 1 \tag{5.16}$$

we assumed that the average of all the deviations was zero

$$\{\hat{\varepsilon}_\beta\} = 0, \quad \{\hat{c}_\beta\} = 0, \quad \{\hat{\mathbf{D}}^*\} = 0 \tag{5.17}$$

In retrospect it seems clear that we should have dealt with the velocity in terms of only $\langle \mathbf{v}_\beta \rangle$ and $\{\langle \mathbf{v}_\beta \rangle\}$; however, we chose to work with the intrinsic phase average, $\langle \mathbf{v}_\beta \rangle^\beta$, which is related to $\langle \mathbf{v}_\beta \rangle$ by

$$\langle \mathbf{v}_\beta \rangle = \varepsilon_\beta \langle \mathbf{v}_\beta \rangle^\beta \tag{5.18}$$

Thus, in addition to Eqs 5.13–5.15, we also used

$$\langle \mathbf{v}_\beta \rangle^\beta = \{\langle \mathbf{v}_\beta \rangle^\beta\} + \hat{\mathbf{v}}_\beta \tag{5.19}$$

and this is not the decomposition used by Quintard and Whitaker (1987, Section 3) in their study of single-phase flow in heterogeneous porous media. One must remember that Eq. 5.19 is a *definition* of the large-scale spatial deviation velocity $\hat{\mathbf{v}}_\beta$, thus there is no question of its correctness. However, there is a question of efficacy, and it seems clear at this time that one should avoid the intrinsic phase average velocity $\langle \mathbf{v}_\beta \rangle^\beta$ *at this level* and make use of decompositions of the form given by Eq. 2.16b.

When Eqs 5.13–5.19 are used in Eq. 5.12, the large-scale averaged transport equation takes the form

$$\begin{aligned}
&\{\varepsilon_\beta\} \frac{\partial}{\partial t} \{\langle c_\beta \rangle^\beta\} + \left[\{\varepsilon_\beta\}\{\langle \mathbf{v}_\beta \rangle^\beta\} + \{\hat{\varepsilon}_\beta \hat{\mathbf{v}}_\beta\}\right] \cdot \nabla \{\langle c_\beta \rangle^\beta\} \\
&+ \frac{\partial}{\partial t} \{\hat{\varepsilon}_\beta \hat{c}_\beta\} + \nabla \cdot \left[\{\hat{\varepsilon}_\beta \hat{c}_\beta\}\{\langle \mathbf{v}_\beta \rangle^\beta\} + \{\varepsilon_\beta\}\{\hat{\mathbf{v}}_\beta \hat{c}_\beta\} + \{\hat{\varepsilon}_\beta \hat{\mathbf{v}}_\beta \hat{c}_\beta\}\right] \\
&= \nabla \cdot \left[(\{\varepsilon_\beta\}\{\mathbf{D}^*\} + \{\hat{\varepsilon}_\beta \hat{\mathbf{D}}\}) \cdot \nabla \{\langle c_\beta \rangle^\beta\}\right] \\
&+ \nabla \cdot \left[\{\varepsilon_\beta\}\{\hat{\mathbf{D}} \cdot \nabla \hat{c}_\beta\} + \{\mathbf{D}^*\} \cdot \{\hat{\varepsilon}_\beta \nabla \hat{c}_\beta\} + \{\hat{\varepsilon}_\beta \hat{\mathbf{D}} \cdot \nabla \hat{c}_\beta\}\right]
\end{aligned} \tag{5.20}$$

This represents Eq. 51 of our original study (Plumb and Whitaker, 1988a), and a little thought will indicate that it could be simplified by the use of

$$\{\langle \mathbf{v}_\beta \rangle\} = \{\varepsilon_\beta\}\{\langle \mathbf{v}_\beta \rangle^\beta\} + \{\hat{\varepsilon}_\beta \hat{\mathbf{v}}_\beta\} \tag{5.21}$$

The development of the closure problem for $\hat{c}_\beta$ follows the type of analysis given in Sections 3 and 7 in the previous chapter, and Section 3 in this chapter, and it leads to a representation for $\hat{c}_\beta$ of the form

$$\hat{c}_\beta = \mathbf{F} \cdot \nabla\{\langle c_\beta \rangle^\beta\} + S \frac{\partial}{\partial t}\{\langle c_\beta \rangle^\beta\} \tag{5.22}$$

The closure variables, **F** and S, must satisfy the following two boundary value problems.

Problem 1

$$\varepsilon_\beta\left(\frac{\partial \mathbf{F}}{\partial t}\right) + \varepsilon_\beta \langle \mathbf{v}_\beta \rangle^\beta \cdot \nabla \mathbf{F} + \{\varepsilon_\beta\}\hat{\mathbf{v}}_\beta + \hat{\varepsilon}_\beta \langle \mathbf{v}_\beta \rangle^\beta - \{\hat{\varepsilon}_\beta \hat{\mathbf{v}}_\beta\}$$

$$-\left\{\hat{\varepsilon}_\beta \frac{\partial \mathbf{F}}{\partial t}\right\} = \nabla \cdot (\varepsilon_\beta \mathbf{D}^* \cdot \nabla \mathbf{F}) + \nabla \cdot (\hat{\varepsilon}_\beta \mathbf{D}^* + \{\varepsilon_\beta\}\mathbf{D}) \tag{5.23a}$$

$$\mathbf{F}(\mathbf{r} + \boldsymbol{\ell}_i) = \mathbf{F}(\mathbf{r}), \quad i = 1, 2, 3 \tag{5.23b}$$

$$\{\mathbf{F}\} = 0 \tag{5.23c}$$

Problem 2

$$\varepsilon_\beta\left(\frac{\partial S}{\partial t}\right) + \varepsilon_\beta \langle \mathbf{v}_\beta \rangle \cdot \nabla S + \hat{\varepsilon}_\beta - \left\{\hat{\varepsilon}_\beta \frac{\partial S}{\partial t}\right\} = \nabla \cdot (\varepsilon_\beta \mathbf{D}^* \cdot \nabla S) \tag{5.24a}$$

$$S(\mathbf{r} + \boldsymbol{\ell}_i) = S(\mathbf{r}), \quad i = 1, 2, 3 \tag{5.24b}$$

$$\{S\} = 0 \tag{5.24c}$$

These two problems represent the transient form of the original closure problems, and when

$$\frac{D^* t}{\ell_{\mathrm{H}}^2} \gg 1 \tag{5.25}$$

one can argue that the **F** and S fields are quasi-steady. Here one should think of ℓ_{H} as the largest of either ℓ_ω or ℓ_η.

The closed form of Eq. 5.20 can be expressed as

$$\{\varepsilon_\beta\}\frac{\partial}{\partial t}\{\langle c_\beta\rangle^\beta\} + \{\langle \mathbf{v}_\beta\rangle\}\cdot\nabla\{\langle c_\beta\rangle^\beta\} + \{\hat{\varepsilon}_\beta S\}\frac{\partial^2}{\partial t^2}\{\langle c_\beta\rangle^\beta\}$$

$$+ \{\varepsilon_\beta\}\mathbf{D}_t\cdot\nabla\frac{\partial}{\partial t}\{\langle c_\beta\rangle^\beta\} = \nabla\cdot\left(\{\varepsilon_\beta\}\mathbf{D}^{**}\cdot\nabla\{\langle c_\beta\rangle^\beta\}\right) \tag{5.26}$$

in which $\mathbf{D}_t$ is referred to as the *skewness vector*. Neither this mixed derivative term, nor the term involving the second derivative with respect to time, appears in the η- and ω-region transport equations given by Eqs 4.4 and 4.14. The origin of these two terms is in the spatial variation of the porosity which is represented by $\hat{\varepsilon}_\beta$, and it is fairly easy to prove that

$$\mathbf{D}_t = 0, \quad \{\hat{\varepsilon}_\beta S\} = 0 \tag{5.27}$$

when $\hat{\varepsilon}_\beta$ is zero. In our two-equation model we assumed that $\varepsilon_{\beta\eta}$ and $\varepsilon_{\beta\omega}$ were constant within the η- and ω-regions, respectively, and because of this the two non-traditional terms in Eq. 5.26 did not appear in Eqs 4.4 and 4.14.

Extensive solutions of the quasi-steady closure problem were presented in our original study (Plumb and Whitaker, 1988b); however, only one direct comparison between theory and experiment was given. This was associated with the experimental study of Fattah and Hoopes (1985) for passive dispersion in stratified systems. The comparison between theory and experiment is shown in Fig. 3. The predicted results are approximately one order of magnitude larger than the empirical representation of the data given by Fattah and Hoopes as

$$D^{**}_{xx} = 0.82\{\langle v_\beta\rangle^\beta\}^{1.22} \tag{5.28}$$

Here the velocity is in centimetres per second and the dispersion coefficient is in square centimetres per second.

In order to solve the quasi-steady form of the closure problem given by Eqs 5.23, we made use of the experimental values of D^*_{yy} shown in Fig. 13 of Chapter V. Obviously there are errors associated with our solution of the closure problem that are related to the scatter of the data shown in Fig. 13, and at the lowest velocity studied by Fattah and Hoopes this error in D^*_{yy} *could be* responsible for the difference between theory and experiment. The increasing disparity between theory and experiment at higher values of the velocity is reminiscent of the behaviour illustrated in Figs 11 and 12 of Chapter V. In our original discussion of this problem, we considered the constraints imposed by Eqs 8.15 of Chapter V and 5.25 of this chapter. Our conclusion was that detailed studies of the transient closure problems were needed, and this situation remains unchanged.

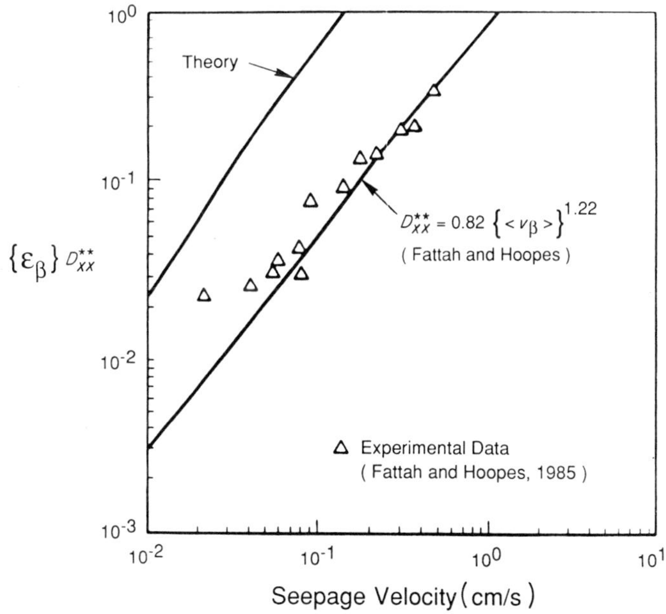

Figure 3. Comparison between theory and experiment for a stratified porous medium.

6. Conclusions

In this work we have considered three levels of averaging: small-scale averaging associated with the process of diffusion and adsorption in porous particles, local-volume averaging associated with convection and diffusion in a porous medium composed of porous particles, and large-scale averaging associated with the spatial smoothing of the effects of local heterogeneities. This type of analysis indicates how physical phenomena that occur at one length scale are manifested in the transport equations at successively higher length scales. It seems clear from this study that the form of the spatially smoothed transport equations becomes increasingly complex with increasing length scale. This complexity can be suppressed by constraining the analysis to one-equation models; however, this is often done in an *ad hoc* manner with no regard for the appropriate length- and time-scale constraints. This is usually accompanied by experimental measurements and adjustable coefficients, and these coefficients may often be nothing more than site-specific, process-specific empirical parameters. The approach taken in this article is that effective transport coefficients should be predicted theoretically and experiments should be performed to test the theory. This places a heavy burden on both the experimentalist and the theoretician.

7. Nomenclature

The equation in which each symbol first appears is given in parentheses. When appropriate, the figure number is given instead of the equation number.

$A_{\eta\omega}$	$A_{\omega\eta}$, the η–ω inter-region area contained within the averaging volume $\mathscr{V}_\infty$ (2.13), m^2.
$\langle c\rangle_\eta$	intrinsic spatial average concentration for the β–σ system in the η-region (2.1), kg mol/m^3.
$\langle c\rangle_\omega$	intrinsic spatial average concentration for the β–σ system in the ω-region (2.2), kg mol/m^3.
$\{\langle c\rangle_\eta\}$	large-scale region average concentration for the η-region (2.5a), kg mol/m^3.
$\{\langle c\rangle_\eta\}^\eta$	large-scale intrinsic region average concentration for the η-region (2.5b), kg mol/m^3.
$\hat{c}_\eta$	$\langle c\rangle_\eta - \{\langle c\rangle_\eta\}^\eta$, η-region spatial deviation concentration for the β–σ system (2.16a), kg mol/m^3.
$\{\langle c\rangle_\omega\}^\omega$	large-scale intrinsic region average concentration for the ω-region (2.22), kg mol/m^3.
$\hat{c}_\omega$	$\langle c\rangle_\omega - \{\langle c\rangle_\omega\}^\omega$, ω-region spatial deviation concentration (2.22), kg mol/m^3.
$\mathbf{D}_\eta^*$	total dispersivity tensor for the β–σ system in the η-region (2.1), m^2/s.
$\mathbf{D}_\omega^*$	total dispersivity tensor for the β–σ system in the ω-region (2.3), m^2/s.
D_ω^*	order of magnitude estimate for $\mathbf{D}_\omega^*$ (3.24), m^2/s.
D_η^*	order of magnitude estimate for $\mathbf{D}_\eta^*$ (3.24), m^2/s.
$\mathbf{D}_{\eta\eta}^{**}$	the η-region total dispersivity tensor (3.9), m^2/s.
$\mathbf{D}_{\eta\omega}^{**}$	the η-region coupling dispersion tensor (4.10), m^2/s.
$\mathbf{D}_{\omega\omega}^{**}$	the ω-region total dispersivity tensor (4.19), m^2/s.
$\mathbf{D}_{\omega\eta}^{**}$	the ω-region coupling dispersion tensor (4.20), m^2/s.
$\hat{\mathbf{D}}$	$\mathbf{D}^* - \{\mathbf{D}^*\}$, spatial deviation dispersion tensor (5.15), m^2/s.
$\mathbf{D}_t$	the skewness vector (5.26), m.
$\mathbf{F}_\eta$	the vector field that maps $\nabla\{\langle c\rangle_\eta\}^\eta$ onto $\hat{c}_\eta$ (3.19), m.
$\mathbf{F}_\omega$	the vector field that maps $\nabla\{\langle c\rangle_\eta\}^\eta$ onto $\hat{c}_\omega$ (3.20), m.
$\mathbf{F}$	the vector field that maps $\nabla\{\langle c_\beta\rangle^\beta\}$ onto $\hat{c}_\beta$ (5.22), m.
$\mathbf{G}_\eta$	the vector field that maps $\nabla\{\langle c\rangle_\omega\}^\omega$ onto $\hat{c}_\eta$ (3.19), m.
$\mathbf{G}_\omega$	the vector field that maps $\nabla\{\langle c\rangle_\omega\}^\omega$ onto $\hat{c}_\omega$ (3.20), m.
$\mathbf{I}$	unit tensor.
$\mathscr{K}_\eta$	dimensionless equilibrium coefficient for the β–σ system in the η-region (2.1).
$\mathscr{K}_\omega$	dimensionless equilibrium coefficient for the β–σ system in the ω-region (2.4).

$k_{\eta\omega}$ the η-region exchange coefficient (4.11), m/s.
$k_{\omega\eta}$ the ω-region exchange coefficient (4.14), m/s.
ℓ_ω characteristic length for the ω-region (Fig. 1), m.
ℓ_η characteristic length for the η-region (Fig. 1), m.
$\boldsymbol{\ell}_i$ lattice vectors describing a spatially periodic porous medium (3.10), m.
$\mathscr{L}$ characteristic length for large-scale averaged quantities (5.16), m.
$\mathscr{L}_\eta$ characteristic length for large-scale averaged quantities associated with the η-region (2.18), m.
ℓ_{H} characteristic length of a local heterogeneity (5.25), m.
$\mathbf{n}_{\eta\omega}$ $-\mathbf{n}_{\omega\eta}$, outwardly directed unit normal vector pointing from the η-region toward the ω-region (2.13).
$\mathbf{r}$ position vector (3.10), m.
R_0 radius of the ω–η system averaging volume $\mathscr{V}_\infty$ (Fig. 1), m.
R_η the scalar field that maps $(\{\langle c\rangle_\omega\}^\omega - \{\langle c\rangle_\eta\}^\eta)$ onto $\hat{c}_\eta$ (3.19).
R_ω the scalar field that maps $(\{\langle c\rangle_\omega\}^\omega - \{\langle c\rangle_\eta\}^\eta)$ onto $\hat{c}_\omega$ (3.20).
S the scalar field that maps $\partial\{\langle c_\beta\rangle^\beta\}/\partial t$ onto $\hat{c}_\beta$ (5.22), s.
t time (2.1), s.
$\mathbf{u}_{\eta\eta}$ a pseudo velocity vector field in the η-region (4.4), m/s.
$\mathbf{u}_{\eta\omega}$ a coupling velocity vector field in the η region (4.4), m/s.
$\mathbf{u}_{\omega\omega}$ a pseudo velocity vector field in the ω-region (4.14), m/s.
$\mathbf{u}_{\omega\eta}$ a coupling velocity vector field in the ω-region (4.14), m/s.
$\mathscr{V}_\infty$ large-scale averaging volume (Fig. 1), m^3.
$\langle \mathbf{v}_\beta\rangle_\eta$ the phase average velocity in the η-region (2.1), m/s.
$\langle \mathbf{v}_\beta\rangle_\omega$ the phase average velocity in the ω-region (2.3), m/s.
$\{\langle \mathbf{v}_\beta\rangle_\eta\}^\eta$ large-scale intrinsic region average velocity for the η-region (2.16b), m/s.
$\{\langle \mathbf{v}_\beta\rangle_\omega\}^\omega$ large-scale intrinsic region average velocity for the ω-region (2.22), m/s.
$\hat{\mathbf{v}}_{\beta\eta}$ $\langle \mathbf{v}_\beta\rangle_\eta - \{\langle \mathbf{v}_\beta\rangle_\eta\}^\eta$, the η-region spatial deviation velocity (2.16b), m/s.
$\hat{\mathbf{v}}_{\beta\omega}$ $\langle \mathbf{v}_\beta\rangle_\omega - \{\langle \mathbf{v}_\beta\rangle_\omega\}^\omega$, the ω-region spatial deviation velocity (2.22), m/s.
$\{\langle \mathbf{v}_\beta\rangle\}$ $\phi_\eta\{\langle \mathbf{v}_\beta\rangle_\eta\}^\eta + \phi_\omega\{\langle \mathbf{v}_\beta\rangle_\omega\}^\omega$, large-scale intrinsic spatial average velocity (2.24), m/s.
V_η volume of the η-region contained within $\mathscr{V}_\infty$ (2.5), m^3.
$\mathbf{v}_{\eta\eta}$ a pseudo velocity vector field in the η-region (4.4), m/s.
$\mathbf{v}_{\eta\omega}$ a coupling velocity vector field in the η-region (4.4), m/s.
$\mathbf{v}_{\omega\omega}$ a pseudo velocity vector field in the ω-region (4.14), m/s.
$\mathbf{v}_{\omega\eta}$ a coupling velocity vector field in the ω-region (4.14), m/s.
V_ω volume of the ω-region contained within $\mathscr{V}_\infty$, m^3.
$\varepsilon_{\beta\eta}$ the β-phase volume fraction in the η-region (2.1).
$\varepsilon_{\beta\omega}$ the β-phase volume fraction in the ω-region (2.4).

$\hat{\varepsilon}_\beta$ large-scale spatial deviation volume fraction (5.13).
ϕ_η volume fraction of the η-region (2.6).
ϕ_ω volume fraction of the ω-region (2.22).

Acknowledgement

This work was supported in part (S.W.) by NSF Grant CBT 8812870.

References

Eidsath, A., Carbonell, R.G., Whitaker, S. and Hermann, L.R. (1983). Dispersion in pulsed systems III. Comparison between theory and experiments for packed beds, *Chem. Engng Sci.* **38**, 1803–1816.

Fattah, Q.N. and Hoopes, J.A. (1985). Dispersion in anisotropic homogeneous porous media, *J. Hydraul. Eng.* **111**, 810–827.

Crapiste, G.H., Rotstein, E. and Whitaker, S. (1986). A general closure scheme for the method of volume averaging, *Chem. Engng. Sci.* **41**, 227–235.

Kim, J.-H., Ochoa, J.A. and Whitaker, S. (1987). Diffusion in anisotropic porous media, *Transport in Porous Media* **2**, 327–356.

Plumb, O.A. and Whitaker, S. (1988a). Dispersion in heterogeneous porous media I: Local volume averaging and large-scale averaging, *Water Resour. Res.* **24**, 913–926.

Plumb, O.A. and Whitaker, S. (1988b). Dispersion in heterogeneous porous media II: Predictions for stratified and two-dimensional spatially periodic systems, *Water Resour. Res.* **24**, 927–938.

Quintard, M. and Whitaker, S. (1987). Ecoulement monophasique en milieu poreux: effet des Hétérogénéités locales, *J. Mécanique théorique et appliquée* **6**, 691–726.

Quintard, M. and Whitaker, S. (1988). Two-phase flow in heterogeneous porous media: The method of large-scale averaging, *Transport in Porous Media* **3**, 357–413.

Quintard, M. and Whitaker, S. (1990). Two-phase flow in heterogeneous porous media I: The influence of large spatial and temporal gradients, *Transport in Porous Media* (in press).

Whitaker, S. (1989). Heat transfer in catalytic packed bed reactors. *In* 'Handbook of Heat and Mass Transfer' (N.P. Cheremisinoff, ed.), Vol. 3, Ch. 10. Gulf Publishers, NJ.

Zanotti, F. and Carbonell, R.G. (1984). Development of transport equations for multiphase systems III: Applications to heat transfer in packed beds, *Chem. Engng. Sci.* **39**, 299–311.

Chapter VII

Dual Porosity Models for Flow in Naturally Fractured Reservoirs

Jim Douglas, Jr and Todd Arbogast
Department of Mathematics, Purdue University, West Lafayette, IN 47907, USA

1. Introduction

1.1 Naturally fractured reservoirs

Within a naturally fractured reservoir there is an interconnected system of fracture planes dividing the porous rock, which will be called the *matrix*, into a collection of blocks. This is somewhat of an idealization, but it is sufficient for the simulation of naturally fractured reservoirs. The fractures, while very thin, have a profound effect on the flow of fluids within the reservoir. Most of the fluid resides in the matrix, where it moves very slowly. When fluid reaches the surface of a matrix block and enters the fractures, it flows comparatively quickly, since the fractures form paths of high permeability. Thus, a naturally fractured reservoir has not a two- but a *three*-scale hierarchy. The smallest scale, of the order of millimetres, is that of the pores. The intermediate and new scale is that of the fracture spacing, which is of the order of metres. Finally, the reservoir length is of the order of kilometres.

In any practical spatial discretization of the reservoir, there will be at least a few matrix blocks in every grid cell and usually many. We cannot, and fortunately we need not, compute the flow of fluids within the individual fractures. We consider dual porosity models which enable us to simulate the flow in naturally fractured reservoirs in a fashion that is both computationally tractable and sufficiently precise for practical purposes.

We begin by recalling some ideas related to single porosity models.

DYNAMICS OF FLUIDS IN HIERARCHICAL POROUS MEDIA
ISBN 0 12-200260-1

1.2 A single porosity model

In a porous medium, define the rock properties porosity and permeability at each point, as usual, as the pore void space per unit bulk volume and the coefficient in Darcy's law (see Eq. 2.2 below) describing the rock's resistance to flow, respectively; see Collins, 1961; Scheidegger, 1974; Bear, 1972; Peaceman, 1977; and Chavent and Jaffré, 1986, as general references on porous media flow. A fractured reservoir could be modelled by allowing the porosity and permeability to vary rapidly and discontinuously over the reservoir; both of these quantities are dramatically greater in the fractures than in the porous rock. However, the computational and data requirements of treating such a model would be too great to approximate the flow in the entire reservoir, and this model must be rejected for practical simulation.

1.3 A dual porosity model

As an alternative, one might try to avoid the discontinuous nature of the porosity and permeability by replacing them locally by their average values. While such a simulation could be done as in the unfractured case, the result would be unsatisfactory, in that correlation to physical observations would not be obtained. The interchange of fluid between the matrix and the fractures must be modeled. The usual technique, due to Pirson (1953), Barenblatt *et al.* (1960) and Warren and Root (1963), is described below.

It has been observed that naturally fractured reservoirs behave as if they possessed two porous structures rather than one (Barenblatt *et al.*, 1960; Warren and Root, 1963). The system of fractures is on a much finer scale than the reservoir as a whole; hence, it can be viewed as a porous structure itself (see Fig. 1). The fractures form the 'void spaces' while the matrix blocks play the

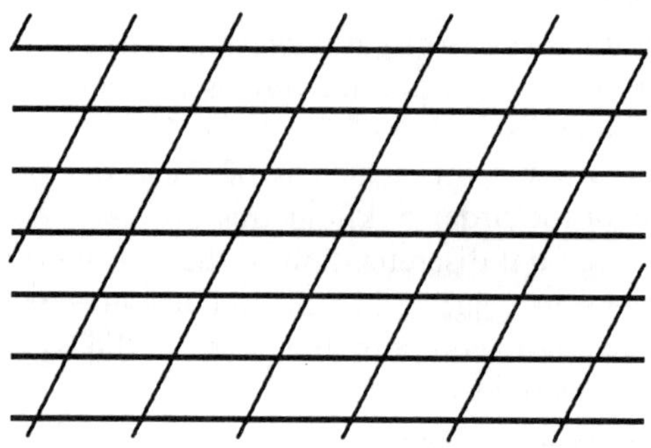

Figure 1. Cross-section of a naturally fractured porous medium, depicting the fracture planes.

role of the 'solid rock'. Of course, there is a difference from the usual porous system: the 'solid part' is in itself permeable.

On the finest scale, the matrix porous structure has a porosity ϕ and a (tensor) permeability $\mathbf{k}$ defined in the usual way, and they are of the order of a tenth and a few millidarcies to a darcy, respectively. (6.3 darcies = 1 ft^2 centipoise/psi/day). On the scale of the fracture system, we can define a fracture porosity Φ as the fracture void space per unit bulk volume, so that Φ is quite small, perhaps of the order of a hundredth. We can also define a fracture permeability tensor $\mathbf{K}$ as the coefficient in the Darcy law that represents resistance to fluid flow *completely within the fracture system*; that is, $\mathbf{K}$ is defined by considering the matrix to be impermeable (Barenblatt *et al.*, 1960; Warren and Root, 1963). It is quite large in magnitude, perhaps of the order of several darcies (Warren and Root, 1963; Kazemi, 1969; de Swaan, 1976).

It is well known how to model the flow of fluids in a single porous structure; a dual porosity model is much more complicated. The main conceptual difficulty lies in the fact that, since the fracture system is viewed as a porous medium, *both* matrix and fracture flow are defined at each point of the matrix. Consequently, we see fluid flow as a combination of general macroscopic motion over the reservoir and flow within matrix blocks, not as actual motion around the matrix blocks (Fig. 2). Since flow in the fractures is much more rapid than in the matrix, we shall assume that fluid does not flow directly from one matrix block to another. Rather, it first flows into the fracture system, and then it can pass into another block or remain in the fractures. We shall determine whether any element of fluid belongs to the matrix or to the fractures by considering the reservoir to be covered by two copies, or 'sheets', equal in extent to the original reservoir. One sheet, which we call Ω, contains the fracture flow. The matrix flow is on the other sheet, called Ω_m; we denote the ith matrix block by Ω_i and assume them to be disjoint, so that $\cup\Omega_i = \Omega_m \subset \Omega$. As a consequence of the assumption that fluid flows from the matrix into the fracture system and vice versa, fluid flows from one sheet to the other. It is this matrix–fracture interaction on the scale of the fracture spacing that must be carefully modeled.

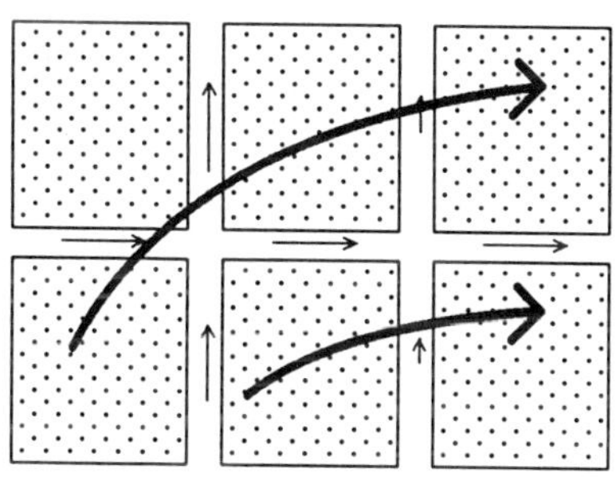

Figure 2. The macroscopic fracture flow.

Its effect on the fracture system will be averaged over the size of an individual matrix block, which is comparable to the fracture spacing; however, the interaction must be treated on a smaller scale as far as the matrix is concerned.

The equations that describe the flow in the fracture system will contain a source term that represents the flow of fluid from the matrix to the fractures; this term will be macroscopically distributed over the entire reservoir, as is the fracture flow itself. The definition of the term will be in terms of quantities that are on the scale of the matrix blocks.

The matrix equations will be effected on the scale of the fracture spacing. In particular, fluid flows into or out of an individual matrix block only through its surface. Such an interaction should be modelled as a boundary condition on the flow equation; fracture quantities will be used to define these boundary conditions.

1.4 Some remarks on the derivation of dual porosity models

The models presented in this paper will be derived first on the basis of physical intuition by applying the dual porosity idea of the previous subsection to define the equations of flow and then by the mathematical theory of homogenization (Bensoussan *et al.*, 1978; Sanchez-Palencia, 1980). This will give us both greater confidence that our physical intuition of the model is correct and a better understanding of the resulting model; in particular, we shall derive an explicit expression for the fracture system's permeability tensor **K**. Homogenization will be used only in a formal sense herein; that is, a rigorous convergence argument will not be carried out.

1.5 A summary of the results

In Section 2 we model single-phase flow of a liquid of constant compressibility in a naturally fractured reservoir intuitively by incorporating the physics of such flow into the dual porosity concept. After giving some general background on homogenization theory, we re-derive the model. There are some subtle differences in the models obtained through the two derivations, but none of any real practical concern. We conclude the section by mentioning some of the mathematical properties of the model, including existence and uniqueness of the solution of the differential system.

In Section 3, we take up a study of the flow of two immiscible fluids, such as oil and water, under the assumption of incompressibility. Again, we first derive a model on physical grounds and then re-derive it from homogenization. Actually, we present three models for immiscible flow, varying in physical and mathematical complexity and therefore in the amount of computational effort needed to approximate the solution of them.

We present in Section 4 some relatively simple finite difference discretization procedures in a computationally tractable form achieved by separating the matrix calculations from those of the fracture system. Both the single phase and immiscible cases are discussed. In Section 5 we present the results of some simulations of immiscible flow based on implementing the algorithms of the fourth section which compare and contrast the predictions of the three immiscible models and an unfractured reservoir model. Finally, in Section 6, we consider briefly some models generalizing those presented herein.

2. Single-phase flow

2.1 Introduction

Consider the flow of a fluid of constant compressibility c in the reservoir; that is, a fluid that satisfies the equation of state

$$d\rho = c\rho \, dp \tag{2.1}$$

where ρ is the density of the fluid and p is the pressure. Let μ be its viscosity. Since ρ is an exponential function of p, flow can be modelled in terms of either variable, but density is the more convenient choice. Let $\rho_f(x, t)$ denote the density of the fluid in the fracture system and $\rho_m(x, t)$ that in the matrix.

For expositional and computational convenience, we shall represent wells as external source (or sink) terms of the form $q_{ext}(x, t)$; they could also be represented by a boundary condition.

The equations describing single-phase flow in a single porosity system are well understood. Darcy's law relates the macroscopic Darcy velocity $v(x, t)$ to the pressure:

$$\mathbf{v} = -\frac{\mathbf{k}}{\mu}(\nabla p - \mathbf{g}\rho) \tag{2.2}$$

where $\mathbf{g}$ is the gravitational, downward-pointing, constant vector. Conservation of mass states that

$$\phi \frac{\partial \rho}{\partial t} + \nabla \cdot (\rho \mathbf{v}) = q_{ext} \tag{2.3}$$

Then, the equation of state (2.1) implies that

$$\phi \frac{\partial \rho}{\partial t} - \nabla \cdot \left[\frac{\mathbf{k}}{\mu c}(\nabla \rho - c\mathbf{g}\rho^2) \right] = q_{ext} \tag{2.4}$$

The first successful attempts at modelling single-phase flow in naturally fractured reservoirs were made in the early 1960s by Barenblatt *et al.* (1960) and Warren and Root (1963); in each of these models a 'quasi-steady state' assumption was made about the matrix–fracture interaction. Models that avoided this assumption were presented later by Kazemi (1969) and de Swaan (1976); unfortunately, these new models were not truly of the dual-porosity type, so that they were difficult to manage computationally. Recently, a general form of the single phase model has appeared in Arbogast (1989b) and Douglas *et al.* (1987); it is this model that we consider below.

2.2 Modelling by physical arguments

Equation 2.4 is the starting point for describing the flow in both the fracture and matrix systems, since each is considered to be a porous medium. As mentioned in the introduction, the equations for the fracture system require a macroscopically distributed matrix source term $q_{\mathrm{m}}(x, t)$ to represent flow from the matrix to the fractures; it is defined below in Eq. 2.10. Hence, flow in the fractures is described by

$$\Phi \frac{\partial \rho_{\mathrm{f}}}{\partial t} - \nabla \cdot \left[\frac{\mathbf{K}}{\mu c} (\nabla \rho_{\mathrm{f}} - c\mathbf{g}\rho_{\mathrm{f}}^2) \right] = q_{\mathrm{ext}} + q_{\mathrm{m}} \quad \text{for } x \in \Omega, \quad t > 0 \tag{2.5}$$

The matrix equations are not special. On each block Ω_i,

$$\phi \frac{\partial \rho_{\mathrm{m}}}{\partial t} - \nabla \cdot \left[\frac{\mathbf{k}}{\mu c} \left(\nabla \rho_{\mathrm{m}} - c\mathbf{g}\rho_{\mathrm{m}}^2 \right) \right] = 0 \quad \text{for } x \in \Omega_i, \quad t > 0 \tag{2.6}$$

We have assumed that the wells interact only with the fracture system; this assumption is based on the fact that flow is much more rapid in the fracture system than in the matrix. Numerical experiments have shown that this assumption is reasonable (see Kazemi, 1969).

Recall that the matrix sees the fractures only at the surfaces of the blocks. There, we must enforce continuity of pressure (or density) in some way. It is not immediately obvious how to do this, since the fracture system has been macroscopically averaged on a scale comparable to the size of the matrix blocks, but the flow in the fractures around an individual matrix block being much faster than the internal matrix flow allows us to assume that the fracture flow essentially reaches equilibrium with respect to the matrix flow. Then, at each time our boundary condition is such that the matrix pressure is constant over the surface of the block; which pressure value to take in the differential model is not uniquely determined, though almost any reasonable choice will yield the same value in discrete form. It is mathematically convenient to take, for each

i, the boundary condition

$$\rho_{\mathrm{m}}(x,t) = \frac{1}{|\hat{\Omega}_i|}\int_{\hat{\Omega}_i} \rho_{\mathrm{f}}(\xi,t)\,\mathrm{d}\xi \quad \text{for } x\in\partial\Omega_i, \quad t>0, \tag{2.7}$$

where $\hat{\Omega}_i$ is Ω_i together with half of the surrounding fractures, $|\hat{\Omega}_i|$ denotes the volume of $\hat{\Omega}_i$, and '∂' denotes the boundary of the set. Thus, at each time t, ρ_{m} is constant on $\partial\Omega_i$.

The matrix source term can be defined as follows. The total mass of fluid leaving the ith matrix block per unit time is

$$-\int_{\partial\Omega_i}\left[\frac{\mathbf{k}}{\mu c}(\nabla\rho_{\mathrm{m}} - c\mathbf{g}\rho_{\mathrm{m}}^2)\right]\cdot\mathbf{n}\,\mathrm{d}a(x)$$

where $\mathbf{n}$ is the outer normal unit vector to the surface. The divergence theorem then implies that

$$-\int_{\partial\Omega_i}\left[\frac{\mathbf{k}}{\mu c}(\nabla\rho_{\mathrm{m}} - c\mathbf{g}\rho_{\mathrm{m}}^2)\right]\cdot\mathbf{n}\,\mathrm{d}a(x) = -\int_{\Omega_i}\nabla\cdot\left[\frac{\mathbf{k}}{\mu c}(\nabla\rho_{\mathrm{m}} - c\mathbf{g}\rho_{\mathrm{m}}^2)\right]\mathrm{d}x$$

$$= -\int_{\Omega_i}\phi\,\frac{\partial\rho_{\mathrm{m}}}{\partial t}\,\mathrm{d}x \tag{2.8}$$

where the last equality follows from Eq. 2.6. This fluid must be distributed into the fracture system in the vicinity of the block. Again, the exact way this is done will have little effect on the computations. It is reasonable to spread it out evenly over $\hat{\Omega}_i$; hence, let

$$q_{\mathrm{m},i}(t) = -\frac{1}{|\hat{\Omega}_i|}\int_{\Omega_i}\phi\,\frac{\partial\rho_{\mathrm{m}}}{\partial t}(\xi,t)\,\mathrm{d}\xi \tag{2.9}$$

and let $\chi_i(x)$ be the characteristic function of $\hat{\Omega}_i$:

$$\chi_i(x) = \begin{cases} 1 & \text{for } x\in\hat{\Omega}_i \\ 0 & \text{for } x\notin\hat{\Omega}_i \end{cases}$$

Finally, let

$$q_{\mathrm{m}}(x,t) = \sum_i q_{\mathrm{m},i}(t)\chi_i(x) \quad \text{for } x\in\Omega, \quad t>0 \tag{2.10}$$

The model is completed by specifying an outer boundary condition and the initial densities. For example, we might assume that there is no flow across $\partial\Omega$. Then,

$$\left[\frac{\mathbf{K}}{\mu c}(\nabla\rho_{\mathrm{f}} - c\mathbf{g}\rho_{\mathrm{f}}^2)\right]\cdot\mathbf{n} = 0 \quad \text{for } x \in \partial\Omega, \quad t > 0 \tag{2.11}$$

Let the initial values be

$$\rho_{\mathrm{f}}(x, 0) = \rho_{\mathrm{init,f}}(x) \quad \text{for } x \in \Omega \tag{2.12}$$

$$\rho_{\mathrm{m}}(x, 0) = \rho_{\mathrm{init,m}}(x) \quad \text{for } x \in \Omega_{\mathrm{m}} \tag{2.13}$$

Most likely,

$$\rho_{\mathrm{init,m}} = \frac{1}{|\hat{\Omega}_i|}\int_{\Omega_i} \rho_{\mathrm{init,f}}(\xi)\,\mathrm{d}\xi \quad \text{for } x \in \Omega_i \tag{2.14}$$

so that the fluids are in equilibrium initially. For consistency, assume that Eqs 2.7 and 2.11 hold for $t = 0$.

2.3 A brief description of homogenization

In order to re-derive the model by homogenization let us recall some of the main ideas of this theory as presented by Bensoussan *et al.* (1978), Sanchez-Palencia (1980) and Ene (Chapter VIII, this volume). Suppose that we are given a physical system with a periodic fine structure. In theory, the governing equations, called the *microscopic* model, of that system can be derived from physics; however the model will capture far more detailed information than we care to know (or far more than we care to pay for in computation). Let us represent this schematically. If $\theta(x)$ is some quantity of physical relevance, it may have a graph such as that shown in Fig. 3, with θ oscillating about some

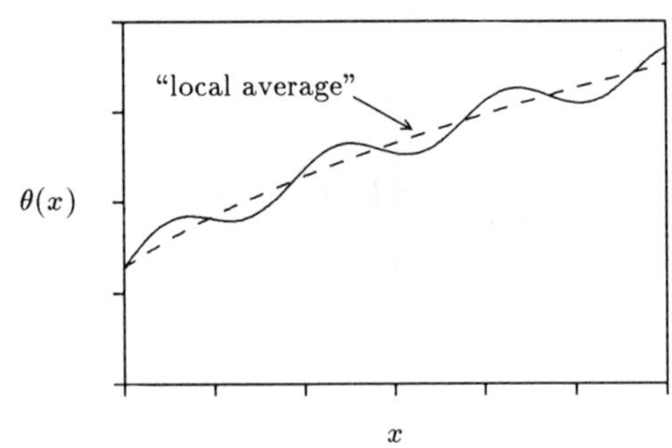

Figure 3. The 'local average' of a function θ.

average value as it varies from point to point within the fine structure. It may suffice to know only this general trend (or 'local average') of θ; homogenization is a mathematical procedure that seeks to derive an equation (or set of equations) for the general trend of θ by constructing a limit of the microscopic model as the period in the periodic structure tends to zero. These equations are then called the *macroscopic* model. Since this model is less detailed, it is usually the case that it is easier to approximate numerically than the original microscopic model for θ.

2.4 Modelling by homogenization

The model of Section 2.2 will be re-derived by homogenization, following Arbogast *et al.* (1989a). Darcy's law is assumed to hold on the smallest (i.e. pore) scale. It is at the scale of the fracture spacing that the problem needs explanation.

Idealize the reservoir by assuming that the fractures are from three sets of parallel, equally spaced planes (see Fig. 4), so that all matrix blocks are identical and the reservoir has a periodic structure. Let Q be a periodic, or *cell*, of this structure, containing a matrix block Q_{m} and half of the surrounding fractures Q_{f} (Fig. 5). Each Ω_i is then some translate of Q.

The microscopic model, which for simplicity will be taken to have all physical parameters as constants, is given by a single porosity system with discontinuous porosity and permeability as described in Section 1.2 above and contains all of the relevant physics, and it does so without requiring any intuition about the dual porosity concept. Let Φ^* be the 'porosity' of an individual fracture; that is, the fracture porosity defined on a finer scale than the fracture width. It is a little less than one, since there is always rock detritus in the fracture. The permeability of an individual fracture is very large and should be a scalar, denoted by K^*; it is defined on the scale of the pores, not on the scale of the fracture spacing, as is $\mathbf{K}$. From Section 2.1 above, Eq. 2.4 is posed over Ω;

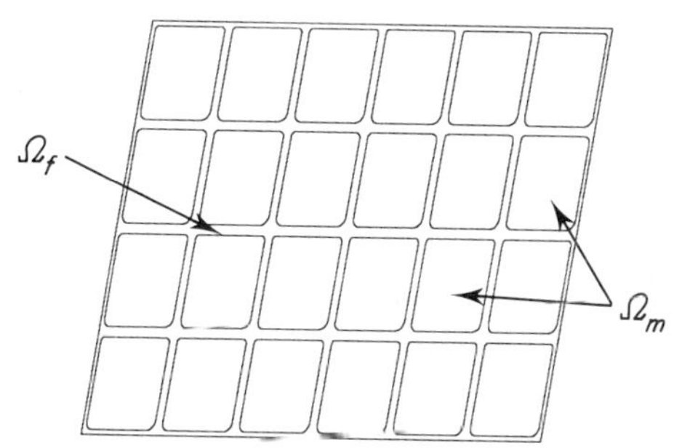

Figure 4. The periodic reservoir Ω.

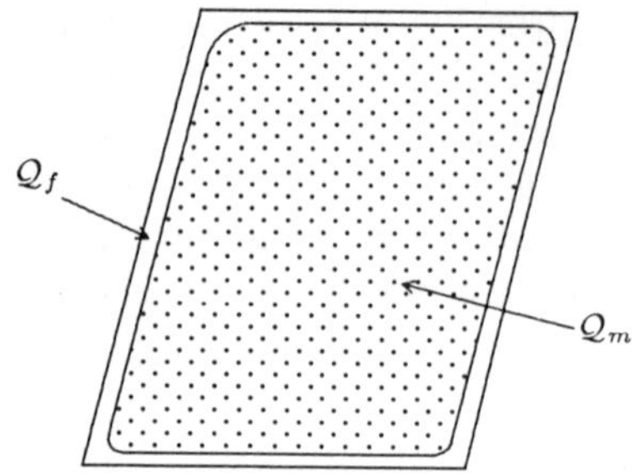

Figure 5. The reservoir cell Q.

however, for clarity we write this on the two parts of the domain separately:

$$\Phi^* \frac{\partial \rho_f}{\partial t} - \nabla \cdot \left[\frac{K^*}{\mu c} (\nabla \rho_f - c\mathbf{g}\rho_f^2) \right] = q_{ext} \quad \text{for } x \in \Omega_f, \quad t > 0 \tag{2.15}$$

$$\phi \frac{\partial \rho_m}{\partial t} - \nabla \cdot \left[\frac{\mathbf{k}}{\mu c} (\nabla \rho_m - c\mathbf{g}\rho_m^2) \right] = q_{ext} \quad \text{for } x \in \Omega_m, \quad t > 0 \tag{2.16}$$

On the interface $\partial\Omega_m$ between the two domains, impose continuity of pressure (i.e. density) and continuity of mass flux between the two domains. Hence,

$$\rho_m = \rho_f \quad \text{for } x \in \partial\Omega_m, \quad t > 0 \tag{2.17}$$

$$\left[\frac{K^*}{\mu c} (\nabla \rho_f - c\mathbf{g}\rho_f^2) \right] \cdot \mathbf{n} = \left[\frac{\mathbf{k}}{\mu c} (\nabla \rho_m - c\mathbf{g}\rho_m^2) \right] \cdot \mathbf{n} \quad \text{for } x \in \partial\Omega_m, \quad t > 0 \tag{2.18}$$

The fine structure that must be homogenized is the nearly periodic behaviour of the system on the scale of the fracture spacing. Since this behaviour is irrelevant in the evaluation of the overall flow in the reservoir, we consider the limit as the matrix blocks become small. To quantify this shrinkage, let ε be a parameter such that $0 < \varepsilon \leqslant 1$. When $\varepsilon = 1$, we have the original microscopic model. For $\varepsilon < 1$, we pose a fractured reservoir that is identical to the original one except that the fracture planes are assumed located at ε times the original distance apart. More explicitly, let the reservoir be composed of cells of size εQ (Fig. 6), and embed the microscopic model ($\varepsilon = 1$) into a family of models (for $\varepsilon < 1$). To this end, for each ε, let Ω_m^ε be the matrix part of the reservoir and Ω_f^ε the fracture part. (Ignore the outer boundary of the reservoir; a rigorous homogenization could not, but the point here is to understand the flow in the interior of the reservoir, since Eq. 2.11 holds on the outer boundary.)

Then, Eqs 2.15–2.18 should hold, though it is necessary to scale the physical

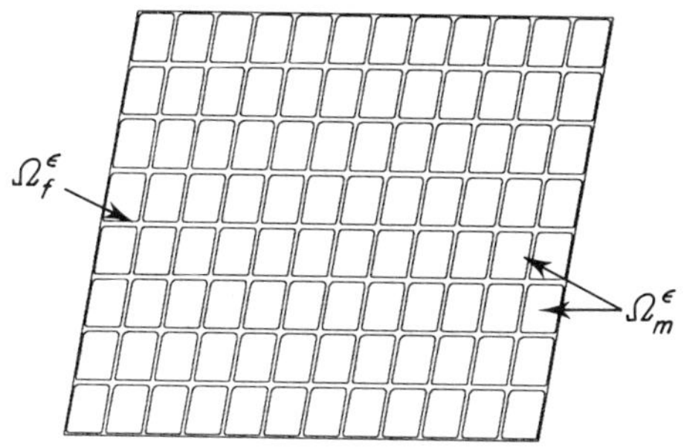

Figure 6. The reservoir Ω^ε for $\varepsilon = \frac{1}{2}$.

parameters in the equations in terms of ε, as can be seen in two different ways. First, if no scaling were done, the homogenization process would produce a single porosity model with averaged coefficients, which has already been pointed out to be inadequate. Second, proper scaling can in some sense preserve matrix-to-fracture mass flow as $\varepsilon \to 0$. This mass flux is

$$\int_{\partial\Omega_{\mathrm{m}}^{\varepsilon}} \left[\frac{\mathbf{k}}{\mu c} (\nabla \rho_{\mathrm{m}} - c\mathbf{g}\rho_{\mathrm{m}}^2) \right] \cdot \mathbf{n} \, da(x) = \int_{\Omega_{\mathrm{m}}^{\varepsilon}} \nabla \cdot \left[\frac{\mathbf{k}}{\mu c} (\nabla \rho_{\mathrm{m}} - c\mathbf{g}\rho_{\mathrm{m}}^2) \right] \mathrm{d}x \quad (2.19)$$

On an individual ε-matrix block $\Omega_i = \varepsilon Q_{\mathrm{m}}$, a change of variables shows that

$$\int_{\varepsilon Q_{\mathrm{m}}} \nabla \cdot \left[\frac{\mathbf{k}}{\mu c} (\nabla \rho_{\mathrm{m}} - c\mathbf{g}\rho_{\mathrm{m}}^2) \right] \mathrm{d}x = \int_{Q_{\mathrm{m}}} \varepsilon^{-1} \nabla \cdot \left[\frac{\mathbf{k}}{\mu c} (\varepsilon^{-1} \nabla \rho_{\mathrm{m}} - c\mathbf{g}\rho_{\mathrm{m}}^2) \right] \varepsilon^3 \, \mathrm{d}x \quad (2.20)$$

Adding these over the matrix blocks (ε-cells) leads to the total matrix–fracture flow

$$\sum_{\varepsilon\text{-cells}} \int_{Q_{\mathrm{m}}} \varepsilon^{-1} \nabla \cdot \left[\frac{\mathbf{k}}{\mu c} (\varepsilon^{-1} \nabla \rho_{\mathrm{m}} - c\mathbf{g}\rho_{\mathrm{m}}^2) \right] \varepsilon^3 \, \mathrm{d}x$$

(In the above the notation has been abused slightly; the ε-cells should have been translated over the reservoir.) Since there are of the order of ε^{-3} cells in $\Omega_{\mathrm{m}}^{\varepsilon}$, this quantity will diverge as $\varepsilon \to 0$ unless $\mathbf{k}$ is scaled by ε^2. It also appears that gravity should be compensated by ε^{-1}, though this is not essential.

This scaling has the effect of making the matrix progressively less permeable as $\varepsilon \to 0$. However, the blocks being smaller implies that more of the fluid in the block is near the surface, so that the scaling prevents the model from reducing the time needed for fluid within the matrix to reach the fractures (as $\varepsilon \to 0$). This is the crucial physical process that must be modelled carefully.

Thus, modulo scale factors, each of the ε-models is identical to that in Eqs 2.15–2.18 above. To avoid confusion as to the value of ε, let $\rho_{\rm f}^{\varepsilon}$ and $\rho_{\rm m}^{\varepsilon}$ denote the solution to the ε-model:

$$\Phi^* \frac{\partial \rho_{\rm f}^{\varepsilon}}{\partial t} - \nabla \cdot \left[\frac{K^*}{\mu c} (\nabla \rho_{\rm f}^{\varepsilon} - c\mathbf{g}(\rho_{\rm f}^{\varepsilon})^2) \right] = q_{\rm ext} \quad \text{for } x \in \Omega_{\rm f}^{\varepsilon}, \quad t > 0 \tag{2.21}$$

$$\phi \frac{\partial \rho_{\rm m}^{\varepsilon}}{\partial t} - \nabla \cdot \left[\frac{\varepsilon^2 \mathbf{k}}{\mu c} (\nabla \rho_{\rm m}^{\varepsilon} - \varepsilon^{-1} c\mathbf{g}(\rho_{\rm m}^{\varepsilon})^2) \right] = q_{\rm ext} \quad \text{for } x \in \Omega_{\rm m}^{\varepsilon}, \quad t > 0 \tag{2.22}$$

$$\rho_{\rm m}^{\varepsilon} = \rho_{\rm f}^{\varepsilon} \quad \text{for } x \in \partial\Omega_{\rm m}^{\varepsilon}, \quad t > 0, \tag{2.23}$$

$$\left[\frac{K^*}{\mu c} (\nabla \rho_{\rm f}^{\varepsilon} - c\mathbf{g}(\rho_{\rm f}^{\varepsilon})^2) \right] \cdot \mathbf{n} = \left[\frac{\varepsilon^2 \mathbf{k}}{\mu c} (\nabla \rho_{\rm m}^{\varepsilon} - \varepsilon^{-1} c\mathbf{g}(\rho_{\rm m}^{\varepsilon})^2) \right] \cdot \mathbf{n} \quad \text{for } x \in \partial\Omega_{\rm m}^{\varepsilon}, \quad t > 0 \tag{2.24}$$

We now perform homogenization heuristically. Assume that any point is described by two variables: $x \in \Omega$ giving the general location of the point and $y \in Q$ the location of the point within the ε-cell εQ. Clearly, x and y are related by the scale ε:

$$\varepsilon y \sim x$$

(up to translation). As a consequence, the gradient operator becomes

$$\nabla \sim \frac{1}{\varepsilon} \nabla_y + \nabla_x, \tag{2.25}$$

with ∇_y and ∇_x being the gradient operators with respect to y and x, respectively.

Assume that the solution behaves as if it were a function of these two space variables (and time) and that it can be expanded in a power series in the scale parameter ε; that is, assume that the oscillations in the solution have wavelengths given by powers of ε. Hence,

$$\rho_{\rm f}^{\varepsilon}(x, t) \sim \sum_{k=0}^{\infty} \varepsilon^k \rho_{\rm f}^k(x, y, t) \quad \text{for } x \in \Omega, \quad y \in Q_{\rm f}, \quad t > 0 \tag{2.26}$$

$$\rho_{\rm m}^{\varepsilon}(x, t) \sim \sum_{k=0}^{\infty} \varepsilon^k \rho_{\rm m}^k(x, y, t) \quad \text{for } x \in \Omega, \quad y \in Q_{\rm m}, \quad t > 0 \tag{2.27}$$

Also, assume explicitly that the $\rho_{\rm f}^k$ are $Q_{\rm f}$-periodic in y.

If we substitute the series Eqs 2.26 and 2.27 into the equations of our model (Eqs 2.21–2.24), expand the gradient according to Eq. 2.25, and collect terms with like powers of ε, then from Eq. 2.21 we obtain three equations for the ε^{-2}, ε^{-1}, and ε^{0} terms when $x \in \Omega$, $y \in Q_{\mathrm{f}}$, and $t > 0$:

$$-\nabla_y \cdot \left[\frac{K^*}{\mu c} \nabla_y \rho_{\mathrm{f}}^0 \right] = 0, \tag{2.28}$$

$$-\nabla_y \cdot \left[\frac{K^*}{\mu c} (\nabla_y \rho_{\mathrm{f}}^1 + \nabla_x \rho_{\mathrm{f}}^0 - c\mathbf{g}(\rho_{\mathrm{f}}^0)^2) \right] - \nabla_x \cdot \left[\frac{K^*}{\mu c} \nabla_y \rho_{\mathrm{f}}^0 \right] = 0 \tag{2.29}$$

$$\Phi^* \frac{\partial \rho_{\mathrm{f}}^0}{\partial t} - \nabla_y \cdot \left[\frac{K^*}{\mu c} (\nabla_y \rho_{\mathrm{f}}^2 + \nabla_x \rho_{\mathrm{f}}^1 - 2c\mathbf{g}\rho_{\mathrm{f}}^0 \rho_{\mathrm{f}}^1) \right]$$
$$- \nabla_x \cdot \left[\frac{K^*}{\mu c} (\nabla_y \rho_{\mathrm{f}}^1 + \nabla_x \rho_{\mathrm{f}}^0 - c\mathbf{g}(\rho_{\mathrm{f}}^0)^2) \right] = q_{\mathrm{ext}} \tag{2.30}$$

The first equations, for ε^0, from Eqs 2.22 and 2.33 are

$$\phi \frac{\partial \rho_{\mathrm{m}}^0}{\partial t} - \nabla_y \cdot \left[\frac{\mathbf{k}}{\mu c} (\nabla_y \rho_{\mathrm{m}}^0 - c\mathbf{g}(\rho_{\mathrm{m}}^0)^2) \right] = q_{\mathrm{ext}} \quad \text{for } x \in \Omega, \quad y \in Q_{\mathrm{m}}, \quad t > 0 \tag{2.31}$$

$$\rho_{\mathrm{m}}^0 = \rho_{\mathrm{f}}^0 \quad \text{for } x \in \Omega, \quad y \in \partial Q_{\mathrm{m}}, \quad t > 0 \tag{2.32}$$

since points on $\partial\Omega_{\mathrm{m}}^{\varepsilon}$ are described globally by $x \in \Omega$ and locally by $y \in \partial Q_{\mathrm{m}}$. The ε^{-1}, ε^0, and ε^1 equations of Eq. 2.24 for $x \in \Omega$, $y \in \partial Q_{\mathrm{m}}$, and $t > 0$ are

$$\left[\frac{K^*}{\mu c} \nabla_y \rho_{\mathrm{f}}^0 \right] \cdot \mathbf{n} = 0 \tag{2.33}$$

$$\left[\frac{K^*}{\mu c} (\nabla_y \rho_{\mathrm{f}}^1 + \nabla_x \rho_{\mathrm{f}}^0 - c\mathbf{g}(\rho_{\mathrm{f}}^0)^2) \right] \cdot \mathbf{n} = 0 \tag{2.34}$$

$$\left[\frac{K^*}{\mu c} (\nabla_y \rho_{\mathrm{f}}^2 + \nabla_x \rho_{\mathrm{f}}^1 - 2c\mathbf{g}\rho_{\mathrm{f}}^0 \rho_{\mathrm{f}}^1) \right] \cdot \mathbf{n} = \left[\frac{\mathbf{k}}{\mu c} (\nabla_y \rho_{\mathrm{m}}^0 - c\mathbf{g}(\rho_{\mathrm{m}}^0)^2) \right] \cdot \mathbf{n} \tag{2.35}$$

No other equations need be considered. As $\varepsilon \to 0$, $\rho_{\mathrm{f}}^{\varepsilon} \to \rho_{\mathrm{f}}^0$ and $\rho_{\mathrm{m}}^{\varepsilon} \to \rho_{\mathrm{m}}^0$. These limits are the 'locally averaged' functions and the equations above combine to give a system in them. Equations 2.28 and 2.33 together form an elliptic system for ρ_f^0 in terms of the y-variables; since its solution vanishes as a function of y, it follows that

$$\rho_f^0 = \rho_f^0(x,t) \quad \text{only} \tag{2.36}$$

This corresponds to our intuition: the local average of ρ_f does not oscillate. Thus, all terms containing $\nabla_y \rho_f^0$ vanish.

Next, the Eqs 2.29 and 2.34 form an elliptic system in y for ρ_f^1 that can be solved for ρ_f^1 in terms of ρ_f^0. To this end, define $\omega_j(y)$ for $j = 1, 2, 3$ as the Q_f-periodic solution of

$$\nabla_y^2 \omega_j = 0 \quad \text{for } y \in Q_f \tag{2.37}$$

$$\nabla_y \omega_j \cdot \mathbf{n} = -\mathbf{e}_j \cdot \mathbf{n} = -\mathbf{n}_j \quad \text{for } y \in \partial Q_m \tag{2.38}$$

where $\mathbf{e}_j$ is the unit vector in the jth direction. Then, by Eq. 2.36,

$$\rho_f^1(x,y,t) = \sum_{j=1}^{3} \omega_j(y) \left[\frac{\partial \rho_f^0}{\partial x_j}(x,t) - cg_j(\rho_f^0)^2 \right] \tag{2.39}$$

(Actually Eq. 2.39 holds up to an arbitrary, additive function of x and t. Since only $\nabla_y \rho_f^1$ occurs below, this ambiguity is unimportant.)

We analyse Eq. 2.30 next. Locally average it by integrating it over Q_f and dividing the result by $|Q|$ to remove the y-variable; then

$$\frac{|Q_f|}{|Q|} \Phi^* \frac{\partial \rho_f^0}{\partial t} - \frac{1}{|Q|} \int_{Q_f} \nabla_y \cdot \left[\frac{K^*}{\mu c} (\nabla_y \rho_f^2 + \nabla_x \rho_f^1 - 2c\mathbf{g}\rho_f^0 \rho_f^1) \right] \mathrm{d}y$$

$$- \frac{1}{|Q|} \int_{Q_f} \nabla_x \cdot \left[\frac{K^*}{\mu c} (\nabla_y \rho_f^1 + \nabla_x \rho_f^0 - c\mathbf{g}(\rho_f^0)^2) \right] \mathrm{d}y = \frac{|Q_f|}{|Q|} q_{\text{ext}} \tag{2.40}$$

Apply the divergence theorem to the first integral above, use Eq. 2.35, make a second application of the divergence theorem, and use Eq. 2.31 to see that

$$-\int_{Q_f} \nabla_y \cdot \left[\frac{K^*}{\mu c} (\nabla_y \rho_f^2 + \nabla_x \rho_f^1 - 2c\mathbf{g}\rho_f^0 \rho_f^1) \right] dy$$

$$= -\int_{\partial Q_f} \left[\frac{K^*}{\mu} (\nabla_y \rho_f^2 + \nabla_x \rho_f^1 - 2c\mathbf{g}\rho_f^0 \rho_f^1) \right] \cdot \mathbf{n}\, da(y)$$

$$= \int_{\partial Q_m} \left[\frac{\mathbf{k}}{\mu c} (\nabla_y \rho_m^0 - c\mathbf{g}(\rho_m^0)^2) \right] \cdot \mathbf{n}\, da(y)$$

$$= \int_{Q_m} \nabla_y \cdot \left[\frac{\mathbf{k}}{\mu c} (\nabla_y \rho_m^0 - c\mathbf{g}(\rho_m^0)^2) \right] dy$$

$$= \int_{Q_m} (\phi \rho_{m,t}^0 - q_{ext})\, dy \tag{2.41}$$

We have used periodicity to see that no contribution arises from the integral over ∂Q, the outer boundary of Q_f, and the fact that the outer normal to ∂Q_f is opposite to that of ∂Q_m. Since Eq. 2.35 established the continuity of the mass flux, we have found the matrix source term. The second integral in Eq. 2.40 is evaluated using Eq. 2.39; its integrand becomes (with implied summation)

$$\nabla_x \cdot \left[\frac{K^*}{\mu c} (\nabla_y \rho_f^1 + \nabla_x \rho_f^0 - c\mathbf{g}(\rho_f^0)^2) \right]$$

$$= \frac{\partial}{\partial x_k} \left[\frac{K^*}{\mu c} \left(\frac{\partial \omega_j}{\partial y_k} \left(\frac{\partial \rho_f^0}{\partial x_j} - c g_j \rho_f^0 \right) + \frac{\partial \rho_f^0}{\partial x_k} - c g_k (\rho_f^0)^2 \right) \right]$$

$$= \frac{\partial}{\partial x_k} \left[\frac{K^*}{\mu c} \left(\frac{\partial \omega_j}{\partial y_k} + \delta_{jk} \right) \left(\frac{\partial \rho_f^0}{\partial x_j} - c g_j \rho_f^0 \right) \right] \tag{2.42}$$

where δ_{jk} is the kronecker symbol. We can now *define* the fracture permeability tensor $\mathbf{K} = \{K_{ij}\}$:

$$K_{ij} = \frac{1}{|Q|} \left\{ \int_{Q_f} \frac{\partial \omega_j}{\partial y_k}\, dy + |Q_f|\, \delta_{ij} \right\} \tag{2.43}$$

Note that $\mathbf{K}$ reflects the geometry of the fracture system through the functions ω_j.

Finally, we have the equation for ρ_{f}^{0}:

$$\Phi \frac{\partial \rho_{\mathrm{f}}^{0}}{\partial t} - \nabla_{x} \cdot \left[\frac{\mathbf{K}}{\mu c} (\nabla_{x} \rho_{\mathrm{f}}^{0} - c\mathbf{g}(\rho_{\mathrm{f}}^{0})^{2}) \right] = q_{\mathrm{ext}} + q_{\mathrm{m}}, \tag{2.44}$$

where $\Phi = |Q_{\mathrm{f}}| \Phi^{*} / |Q|$ is clearly the macroscopic fracture porosity and

$$q_{\mathrm{m}} = -\frac{1}{|Q|} \int_{Q_{\mathrm{m}}} \phi \frac{\partial \rho_{\mathrm{m}}^{0}}{\partial t} \, \mathrm{d}y \tag{2.45}$$

With Eqs 2.31 and 2.32 (and the outer boundary condition (Eq. 2.11) and the initial conditions (Eqs 2.12 and 2.13)), we have a complete model for the flow.

We should make a few brief remarks. First, the necessary requirement for physical relevance that **K** be symmetric and positive definite can be demonstrated (see Arbogast *et al.*, 1989a). Second, the formal arguments leading to convergence of the model as $\varepsilon \to 0$ can be made rigorous, at least if the gravity term is linearized as described in Section 2.6 below. Again, see Arbogast *et al.* (1989a) for details. The argument there gives a somewhat less physically clear picture than that given by the formalism here, but the mathematical proof confirms that the formal picture is correct.

2.5 A comparison of the two versions of the model

The only noticeable difference between the two versions of the model is that the external source term appears in the matrix equations of the homogenized model but not in the physically defined one. This term was left out of the latter version deliberately; it is not very significant and it would be appropriate to delete it from the homogenized version.

The other differences are subtler. The physically defined version of the model has a finite number of finite-sized matrix blocks, whereas the homogenized version has an infinite number of infinitely small matrix blocks, one for each $x \in \Omega$. As a consequence, the matrix source term is defined in the former version by a finite sum; it is an integral in the latter version. This difference has no consequence in practice, as we must restrict to a finite number of blocks to approximate the solution numerically.

What should be emphasized is that the form of the equations is the same. In particular, the matrix source term is defined to be the total amount of fluid leaving the matrix blocks per unit volume. Also, the boundary value on each block is constant in space, being explicitly a local average when the block has some size and being simply the value at the point where the block is when it is infinitely small. Again, in practice, these conditions are the same.

2.6 Some mathematical properties of the model

The single phase model is mathematically relatively simple. Further, the system of differential equations can be made linear by an approximation of the terms that contain the gravitational vector. This is frequently done, since the densities of the fluids do not change greatly in any reasonable simulation. So, let ρ_{ref} be a constant reference density, and approximate ρ^2 by

$$\rho^2 = [(\rho - \rho_{\text{ref}}) + \rho_{\text{ref}}]^2 \approx \rho_{\text{ref}}(2\rho - \rho_{\text{ref}}) \tag{2.46}$$

in Eqs 2.5, 2.6, and 2.11, and also in Eqs 2.31 and 2.44.

With this modification (and with Eq. 2.14), it has been shown (Arbogast, 1989b; Arbogast *et al.*, 1989a) that there exists unique solution $\{\rho_{\text{f}}, \rho_{\text{m}}\}$ to the differential system of either version of the model. Moreover, the solution depends continuously on the data q_{ext}, $\rho_{\text{init,f}}$ and ρ_{ref}. Thus, a small change (or error) in the data produces only a small change in the solution, as measured in appropriate Sobolev spaces.

3. Two-phase immiscible flow

3.1 Introduction

We consider saturated, two-phase, incompressible, immiscible flow, the phases being o (oil) and w (water), with densities and viscosities ρ_α and μ_α, $\alpha = \text{o, w}$, respectively. We begin by recalling the equations that govern such flow in a single porosity system. Let $s(x, t)$ denote the w-saturation, so that the o-phase has saturation $1 - s$. Let $p_\alpha(x, t)$, $\alpha = \text{o, w}$, represent the pressure in the α-phase, and denote the capillary pressure between the two phases by

$$p_{\text{c}}(s) = p_{\text{o}} - p_{\text{w}} \tag{3.1}$$

where, as usual, $p_{\text{c}}(s)$ is assumed to be a function of s only. With w being the wetting phase, $p_{\text{c}}(s)$ typically is a decreasing function of s, as shown in Fig. 7. It becomes infinite as the saturation tends to the residual water saturation $s_{\min}$, and it is zero at the residual oil saturation corresponding to $s = s_{\max}$.

The presence of one phase interferes with the flow of the other. This is quantified by defining relative permeability functions $k_{\text{r}\alpha}(s)$, $\alpha = \text{o, w}$, as functions of the saturation (see Fig. 8 for typical examples). Usually, $k_{\text{rw}}(s_{\min}) = k_{\text{ro}}(s_{\max}) = 0$. Finally, let $\mathbf{k}(x)$ be the absolute permeability, so that $\mathbf{k}k_{\text{r}\alpha}$ is the permeability of the rock to the α-phase at the point x with saturation s.

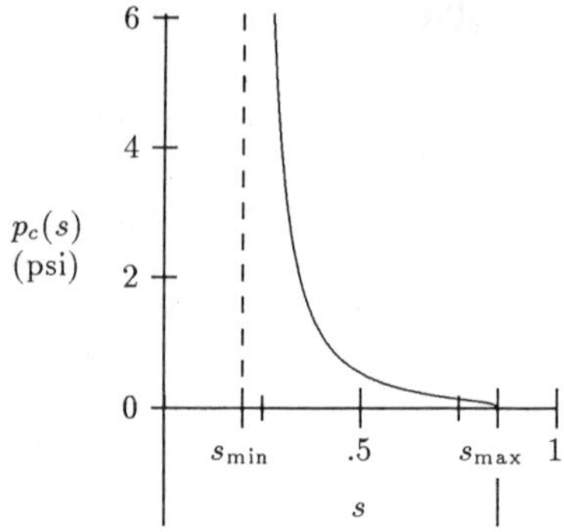

Figure 7. A typical capillary pressure function.

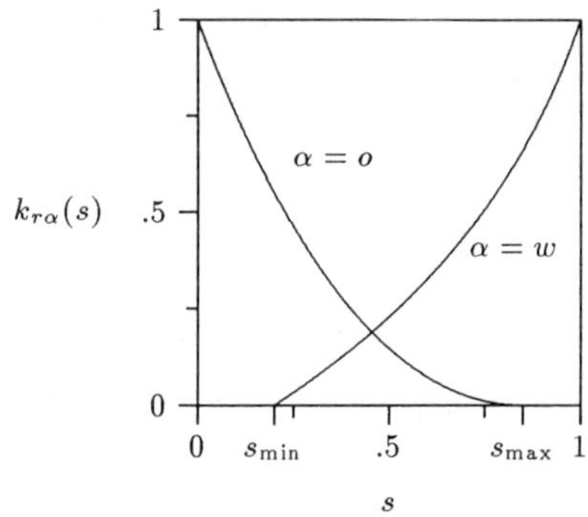

Figure 8. Typical relative permeability functions.

It is more convenient to work with the potentials

$$\psi_\alpha = p_\alpha - \rho_\alpha g z, \quad \alpha = \mathrm{o}, \mathrm{w} \tag{3.2}$$

than pressures; from here on g is the gravitational constant (not vector) and $z(x)$ is the depth. Darcy's law for the volumetric flow rates in two-phase flow takes the form

$$\mathbf{v}_\alpha = -\lambda_\alpha(s)\nabla\psi_\alpha, \quad \alpha = \mathrm{o}, \mathrm{w}, \tag{3.3}$$

where the phase mobilities are defined by

$$\boldsymbol{\lambda}_\mathrm{a}(s) = \frac{\mathbf{k}k_{\mathrm{r}\alpha}(s)}{\mu_\alpha}, \quad \alpha = \mathrm{o}, \mathrm{w} \tag{3.4}$$

Incompressibility and conservation of mass (or, equivalently, volume) imply that

$$\phi \frac{\partial s}{\partial t} + \nabla \cdot \mathbf{v}_{\mathrm{w}} = q_{\mathrm{ext,w}}, \tag{3.5}$$

$$-\phi \frac{\partial s}{\partial t} + \nabla \cdot \mathbf{v}_{\mathrm{o}} = q_{\mathrm{ext,o}}, \tag{3.6}$$

where $q_{\mathrm{ext},\alpha}$ is the external volumetric α-source.

It is convenient to define a 'capillary potential',

$$\psi_{\mathrm{c}} = \psi_{\mathrm{o}} - \psi_{\mathrm{w}} = p_{\mathrm{c}}(s) - (\rho_{\mathrm{o}} - \rho_{\mathrm{w}})gz \tag{3.7}$$

and to use it and ψ_{w} as the primary dependent variables. Then,

$$s = p_{\mathrm{c}}^{-1}(\psi_{\mathrm{c}} + (\rho_{\mathrm{o}} - \rho_{\mathrm{w}})gz) \tag{3.8}$$

is defined from ψ_{c} and

$$\mathbf{v}_{\mathrm{o}} = -\lambda_{\mathrm{o}}(s)\nabla(\psi_{\mathrm{w}} + \psi_{\mathrm{c}}) \tag{3.9}$$

It is frequently useful to define a total volumetric flow rate and a total mobility by

$$\mathbf{v} = \mathbf{v}_{\mathrm{o}} + \mathbf{v}_{\mathrm{w}} = -\lambda(s)\nabla\psi_{\mathrm{w}} - \lambda_{\mathrm{o}}(s)\nabla\psi_{\mathrm{c}} \tag{3.10}$$

$$\lambda(s) = \lambda_{\mathrm{w}}(s) + \lambda_{\mathrm{o}}(s) \tag{3.11}$$

respectively, and to add (3.5) and (3.6) to obtain a *pressure* equation,

$$\nabla \cdot \mathbf{v} = q_{\mathrm{ext}} \tag{3.12}$$

where $q_{\mathrm{ext}} = q_{\mathrm{ext,o}} + q_{\mathrm{ext,w}}$ is the total external volumetric source. Then, Eq. 3.5 is called the *saturation* equation.

In summary, the basic equations for describing two-phase, incompressible, immiscible flow in a single porosity system are given by Eqs 3.3, 3.5, 3.8, and either Eqs 3.6 and 3.9 or Eqs 3.10–3.12.

3.2 Modelling by physical arguments

There exists an extensive literature on the modelling of immiscible flow in naturally fractured reservoirs; from it we cite de Swaan (1978), Thomas *et al.*

(1983), Kazemi and Gilman (1983, 1987), Gilman (1986), and Sonier *et al.* (1986). Most of these papers consider models that effectively define the matrix–fracture interaction by introducing various *ad hoc* parameters; the rest do not incorporate the matrix boundary condition in any general way. Herein we consider models (Douglas *et al.*, 1987, 1989; Arbogast *et al.*, 1988, 1989b; Douglas, 1989) that treat the interaction explicitly through boundary conditions on the matrix blocks.

We denote fracture quantities by upper case letters and matrix quantities by corresponding lower case letters.

Capillary pressure and relative permeability functions are somewhat different in the fractures than in the matrix blocks. Generally, one assumes that the fractures are essentially like spaces between two parallel planes and that $S_{\min} = 0$ and $S_{\max} = 1$. Typical examples are shown in Fig. 9.

There are matrix source terms $q_{\mathrm{m},\alpha}$, $\alpha = \mathrm{o}, \mathrm{w}$, for each of the phases. The saturation, pressure, and capillary equations in the fracture system can be written as

$$\Phi \frac{\partial S}{\partial t} - \nabla \cdot [\Lambda_{\mathrm{w}}(S) \nabla \Psi_{\mathrm{w}}] = q_{\mathrm{ext,w}} + q_{\mathrm{m,w}} \quad \text{for } x \in \Omega, \quad t > 0 \tag{3.13}$$

$$-\nabla \cdot [\Lambda(S) \nabla \Psi_{\mathrm{w}} + \Lambda_{\mathrm{o}}(S) \nabla \Psi_{\mathrm{c}}] = q_{\mathrm{ext}} \quad \text{for } x \in \Omega, \quad t > 0 \tag{3.14}$$

$$S = P_{\mathrm{c}}^{-1}(\Psi_{\mathrm{c}} + (\rho_{\mathrm{o}} - \rho_{\mathrm{w}}) g z) \tag{3.15}$$

since incompressibility requires that $q_{\mathrm{m,o}} + q_{\mathrm{m,w}} = 0$. The equations on the block

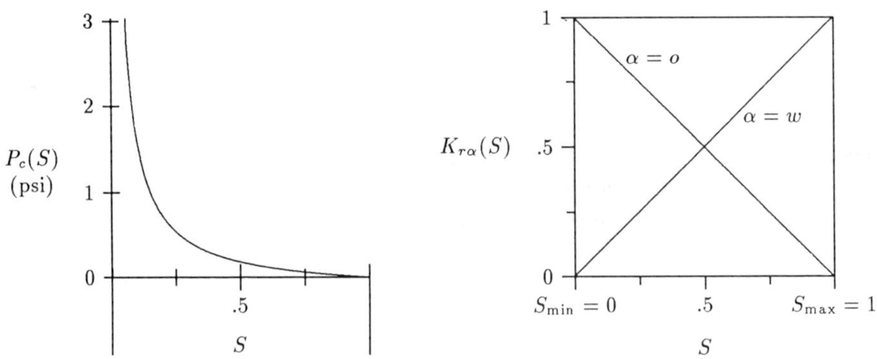

Capillary pressure. Relative permeabilities.

Figure 9. Typical fracture functions.

Ω_i are

$$\phi \frac{\partial s}{\partial t} - \nabla \cdot [\boldsymbol{\lambda}_{\mathrm{w}}(s) \nabla \psi_{\mathrm{w}}] = 0 \quad \text{for } x \in \Omega_i, \quad t > 0 \tag{3.16}$$

$$-\nabla \cdot [\boldsymbol{\lambda}(s) \nabla \psi_{\mathrm{w}} + \boldsymbol{\lambda}_{\mathrm{o}}(s) \nabla \psi_{\mathrm{c}}] = 0 \quad \text{for } x \in \Omega_i, \quad t > 0 \tag{3.17}$$

$$s = p_{\mathrm{c}}^{-1}(\psi_{\mathrm{c}} + (\rho_{\mathrm{o}} - \rho_{\mathrm{w}})gz) \tag{3.18}$$

where again we assume that the external sources affect the fracture system only. The boundary conditions for the matrix problems are given by requiring continuity of the potentials:

$$\psi_{\mathrm{w}}(x, t) = \frac{1}{|\hat{\Omega}_i|} \int_{\hat{\Omega}_i} \Psi_{\mathrm{w}}(\xi, t) \, \mathrm{d}\xi \quad \text{for } x \in \partial\Omega_i, \quad t > 0 \tag{3.19}$$

and

$$\psi_{\mathrm{c}}(x, t) = \frac{1}{|\hat{\Omega}_i|} \int_{\hat{\Omega}_i} \Psi_{\mathrm{c}}(\xi, t) \, \mathrm{d}\xi \quad \text{for } x \in \partial\Omega_i, \quad t > 0 \tag{3.20}$$

The matrix source terms are defined in a fashion analogous to the single phase case. The volume of the w-fluid leaving the ith block is

$$\int_{\partial\Omega_i} \mathbf{v}_{\mathrm{w}} \cdot \mathbf{n} \, \mathrm{d}a(x) = \int_{\Omega_i} \nabla \cdot \mathbf{v}_{\mathrm{w}} \, \mathrm{d}x = -\int_{\Omega_i} \phi \frac{\partial s}{\partial t} \, \mathrm{d}x \tag{3.21}$$

so, let

$$q_{\mathrm{m,w},i}(t) = -\frac{1}{|\hat{\Omega}_i|} \int_{\Omega_i} \phi \frac{\partial s}{\partial t} \, \mathrm{d}x \tag{3.22}$$

and

$$q_{\mathrm{m,w}} = \sum_i q_{\mathrm{m,w},i}(t) \chi_i(x) \quad \text{for } x \in \Omega, \quad t > 0 \tag{3.23}$$

We complete the model by specifying the external boundary conditions and the initial conditions for the system. For the case of no flow across the external boundary,

$$\boldsymbol{\Lambda}_\alpha(s) \nabla \Psi_\alpha \cdot \mathbf{n} = 0 \quad \text{for } x \in \partial\Omega, \quad t > 0, \quad \alpha = \mathrm{o, w, c} \tag{3.24}$$

Initial saturations (i.e. capillary potentials) must be specified:

$$\Psi_c(x, 0) = \Psi_{\text{init},c}(x) \quad \text{for } x \in \Omega \tag{3.25}$$

$$\psi_c(x, 0) = \psi_{\text{init},c}(x) \quad \text{for } x \in \Omega_m \tag{3.26}$$

To be consistent, Eqs 3.19, 3.20, and 3.24 should hold when $t = 0$.

3.3 Modelling by homogenization

Recall that Φ^* and K^* are the porosity and absolute permeability of the fractures on the scale of the pores, and let $\Lambda^*_\alpha(S) = K^* K_{r\alpha}(S)/\mu_\alpha$, $\alpha = \text{o, w}$, and $\Lambda^*(S) = \Lambda^*_o(S) + \Lambda^*_w(S)$. Again for simplicity, assume that the coefficients do not depend explicitly on space and time. We do *not*, however, assume them independent of saturation; this dependence is a crucial aspect of immiscible flow.

The scaled microscopic model for each $\varepsilon > 0$ is best defined in terms of Eqs 3.6 and 3.9, rather than Eqs 3.10–3.12. To exploit the natural symmetry of the equations, let $\sigma_w = 1$ and $\sigma_o = -1$. Recall that $\Psi_o = \Psi_c + \Psi_w$ and $\psi_o = \psi_c + \psi_w$, and note that

$$\frac{\partial S}{\partial t} = P_c'(S)^{-1} \frac{\partial \Psi_c}{\partial t} \quad \text{and} \quad \frac{\partial s}{\partial t} = p_c'(s)^{-1} \frac{\partial \psi_c}{\partial t}$$

Then, in the fractures,

$$\sigma_\alpha \Phi^* P_c'(S)^{-1} \frac{\partial \Psi^\varepsilon_c}{\partial t} - \nabla \cdot [\Lambda^*_\alpha(S) \nabla \Psi^*_\alpha] = q_{\text{ext},\alpha} \quad \text{for } x \in \Omega^\varepsilon_f, \quad t > 0, \quad \alpha = \text{o, w} \tag{3.27}$$

$$S = P_c^{-1}(\Psi^\varepsilon_c + (\rho_o - \rho_w)gz) \tag{3.28}$$

and, in the matrix,

$$\sigma_\alpha \phi p_c'(s)^{-1} \frac{\partial \psi_c}{\partial t} - \varepsilon^2 \nabla \cdot [\lambda_\alpha(s) \nabla \psi^\varepsilon_\alpha] = q_{\text{ext},\alpha} \quad \text{for } x \in \Omega^\varepsilon_m, \quad t > 0, \quad \alpha = \text{o, w} \tag{3.29}$$

$$s = p_c^{-1}(\psi^\varepsilon_c + (\rho_o - \rho_w)gz) \tag{3.30}$$

On the matrix–fracture interface, we require continuity of the pressures and the phase volumetric fluxes; that is,

$$\psi_\alpha^\varepsilon(x,t) = \Psi_\alpha^\varepsilon(x,t) \qquad \text{for } x \in \partial\Omega_m^\varepsilon, \quad t > 0, \quad \alpha = \text{o, w}, \tag{3.31}$$

$$[\Lambda_\alpha^*(S)\nabla\Psi_\alpha^\varepsilon]\cdot\mathbf{n} = \varepsilon^2[\boldsymbol{\lambda}_\alpha(s)\nabla\psi_\alpha^\varepsilon]\cdot\mathbf{n} \quad \text{for } x \in \partial\Omega_m^\varepsilon, \quad t > 0, \quad \alpha = \text{o, w} \tag{3.32}$$

The correctness of the scaling follows as in the previous section. Again, were it absent, a single porosity model would result from homogenization (Bourgeat, 1986; Amaziane and Bourgeat, 1988). Also, the total matrix–fracture w-flow is

$$\sum_{\varepsilon\text{-cells}} \int_{\varepsilon\partial Q_m} v_w^\varepsilon \cdot \mathbf{n}\, da(x) = \sum_{\varepsilon\text{-cells}} \int_{\varepsilon Q_m} \nabla\cdot\mathbf{v}_w^\varepsilon\, dx = \sum_{\varepsilon\text{-cells}} \int_{\varepsilon Q_m} \nabla\cdot\left[\boldsymbol{\lambda}_w(s)\nabla\psi_w^\varepsilon\right] dx$$

$$= \sum_{\varepsilon\text{-cells}} \int_{Q_m} \varepsilon^{-2}\nabla\cdot[\boldsymbol{\lambda}_w(s)\nabla\psi_w^\varepsilon]\varepsilon^3\, dx \tag{3.33}$$

Since there are on the order of ε^{-3} cells, $\lambda_w(s)$ must be scaled by ε^2 to preserve the flow from matrix to fractures as $\varepsilon \to 0$. Similarly, $\boldsymbol{\lambda}_o(s)$ (and hence $\boldsymbol{\lambda}(s)$) must be scaled by ε^2. The definition of the potentials, Eq. 3.2, implies that gravity has been scaled by ε^{-1} implicitly. If we had chosen to use the pressure variables, we would have to do this explicitly.

We again assume that every point in the reservoir is described by the general location $x \in \Omega$ and by the specific location $y \in Q$ and that the solution can be expanded in power series in ε for $\alpha = \text{w, c}$ (with implied summation):

$$\Psi_\alpha^\varepsilon(x,t) \sim \varepsilon^k \Psi_\alpha^k(x,y,t) \tag{3.34}$$

$$\psi_\alpha^\varepsilon(x,t) \sim \varepsilon^k \psi_\alpha^k(x,y,t) \tag{3.35}$$

As before, Ψ_w^k and Ψ_c^k are taken to be Q_f-periodic in y. If we set $\Psi_o^k = \Psi_c^k + \Psi_w^k$ and $\psi_o^k = \psi_c^k + \psi_w^k$, then Eqs 3.34 and 3.35 hold with $\alpha = \text{o}$.

We shall expand saturation-dependent quantities in powers of ε. If θ is such a quantity, then, with '$\circ$' denoting composition of functions,

$$\theta(s) = \theta(p_c^{-1}(\psi_c^\varepsilon + (\rho_o - \rho_w)gz))$$

$$= (\theta \circ p_c^{-1})(\psi_c^\varepsilon + (\rho_o - \rho_w)gz)$$

$$= (\theta \circ p_c^{-1})(\psi_c^0 + (\rho_o - \rho_w)gz) + (\theta \circ p_c^{-1})'(\sigma)(\psi_c^\varepsilon - \psi_c^0)$$

$$- \theta(p_c^{-1}(\psi_c^0 + (\rho_o - \rho_w)gz)) + \varepsilon\hat{\theta}^1 + \varepsilon^2\hat{\theta}^2 + \cdots \tag{3.36}$$

for some σ and $\hat{\theta}^1, \hat{\theta}^2, \ldots$. Set

$$s^0 = p_c^{-1}(\psi_c^0 + (\rho_o - \rho_w)gz) \tag{3.37}$$

so that

$$\theta(s) = \theta(s^0) + \varepsilon^k \hat{\theta}^k \tag{3.38}$$

Similarly, let

$$S^0 = P_c^{-1}(\Psi_c^0 + (\rho_o - \rho_w)gz) \tag{3.39}$$

analogous expansions to those of Eq. 3.38 hold for functions of S.

Now substitute Eqs 3.34 and 3.35 into the microscopic model and expand the gradient according to Eq. 2.25 and the functions of saturations by Eq. 3.38 and collect terms by powers of ε. For $x \in \Omega$, $y \in Q_f$, and $t > 0$, it follows from Eq. 3.27 for $\alpha = $ o, w, that

$$-\nabla_y \cdot [\Lambda_\alpha^*(S^0)\nabla_y \Psi_\alpha^0] = 0 \tag{3.40}$$

$$-\nabla_y \cdot [\Lambda_\alpha^*(S^0)(\nabla_y \Psi_\alpha^1 + \nabla_x \Psi_\alpha^0) + \hat{\Lambda}_\alpha^{*,1}\nabla_y \Psi_\alpha^0] = 0 \tag{3.41}$$

$$\sigma_\alpha \Phi^* P_c'(S)^{-1} \frac{\partial \Psi_c^0}{\partial t} - \nabla_y \cdot [\Lambda_\alpha^*(S^0)(\nabla_y \Psi_\alpha^2 + \nabla_x \Psi_\alpha^1) + \hat{\Lambda}_\alpha^{*,1}(\nabla_y \Psi_\alpha^1 + \nabla_x \Psi_\alpha^0)$$

$$+ \hat{\Lambda}_\alpha^{*,2}\nabla_y \Psi_\alpha^0] - \nabla_x \cdot [\Lambda_\alpha^*(S^0)(\nabla_y \Psi_\alpha^1 + \nabla_x \Psi_\alpha^0) + \hat{\Lambda}_\alpha^{*,1}\nabla_y \Psi_\alpha^0] = q_{\text{ext},\alpha} \tag{3.42}$$

Also, from Eqs 3.29 and 3.31,

$$\sigma_\alpha \phi p_c'(s)^{-1} \frac{\partial \psi_c^0}{\partial t} - \nabla_y \cdot [\lambda_\alpha(s^0)\nabla_y \psi_\alpha^0] = q_{\text{ext},\alpha}$$

$$\text{for } x \in \Omega, \quad y \in Q_m, \quad t > 0, \quad \alpha = \text{o, w} \tag{3.43}$$

$$\psi_\alpha^0 = \Psi_\alpha^0 \quad \text{for } x \in \Omega, \quad y \in \partial Q_m, \quad t > 0, \quad \alpha = \text{o, w} \tag{3.44}$$

From Eq. 3.32 for $x \in \Omega$, $y \in \partial Q_m$, $t > 0$, $\alpha = $ o, w, we obtain the relations

$$[\Lambda_\alpha^*(S^0)\nabla_y \Psi_\alpha^0] \cdot \mathbf{n} = 0, \tag{3.45}$$

$$[\Lambda_\alpha^*(S^0)(\nabla_y \Psi_\alpha^1 + \nabla_x \Psi_\alpha^0) + \hat{\Lambda}_\alpha^{*,1}\nabla_y \Psi_\alpha^0] \cdot \mathbf{n} = 0, \tag{3.46}$$

$$[\Lambda_\alpha^*(S^0)(\nabla_y\Psi_\alpha^2 + \nabla_x\Psi_\alpha^1) + \hat{\Lambda}_\alpha^{*,1}(\nabla_y\Psi_\alpha^1 + \nabla_x\Psi_\alpha^0) + \hat{\Lambda}_\alpha^{*,2}\nabla_y\Psi_\alpha^0]\cdot\mathbf{n}$$

$$= [\lambda_\alpha(s^0)\nabla_y\psi_\alpha^0]\cdot\mathbf{n} \tag{3.47}$$

The analysis of these equations is similar to that in Section 2.4 above. First, Eqs 3.40 and 3.45 give homogeneous elliptic systems for the Ψ_α^0, from which we can conclude that Ψ_α^0 is independent of y:

$$\Psi_\alpha^0 = \Psi_\alpha^0(x, t) \quad \text{only, } \alpha = \text{o, w} \tag{3.48}$$

This is not as easy to see as in Section 2.4; the system is degenerate, since $\Lambda_w(S^0)$ is zero if $S^0 = S_{min}$ and $\Lambda_o(S^0)$ is zero if $S^0 = S_{max}$. Multiply Eq. 3.40 by Ψ_α^0 and integrate over $y \in Q_f$; then the divergence theorem, Eq. 3.45, and periodicity in y over ∂Q show that

$$-\int_{Q_f} \nabla_y\cdot[\Lambda_\alpha^*(S^0)\nabla_y\Psi_\alpha^0]\Psi_\alpha^0\,dy = \int_{Q_f} \Lambda_\alpha^*(S^0)|\nabla_y\Psi_\alpha^0|^2\,dx = 0 \tag{3.49}$$

Since the integrand is non-negative, it must be zero; consequently, one of the following must hold:

(i) $S^0 = S_{min}$ and $\Psi_o^0 = \Psi_o^0(x, t)$ only;
(ii) $S^0 = S_{max}$ and $\Psi_w^0 = \Psi_w^0(x, t)$ only;
(iii) Eq. 3.48 holds.

But Eq. 3.39 implies that if any two of S^0, Ψ_o^0, and Ψ_w^0 are independent of y, so is the other one. Thus, Eq. 3.48 holds and

$$S^0 = S^0(x, t) \text{ only} \tag{3.50}$$

Now, all terms containing $\nabla_y\Psi_\alpha^0$ or $\nabla_y S^0$ drop out.

Next, Eqs 3.41 and 3.46 can be used to write Ψ_α^1 in terms of Ψ_α^0. If $\Lambda_\alpha^*(S^0) \neq 0$,

$$\Psi_\alpha^1(x, y, t) = \sum_{j=1}^{3} \omega_j(y)\frac{\partial\Psi_\alpha^0}{\partial x_j}(x, t) \tag{3.51}$$

up to an additive function of x and t; the functions $\omega_j(y)$ are those defined in Eqs 2.37 and 2.38. Since only $\Lambda_\alpha^*(S^0)\nabla_y\Psi_\alpha^1$ is used below, the ambiguities in the value of Ψ_α^1 are irrelevant.

If Eq. 3.42 is locally averaged, it follows that

$$\sigma_\alpha \Phi P_c'(S)^{-1} \frac{\partial \Psi_c^0}{\partial t} - \frac{1}{|Q|} \int_{Q_f} \nabla_y \cdot [\Lambda_\alpha^*(S^0)(\nabla_y \Psi_\alpha^2 + \nabla_x \Psi_\alpha^1)$$

$$+ \hat{\Lambda}_\alpha^{*,1}(\nabla_y \Psi_\alpha^1 + \nabla_x \Psi_\alpha^0)] \, dy - \frac{1}{|Q|} \int_{Q_f} \nabla_x \cdot [\Lambda_\alpha^*(S^0)(\nabla_y \Psi_\alpha^1 + \nabla_x \Psi_\alpha^0)] \, dy$$

$$= \frac{|Q_f|}{|Q|} q_{\text{ext},\alpha} \tag{3.52}$$

with the fracture porosity Φ being correctly defined earlier. Transform the first integral by use of the divergence theorem (twice) and Eq. 3.47; then,

$$- \int_{Q_f} \nabla_y \cdot [\Lambda_\alpha^*(S^0)(\nabla_y \Psi_\alpha^2 + \nabla_x \Psi_\alpha^1) + \hat{\Lambda}_\alpha^{*,1}(\nabla_y \Psi_\alpha^1 + \nabla_x \Psi_\alpha^0)] \, dy$$

$$= - \int_{\partial Q_f} [\Lambda_\alpha^*(S^0)(\nabla_y \Psi_\alpha^2 + \nabla_x \Psi_\alpha^1) + \hat{\Lambda}_\alpha^{*,1}(\nabla_y \Psi_\alpha^1 + \nabla_x \Psi_\alpha^0)] \cdot \mathbf{n} \, da(y)$$

$$= \int_{\partial Q_m} [\lambda_\alpha(s^0) \nabla_y \psi_\alpha^0] \cdot \mathbf{n} \, da(y)$$

$$= \int_{Q_m} \nabla_y \cdot [\lambda_\alpha(s^0) \nabla_y \psi_\alpha^0] \, dy$$

$$= \int_{Q_m} \left[\sigma_\alpha \phi p_c'(s)^{-1} \frac{\partial \psi_c^0}{\partial t} - q_{\text{ext},\alpha} \right] dy \tag{3.53}$$

again the part of the second integral above over ∂Q vanishes, and the outer normal to ∂Q_f is opposite of that to ∂Q_m; Eq. 3.43 has been used in the last equality above.

The second integrand in Eq. 3.52 is simplified with Eq. 3.51 as follows (with summation implied):

$$\nabla_x \cdot [\Lambda_\alpha^*(S^0)(\nabla_y \Psi_\alpha^1 + \nabla_x \Psi_\alpha^0)] = \frac{\partial}{\partial x_k} \left[\Lambda_\alpha^*(S^0) \left(\frac{\partial \omega_j}{\partial y_k} \frac{\partial \Psi_\alpha^0}{\partial x_j} + \frac{\partial \Psi_\alpha^0}{\partial x_k} \right) \right]$$

$$= \frac{\partial}{\partial x_k} \left[\Lambda_\alpha^*(S^0) \left(\frac{\partial \omega_j}{\partial y_k} + \delta_{kj} \right) \frac{\partial \Psi_\alpha^0}{\partial x_j} \right] \tag{3.54}$$

Define $\mathbf{K}$ as in Eq. 2.43 and let

$$\Lambda_\alpha(S^0) = \frac{\mathbf{K}K_{r\alpha}(S^0)}{\mu_\alpha}, \quad \alpha = \text{o, w} \tag{3.55}$$

Finally, set

$$q_{m,\alpha} = -\frac{\sigma_\alpha}{|Q|}\int_{Q_m} \phi p_c'(s)^{-1}\frac{\partial \psi_c^0}{\partial t}\,dy = -\frac{\sigma_\alpha}{|Q|}\int_{Q_m} \phi \frac{\partial s^0}{\partial t}\,dy \tag{3.56}$$

and combine Eqs 3.39 and 3.52–3.54 to obtain the equations

$$\sigma_\alpha \Phi \frac{\partial S^0}{\partial t} - \nabla\cdot[\Lambda_\alpha(S^0)\nabla\Psi_\alpha^0] = q_{\text{ext},\alpha} + q_{m,\alpha} \quad \text{for } x\in\Omega, \quad t>0, \quad \alpha = \text{o, w} \tag{3.57}$$

The homogenized version consists of Eqs 3.56, 3.57, 3.39, 3.43, 3.37 and 3.44 together with the initial and boundary conditions corresponding to Eqs 3.24–3.26. We prefer to write the model in terms of pressure and saturation equations. Add the Eqs 3.57 for $\alpha = \text{o, w}$ to obtain the pressure equation

$$-\nabla_x\cdot[\Lambda(S^0)\nabla_x\Psi_w^0 + \Lambda_o(S^0)\nabla_x\Psi_c^0] = q_{\text{ext}} \quad \text{for } x\in\Omega, \quad t>0 \tag{3.58}$$

where $\Lambda(S^0) = \Lambda_o(S^0) + \Lambda_w(S^0)$, to replace Eq. 3.57 for $\alpha = \text{o}$; the remaining Eq. 3.57 for $\alpha = \text{w}$ is the saturation equation. Similarly, replace Eq. 3.43 for $\alpha = \text{o}$ by

$$\nabla_y\cdot[\lambda(s^0)\nabla_y\psi_w^0 + \lambda_o(s^0)\nabla_y\psi_c^0] = 0 \quad \text{for } x\in\Omega, \quad y\in Q_m, \quad t>0 \tag{3.59}$$

where $\lambda(s^0) = \lambda_o(s^0) + \lambda_w(s^0)$. (We take Eq. 3.44 to hold for $\alpha = \text{c}$, also.)

The two versions of the model again have subtle differences, but none of any practical concern. The key features are reflected in both; namely, that the boundary conditions on the matrix problems are constant in space (with respect to the matrix variables) and that the matrix source term is defined as the average total flow out of the matrix blocks.

3.4 Neglecting gravity in the matrix

When the matrix blocks are quite small, it seems reasonable for some simulations to neglect gravity in the matrix. In that case, the amount of computation required to simulate the flow can be significantly reduced, since the matrix pressure equations drop out. To see this, write the matrix equations in terms of a so-called 'global pressure' variable (Antoncev, 1972; Chavent, 1976) defined by

$$p(x,t) = \tfrac{1}{2}(p_o + p_w) + \tfrac{1}{2}\int_0^{p_c(s)} \left(\frac{\lambda_o - \lambda_w}{\lambda}\right)(p_c^{-1}(\xi))\,d\xi$$

$$= p_w + \int_0^{p_c(s)} \left(\frac{\lambda_o}{\lambda}\right)(p_c^{-1}(\xi))\,d\xi \tag{3.60}$$

so that

$$\lambda(s)\nabla p = \lambda(s)\nabla p_w + \lambda_o(s)\nabla p_c \tag{3.61}$$

If the gravity terms are deleted in the matrix equations, matrix potentials are the same as matrix pressures, so that $\lambda(s)\nabla p$ is the total volumetric flow rate v (see Eq. 3.10), and Eq. 3.61 becomes

$$\nabla\cdot(\lambda(s)\nabla p) = 0 \tag{3.62}$$

The boundary conditions Eqs 3.19 and 3.20 state that the pressures p_w and p_c are constant over $\partial\Omega_i$ for each i; consequently, the same is true of p. Hence, Eq. 3.62 is an elliptic problem with constant boundary values, so that p is this constant. Then Eq. 3.62 implies that

$$\lambda_w(s)\nabla p_w = -\frac{\lambda_w(s)\lambda_o(s)}{\lambda(s)}\nabla p_c, \tag{3.63}$$

and Eq. 3.16 can be rewritten in terms of s to obtain

$$\phi\frac{\partial s}{\partial t} - \nabla\cdot\left[\frac{\lambda_w(s)\lambda_o(s)}{\lambda(s)}p_c'(s)\nabla s\right] = 0 \quad \text{for } x\in\Omega_i, \quad t > 0 \tag{3.64}$$

For emphasis, we record the boundary condition below:

$$s(x,t) = p_c^{-1}\left(\frac{1}{|\hat{\Omega}_i|}\int_{\hat{\Omega}_i}\Psi_c(\xi,t)\,d\xi + (\rho_w - \rho_o)gz\right) \quad \text{for } x\in\partial\Omega_i, \quad t > 0 \tag{3.65}$$

These two equations and a setting of the initial saturations replace the matrix equations of the previous model.

This model can be derived from homogenization theory by not compensating gravity, so that the matrix gravity terms tend to zero with ε.

3.5 A limit model

Sometimes the fracturing is sufficiently extensive that the matrix blocks can be considered to be small enough that the fluid pressures throughout a block can be assumed in equilibrium with those in the surrounding fractures; thus, the time taken by matrix fluids to flow to the surfaces of the blocks is completely neglected. In this case the dual porosity model can be treated as a single porosity model derivable as a simple (*not* homogenization) limit of the model of the previous subsection as the fracture spacing $\varepsilon \to 0$. The validity of the limit model is not limited to fractures having the regular geometric form assumed in the models discussed so far; the size alone of the blocks is the determining factor.

As $\varepsilon \to 0$, the boundary conditions (Eq. 3.65) dominate the differential equations (Eq. 3.64) in a block, and Eq. 3.65 holds in the limit on the block located at each $x \in \Omega$:

$$s(x, t) = p_c^{-1}(\Psi_c(x, t) + (\rho_w - \rho_o)gz) \quad \text{for } x \in \Omega, \quad t > 0 \tag{3.66}$$

where we use Lebesgue's theorem on the differentiation of the integral (Stein, 1970) to see that, for fixed $x \in \varepsilon\hat{\Omega}_i$,

$$\lim_{\varepsilon \to 0} \frac{1}{|\varepsilon\hat{\Omega}_i|} \int_{\varepsilon\hat{\Omega}_i} \Psi_c(\xi, t)\, d\xi = \Psi_c(x, t) \tag{3.67}$$

Thus, the entire matrix problem is replaced by Eq. 3.65. A typical example of this very important relation is given in Fig. 10. The matrix source term defined

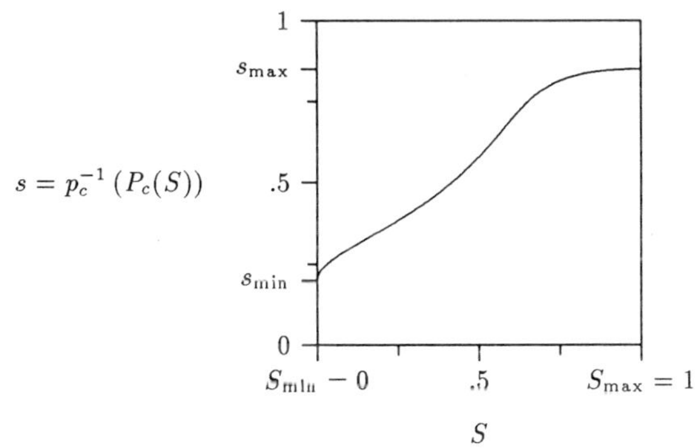

Figure 10. A typical matrix-fracture capillary pressure relation.

by Eqs 3.22 and 3.23 becomes

$$q_{\mathrm{m,w}}(x,t) = -\lim_{\varepsilon\to 0}\sum_i \left(\frac{1}{|\varepsilon\hat{\Omega}_i|}\int_{\varepsilon\Omega_i}\phi\,\frac{\partial s}{\partial t}(\xi,t)\,\mathrm{d}\xi\right)\chi_i(x)$$

$$= -\left(\frac{|Q_{\mathrm{m}}|}{|Q|}\,\phi\right)\frac{\partial s}{\partial t}(x,t) \quad \text{for } x\in\Omega,\quad t>0 \tag{3.68}$$

again by Lebesgue's theorem.

Then, Eq. 3.13 reduces to

$$\Phi\,\frac{\partial S}{\partial t} + \hat{\phi}\,\frac{\partial s}{\partial t} - \nabla\cdot[\Lambda_{\mathrm{w}}(S)\nabla\Psi_{\mathrm{w}}] = q_{\mathrm{ext,w}} \quad \text{for } x\in\Omega,\quad t>0 \tag{3.69}$$

with

$$\hat{\phi} = \frac{|Q_{\mathrm{m}}|}{|Q|}\,\phi \tag{3.70}$$

This model can be interpreted as a single porosity system with a saturation-dependent porosity for computational purposes; however, physically there remains a block associated with each point x.

4. Some discretization techniques

4.1 General remarks

Finite difference procedures for approximating the solutions to the models for single-phase or two-phase flow will be discussed, with emphasis placed on solving the matrix equations independently of the fracture equations. This is important practically, for, if only one matrix block per fracture node were employed in the computational model and if each matrix block were simulated using only one node, the size of the algebraic system would double over that in an unfractured reservoir simulation. If Gaussian elimination is used to solve any resulting large linear system, the number of arithmetic operations needed would be four to eight times that of the unfractured situation. Normally, there will be more than one matrix node per fracture node.

The simplest way to treat the fracture and matrix systems separately is to consider them sequentially. Solve first, say, for the state in the matrix blocks at the next time level, taking the matrix boundary values to be defined by the fracture data at the current time level. Then advance the fracture solution, using the new matrix data to define the matrix source term. Unfortunately, such a procedure is doomed to failure, since a small change in the boundary values can cause flow of a significant volume of fluid in comparison to the volume of the fractures. Consequently, it is easy to set up numerical oscillations: the matrix absorbs more fluid from the surrounding fractures on one step than can be resident there and then returns more to the fractures on the next step than their total volume. The matrix–fracture interaction must be handled *implicitly*, which normally would imply simultaneous solution. However, this interaction can be computed implicitly and separately by a particular linearization of the matrix problems to be made precise below. The final procedure requires solution of a large linear or non-linear system, corresponding to the fracture equations, and many small linear systems, each corresponding to a matrix block approximated by one to a few dozen nodes. The fracture system involves about as many nodes as are used in simulation of an unfractured reservoir.

However the algebraic equations generated on each time step are solved, the separation procedure detailed below will lead to more efficient computation, but there is an added advantage of the separation when it is carried out on a parallel computing device, since matrix blocks interact only through the fracture system and do not directly affect each other. As a consequence, it is natural to solve the matrix problems in parallel, and little communication of data between processors is required. The (non-linear) algebraic equations in the fractures can be treated either by domain decomposition techniques to solve the fracture equations in parallel or on a large vector processor. An additional possibility is to use a network consisting of a large vector computer and a set of smaller workstations. The fracture calculation can be performed on the large computer, and the matrix problems can be apportioned over the entire network (Douglas *et al.*, 1987; Arbogast, 1989b).

4.2 Some finite difference notation

Discretize the time variable by choosing $t^0, t^1, t^2, \ldots, t^N$ such that $0 = t^0 < t^1 < t^2 < \cdots < t^N$ and set $\Delta t^n = t^n - t^{n-1}$ for $n = 1, 2, \ldots, N$.

We shall discretize the space variables by defining grids over Ω and over each matrix block Ω_i. In a finite difference context, it is simplest to consider Ω and each Ω_i to be rectangular parallelepipeds; more general domains can be treated by either finite difference or finite element techniques quite analogous to the finite difference methods to be described herein. Suppose that $\Omega = [0, D_1] \times [0, D_2] \times [0, D_3]$. Then, divide each D_j into N_j intervals, which for simplicity we take to be of equal size $H_j = D_j/N_j, j = 1, 2, 3$. Thus, the fracture grid is the lattice of points

$$G_f = \{x_L : L = (L_1, L_2, L_3), \quad x_L = (L_1 H_1, L_2 H_2, L_3 H_3), \\ \text{and } L_j = 0, 1, 2, \ldots, N_j, \quad j = 1, 2, 3\}$$

We shall assume for notational convenience that the matrix blocks are all of the same size and consider a grid defined on the representative matrix block Q_{m}. Let h_j and n_j be analogous to H_j and N_j and set

$$G_{\mathrm{m}} = \{ y_l : l = (l_1, l_2, l_3), \quad y_l = (l_1 h_1, l_2 h_2, l_3 h_3), \quad \text{and } l_j = 0, 1, 2, \ldots, n_j, \quad j = 1, 2, 3 \}$$

Also, let

$$\mathring{G}_{\mathrm{m}} = \{ y_l : l = (l_1, l_2, l_3), \quad y_l = (l_1 h_1, l_2 h_2, l_3 h_3), \quad \text{and } l_j = 1, 2, \ldots, n_j - 1, \quad j = 1, 2. 3 \}$$

indicate the interior nodes and $\partial G_{\mathrm{m}} = G_{\mathrm{m}} \setminus \mathring{G}_{\mathrm{m}}$ the boundary nodes. (In practice the matrix grid should be graded to place more nodes near the surface of the block, as discussed in Douglas *et al.* (1989). Also, advantage should be taken of the symmetry of the solution on a matrix block to allow the solution to be computed only at necessary nodes.)

We wish to approximate the solutions to the models at the time levels t^n. A function Θ related to the fracture system will be approximated at each point $x_L \in G_{\mathrm{f}}$; this approximation will be denoted by

$$\Theta_L^n \approx \Theta(x_L, t^n)$$

Each H_j will be larger than the spacing of the fracture planes. Only those blocks which sit over the points of G_{f} will enter the discretization process. For a function θ associated with the block at the point $x_L \in G_{\mathrm{f}}$, we denote the approximation to θ at $y_l \in G_{\mathrm{m}}$ by

$$\theta_{L,l}^n \approx \theta(x_{L,l}, t^n)$$

where $x_{L,l} = x_L - y_l$ (we assume that a top corner of the block is x_L).

Time derivatives will be discretized by backward Euler approximations and spatial derivatives by standard seven-point differences. Grid points adjacent to a given point x_L are denoted by $x_{L \pm e_j}$, where $e_1 = (1, 0, 0)$, $e_2 = (0, 1, 0)$, and $e_3 = (0, 0, 1)$. Points half-way between x_L and $x_{L \pm e_j}$ are denoted by $x_{L \pm \frac{1}{2} e_j}$. A similar notation will be employed for points adjacent to $x_{L,l}$.

4.3 Single-phase flow

The discretization of either version of the single-phase model produces the same discrete model. Though it could easily be included using the linearization of

Eq. 2.46 (Douglas *et al.*, 1987; Arbogast, 1989b), gravity will be neglected both in the matrix blocks and in the fractures; moreover, the permeabilities will be taken to be scalar.

Let the spatial derivatives in Eq. 4.3 below be approximated by a standard seven-point difference operator,

$$\nabla_{h,L,l}\cdot\left[\frac{k}{\mu c}\nabla_{h,L,l}\rho_{\mathrm{m}}^{n}\right]=\sum_{j=1}^{3}\frac{1}{h_j^2}\left\{\frac{k(x_{L,l+\frac{1}{2}e_j})}{\mu c}(\rho_{\mathrm{m},L,l+e_j}^{n}-\rho_{\mathrm{m},L,l}^{n})\right.$$

$$\left.-\frac{k(x_{L,l-\frac{1}{2}e_j})}{\mu c}(\rho_{\mathrm{m},L,l}^{n}-\rho_{\mathrm{m},L,l-e_j}^{n})\right\} \tag{4.1}$$

and the spatial operator in Eq. 4.6 analogously. The numerical algorithm will be described below in four parts:

(i) *Initialization* at time t^0. For each L and l, let

$$\rho_{\mathrm{f},L}^{0}=\rho_{\mathrm{init,f}}(x_L)\quad\text{and}\quad\rho_{\mathrm{m},L,l}^{0}=\rho_{\mathrm{init,m}}(x_{L,l}) \tag{4.2}$$

(ii) *A time step in the matrix system.* For each L, l, and $n\geqslant 1$,

$$\phi(x_{L,l})\frac{\rho_{\mathrm{m},L,l}^{n}-\rho_{\mathrm{m},L,l}^{n-1}}{\Delta t^n}-\nabla_{h,L,l}\cdot\left[\frac{k}{\mu c}\nabla_{h,L,l}\rho_{\mathrm{m}}^{n}\right]=0\quad\text{if } y_l\in\mathring{G}_{\mathrm{m}} \tag{4.3}$$

$$\rho_{\mathrm{m},L,l}^{n}=\rho_{\mathrm{f},L}^{n}\quad\text{if } y_l\in\partial G_{\mathrm{m}} \tag{4.4}$$

(iii) *The matrix source term.* For each L and $n\geqslant 1$,

$$q_{\mathrm{m},L}^{n}=-\frac{1}{|Q|}\sum_{l}\phi(x_{L,l})\frac{\rho_{\mathrm{m},L,l}^{n}-\rho_{\mathrm{m},L,l}^{n-1}}{\Delta t^n}h_1h_2h_3 \tag{4.5}$$

(iv) *A time step in the fracture system.* For each L and $n\geqslant 1$,

$$\Phi(x_L)\frac{\rho_{\mathrm{f},L}^{n}-\rho_{\mathrm{f},L}^{n-1}}{\Delta t^n}-\nabla_{h,L}\cdot\left[\frac{K}{\mu c}\nabla_{h,L}\rho_{\mathrm{f}}^{n}\right]=q_{\mathrm{ext}}(x_L,t^n)+q_{\mathrm{m},L}^{n} \tag{4.6}$$

where $\rho_{\mathrm{f},L\pm e_j}^{n}$ is defined outside of G_{f} by reflection:

$$\rho_{\mathrm{f},L\pm e_j}^{n}=\rho_{\mathrm{f},L\mp e_j}^{n}\quad\text{for } x_{L\pm e_j}\notin G_{\mathrm{f}} \tag{4.7}$$

The algorithm is completely implicit, and Eq. 4.4 as stated requires the simultaneous solution of (ii)–(iv). This can be rectified by changing the way in which it is implemented. The key idea is to recognize that Eq. 4.3 is linear in $\rho^n_{\mathrm{m},L,l}$ and can be solved by finding any particular solution of Eq. 4.3 without requiring satisfaction of Eq. 4.4 and adding a solution to the problem not containing the term $\rho^{n-1}_{\mathrm{m},L,l}$ but satisfying proper boundary values. The simplest way to do this is to solve for $\bar{\rho}^n_{\mathrm{m},L,l}$ such that

$$\phi(x_{L,l})\frac{\bar{\rho}^n_{\mathrm{m},L,l}-\rho^{n-1}_{\mathrm{m},L,l}}{\Delta t^n}-\nabla_{h,L,l}\cdot\left[\frac{k}{\mu c}\nabla_{h,L,l}\bar{\rho}^n_{\mathrm{m}}\right]=0,\quad y_l\in\mathring{G}_{\mathrm{m}}, \tag{4.8}$$

$$\rho^n_{\mathrm{m},L,l}=\rho^{n-1}_{\mathrm{f},L},\quad y_l\in\partial G_{\mathrm{m}}, \tag{4.9}$$

and then to solve for $\breve{\rho}^n_{\mathrm{m},L,l}$ satisfying

$$\phi(x_{L,l})\frac{\breve{\rho}^n_{\mathrm{m},L,l}}{\Delta t^n}-\nabla_{h,L,l}\cdot\left[\frac{k}{\mu c}\nabla_{h,L,l}\breve{\rho}^n_{\mathrm{m}}\right]=0\quad \text{if } y_l\in\mathring{G}_{\mathrm{m}}, \tag{4.10}$$

$$\breve{\rho}^n_{\mathrm{m},L,l}=1\quad \text{if } y_l\in\partial G_{\mathrm{m}} \tag{4.11}$$

Both of these problems can be solved without knowing $\rho^n_{\mathrm{f},L}$. Clearly,

$$\rho^n_{\mathrm{m},L,l}=\bar{\rho}^n_{\mathrm{m},L,l}+(\rho^n_{\mathrm{f},L}-\rho^{n-1}_{\mathrm{f},L})\breve{\rho}^n_{\mathrm{m},L,l} \tag{4.12}$$

however, $\rho^n_{\mathrm{m},L,l}$ cannot be evaluated until $\rho^n_{\mathrm{f},L}$ is obtained from (iv).

Step (iii) can be implemented so as to define $q^n_{\mathrm{m},L}$ implicitly in terms of $\rho^n_{\mathrm{f},L}$:

$$\begin{aligned} q^n_{\mathrm{m},L} = {} & -\frac{1}{|Q|}\sum_l \phi(x_{L,l})\frac{\bar{\rho}^n_{\mathrm{m},L,l}-\rho^{n-1}_{\mathrm{m},L,l}}{\Delta t^n}h_1h_2h_3 \\ & -\frac{\rho^n_{\mathrm{f},L}-\rho^{n-1}_{\mathrm{f},L}}{\Delta t^n}\frac{1}{|Q|}\sum_l \phi(x_{L,l})\breve{\rho}^n_{\mathrm{m},L,l}h_1h_2h_3 \end{aligned} \tag{4.13}$$

Now, $\rho^n_{\mathrm{f},L}$ can be found from (iv), and the matrix solution can be updated using Eq. 4.12. The time step is finished. (Note that in this *linear* problem it is necessary to find $\breve{\rho}^n_{\mathrm{m},L,l}$ only once and not on each time step; the last sum in Eq. 4.13 need be evaluated just once, as well.)

The modified algorithm can be understood in physical terms. The problem given by Eqs 4.8 and 4.9 determines the flow of matrix fluid that results when a fracture system response is temporarily suspended; that is, when the boundary values on ∂G_{m} remain unchanged during the time step. The next relations, Eqs 4.10 and 4.11, determine the matrix flow that results from a unit change in the fracture density. The proper multiple of the latter solution added to the former is the complete solution to Eqs 4.3 and 4.4, as given in Eq. 4.12.

A finite element version of this algorithm appears in Arbogast (1989b) and Douglas *et al.* (1987). It was also shown in Arbogast (1989b) that the solutions of the finite element scheme converge to the solution of the differential model at the optimal rate as the Δt^n's and the grid spacings of G_f and G_m tend to zero.

4.4 Immiscible flow

The discretization of the model of Sections 3.2 and 3.3 leads to an algorithm that is noticeably more complicated than the one above; however, the main ideas are quite similar. The matrix equations will be completely linearized, but *not* the fracture equations. A Newton iteration will be used to solve the non-linear equations of the discretized fracture system; the discrete matrix system is directly solvable. The five parts of the algorithm again uncouple the calculations related to the matrix blocks from those of the fracture calculation:

(i) *Initialization.* For each L and l,

$$\Psi^0_{c,L} = \Psi_{\text{init,c}}(x_L) \tag{4.14}$$

$$\psi^0_{c,L,l} = \psi_{\text{init,c},L,l}(x_{L,l}) \tag{4.15}$$

$$S^0_L = P_c^{-1}(\Psi^0_{c,L} + (\rho_o - \rho_w)gz(x_L)) \tag{4.16}$$

$$s^0_{L,l} = p_c^{-1}(\psi^0_{c,L,l} + (\rho_o - \rho_w)gz(x_{L,l})) \tag{4.17}$$

The initial water potentials can be determined by solving Eqs 4.15 and 4.43 below. The Newton procedure will require an initial guess at the solution, so that it is convenient that $\Psi^0_{w,L}$ be found; the initial matrix water potential is of little interest.

(ii) *Matrix system.* For each L, l, and for $n \geqslant 1$, find $\{\bar{\psi}^n_{c,L,l}, \bar{\psi}^n_{w,L,l}\}$ by solving

$$\phi(x_{L,l})\frac{\bar{\psi}^n_{c,L,l} - \psi^{n-1}_{c,L,l}}{p'_c(s^{n-1})\Delta t^n} - \nabla_{h,L,l}\cdot[\lambda_w(s^{n-1})\nabla_{h,L,l}\bar{\psi}^n_w] = 0 \quad \text{if } y_l \in \mathring{G}_m, \tag{4.18}$$

$$-\nabla_{h,L,l}\cdot[\lambda(s^{n-1})\nabla_{h,L,l}\bar{\psi}^n_w + \lambda_o(s^{n-1})\nabla_{h,L,l}\psi^n_c] = 0 \quad \text{if } y_l \in \mathring{G}_m \tag{4.19}$$

$$\bar{\psi}^n_{c,L,l} = \Psi^{n-1}_{c,L} \quad \text{if } y_l \in \partial G_m, \tag{4.20}$$

$$\bar{\psi}^n_{w,L,l} = \Psi^{n-1}_{w,L} \quad \text{if } y_l \in \partial G_m, \tag{4.21}$$

and determine $\{\breve{\psi}^n_{\mathrm{c},L,l}, \breve{\psi}^n_{\mathrm{w},L,l}\}$ and $\{\hat{\psi}^n_{\mathrm{c},L,l}, \hat{\psi}^n_{\mathrm{w},L,l}\}$ by solving

$$\phi(x_{L,l}) \frac{\breve{\psi}^n_{\mathrm{c},L,l}}{p'_{\mathrm{c}}(s^{n-1})\Delta t^n} - \nabla_{h,L,l} \cdot [\lambda_{\mathrm{w}}(s^{n-1})\nabla_{h,L,l}\breve{\psi}^n_{\mathrm{w}}] = 0 \quad \text{if } y_l \in \mathring{G}_{\mathrm{m}}, \tag{4.22}$$

$$-\nabla_{h,L,l} \cdot [\lambda(s^{n-1})\nabla_{h,L,l}\breve{\psi}^n_{\mathrm{w}} + \lambda_{\mathrm{o}}(s^{n-1})\nabla_{h,L,l}\breve{\psi}^n_{\mathrm{c}}] = 0 \quad \text{if } y_l \in \mathring{G}_{\mathrm{m}}, \tag{4.23}$$

$$\breve{\psi}^n_{\mathrm{c},L,l} = 1 \quad \text{if } y_l \in \partial G_{\mathrm{m}}, \tag{4.24}$$

$$\breve{\psi}^n_{\mathrm{w},L,l} = 0 \quad \text{if } y_l \in \partial G_{\mathrm{m}}, \tag{4.25}$$

and

$$\phi(x_{L,l}) \frac{\hat{\psi}^n_{\mathrm{c},L,l}}{p'_{\mathrm{c}}(s^{n-1})\Delta t^n} - \nabla_{h,L,l} \cdot [\lambda_{\mathrm{w}}(s^{n-1})\nabla_{h,L,l}\hat{\psi}^n_{\mathrm{w}}] = 0 \quad \text{if } y_l \in \mathring{G}_{\mathrm{m}}, \tag{4.26}$$

$$-\nabla_{h,L,l} \cdot [\lambda(s^{n-1})\nabla_{h,L,l}\hat{\psi}^n_{\mathrm{w}} + \lambda_{\mathrm{o}}(s^{n-1})\nabla_{h,L,l}\hat{\psi}^n_{\mathrm{c}}] = 0 \quad \text{if } y_l \in \mathring{G}_{\mathrm{m}}, \tag{4.27}$$

$$\hat{\psi}^n_{\mathrm{c},L,l} = 0 \quad \text{if } y_l \in \partial G_{\mathrm{m}}, \tag{4.28}$$

$$\hat{\psi}^n_{\mathrm{w},L,l} = 1 \quad \text{if } y_l \in \partial G_{\mathrm{m}}, \tag{4.29}$$

where

$$\begin{aligned}\nabla_{h,L,l} \cdot [\lambda_\alpha(s^{n-1})\nabla_{h,L,l}\psi^n] = \sum_{j=1}^{3} \frac{1}{h_j^2} \Bigg\{ &\lambda_{\mathrm{a}}\left(\frac{s^{n-1}_{L,l+e_j} + s^{n-1}_{L,l}}{2}\right)(\psi^n_{L,l+e_j} - \psi^n_{L,l}) \\ &- \lambda_\alpha\left(\frac{s^{n-1}_{L,l} + s^{n-1}_{L,l-e_j}}{2}\right)(\psi^n_{L,l} - \psi^n_{L,l-e_j})\Bigg\}\end{aligned} \tag{4.30}$$

Note that these equations are indeed linear, since the mobilities and p'_{c} are evaluated at the previous time level. The matrix potentials $\psi^n_{\mathrm{c},L,l}$ and $\psi^n_{\mathrm{w},L,l}$ are defined below in Eqs 4.39 and 4.40; they satisfy the expected equations, namely Eqs 4.42–4.45. Equations 4.18–4.21 define a particular solution to the linear equations, while Eqs 4.22–4.25 and Eqs 4.26–4.29 give solutions to the homogeneous problem which describe unit changes in the boundary conditions.

(iii) *The matrix source term.* For each L and $n \geqslant 1$,

$$\tilde{\psi}^n_{\mathrm{w},L,l} = \bar{\psi}^n_{\mathrm{w},L,l} + (\Psi^n_{\mathrm{c},L} - \Psi^{n-1}_{\mathrm{c},L})\breve{\psi}^n_{\mathrm{w},L,l} + (\Psi^n_{\mathrm{w},L} - \Psi^{n-1}_{\mathrm{w},L})\hat{\psi}^n_{\mathrm{w},L,l} \quad (4.31)$$

$$\tilde{s}^n_{L,l} = p_{\mathrm{c}}^{-1}(\tilde{\psi}^n_{\mathrm{w},L,l} + (\rho_{\mathrm{o}} - \rho_{\mathrm{w}})gz(x_{L,l})) \quad (4.32)$$

$$q^n_{\mathrm{m,w},L} = -\frac{1}{|Q|}\sum_l \phi(x_{L,l})\frac{\tilde{s}^n_{L,l} - s^{n-1}_{L,l}}{\Delta t^n} h_1 h_2 h_3 \quad (4.33)$$

The quantity $q^n_{\mathrm{m,w},L}$ is given implicitly in terms of the fracture potentials at the nth time level; however, in view of Eqs 4.40 and 4.41 below, Eq. 4.33 is clearly a discretization of Eq. 3.22.

(iv) *Fracture system.* (This is the most critical system to solve well.) For each L and $n \geqslant 1$, solve the following non-linear system of equations for $\Psi^n_{\mathrm{c},L}$, $\Psi^n_{\mathrm{w},L}$, S^n_L and $q^n_{\mathrm{m,w},L}$:

$$\Phi(x_L)\frac{S^n_L - S^{n-1}_L}{\Delta t^n} - \sum_{j=1}^{3}\frac{1}{H_j^2}\{\Lambda^n_{\mathrm{w},L,l}(\Psi^n_{\mathrm{w},L+e_j} - \Psi^n_{\mathrm{w},L}) - \Lambda^n_{\mathrm{w},L-e_j,j}(\Psi^n_{\mathrm{w},L} - \Psi^n_{\mathrm{w},L-e_j})\} = q_{\mathrm{ext,w}}(x_L, t^n) + q^n_{\mathrm{m,w},L}, \quad (4.34)$$

$$-\sum_{j=1}^{3}\frac{1}{H_j^2}\{\Lambda^n_{L,j}(\Psi^n_{\mathrm{w},L+e_j} - \Psi^n_{\mathrm{w},L}) - \Lambda^n_{L-e_j,j}(\Psi^n_{\mathrm{w},L} - \Psi^n_{\mathrm{w},L-e_j}) + \Lambda^n_{\mathrm{o},L,j}(\Psi^n_{\mathrm{c},L+e_j} - \Psi^n_{\mathrm{c},L}) - \Lambda^n_{\mathrm{o},L-e_j,j}(\Psi^n_{\mathrm{c},L} - \psi^n_{\mathrm{c},L-e_j})\} = q_{\mathrm{ext}}(x_L, t^n) \quad (4.35)$$

$$S^n_L = P_{\mathrm{c}}^{-1}(\Psi^n_{\mathrm{c},L} + (\rho_{\mathrm{o}} - \rho_{\mathrm{w}})gz(x_L)) \quad (4.36)$$

where again the no-flow boundary conditions of Eq. 3.24 are imposed by reflection:

$$\Psi^n_{\alpha,L\pm e_j} = \Psi^n_{\alpha,L\mp e_j}, \quad \alpha = \mathrm{w}, \mathrm{c} \quad (4.37)$$

if $x_{L\pm e_j}$ is outside the reservoir. The mobilities should be upstream weighted:

$$\Lambda^n_{\alpha,L,j} = \begin{cases} \Lambda_\alpha(S^n_{L+e_j}), & \text{if } \Psi^n_{\alpha,L} < \Psi^n_{\alpha,L+e_j}, \\ \Lambda_\alpha(S^n_L), & \text{otherwise,} \end{cases} \tag{4.38}$$

for $\alpha = \text{o}, \text{w}$ (where $\Psi^n_{\text{o},L} = \Psi^n_{\text{c},L} + \Psi^n_{\text{w},L}$). This is intended to prevent numerical instabilities (Peaceman, 1977; Douglas *et al.*, 1989).

Note that Eqs 4.31–4.38 form a completely implicit procedure for the fracture quantities and the boundary values on the blocks. Indeed, the equations can be interpreted as a system of non-linear equations in the fracture quantities alone, and a variant of Newton's method can be used to obtain the solution; this is not completely straightforward, because of sharp corners at S_{max} or s_{max} in the graphs of the capillary functions (Figs 7 and 9). Since the rate of convergence in a Newton iteration is related to the values of the second derivatives (which fail to exist at the corners of the capillary curves), it is necessary to modify the algorithm so as to aid the convergence by, for example, requiring the solution Ψ_c to take at least two iterations to cross a value slightly above the critical value $\Psi_{\text{c},L} = P_\text{c}(S_{\text{max}}) - (\rho_\text{o} - \rho_\text{w})gz = -(\rho_\text{o} - \rho_\text{w})gz$ at any applicable point. It is also necessary to recognize that $P'_\text{c}(S_{\text{max}}) = p'_\text{c}(s_{\text{max}}) = -\infty$ (or a large negative number in practice; see Douglas *et al.*, 1989, for details).

(v) *Matrix update*. For each L, l, and $n \geqslant 1$, let

$$\psi^n_{\text{c},L,l} = \bar{\psi}^n_{\text{c},L,l} + (\Psi^n_{\text{c},L,l} - \Psi^{n-1}_{\text{c},L,l})\breve{\psi}^n_{\text{c},L,l} + (\Psi^n_{\text{w},L,l} - \Psi^{n-1}_{\text{w},L,l})\hat{\psi}^n_{\text{c},L,l} \tag{4.39}$$

$$\psi^n_{\text{w},L,l} = \bar{\psi}^n_{\text{w},L,l} + (\Psi^n_{\text{c},L,l} - \Psi^{n-1}_{\text{c},L,l})\breve{\psi}^n_{\text{w},L,l} + (\Psi^n_{\text{w},L,l} - \Psi^{n-1}_{\text{w},L,l})\hat{\psi}^n_{\text{w},L,l} \tag{4.40}$$

$$s^n_{L,l} = p_\text{c}^{-1}(\psi^n_{\text{c},L,l} + (\rho_\text{o} - \rho_\text{w})gz(x_{L,l})) \tag{4.41}$$

This completes the time step.

The above algorithm can be implemented sequentially. The following discrete matrix problem has been solved:

$$\phi(x_{L,l})\frac{\psi^n_{\text{c},L,l} - \psi^{n-1}_{\text{c},L,l}}{p'_\text{c}(s^{n-1})\Delta t^n} - \nabla_{h,L,l}\cdot[\lambda_\text{w}(s^{n-1})\nabla_{h,L,l}\psi^n_\text{w}] = 0 \quad \text{if } y_l \in \mathring{G}_\text{m} \tag{4.42}$$

$$-\nabla_{h,L,l}\cdot[\lambda(s^{n-1})\nabla_{h,L,l}\psi^n_\text{w} + \lambda_\text{o}(s^{n-1})\nabla_{h,L,l}\psi^n_\text{c}] = 0 \quad \text{if } y_l \in \mathring{G}_\text{m} \tag{4.43}$$

$$\psi^n_{\text{c},L,l} = \Psi^n_{\text{c},L} \quad \text{if } y_l \in \partial G_\text{m} \tag{4.44}$$

$$\psi^n_{\mathrm{w},L,l} = \Psi^n_{\mathrm{w},L} \quad \text{if } y_l \in \partial G_{\mathrm{m}} \tag{4.45}$$

Assuming that the wetting fluid is the denser, it should be noted that the block associated with the fracture point x_L is interpreted to lie below x_L for imbibition and above for drainage; otherwise, fluid is trapped by the numerical simulation as P_c tends to zero. We have considered the case of imbibition.

It should be remarked that the discretization of the fracture system conserves mass, except for inexactness caused by not carrying out the Newton iteration to complete convergence. The linearized matrix problems do not; however, there is no *net* mass balance error, as any fluid that leaves the matrix is transmitted to the fracture system. There is a small error in the time at which fluid is transferred from one system to the other.

4.5 Simplified models for immiscible flow

A similar procedure to that presented in the last subsection can be given for the models of Sections 3.4 and 3.5. We describe the discretization of the model that neglects gravity in the matrix first; the limit model requires only a simple modification of this algorithm.

The initialization and fracture system steps (i.e. (i) and (iv)) of the previous algorithm remain unchanged; steps (ii), (iii) and (v) become:

(ii) *Matrix system.* For each L, l, and $n \geqslant 1$, determine $\bar{s}^n_{L,l}$ and $\check{s}^n_{L,l}$ by solving

$$\phi(x_{L,l})\frac{\bar{s}^n_{L,l} - s^{n-1}_{L,l}}{\Delta t^n} - \nabla_{h,L,l}\cdot\left[\frac{\lambda_{\mathrm{w}}(s^{n-1})\lambda_{\mathrm{o}}(s^{n-1})}{\lambda(s^{n-1})}p_{\mathrm{c}}'(s^{n-1})\nabla_{h,L,l}\bar{s}^n\right] = 0$$

$$\text{if } y_l \in \mathring{G}_{\mathrm{m}}, \tag{4.46}$$

$$\bar{s}^n_{L,l} = p_{\mathrm{c}}^{-1}(\Psi^{n-1}_{\mathrm{c},L} + (\rho_{\mathrm{o}} - \rho_{\mathrm{w}})gz(x_{L,l})) \quad \text{if } y_l \in \partial G_{\mathrm{m}}, \tag{4.47}$$

and

$$\phi(x_{L,l})\frac{\check{s}^n_{L,l}}{\Delta t^n} - \nabla_{h,L,l}\cdot\left[\frac{\lambda_{\mathrm{w}}(s^{n-1})\lambda_{\mathrm{o}}(s^{n-1})}{\lambda(s^{n-1})}p_{\mathrm{c}}'(s^{n-1})\nabla_{h,L,l}\check{s}^n\right] = 0$$

$$\text{if } y_l \in \mathring{G}_{\mathrm{m}}, \tag{4.48}$$

$$\check{s}_{L,l} = 1 \quad \text{if } y_l \in \partial G_{\mathrm{m}} \tag{4.49}$$

Here, the spatial derivatives are approximated as in Eq. 4.30 above.

(iii) *The matrix source term.* For each L and $n \geqslant 1$, evaluate the matrix source term $q^n_{\mathrm{m,w},L}$ by Eq. 4.33, where now

$$\tilde{s}^n_{L,l} = \bar{s}^n_{L,l} + [p_c^{-1}(\Psi^n_{c,L} + (\rho_o - \rho_w)gz(x_L)) - p_c^{-1}(\Psi^{n-1}_{c,L} + (\rho_o - \rho_w)gz(x_L))]\bar{s}^n_{L,l} \tag{4.50}$$

(v) *Matrix update.* The matrix saturation $s^n_{L,l}$ is equal to $\tilde{s}^n_{L,l}$ given by Eq. 4.50. It should be clear that the matrix problem has solved the equations

$$\phi(x_{L,l})\frac{s^n_{L,l} - s^{n-1}_{L,l}}{\Delta t^n} - \nabla_{h,L,l}\cdot\left[\frac{\lambda_w(s^{n-1})\lambda_o(s^{n-1})}{\lambda(s^{n-1})}p_c'(s^{n-1})\nabla_{h,L,l}s^n\right] = 0 \quad \text{if } y_l \in \mathring{G}_m, \tag{4.51}$$

$$s^n_{L,l} = p_c^{-1}(\Psi^n_{c,L} + (\rho_o - \rho_w)gz(x_L)) \quad \text{if } y_l \in \partial G_m. \tag{4.52}$$

A finite element algorithm similar to the one above has been shown by Arbogast *et al.* (1988) to be convergent under the assumption that the mobilities do not degenerate to zero.

The limit model is the restriction of the above algorithm that results from not allowing any of the matrix nodes to lie inside the blocks; that is, $\mathring{G}_m = \varnothing$. Then Eq. 4.52 holds for all matrix nodes, and a single linear system must be solved (the one of (iv)).

5. Some computational results

We present recovery curves for some simulations of petroleum reservoir water-flooding. Though these curves give only a gross indication of the flow of fluids within the reservoir, they are sufficient to illustrate important features of naturally fractured reservoir simulation.

In all cases, the reservoir is assumed horizontal and rectangular, with height 10 m and length 300 m. For computational simplicity, the reservoir is assumed to be uniform in the other direction; consequently, the fracture calculations are two-dimensional over Ω, though the matrix calculations must remain three-dimensional over Q. Fracture porosity and permeability are assumed to be 0.01 and 1 darcy, respectively. The corresponding quantities for the matrix are 0.2 and 5 millidarcies. Initially, the reservoir contains 76% oil ($\mu_o = 0.5$ cP and $\rho_o = 0.7\ \mathrm{g/cm^3}$) and 24% water ($\mu_w = 2$ cP and $\rho_w = 1\ \mathrm{g/cm^3}$). Water is injected uniformly into the reservoir along one end at a constant rate of one pore-volume every five years. Oil and water are produced at the top of the other end. The recovery curve is given as the graph of the cumulative amount of oil produced versus the cumulative amount of water injected. The residual water level in the matrix is 0.20, while the residual oil level is 0.15; hence, there is a maximum recovery of 0.65 pore-volumes. In our simulations, the only reservoir parameter that is varied is the size of the matrix blocks, which we take to be cubes.

For the spatial discretization, we took 40 nodes in the horizontal direction and 10 in the vertical direction over Ω. In the matrix block Q, we took the equivalent of 16 interior nodes in the full model (Sections 3.2–3.3) simulations (four nodes on each of four horizontal planes), and we took the equivalent of 27 interior nodes in the simulations using the model that neglects gravity in the matrix (Section 3.4). The time step varied from one day initially to twenty days near the end of the simulations. (An entire simulation was for about twenty years.)

First we compare a fractured reservoir to an unfractured one, with the unfractured one possessing the matrix properties of the fractured one, except for some minor modifications to account for the fracture space. The numerical procedure for the unfractured case is a straightforward modification (consisting of deletions) of the procedure for the fractures presented in Section 4.4. The fractured reservoir has blocks of sidelength 200 cm, and the simulation is given for the full model. From Fig. 11, we see that indeed there is a significant difference in response between the two reservoirs. Better sweep efficiency results in the unfractured reservoir, though it takes a much higher pressure gradient across the reservoir to obtain this recovery. The curve for the unfractured reservoir has a fairly obvious 'breakthrough' at about 0.5 pore-volumes injected, beyond which recovery drops dramatically; on the other hand, the recovery curve for the fractured reservoir varies much more gently.

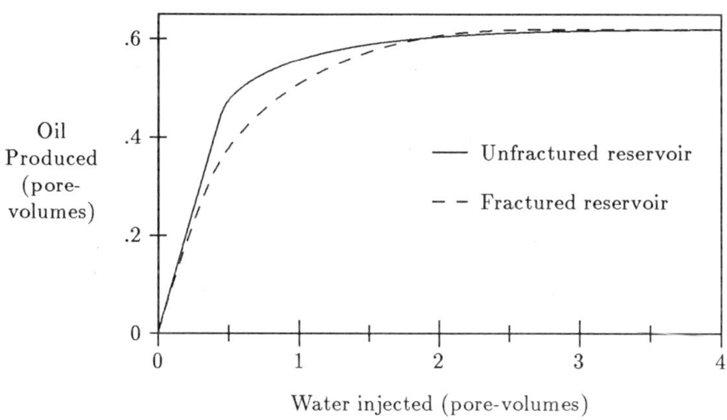

Figure 11. A comparison of the production data for an unfractured reservoir and for a naturally fractured reservoir with matrix blocks of sidelength 200 cm.

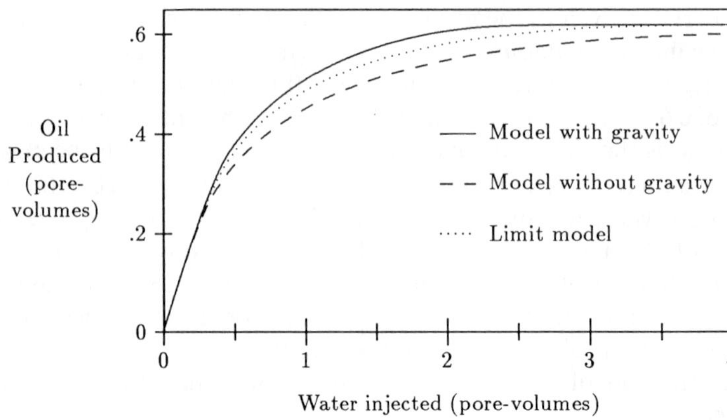

Figure 12. A comparison of the production data for the various immiscible models for a reservoir with matrix blocks of sidelength 200 cm.

Next, we look at the three immiscible models of this presentation. As can be seen in Fig. 12, these models predict significantly different recoveries. The full model predicts the greatest sweep efficiency. Surprisingly, in this series of simulations the limit model (Section 3.5) seems to give a better prediction than the model which simply neglects gravity in the matrix (see Douglas *et al.* (1989) for additional simulations of the simplified models).

Finally, the effect of matrix block size on the full model is considered in Fig. 13. Sweep efficiency increases with increasing block size. It can also be seen that, as the block size tends to zero, the recovery curves of the simulations tend to that of the limit model (where the block size is 0 cm). If we continue reducing the block size below 50 cm, the recovery curves become almost indistinguishable from that for the limit model at about 10 cm.

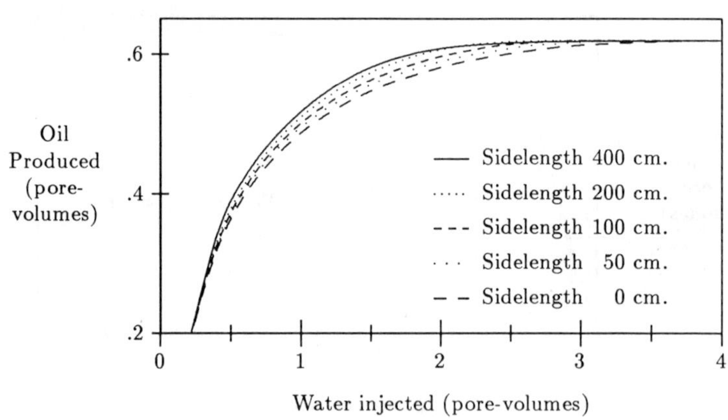

Figure 13. The effect of matrix block size on the production data, and convergence to the limit model.

6. Some other models

Many other dual porosity models can be defined to incorporate various physical phenomena. We briefly mention a few of these in this section.

Miscible flows can be considered; a simple dual porosity model for such appears in Arbogast *et al.* (1989b). It is similar to the models presented above. A new feature is that a continuity of component mass concentration is imposed at the surfaces of each matrix block in addition to the imposition of continuity of pressure. This model is not completely satisfactory for simulating miscible displacement; the 'first-order' model discussed below provides a much more realistic simulation of that physical process; see Arbogast (1989a).

The immiscible water-flooding problem is sensitive to gravitational segregation of water and oil. If the matrix block height is large, it may be important to model this effect. It was incorporated somewhat crudely in the model of Section 3, since the block height was considered to be small (infinitely so, in the homogenized version). We assumed that at each time a potential equilibrium exists from the top to the bottom of the block as its surface. The block height can be considered to be large and only the width and depth to be small. Homogenization will produce a model in which the blocks have the original height but no width or depth. Then, the full vertical variation in the potentials is reflected in the boundary condition for the matrix problems (see Arbogast *et al.* (1989b) and Douglas (1989) for additional information).

In the models treated in this presentation, the matrix boundary conditions have always imposed continuity between matrix and fracture pressures. This means that the 'capillary end effect' of Richardson *et al.* (1952), Douglas *et al.* (1958) and Collins (1961) has been imposed without exception in the water-flooding models; Russell (1988) has considered the possibility that flow in the fractures is sufficiently rapid that there is not enough time for equilibrium to be reached between the matrix and fracture potentials, and he has proposed replacing the strict continuity of oil pressure between the matrix and fracture systems by a 'third-type' or Robin boundary condition. In our notation (see Section 3), this would take the form

$$-\lambda_{\mathrm{o}}|p_{\mathrm{c}}'(s)|\nabla s\cdot v = C\left(p_{\mathrm{c}}(s) - \frac{1}{|\hat{\Omega}_i|}\int_{\hat{\Omega}_i} P_{\mathrm{c}}(S(\xi,t))\,\mathrm{d}\xi\right) \quad \text{for } x\in\partial\Omega_i,\quad t>0 \tag{6.1}$$

where C is some empirical constant. The effect of such a boundary condition is to permit a delay in the imbibition of water into the matrix blocks.

Finally, we mention a class of models that might be referred to as 'first-order' models. These models are not strictly derived from homogenization theory, since that theory assumes that the matrix block size tends to zero. The leading terms in these models are derived from homogenization (giving the 'zeroth-order' model), but an attempt is made to take into account the actual finite size of the matrix blocks by assuming that the boundary values on each matrix block vary linearly in space at any given time, rather than being constant in space. Such a model has been considered for single-component, single-phase flow in Arbogast (1988) and for two-component miscible flow in Arbogast (1989a).

Acknowledgement

This work was supported in part by the National Science Foundation.

References

Amaziane, B. and Bourgeat, A. (1988). Effective behaviour of two-phase flow in heterogeneous reservoir. *In* 'Numerical Simulation in Oil Recovery' (M.F. Wheeler, ed.), The IMA Volumes in Mathematics and its Applications, Vol. 11, pp. 1–22. Springer-Verlag, Berlin and New York.

Antoncev, S.N. (1972). On the solvability of boundary value problems for degenerate two-phase porous flow equations, *Dinamika Splošnoĭ Sredy vyp.* **10**, 28–53 (in Russian).

Arbogast, T. (1988). The double porosity model for single phase flow in naturally fractured reservoirs. *In* 'Numerical Simulation in Oil Recovery' (M.F. Wheeler, ed.), The IMA Volumes in Mathematics and its Applications, Vol. 11, pp. 23–45. Springer-Verlag, Berlin and New York.

Arbogast, T. (1989a). On the simulation of incompressible, miscible displacement in a naturally fractured petroleum reservoir, *R.A.I.R.O. Modél Math. Anal. Numér.* **23**, to be published.

Arbogast, T. (1989b). Analysis of the simulation of single phase flow through a naturally fractured reservoir, *SIAM J. Numer. Anal.* **26**, 12–29.

Arbogast, T., Douglas, J., Jr. and Santos, J.E. (1988). Two-phase immiscible flow in naturally fractured reservoirs. *In* 'Numerical Simulation in Oil Recovery', (M.F. Wheeler, ed.), The IMA Volumes in Mathematics and its Applications, Vol. 11, pp. 47–66. Springer-Verlag, Berlin and New York.

Arbogast, T., Douglas, J., Jr. and Hornung, U. (1989a). Derivation of the double porosity model of single phase flow via homogenization theory, to be published.

Arbogast, T., Douglas, J., Jr. and Hornung, U. (1989b). Modeling of naturally fractured reservoirs by formal homogenization techniques, to be published.

Barenblatt, G.I., Zheltov, I.P. and Kochina, I.N. (1960). Basic concepts in the theory of seepage of homogeneous liquids in fissured rocks [strata], *Prikl. Mat. Mekh.* **24**, 852–864; *J. Appl. Math. Mech.* **24**, 1286–1303.

Bear, J. (1972). 'Dynamics of Fluids in Porous Media'. Elsevier, New York.

Bensoussan, A., Lions, J.L. and Papanicolaou, G. (1978). 'Asymptotic Analysis for Periodic Structures'. North-Holland, Amsterdam.

Bourgeat, A. (1986). Homogenization of two phase flow equations, *Proc. Symp. Pure Math.* **45**, 157–163.

Chavent, G. (1976). A new formulation of diphasic incompressible flows in porous media. *In* 'Applications of Methods of Functional Analysis to Problems in Mechanics', Lecture Notes in Mathematics, Vol. 503. Springer-Verlag, Berlin and New York.

Chavent, G. and Jaffré, J. (1986). 'Mathematical Models and Finite Elements for Reservoir Simulation'. North-Holland, Amsterdam.

Collins, R.E. (1961). 'Flow of Fluids Through Porous Materials'. Reinhold, New York.

Douglas, J., Jr. (1989). Three models for waterflooding in a naturally fractured petroleum reservoir, to be published.

Douglas, J., Jr., Arbogast, T. and Paes Leme, P.J. (1989). Two models for the waterflooding of naturally fractured reservoirs. *Proc. Tenth SPE Symposium on Reservoir Simulation*, Paper SPE 18 425, pp. 219–225. Society of Petroleum Engineers, Dallas, TX.

Douglas, J., Jr., Blair, P.M. and Wagner, R.J. (1958). Calculation of linear waterflood behavior including the effects of capillary pressure, *Trans. AIME* **213**, 96–104.

Douglas, J., Jr., Paes Leme, P.J., Arbogast, T. and Schmitt, T. (1987). Simulation of flow in naturally fractured reservoirs. *Proc. Ninth SPE Symposium on Reservoir Simulation*, Paper SPE 16019, pp. 271–279. Society of Petroleum Engineers, Dallas, TX.

Ene, H.I. (1990). Application of the homogenization method to transport in porous media. *In* 'Dynamics of Fluids in Hierarchical Porous Media', (J.H. Cushman, ed.), pp. 223–241. Academic Press, London.

Gilman, J.R. (1986). An efficient finite-difference method for simulating phase segregation in the matrix blocks in double-porosity reservoirs, *Soc. Petrol. Eng. J.* **26**, 403–413.

Kazemi, H. (1969). Pressure transient analysis of naturally fractured reservoirs with uniform fracture distribution, *Soc. Petrol. Eng. J.* **9**, 451–462.

Kazemi, H. and Gilman, J.R. (1983). Improvements in simulation of naturally fractured reservoirs, *Soc. Petrol. Eng. J.* **23**, 695–707.

Kazemi, H. and Gilman, J.R. (1987). Improved calculations for viscous and gravity displacement in matrix blocks in dual-porosity simulators. *Proc. Ninth SPE Symposium on Reservoir Simulation*, Paper SPE 16010, pp. 193–208. Society of Petroleum Engineers, Dallas, TX.

Peaceman, D.W. (1977). 'Fundamentals of Numerical Reservoir Simulation'. Elsevier, New York.

Pirson, S.J. (1953). Performance of fractured oil reservoirs, *Bull. Am. Assoc. Petrol. Geologists* **37**, 232–244.

Richardson, J.G., Kerver, J.K., Hafford, J.A. and Osoba, J.S. (1952). Laboratory determination of relative permeability, *Trans. AIME* **195**, 187–196.

Russell, T.F. (1988). Formulation of a model accounting for capillary non-equilibrium in naturally fractured subsurface formations. *Proc. Fourth Wyoming Enhanced Oil Recovery Symposium* (R.E. Ewing and D. Copeland, eds), pp. 103–114. University of Wyoming, Laramie, WY.

Sanchez-Palencia, E. (1980). 'Non-homogeneous Media and Vibration Theory', Lecture Notes in Physics, Vol. 127. Springer-Verlag, Berlin and New York.

Scheidegger, A.E. (1974). 'The Physics of Flow Through Porous Media'. University of Toronto Press, Toronto.

Sonier, F., Souillard, P. and Blaskovich, F.T. (1986). Numerical simulation of naturally fractured reservoirs. *Proc. 61st Annual Technical Conference and Exhibition of the Society of Petroleum Engineers, Dallas, Texas*, Paper SPE 15 627. Society of Petroleum Engineers, Dallas, TX.

Stein, E.M. (1970). 'Singular Integrals and Differentiability Properties of Functions'. Princeton University Press, Princeton, NJ.

de Swaan, A. (1976). Analytic solutions for determining naturally fractured reservoir properties by well testing, *Soc. Petrol. Eng. J.* **16**, 117–122.

de Swaan, A. (1978). Theory of waterflooding in fractured reservoirs, *Soc. Petrol. Eng. J.* **18**, 117–122.

Thomas, L.K., Dixon, T.N. and Pierson, R.G. (1983). Fractured reservoir simulation, *Soc. Petrol. Eng. J.* **23**, 42–54.

Warren, J.E. and Root, P.J. (1963). The behaviour of naturally fractured reservoirs, *Soc. Petrol. Eng. J.* **3**, 245–255.

Chapter VIII

Application of the Homogenization Method to Transport in Porous Media

Horia I. Ene
INCREST, Department of Mathematics, Bd. Pacii 220, 79622 Bucharest, Romania
(Present address: University of Bucharest, Faculty of Mathematics, Mechanics Department, St. Academici 14, 70109 Bucharest, Romania)

1. Introduction

A porous medium may have a microstructure on a scale much smaller than the macroscopic scale which is of interest. Classically, phenomena arising at the macroscopic level are described by taking mean values of the microscopic quantities. There are several methods for doing this. Of prime interest in this chapter is the homogenization method. A more descriptive name of this method is 'an asymptotic method for the study of periodic media'. Such a name is long, but it contains more detail concerning the idea of taking mean values for periodic structures.

What are the reasons for using a periodic model of a porous medium? The major reason for this choice is that for such a medium we have at our disposal a good mathematical technique which permits us to prove the existence and uniqueness of solutions, to construct the homogenization operator and the homogenization coefficients, and so on. There is another important reason: by use of this method it is possible to obtain new results concerning the macroscopic equations arising in flow through porous media, to indicate the boundary conditions, or to make more clear the dimensionless parameters which are significant in such phenomena. The present chapter deals with this aspect. The reader interested in mathematical aspects of the method is referred to the excellent books of Bensoussan *et al.* (1978) and Sanchez-Palencia (1980).

A more detailed presentation of the problem may be found in the review paper by Levy (1987) or in the book by Ene and Polisevski (1987).

DYNAMICS OF FLUIDS IN HIERARCHICAL POROUS MEDIA
ISBN 0-12-200260-1

2. The homogenization method

2.1 Periodic media

If we consider a medium with fine periodic structure with period l and if we let L denote a characteristic length of the macroscopic domain, we may define a natural small parameter ε by l/L. In studies of the mechanical behaviour of such media, it is clear that the quantities describing the phenomena depend on ε. The asymptotic process $\varepsilon \to 0$ represents the transition from the microscopic to macroscopic. To make precise the fact that the medium varies rapidly on the microscopic scale l and slowly on the macroscopic scale L, we assume that every property of the medium is of the form $f(\mathbf{x}, \mathbf{y})$ with $\mathbf{y} = \mathbf{x}/\varepsilon$. We denote by $\mathbf{x} = (x_1, x_2, x_3)$ the position vector of a point in Cartesian coordinates and by $\mathbf{y} = (y_1, y_2, y_3)$ the vector of stretched coordinates. The period may be considered homotetic with the ratio ε of a basic period Y in the stretched coordinates $\mathbf{y} = \mathbf{x}/\varepsilon$. The periodicity of the medium is εY. Now, because $f(\mathbf{x}, \mathbf{y})$ has period Y with respect to the $\mathbf{y}$ variable, the function f of the $\mathbf{y}$ variable takes equal values on the opposite faces of Y. Such a function $f(\mathbf{x}, \mathbf{y})$ with $\mathbf{y} = \mathbf{x}/\varepsilon$ will be called locally periodic.

The unknown field quantity $u^\varepsilon(\mathbf{x})$ will be in the form of a double-scale asymptotic expansion:

$$u^\varepsilon(\mathbf{x}) = u^0(\mathbf{x}, \mathbf{y}) + \varepsilon u^1(\mathbf{x}, \mathbf{y}) + \varepsilon^2 u^2(\mathbf{x}, \mathbf{y}) + \ldots$$

In such expansion, the terms $u^i(\mathbf{x}, \mathbf{y})$ are periodic in the $\mathbf{y}$ variable with the same period as that of the structure. If we introduce a two-scale expansion in the partial differential equations of our problem, and identify equal powers of ε, we obtain equations in $\mathbf{x}$ and $\mathbf{y}$ variables. At this point, it is important to remark that in a periodic microstructure the equations in $\mathbf{y}$ are solvable and, consequently, the equations in $\mathbf{x}$ give us the macroscopic, or homogenized, equations describing the global behaviour of the medium. In fact, this process gives the passage from the microscopic description of the medium to the macroscopic one. The hypothesis of periodicity plays an essential part in obtaining results in a precise form, which may be used as a simplified model for more general situations. In order to illustrate these ideas, we shall try to explain them using as an example the elliptic equation in divergence form.

2.2. Homogenization of the elliptic equation

We consider the simplest boundary-value problem:

$$-\frac{\partial}{\partial x_i}\left(a_{ij}^\varepsilon(\mathbf{x})\frac{\partial u^\varepsilon}{\partial x_j}\right) = f(\mathbf{x}) \quad \text{in } \Omega \tag{2.1}$$

$$u^{\varepsilon}\Big|_{\partial\Omega} = 0 \tag{2.2}$$

where $f(\mathbf{x})$ is a given smooth function defined on Ω and $a_{ij}^{\varepsilon}(\mathbf{x}) = a_{ij}(\mathbf{x}/\varepsilon)$ are smooth real functions, Y-periodic, such that there exists $\gamma > 0$ with $a_{ij}(\mathbf{y})\xi_i\xi_j \geqslant \gamma\xi_i\xi_i$, $\forall \mathbf{y} \in Y$, $\boldsymbol{\xi} \in \mathbb{R}^3$ and $a_{ij}^{\varepsilon} = a_{ji}^{\varepsilon}$. Here Ω is a bounded domain of the space $\mathbb{R}^3$ of coordinates x_i, inhomogeneous in the sense that the coefficients a_{ij} are functions of $\mathbf{x}$.

Equation 2.1 may, for example, represent steady thermal diffusion with u^{ε} the temperature, $f(\mathbf{x})$ the given source term, and the tensor a_{ij}^{ε} being the thermal conductivity. We assume that the material has a locally periodic structure, the basic period Y being a parallelepiped cell.

As we have seen before, we look for an asymptotic expansion of u^{ε} (note that $\varepsilon \ll 1$) in the form:

$$u^{\varepsilon}(\mathbf{x}) = u^0(\mathbf{x}, \mathbf{y}) + \varepsilon u^1(\mathbf{x}, \mathbf{y}) + \varepsilon^2 u^2(\mathbf{x}, \mathbf{y}) + \ldots \tag{2.3}$$

where the functions $u^i(\mathbf{x}, \mathbf{y})$ are Y-periodic in the variable $\mathbf{y}$, $\forall \mathbf{x} \in \Omega$.

The basic idea of the method is to insert Eq. 2.3 in Eq. 2.1 and to identify the powers of ε. In order to do this, we consider first $\mathbf{x}$ and $\mathbf{y}$ as independent variables, and next we replace $\mathbf{y}$ by $\mathbf{x}/\varepsilon$. Now the derivatives may be understood as:

$$\frac{\mathrm{d}}{\mathrm{d}x_i} \to \frac{\partial}{\partial x_i} + \frac{1}{\varepsilon}\frac{\partial}{\partial y_i} \tag{2.4}$$

The total dependence on $\mathbf{x}$ of a function $u(\mathbf{x}, \mathbf{y})$ is obtained directly and through the variable $\mathbf{y}$.

From Eq. 2.1 we have successively:

$$-\frac{\partial}{\partial y_i}\left(a_{ij}(\mathbf{y})\frac{\partial u^0}{\partial y_j}\right) = 0 \tag{2.5}$$

$$-\frac{\partial}{\partial y_i}\left(a_{ij}(\mathbf{y})\frac{\partial u^1}{\partial y_j}\right) - \frac{\partial}{\partial y_i}\left(a_{ij}(\mathbf{y})\frac{\partial u^0}{\partial x_j}\right) - \frac{\partial}{\partial x_i}\left(a_{ij}(\mathbf{y})\frac{\partial u^0}{\partial y_j}\right) = 0 \tag{2.6}$$

$$-\frac{\partial}{\partial y_i}\left(a_{ij}(\mathbf{y})\frac{\partial u^2}{\partial y_j}\right) - \frac{\partial}{\partial y_i}\left(a_{ij}(\mathbf{y})\frac{\partial u^1}{\partial x_j}\right) - \frac{\partial}{\partial x_i}\left(a_{ij}(\mathbf{y})\frac{\partial u^1}{\partial x_j}\right) - \frac{\partial}{\partial x_i}\left(a_{ij}(\mathbf{y})\frac{\partial u^0}{\partial x_j}\right) = f(\mathbf{x}) \tag{2.7}$$

Equation 2.5 is a differential problem for the Y-periodic function $u^0(\mathbf{y})$. But the only periodic solution of Eq. 2.5 is $u^0(\mathbf{x}, \mathbf{y}) = u^0(\mathbf{x})$.

With this result, Eq. 2.6 takes the form:

$$-\frac{\partial}{\partial y_i}\left(a_{ij}(\mathbf{y})\frac{\partial u^1}{\partial y_j}\right)=\frac{\partial u^0}{\partial x_j}\frac{\partial a_{ij}}{\partial y_i} \tag{2.8}$$

This is an equation in $u^1(\mathbf{y})$ if u^0 is known. We look for the solution of Eq. 2.8 in the sense of distributions. By virtue of the linearity of Eq. 2.8, the solution is of the form:

$$u^1(\mathbf{x},\mathbf{y})=\frac{\partial u^0}{\partial x_k}\Gamma^k(\mathbf{y})+c(\mathbf{x}) \tag{2.9}$$

where $\Gamma^k(y)$ is, in the sense of distributions, the periodic solution of an equation of the form:

$$-\frac{\partial}{\partial y_i}\left(a_{ij}(y)\frac{\partial \Gamma^k}{\partial y_j}\right)=\frac{\partial a_{ij}}{\partial y_i} \tag{2.10}$$

with the mean value equal to zero. By mean operator, we understand:

$$\tilde{\phi}=\frac{1}{|Y|}\int_Y \phi(\mathbf{y})\,d\mathbf{y} \tag{2.11}$$

$|Y|$ being the volume of the basic cell Y. In Eq. 2.9, the term $c(\mathbf{x})$ is an additive constant.

Using Eq. 2.9, we have:

$$a_{ij}(\mathbf{y})\left(\frac{\partial u^0}{\partial x_j}+\frac{\partial u^1}{\partial y_j}\right)=\left[a_{ik}(\mathbf{y})+a_{ij}(\mathbf{y})\frac{\partial \Gamma^k}{\partial y_j}\right]\frac{\partial u^0}{\partial x_k} \tag{2.12}$$

and applying the mean operator to Eq. 2.7 we obtain:

$$-\frac{\partial}{\partial x_i}\left(\tilde{a}_{ik}\frac{\partial u^0}{\partial x_k}\right)=f \tag{2.13}$$

where

$$\tilde{a}_{ik}=\left[a_{ik}(\mathbf{y})+a_{ij}(\mathbf{y})\frac{\partial \Gamma^k}{\partial y_j}\right]^{\sim}=\tilde{a}_{ik}+\left(a_{ij}\frac{\partial \Gamma^k}{\partial y_j}\right)^{\sim} \tag{2.14}$$

In this computation, the first two terms from Eq. 2.7 are integrated by parts and

$$\int_Y \frac{\partial}{\partial y_i}\left(a_{ij}\frac{\partial v}{\partial y_j}\right)\mathrm{d}\mathbf{y} = \int_{\partial Y} n_i a_{ij}\frac{\partial v}{\partial y_j}\,\mathrm{d}\mathbf{s} = 0 \tag{2.15}$$

by periodicity.

Equation 2.13 is the homogenized (macroscopic) equation and the coefficients $\tilde{a}_{ik}$ are the homogenized coefficients. Note that they are well determined constants which depend on the local coefficients $a_{ij}(\mathbf{y})$. Also it is interesting to observe that they are not simply the mean values of the microscopic coefficients. Of course, they have the following properties:

$$\tilde{a}_{ik} = \tilde{a}_{ki}$$

$$\tilde{a}_{ik}\xi_i\xi_k \geqslant \gamma\xi_i\xi_i, \quad \gamma > 0, \quad \forall\xi\in\mathbb{R}^3$$

From a mathematical point of view, it is interesting to note that the solutions u^ε of Eq. 2.1 converge, in some sense, when $\varepsilon \to 0$, to the solution u^0 of Eq. 2.13 (see Bensoussan *et al.*, 1978). All previous results remain unchanged if the microscopic coefficients $a_{ij}(\mathbf{y})$ are discontinuous.

It is interesting to observe that the homogenized medium is anisotropic even if at the microscopic level it is isotropic. Indeed, if $a_{ij} = a\delta_{ij}$, from Eq. 2.14 we have for $i \neq k$:

$$\tilde{a}_{ik} = \frac{1}{|Y|}\int_Y a\frac{\partial\Gamma^k}{\partial y_i}\,\mathrm{d}y \tag{2.16}$$

which is not necessarily equal to zero.

3. Darcy's law

We consider the flow of a Newtonian fluid in a domain formed by the cavities of a rigid porous solid, periodically distributed in the space. The simplest representation of such a medium is a porous material formed by tubes parallel to the three directions of the coordinate axis, with a non-empty intersection. The pores thus have a periodic geometric structure, the period of which is small with respect to the other geometric parameters. We are still within the general framework of the homogenization method, but it is necessary to note the problem presented here has periodicity in the geometry of the material, and not necessarily in the coefficients of the equations as in the previous section. Periodic coefficients do occur in the temperature equation, which is considered later.

The asymptotic process and the limit equations may be very different in structure if several small parameters are involved in the problem. For example, if the viscosity coefficient is small, the velocity vector may be large and the non-linear terms may appear in the asymptotic equation leading to a non-linear form of Darcy's law.

Following the initial ideas from Ene and Sanchez-Palencia (1975) we consider a slow flow in a porous body Ω formed by the fluid and solid domains $\Omega_{\varepsilon f}$ and $\Omega_{\varepsilon s}$:

$$\Omega_{\varepsilon f} = \{\mathbf{x} \in \Omega;\quad \mathbf{x} \in \varepsilon Y_f\};\quad \Omega_{\varepsilon s} = \{\mathbf{x} \in \Omega;\quad \mathbf{x} \in \varepsilon Y_s\}$$

where the parallelepipedic period Y of the space variables y_i consists of a fluid part Y_f and a solid part Y_s separated by a smooth boundary $\partial\Omega_{\varepsilon f}$.

If p^ε and $\mathbf{v}^\varepsilon$ denote the pressure and the velocity of the flow, they must satisfy the equations:

$$\nabla \cdot \mathbf{v}^\varepsilon = 0 \tag{3.1}$$

$$0 = -\nabla p^\varepsilon + \mu \Delta \mathbf{v}^\varepsilon + \mathbf{f} \tag{3.2}$$

and the boundary condition:

$$\mathbf{v}^\varepsilon \Big|_{\partial\Omega_{\varepsilon f}} = 0 \tag{3.3}$$

Here we consider only the Stokes equations. For the general case of the Navier–Stokes equations, see Ene and Polisevski (1987).

The only small parameter in the problem is ε. The viscosity coefficient μ and the applied density of forces $\mathbf{f}(\mathbf{x})$ are independent of ε. In order to study the asymptotic process, we consider expansions of the form:

$$\mathbf{v}^\varepsilon(\mathbf{x}) = \varepsilon^2 \mathbf{v}^0(\mathbf{x}, \mathbf{y}) + \varepsilon^3 \mathbf{v}^1(\mathbf{x}, \mathbf{y}) + \ldots \tag{3.4}$$

$$p^\varepsilon(\mathbf{x}) = p^0(\mathbf{x}) + \varepsilon p^1(\mathbf{x}, \mathbf{y}) + \ldots \tag{3.5}$$

where $\mathbf{y} = \mathbf{x}/\varepsilon$ and all the functions are considered to be Y-periodic with respect to the variable $\mathbf{y}$.

In the following we introduce the sub-index $\mathbf{y}$ or $\mathbf{x}$ to specify the partial derivative. From Eq. 3.1 at order ε, Eq. 3.2 at order ε^0, and Eq. 3.3 at order ε^2, we obtain the problem in the $\mathbf{y}$ variable for the Y-periodic functions $\mathbf{v}^0$ and p^1:

$$\nabla_y \cdot \mathbf{v}^0 = 0 \tag{3.6}$$

$$0 = -\nabla_y p^1 + \mu\Delta_y \mathbf{v}^0 + \mathbf{f} - \nabla_x p^0 \tag{3.7}$$

$$\mathbf{v}^0\Big|_{\partial\Omega_{ef}} = 0 \tag{3.8}$$

In this problem, $\mathbf{f} - \nabla_x p^0$ plays the role of a given force. If we denote by $\mathbf{w}^i$ the solution of the local problem, in the weak sense:

$$\nabla_y \cdot \mathbf{w}^i = 0 \tag{3.9}$$

$$0 = -\nabla_y q^i + \mu\Delta_y \mathbf{w}^i + \mathbf{e}_i \tag{3.10}$$

$$\mathbf{w}^i\Big|_{\delta\Omega_{ef}} = 0 \tag{3.11}$$

with $\mathbf{w}^i$ and q^i Y-periodic and $\mathbf{e}_i$ the unit vector in the direction of the y_i axis. By linearity, we can write the solution $\mathbf{v}^0$ of Eqs 3.6, 3.7, and 3.8 in the form:

$$\mathbf{v}^0 = \frac{1}{\mu}(\mathbf{f} - \nabla_x p^0)\mathbf{w}^i \tag{3.12}$$

When we apply the mean operator of Eq. 2.11 to Eq. 3.12, we get Darcy's law:

$$\tilde{v}_j^0 = -\frac{K_{ij}}{\mu}\left(\frac{\partial p^0}{\partial x_i} - f_i\right) \tag{3.13}$$

$$K_{ij} = \tilde{w}_j^i \tag{3.14}$$

and the continuity equation:

$$\nabla_x \cdot \tilde{\mathbf{v}}^0 = 0 \tag{3.15}$$

The constant tensor K_{ij}, defined as the mean value of the component j of the vector $\mathbf{w}^i$, depends only on the geometry of the period Y, and is the usual permeability tensor. It is a symmetric, positive-definite matrix.

If we have a small viscosity and a large velocity in the Navier–Stokes equations, a more general version of Eq. 3.7 takes the form:

$$\rho(\mathbf{v}^0\nabla_y)\mathbf{v}^0 = -\nabla_y p^1 + \mu\Delta_y \mathbf{v}^0 + \mathbf{f} - \nabla_x p^0$$

Consequently, we have a non-linear form of Darcy's law which is difficult to explain. This is the situation for gas flow.

4. Thermal equation

Rather than go through a detailed explanation of the use of the homogenization method for the energy equation in a porous medium, we simply present the results. The details are given in Ene and Polisevski (1987). In the thermal equation, it is important to note that at the microscopic level we have two equations: the complete equations of thermo-convective phenomena in the fluid part of the medium and the conduction equation in the skeleton. Using the same type of asymptotic expansions as in Section 2 we obtain the macroscopic equation for temperature in the form:

$$\rho^0 c_f v_k^0 \frac{\partial T^0}{\partial x_k} = \mu (K^{-1})_{kj} \tilde{v}_j^0 \tilde{v}_k^0 + \frac{\partial}{\partial x_j}\left(\tilde{\chi}_{jk} \frac{\partial T^0}{\partial x_k}\right) \tag{4.1a}$$

where T^0 and ρ^0 are the first terms of the expansions of temperature and density, and the macroscopic thermal conductivity is given by:

$$\tilde{X}_{jk} = \tilde{X}_{ij}(\mathbf{y}) + \left(\chi_{ik}(\mathbf{y}) \frac{\partial \theta_j}{\partial y_k}\right)^{\sim} \tag{4.1b}$$

with θ_j the solution of Eq. 2.10. In fact, $\tilde{\chi}_{jk}$ are the same as $\tilde{a}_{ik}$ from Eq. 2.14.

For non-steady flow, Eq. 4.1 takes the form:

$$[n\rho_f c_f + (1-n)\rho_s c_s] \frac{\partial T^0}{\partial t} + \rho^0 c_f \tilde{v}_k^0 \frac{\partial T^0}{\partial x_k} = \mu (K^{-1})_{kj} \tilde{v}_j^0 \tilde{v}_k^0 + \frac{\partial}{\partial x_j}\left(\tilde{\chi}_{jk} \frac{\partial T^0}{\partial x_k}\right) \tag{4.2}$$

It is important to remark that for natural convection in porous media, the viscous dissipation term $\rho(K^{-1})_{kj}\tilde{v}_j^0 \tilde{v}_k^0$ is negligible and consequently the equation takes the form:

$$[n\rho_f c_f + (1-n)\rho_s c_s] \frac{\partial T^0}{\partial t} + \rho^0 c_f v_k^0 \frac{\partial T^0}{\partial x_k} = \frac{\partial}{\partial x_j}\left(\tilde{\chi}_{jk} \frac{\partial T^0}{\partial x_k}\right) \tag{4.3}$$

If we study diffusion in porous media, the macroscopic equation has the same form as Eq. 4.3 with the diffusion coefficient calculated by Eq. 4.1b.

Equation 4.1b gives us the opportunity to compute the macroscopic coefficients exactly via the solution of an elliptic equation (Eq. 2.10). But we also have the opportunity to estimate macroscopic coefficients such as the elastic coefficients for composite materials (see Tartar, 1981; Ene and Pasa, 1987).

The general estimate tells us that: *the upper bound for the thermal conductivity is given by the arithmetic mean value with respect to x_2 and x_3 coordinates followed* *by the harmonic mean with respect to x_1, and the lower bound is given by taking firstly the harmonic mean with respect to x_1 followed by the arithmetic mean with respect to x_2 and x_3 coordinates.*

Note that such a computation of the upper and lower bound gives us the opportunity to consider anisotropy of the medium. In order to illustrate this, we present a few examples.

If we take a two-dimensional case and use a square cell with unit dimension, and let $a_{ij}^{\varepsilon} = a\delta_{ij}$ where a has constant value α in the square of the length $1/3$ centred in the middle of our cell and 1 in the remainder of the cell, we find

$$\frac{7\alpha + 2}{3(2\alpha + 1)} \leqslant \tilde{\chi} \leqslant \frac{3(2 + \alpha)}{7 + 2\alpha} \tag{4.4}$$

Using the same computation, if we have an elliptic inclusion in our square cell, with the surface equal to $1/4$, and if $\alpha = 10$, we obtain two different bounds:

(1) If the ellipse is oriented along the x_1 direction, say $(0.21)^2 x_1^2 + (0.4)^2 x_2^2 = (0.4)^2(0.21)^2$, we have:

$$0.70 \leqslant \tilde{\chi} \leqslant 2.43 \tag{4.5}$$

(2) If the same ellipse is rotated by an angle equal to $\pi/2$ we obtain:

$$0.74 \leqslant \tilde{\chi} \leqslant 1.53 \tag{4.6}$$

This example illustrates the anisotropy of the medium and its influences on the bounds of the coefficients. If, instead of the ellipse, we take a circle of the same surface, we obtain:

$$0.81 \leqslant \tilde{\chi} \leqslant 1.80 \tag{4.7}$$

5. Darcy's law or Brinkman's law

The situation explained in Section 3 depends strongly on the asymptotic dimensions of the pores. We can also give the asymptotic structure of the solution for the fluid flow of a viscous fluid past an array of fixed obstacles (see Levy, 1981; Sanchez-Palencia, 1982, 1986). In the geometric framework of Section 3 we consider obstacles $\eta\gamma$ in the $\mathbf{y}$ variables, i.e. $\varepsilon\eta\gamma$ in the standard $\mathbf{x}$ variables; the dimension of the period being of order ε as usual. An important parameter for the description of the asymptotic behaviour is:

$$m = \lim \frac{\varepsilon^2}{\eta} \tag{5.1}$$

In fact, our problem has two small parameters ε and η, but the asymptotic behaviour depends on m.

If $\varepsilon \to 0$ and η is fixed ($\eta = 0(1)$), we have the classical homogenization problem of Section 3 and the asymptotic velocity is:

$$v_k^\eta = \varepsilon^2 w_k^i \left(f_i - \frac{\partial p^0}{\partial x_i} \right) \tag{5.2}$$

$$\nabla \cdot \tilde{\mathbf{v}}^\eta = 0 \tag{5.3}$$

or

$$\tilde{v}_k^\eta = K_{ik}^\eta \left(f_i - \frac{\partial p^0}{\partial x_i} \right) \tag{5.4}$$

where K_{ik}^η is the permeability tensor which depends on η.

Now for $\eta \ll 1$, flow without interaction between different obstacles gives a good approximation. This flow is defined in the following way:

$$\Delta_y \mathbf{w}^k = \nabla_x p^k \tag{5.5}$$

$$\nabla_y \cdot \mathbf{w}^k = 0 \tag{5.6}$$

$$\mathbf{w}^k \bigg|_{\partial\gamma} = 0 \tag{5.7}$$

$$\mathbf{w}^k \to \mathbf{e}_k \quad \text{for} \quad |\mathbf{y}| \to \infty \tag{5.8}$$

The corresponding force on the obstacle is

$$T(\mathbf{w}^k) = \mathbf{e}_i \int_{\partial\gamma} \sigma_{ij}^k n_j \, \mathrm{d}s \tag{5.9a}$$

$$\sigma_{ij}^k = -p^k \delta_{ij} + \left(\frac{\partial \mathbf{w}_i^k}{\partial y_j} + \frac{\partial \mathbf{w}_j^k}{\partial y_i} \right) \tag{5.9b}$$

If we introduce the translation tensor H_{ij} of components

$$H_{ij} = T_i(\mathbf{w}^j)$$

the asymptotic structure of the flow is given by:

$$\mathbf{v}(x) \cong \mathbf{f}(\varepsilon) U_k(x) \mathbf{w}^k(y) \tag{5.10a}$$

where

$$\mathbf{f}(\varepsilon) = \begin{cases} \dfrac{\varepsilon^2}{\eta} & \text{if } m = 0 \\ 1 & \text{if } m \text{ infinite or finite} \neq 0 \end{cases} \tag{5.10b}$$

and $\mathbf{U}(x)$ is the solution of:

$$\nabla_x p^0 + \mathbf{H} \cdot \mathbf{U} = \mathbf{f} \tag{5.11a}$$

$$\nabla_x \cdot \mathbf{U} = 0 \tag{5.11b}$$

$$\mathbf{U} \cdot \mathbf{n} \Big|_{\partial\Omega} = 0 \tag{5.11c}$$

if $m = 0$,

$$\Delta_x \mathbf{U} + \nabla_x p^0 + \frac{1}{m} \mathbf{H} \cdot \mathbf{U} = \mathbf{f} \tag{5.12a}$$

$$\nabla_x \cdot \mathbf{U} = 0 \tag{5.12b}$$

$$\mathbf{U} \Big|_{\partial\Omega} = 0 \tag{5.12c}$$

if m is finite $\neq 0$, and

$$-\Delta_x \mathbf{U} + \nabla_x p^0 = \mathbf{f} \tag{5.13a}$$

$$\nabla_x \cdot \mathbf{U} = 0 \tag{5.13b}$$

$$\mathbf{U} \Big|_{\partial\Omega} = 0 \tag{5.13c}$$

if m is infinite.

In fact, the asymptotic behaviour (Eq. 5.10a) contains an order function $\mathbf{f}(\varepsilon)$, a uniform flow $\mathbf{U}(x)$ given by Eqs 5.11–5.13, and the asymptotic structure $\mathbf{w}$ near the obstacles. For $m = \infty$ (i.e. obstacles very small) there is no action of the obstacles on the $\mathbf{U}$ flow, which is the same as in the absence of obstacles (Stoke's flow). For $m = 0(1)$ (i.e. Eq. 5.12), we have Brinkman's equation: the global movement $\mathbf{U}$ takes into account the force of the obstacles on the fluid.

For $m = 0$ (i.e. Eq. 5.11), the global flow is negligibly small with respect to that of the preceeding case and is given by the order function $\mathbf{f}(\varepsilon)$ (Darcy's flow).

6. Fluid flow in a fractured porous medium

A very different problem arises in the case of fractured porous media. In this case it is necessary to consider the motion in fractures and also the motion in the porous blocks (see also Douglas and Arbogast, Chapter VII this volume). A large variety of situations arise as a function of the dimensions of the pores and fractures, and also as a function of the model of flow considered. For example, we may envision a porous medium with a doubly periodic structure εY and $\varepsilon^2 Z$, where the small pores εY represent fractures and the very small pores $\varepsilon^2 Z$ represent the porous blocks. If we consider Stoke's flow in both parts of the body, we obtain a macroscopic equation of Darcy's type with a permeability tensor completely determined by the fracture geometry, as if the blocks were solids (Levy, 1988). These results are of the same type as the classical results of Barenblatt.

A very different situation may appear if we consider fractures of dimension ε and of porous blocks of the dimension $\eta = \eta(\varepsilon)$ with $\eta < \varepsilon$. Consider a porous body Ω formed by fractures and porous domains $\Omega_{\eta F}$ and $\Omega_{\eta P}$:

$$\Omega_{\eta F} = \{\mathbf{x} \in \Omega; \quad \mathbf{x} \in \eta Y_F\}; \quad \Omega_{\eta P} = \{\mathbf{x} \in \Omega; \quad \mathbf{x} \in \eta Y_P\}$$

where the parallelepipedic period Y of the space variables y_i consists of a fractured part Y_F and a porous part Y_P separated by a smooth boundary $\partial\Omega_{\eta F}$. If p and $\mathbf{v}$ denote the pressure and the velocity of the flow, they must satisfy in dimensionless form the equations:

$$-\Delta\mathbf{v} = \mathbf{f} - \nabla p \quad \text{in } \Omega_{\eta F} \tag{6.1}$$

$$k\mathbf{v} = \mathbf{f} - \nabla p \quad \text{in } \Omega_{\eta P} \tag{6.2}$$

$$\nabla \cdot \mathbf{v} = 0 \quad \text{in } \Omega \tag{6.3}$$

and boundary conditions

$$\mathbf{v}\Big|_{\Omega_{\eta F}} \times \mathbf{n} = 0 \tag{6.4}$$

$$[\mathbf{v} \cdot \mathbf{n}] = 0 \tag{6.5}$$

$$[p] = 0 \tag{6.6}$$

where k is the inverse of the dimensionless permeability tensor. Note that in this case we consider the Stoke's flow valid in the fissures and the Darcy's flow in the porous blocks (see Ene and Vernescu, 1985; Vernescu, 1988).

As in Section 5, we may obtain different results:

(i) If $\eta = 0(\varepsilon^3)$, we have at the macroscopic level

$$-\Delta \mathbf{U} + C\mathbf{U} = \mathbf{f} - \nabla_x p^0 \tag{6.7}$$

where

$$C = \frac{\pi}{2} \frac{3k}{1 + 2k} \lim_{\varepsilon \to 0} \frac{\eta}{\varepsilon^3}$$

In this case motion is governed by Brinkman's law.

(ii) If $\eta = 0(\varepsilon^3)$ we have:

$$-\Delta \mathbf{U} = \mathbf{f} - \nabla p^0 \tag{6.8}$$

In this case, the porous blocks are so small that the fluid behaves as if it were a free fluid.

(iii) If $\varepsilon^3 \ll \eta \ll \varepsilon$ we have a Darcy's flow:

$$C\mathbf{U} = \mathbf{f} - \nabla p^0 \tag{6.9}$$

with C given by the same formula as in case (i).

(iv) If $\eta = 0(\varepsilon)$ we obtain a Darcy's law but with a different permeability tensor

$$\tilde{v}_j^0 = -L_{ij}\left(\frac{\partial p^0}{\partial x_i} - f_i\right) \tag{6.10}$$

$$L_{ij} = \tilde{\phi}_j^i \tag{6.11}$$

where $\boldsymbol{\phi}^i$ is the solution of the problem:

$$-\Delta_y \boldsymbol{\phi}^i = \mathbf{e}_i - \nabla_y q^i \quad \text{in } Y_{\mathrm{F}} \tag{6.12}$$

$$k\boldsymbol{\phi}^i = \mathbf{e}_i - \nabla_y q^i \quad \text{in } Y_{\mathrm{P}} \tag{6.13}$$

$$\nabla_y \cdot \boldsymbol{\phi}^i = 0 \qquad \text{in } Y$$

$$\boldsymbol{\phi}^i \times \mathbf{n} = 0, \quad [\boldsymbol{\phi}^i \cdot \mathbf{n}] = 0[q^i] = \text{on } \partial\Omega_{\eta\mathrm{F}} \tag{6.14}$$

and $\boldsymbol{\phi}^i$, q^i are Y periodic functions.

Note that in the last case, the permeability tensor L_{ij} is also symmetric and positive-definite and depends on the porous blocks and on the geometry of the fractures.

To conclude this section, it is important to observe that flow in fractured porous media may be governed by a large variety of equations depending on the scale and the small parameters in the problem.

7. Solid deformable skeleton

In this section we consider a very different problem: the behaviour of a mixture formed by a viscoelastic skeleton and a viscous incompressible fluid. The solid and fluid parts are periodic as in Section 3. We assume the solid skeleton and the fluid phase are each connected and that the viscosity of the fluid is small. We see that Darcy's law holds only in the case of large viscosity and small velocity, or small viscosity and possibly large velocity. More precisely, the small magnitude is of order ε^2 and the larger is of order ε^0. As a consequence of the fact that the displacement vector in the solid part is of order ε^0, we may take the velocity in the fluid part of the same order and the viscosity of the fluid of order ε^2.

In the solid part of the mixture, the equations are

$$\rho_s \frac{\partial^2 u_i}{\partial t^2} - \frac{\partial \sigma^s_{ij}}{\partial x_j} = f_i \tag{7.1}$$

$$\sigma^s_{ij} = a^s_{ijkh} e_{kh}(\mathbf{u}) + b^s_{ijkh} e_{kh}\left(\frac{\partial \mathbf{u}}{\partial t}\right) \tag{7.2a}$$

$$e_{kh}(\mathbf{u}) = \frac{1}{2}\left(\frac{\partial \mathbf{u}_k}{\partial x_h} + \frac{\partial \mathbf{u}_h}{\partial x_k}\right) \tag{7.2b}$$

where $\mathbf{f}$ is the exterior body force and the coefficients a^s_{ijkh}, b^s_{ijkh} satisfy the usual properties of symmetry and positivity.

In the fluid part, the equations are:

$$\rho_f \frac{\partial^2 u_i}{\partial t^2} - \frac{\partial \sigma^f_{ij}}{\partial x_j} = f_i \tag{7.3}$$

$$\sigma^f_{ij} = -p\delta_{ij} + \varepsilon^2 \mu \delta_{ik} \delta_{jh} e_{kh}(\mathbf{v}) \tag{7.4}$$

$$\nabla \cdot \mathbf{v} = 0, \quad \mathbf{v} = \frac{\partial \mathbf{u}}{\partial t} \tag{7.5}$$

At the interface between the solid and fluid, we must have continuity of displacement and stress:

$$[\mathbf{u}] = 0, \quad [\sigma_{ij}n_j] = 0 \tag{7.6}$$

We must add initial and boundary conditions:

$$\mathbf{u} = 0 \quad \text{on } \partial\Omega \tag{7.7}$$

$$\mathbf{u} = \frac{\partial \mathbf{u}}{\partial t} = 0 \quad \text{for } t = 0 \tag{7.8}$$

We look for Y-periodic coefficients in the variables $\mathbf{y} = \mathbf{x}/\varepsilon$:

$$\rho^\varepsilon(\mathbf{x}) = \rho(\mathbf{x}/\varepsilon), \quad a^\varepsilon_{ijkh}(\mathbf{x}) = a_{ijkh}(\mathbf{x}/\varepsilon) \quad \text{and} \quad b^\varepsilon_{ijkh}(\mathbf{x}) = b_{ijkh}(\mathbf{x}/\varepsilon), \quad \text{with:}$$

$$b_{ijkh} = \begin{cases} b^{\mathrm{s}}_{ijkh} & \text{in } \Omega_{\mathrm{s}} \\ \varepsilon^2 \mu \delta_{ik}\delta_{jh} & \text{in } \Omega_{\mathrm{f}} \end{cases}$$

We assume that the appropriate asymptotic expansion in the solid part is analogous to that of viscoelastic mixture and in the fluid part analogous to that of flow through porous media. The first term of the displacement vector expansion in the solid does not depend on $\mathbf{y}$. However, it does in the fluid region. It is thus natural to introduce the relative displacement of the fluid with respect to the solid, $\mathbf{u}^{\mathrm{r}}(\mathbf{x}, \mathbf{y}, t) = \mathbf{u}^0(\mathbf{x}, \mathbf{y}, t) - \mathbf{u}^0(\mathbf{x}, t)$ and, consequently, we search for a two-scale asymptotic expansion in $\Omega_{\varepsilon\mathrm{f}}$ and $\Omega_{\varepsilon\mathrm{s}}$:

$$\mathbf{u}^\varepsilon(\mathbf{x}, t) = \mathbf{u}^0(\mathbf{x}, t) + \mathbf{u}^{\mathrm{r}}(\mathbf{x}, \mathbf{y}, t) + \varepsilon \mathbf{u}^1(\mathbf{x}, \mathbf{y}, t) + \dots \tag{7.9}$$

where $\mathbf{y} = \mathbf{x}/\varepsilon$ and all functions are Y-periodic in $\mathbf{y}$. Note that the vector $\mathbf{u}^{\mathrm{r}}$ is zero on Y_{s} and on $\partial\Omega$. Application of the homogenization technique gives

$$\tilde{v}^{\mathrm{r}}_k(t) = \int_0^t g_{ki}(t-s)\left(f_i - \frac{\partial p^0}{\partial x_i} - \rho \frac{\partial v^0_i}{\partial t} \right)(s)\, \mathrm{d}s \tag{7.10}$$

where $g_{ki}(\xi)$ is a well-defined function of ξ which decreases exponentially as $\xi \to \infty$ and $g_{ki} = g_{ik}$ (see Ene, 1988).

The macroscopic equation of balance of momentum is:

$$\tilde{\rho} \frac{\partial^2 u^0_i}{\partial t^2} + \rho_{\mathrm{f}} \frac{\partial \tilde{v}^{\mathrm{r}}_i}{\partial t} - \frac{\partial \sigma^{\mathrm{T}}_{ij}}{\partial x_j} = f_i \tag{7.11}$$

where the effective (or total) stress tensor is defined by:

$$\sigma_{ij}^{\mathrm{T}} = \tilde{\sigma}_{ij}^{0} - np^{0}\delta_{ij} \tag{7.12}$$

$$\tilde{\sigma}_{ij}^{0} = \alpha_{ijkh}^{0} e_{kh}(\mathbf{u}^{0}) + \alpha_{ijkh}^{1} e_{kh}\left(\frac{\partial \mathbf{u}^{0}}{\partial t}\right) - \alpha_{ij}^{2} p^{0} + \int_{0}^{t} g_{ijkh}(t-s) e_{kh}(\mathbf{u}^{0})(s)\,\mathrm{d}s$$

$$+ \int_{0}^{t} g_{ij}^{*}(t-s) p^{0}(s)\,\mathrm{d}s \tag{7.13}$$

and n is the usual porosity of the medium $n = |Y_{\mathrm{f}}|/|Y|$. The constitutive Eq. 7.13 contains an elastic term α^0, a viscoelastic term with instantaneous memory α^1, a pressure term α^2, and two terms with long memory g and g^*, which are functions of strain and pressure.

Of course these equations are subject to the balance of mass which is of the form

$$n\nabla_x \cdot \mathbf{v}^0 + \nabla_x \cdot \tilde{\mathbf{v}}^{\mathrm{r}} = \frac{1}{|Y|} \int_{Y_{\mathrm{s}}} \nabla_y \cdot \mathbf{v}^1 \,\mathrm{d}y \tag{7.14}$$

The macroscopic motion of the mixture may be described by the displacement vector in the solid $\mathbf{u}^0(\mathbf{x}, t)$, the pore pressure in the fluid $p^0(\mathbf{x}, t)$ and the mean value of the relative velocity $\tilde{\mathbf{v}}^{\mathrm{r}}$. These quantities satisfy Eq. 7.14 (conservation of mass), Eq. 7.10 (Darcy's law), and Eq. 7.11 (conservation of momentum). The effective stress tensor is defined by Eqs 7.12 and 7.13. Note that, in this case, Darcy's law is of integro-differential type.

Important remark: a mixture of a viscous fluid with a linear viscoelastic skeleton gives rise at the macroscopic level to a viscoelastic medium with long memory. In fact, to describe the behaviour of the mixture, Eqs 7.10 and 7.13 illustrate that the history of deformation should be taken into consideration at the macroscopic level. In the particular case of an elastic skeleton, our results reduce to those obtained by Sanchez-Palencia (1980) and Levy (1979).

8. Dimensionless parameters

Let us return to the complete system of equations which describe motion in a porous medium when the skeleton is rigid. If $\mathbf{v}$, p, ρ, T denote the velocity, the pressure, the density, and the temperature, we have:

$$\mathbf{v} = -\frac{\mathbf{k}}{\mu}(\nabla p - \rho \mathbf{f}) \tag{8.1}$$

$$\frac{\partial(n\rho)}{\partial t} + \nabla \cdot (\rho \mathbf{v}) = 0 \tag{8.2}$$

$$\rho = \rho_0[1 - \alpha(T - T_r) + \beta(p - p_r)] \tag{8.3}$$

$$[n\rho_f c_f + (1 - n)\rho_s c_s] \frac{\partial T}{\partial t} + \rho_f c_f \mathbf{v} \cdot \nabla T = \frac{\mu}{k} v^2 + \chi \Delta T \tag{8.4}$$

For natural convection in a porous medium, the viscous dissipation is negligibly small, and instead of Eq. 8.4 we have

$$[n\rho_f c_f + (1 - n)\rho_s c_s] \frac{\partial T}{\partial t} + \rho_f c_f \mathbf{v} \cdot \nabla T = \chi \Delta T \tag{8.5}$$

In order to establish the physical significance of the various terms in these systems of equations, we take a characteristic pore length, l, a characteristic length, L, of the macroscopic domain, a characteristic velocity, U, and a reference temperature, T_r. The small parameter involved in the homogenization technique is defined by $\varepsilon = l/L$ and it is known that the permeability is of the form $\mathbf{k} = \mathbf{k}^* l^2$, with $\mathbf{k}^*$ a dimensionless permeability. It is expedient to introduce the following dimensionless numbers:

the Reynolds number $$\mathrm{Re}_\varepsilon = \frac{U\rho_0 l}{\mu} \tag{8.6}$$

the Prandtl number $$\mathrm{P} = \frac{\mu c}{\chi} \tag{8.7}$$

the temperature number $$\mathrm{S}_\varepsilon = \frac{\mu U^2}{\chi(T - T_r)} \tag{8.8}$$

the Grashof number $$\mathrm{G}_\varepsilon = \frac{g\alpha\rho_0^2 l^3 (T - T_r)}{\mu^2} \tag{8.9}$$

the Rayleigh number $$\mathrm{Ra}_\varepsilon = \mathrm{G}_\varepsilon \mathrm{P} \tag{8.10}$$

Darcy's law holds for $\mathrm{Re}_\varepsilon \sim \mathrm{O}(\varepsilon^{-1})$ or, in an equivalent form,

$$\frac{l^2 \rho_0 U}{\mu L} \sim \mathrm{O}(1) \tag{8.11}$$

Equation 8.4 makes sense for $S_\varepsilon \sim O(\varepsilon^2)$ and $P \sim O(\varepsilon^2)$, or

$$\frac{\mu L^2 U^2}{l^2 \chi (T - T_r)} \sim O(1) \tag{8.12}$$

$$\frac{\mu c L^2}{\chi l^2} \sim O(1) \tag{8.13}$$

If we take into consideration the range of validity of Darcy's law, Eqs 8.12 and 8.13 may be combined in the form:

$$P Re_\varepsilon \sim \varepsilon^{-1} S_\varepsilon \tag{8.14}$$

In the case of natural convection in porous media, Eq. 8.5 gives us $P \cdot Re_\varepsilon \sim O(\varepsilon)$ and $Ra_\varepsilon \sim O(\varepsilon)$, or

$$\frac{U \rho_0 L c}{\chi} \sim O(1) \tag{8.15}$$

$$\frac{\rho_0^2 g \alpha l^2 L c (T - T_r)}{\mu \chi} \sim O(1) \tag{8.16}$$

The Rayleigh number, Eq. 8.10, is defined at the microscopic level with l^3. In the study of natural convection in porous media, it is usual to define the Rayleigh number with two scales as, $l^2 L$ instead of l^3, i.e.

$$Ra = \frac{\rho_0^2 g \alpha l^2 L c (T - T_r)}{\mu \chi} \tag{8.17}$$

From Eqs 8.16 and 8.17 it is clear that the previous condition may be written in the form $Ra \sim O(1)$; this is in agreement with experimental results.

9. Conclusions

Porous media have a natural hierarchy of distinct spatial and temporal scales. In this chapter we have illustrated a method for passing from the pore scale (the microscopic level) to the hydrodynamic scale (the macroscopic level) by using the homogenization method. To describe transport phenomena in porous media we need macroscopic equations of motion. These may be obtained by the homogenization method, and the equations thus obtained may be studied in terms of the homogenized coefficients.

It is clear that various generalizations to Darcy's law must be used to adequately describe fluid flow through certain porous media. Some of the equations were presented here. It is important to note that the periodic model of a porous medium was adopted only because we have a mathematical technique which enables us to deduce the equations in such a model. But the equations obtained are general and may be analysed with the aid of dimensionless parameters.

References

Bensoussan, A., Lions, J.L. and Papanicolaou, G. (1978). 'Asymptotic Analysis for Periodic Structure'. North-Holland, Amsterdam.

Ene, H.I. (1988). The use of the homogenization method to describe the viscoelastic behavior of a porous saturated medium. *In* 'Rock and Soil Rheology' (N. Cristescu and H.I. Ene, eds), Lecture Notes in Earth Sciences, Vol. 14. Springer-Verlag, Berlin and New York.

Ene, H.I. and Pasa, G. (1987). 'The Homogenization Method: Applications to the Theory of Composite Materials'. Ed. Academiei, Bucharest (in Romanian).

Ene, H.I. and Polisevski, D. (1987). 'Thermal Flow in Porous Media'. D. Reidel, Dordrecht, Boston, Lancaster and Tokyo.

Ene, H.I. and Sanchez-Palencia, E. (1975). Equations et phenomenes de surface pour l'ecoulement dans un modele de milieu poreux, *J. Mec.* **14**, 73–108.

Ene, H.I. and Vernescu, B. (1985). Homogenization of a singular perturbation problem, *Rev. Roum. Math. Pure Appl.* **30**, 815–822.

Levy, T. (1979). Propagation of waves in a fluid-saturated porous elastic solid. *Int. J. Engng Sci.* **17**, 1005–1014.

Levy, T. (1981). Loi de Darcy ou loi de Brinkman, *C.R. Acad. Sci. Paris II* **292**, 871–874.

Levy, T. (1987). Fluids in porous media and suspensions. *In* 'Homogenization Techniques for Composite Materials'. (E. Sanchez-Palencia and A. Zaoui, eds), Lecture Notes in Physics, Vol. 272. Springer-Verlag, Berlin and New York.

Levy, T. (1988). Ecoulement d'un fluide dans un milieu poreux fisure, *C.R. Acad. Sci. Paris II* **306**, 1413–1417.

Sanchez-Palencia, E. (1980). 'Non-homogeneous Media and Vibration Theory'. Lecture Notes in Physics, Vol. 127. Springer-Verlag, Berlin and New York.

Sanchez-Palencia, E. (1982). On the asymptotics of the fluid flow past an array of fixed obstacle, *Int. J. Engng Sci.* **20**, 1291–1301.

Sanchez-Palencia, E. (1986). Homogenization in mechanics: A survey of solved and open problems, *Rend. Sem. Mat. Univ. Polit. Torino* **44**, 1–45.

Tartar, L. (1981). 'Estimations des coefficients homogeneisee'. Lecture Notes in Mathematics, Vol. 704, p. 364. Springer Verlag, Berlin and New York.

Vernescu, B. (1988). Asymptotic methods in flow through porous media. Ph.D. Thesis, University of Bucharest.

Chapter IX

Effect of the Structure of the Porous Medium on Unstable Miscible Displacement

D.E. Moissis and M.F. Wheeler
Department of Mathematics, University of Houston, Houston, TX 77005, USA

1. Introduction

Displacement of a fluid in a porous medium by a less viscous fluid is generally unstable; small perturbations in the flow tend to grow with time, producing large protrusions of the displacing fluid into the displaced fluid. These protrusions are called fingers and this type of instability is known as viscous fingering. Viscous fingering dominates a large number of miscible enhanced oil recovery processes, causing a severe reduction of the displacement efficiency. It may also occur in groundwater contaminant flow, causing uneven spread of a contaminant. In both cases the understanding and quantitative prediction of the phenomenon is important.

In flow through a porous medium small flow disturbances are continuously generated, due to heterogeneities of the medium. Moreover, the processes of finger growth and interaction are linked to the heterogeneity of the porous medium and, in particular, to the spatial variation of porosity and permeability. Investigation of the effect of the structure of the porous medium on the displacement is an area of intense research interest. A major part of the problem is developing a realistic description of the medium for numerical simulation. Conditional simulation (Journel and Huijbregts, 1978; Delhomme, 1979; Lake, 1987), involves generation of 'synthetic' porous media that are compatible with the available statistical information. To obtain statistically meaningful results, simulations must be performed for several different 'realizations' of the same porous medium and results must be averaged. A number of statistical techniques that generate realizations of a porous medium from raw statistics are discussed in the literature (Matheron, 1973; Mejia and Rodriguez-Iturbe, 1974; Bartlett, 1975; Mantoglou and Wilson, 1982).

DYNAMICS OF FLUIDS IN HIERARCHICAL POROUS MEDIA
ISBN 0-12-200260-1

In this paper the spatial variation of permeability is described by two parameters, the coefficient of permeability variation CV and the correlation length λ. The former parameter is the ratio of the standard deviation of the permeability distribution (which is assumed log normal) to the mean permeability and represents the degree of heterogeneity of the medium. The correlation length is a measure of the range over which permeability is correlated. The results presented in this paper show the effect of these two parameters and of the viscosity ratio on horizontal unstable miscible displacement. All simulations mentioned are limited to two-dimensional porous media and to a linear (rectangular) geometry.

2. Methodology

2.1 The problem

Consider incompressible miscible displacement in a porous medium having the shape of a rectangular parallelepiped, one of whose dimensions is much smaller than the other two, so that flow can be assumed two-dimensional. The pore space of the medium is initially filled with the resident (or displaced) fluid ($c = 0$). The medium is flooded at the side $x = 0$ with pure invading (or displacing) fluid ($c = 1$). Thus x is the principal flow direction and y is the direction transverse to the flow. The x–y plane is assumed horizontal. Let L_x and L_y be the x and y dimensions of the parallelepiped and let $\Omega = (0, L_x) \times (0, L_y)$. Under the above assumptions, the displacement can be modelled by the following set of equations:

$$u = -\frac{k}{\mu}\frac{\partial p}{\partial x} \tag{2.1}$$

$$v = -\frac{k}{\mu}\frac{\partial p}{\partial y} \tag{2.2}$$

$$\frac{\partial u}{\partial x} + \frac{\partial v}{\partial y} = 0 \tag{2.3}$$

$$\phi\frac{\partial c}{\partial t} + u\frac{\partial c}{\partial x} + v\frac{\partial c}{\partial y} = \nabla\cdot(\mathbf{D}\nabla c) \tag{2.4}$$

Equations 2.1–2.3 can be combined to yield the 'pressure equation'. Equation 2.4 is the 'concentration equation'. **D** is the velocity-dependent dispersion tensor, given by Peaceman (1966):

$$\mathbf{D} = D_{\mathrm{m}}\mathbf{I} + \frac{\alpha_{\mathrm{l}}}{|\mathbf{u}|}\begin{bmatrix} u^2 & uv \\ uv & v^2 \end{bmatrix} + \frac{\alpha_{\mathrm{t}}}{|\mathbf{u}|}\begin{bmatrix} v^2 & -uv \\ -uv & u^2 \end{bmatrix} \tag{2.5}$$

where D_{m} is the molecular diffusivity and α_{l} and α_{t} are the longitudinal and transverse dispersion coefficients respectively. $\mathbf{u} = (u, v)$ is the velocity vector and $|\mathbf{u}| = \sqrt{(u^2 + v^2)}$.

The boundary conditions are:

$$v = 0, \quad \frac{\partial c}{\partial y} = 0, \qquad \text{at } y = 0 \text{ and } y = L_y \text{ (no flow)} \tag{2.6}$$

$$p = p_0(t), \quad c = 1 \qquad \text{at } x = 0 \qquad \text{(inflow)} \tag{2.7}$$

$$p = 0, \quad \phi\frac{\partial c}{\partial t} + \mathbf{u}\cdot\nabla c = 0 \quad \text{at } x = L_x \qquad \text{(outflow)} \tag{2.8}$$

The value of p_0, which is a function of time, is adjusted at each time step, so that the average flow rate per unit area remains constant.

The initial condition is:

$$c(\mathbf{x}, 0) = 0 \tag{2.9}$$

The viscosity μ in Eqs 2.1–2.2 is a function of the concentration c of the invading fluid. The equation of state used in this work is:

$$\frac{\mu}{\mu_{\mathrm{r}}} = [1 + (M^{1/4} - 1)c]^{-4} \tag{2.10}$$

where μ_{r} and μ_{i} are the viscosities of the resident fluid and the invading fluid respectively and $M = \mu_{\mathrm{r}}/\mu_{\mathrm{i}}$.

2.2 Numerical method

The coupled system consisting of the pressure equation (Eqs 2.1–2.3) and the concentration equation (Eq. 2.4) is solved sequentially. The solution of the concentration equation is approximated using a finite-element modified method of characteristics (MMOC). (Russell, 1982; Douglas and Russell, 1982; Russell and Wheeler, 1983; Ewing *et al.*, 1983; Russell *et al.*, 1986.) The time derivative and the convective terms are combined in a directional τ derivative; time stepping is along the characteristics. The characteristics are tracked backwards, using small time steps. The dispersion terms are treated implicitly, using large time steps. Continuous bilinears are employed for the concentration approximation space. The truncation error is $O(\Delta t)c_{\tau\tau} + O(\Delta x)^2$, where Δt is the time step and Δx the spatial grid spacing. Since the derivative $c_{\tau\tau}$ that multiplies the $O(\Delta t)$ term is along the characteristic, it is very small and consequently it makes the first term of the above error estimate very small. The remaining term is of order $(\Delta x)^2$. The method generates very little numerical dispersion (Russell *et al.*, 1986; Moissis, 1988) and, being implicit, allows the use of much larger time steps than explicit procedures (where time step is limited by a CFL condition). Although the method is not conservative, in practice mass is conserved if a reasonably small time step is used.

Because accurate velocities are required for implicit characteristic tracking, the pressure equation is solved with a mixed finite element scheme (MFEM) (Wheeler and Gonzalez, 1984). Equations 2.1–2.3 are not combined, but solved separately, by seeking a finite-element approximation for both pressure and velocity. Next lowest order Raviart–Thomas spaces are used as the approximating spaces. The velocities provided by this method are second-order accurate globally, while velocities at certain superconvergence points are third-order accurate (Nakata *et al.*, 1985). The high order of accuracy of the velocity approximation allows the use of a coarser grid than needed for standard methods. In addition, this method conserves mass in each gridblock.

For completeness we briefly describe the combined MFEM-MMOC numerical formulation. We first establish notation. Let $M_h(Z_h)$ denote the finite dimensional space spanned by continuous (discontinuous) piecewise bilinear functions defined on Ω. Let $V_h = V_h^x \times \tilde{V}_h^y$, where $V_h^x(V_h^y)$ denotes the tensor product space of continuous piecewise quadratics on $[0, L_x]([0, L_y])$ and discontinuous piecewise linears on $[0, L_y]([0, L_x])$ and $\tilde{V}_h^y = V_h^y \cap \{z \mid z(x, 0) = z(x, L_y) = 0\}$. Note that $V_h \subset H(\Omega; \text{div})$, where $H(\Omega; \text{div})$ is the space of vector functions which are square integrable and whose divergence is square integrable. Finally we define the inner product:

$$\langle \psi_1, \psi_2 \rangle = \int_\Omega \psi_1 \psi_2 \, \mathrm{d}x \, \mathrm{d}y \tag{2.11}$$

Let $\Delta t > 0$ and $t^n = n\Delta t$. In the MFE-MMOC formulation we seek an approximation $(P^n, \mathbf{U}^n; C^n)$ in $Z_h \times V_h \times M_h$ to the solution $(p(\mathbf{x}, t^n), \mathbf{u}(\mathbf{x}, t^n); c(\mathbf{x}, t^n))$, as follows:

$$C^0(\mathbf{x}) = \begin{cases} 1, & x < 0 \\ 0, & x > 0 \end{cases} \tag{2.12}$$

For $n \geqslant 0$, the MFE approximation $(P^n, \mathbf{U}^n)$ is defined by:

$$\left.\begin{aligned} &\left\langle \frac{\mu(C^n)}{k}\mathbf{U}^n, \mathbf{v} \right\rangle - \langle P^n, \nabla\cdot\mathbf{v}\rangle = \int_0^{L_y} p_0(y, t^n)v(0, y), \quad && \mathbf{v}\in V_h \\ &\langle \nabla\cdot\mathbf{U}^n, w\rangle = 0 && w\in Z_h \end{aligned}\right\} \tag{2.13}$$

For $n \geqslant 1$, the MMOC approximation C^n is defined by:

$$\left\langle \phi \frac{C^n - \check{C}^{n-1}}{\Delta t}, \chi \right\rangle + \langle \mathbf{D}(E\mathbf{U}^n)\nabla C^n, \nabla\chi\rangle = 0, \quad \chi\in M_h \tag{2.14}$$

with

$$\left.\begin{aligned} C^n(0, y) &= 1 \\ C^n(1, y) &= \check{C}^{n-1}(1, y) \end{aligned}\right\}, \quad y\in(0, L_y) \tag{2.15}$$

where

$$\check{C}^{n-1}(\mathbf{x}) = C^{n-1}\left(\mathbf{x} - \frac{E\tilde{\mathbf{U}}^n(\mathbf{x})}{\phi}\Delta t \right) \tag{2.16}$$

$E\mathbf{U}^n(\mathbf{x})$ is defined as follows:

$$E\mathbf{U}^n = \begin{cases} 2\mathbf{U}^{n-1} - \mathbf{U}^{n-2}, & n > 1 \\ \mathbf{U}^0, & n = 1 \end{cases} \tag{2.17}$$

$E\tilde{\mathbf{U}}^n(\mathbf{x})$ is an approximate average velocity between the times t^{n-1} and t^n, which is computed by segmenting the time step Δt into smaller sub-time steps and using a predictor–corrector procedure to determine the velocity along the characteristic in each sub-time step (Moissis, 1988; Dawson *et al.*, 1989).

The combination of the MFEM and the MMOC has been previously applied to the solution of miscible displacement problems (Russell and Wheeler, 1983; Ewing *et al.*, 1983, Russell *et al.*, 1986; Moissis *et al.*, 1987).

2.3 Generation of permeability fields

The permeability fields of the media used in the simulations presented in this paper were generated in such a way that $z = \ln k$ follows a normal distribution with mean $\bar{z} = 0$, given variance σ^2 and given correlation length λ. The correlation length is defined from the autocorrelation function of the log permeability field (which is assumed stationary)

$$\gamma(u) = \frac{1}{\sigma^2} E\{[z(x+u) - \bar{z}][z(x) - \bar{z}]\} \tag{2.18}$$

as the lag distance u for which $\gamma(u) = 1/e$. All distributions generated in this work are isotropic, so that σ^2 and λ are constants independent of direction. In the following discussion the correlation length λ is assumed scaled by the length of the medium.

In order to study the effects of permeability variations on different scales, it is desired to generate distributions with at least two scales of heterogeneity. The simplest case will involve only two scales. The correlation length will determine the order of the 'longest' of these two scales. The 'shorter' scale will be of the order of the grid that is used in the flow simulations. The method used for the generation of permeability fields is described below. For simplicity in the description, the domain is taken to be the unit square.

A number N of points, which will be called 'pilot points', are selected in the domain. The locations of the pilot points are chosen randomly; each coordinate follows a uniform distribution in $[0, 1]$. To avoid potential aggregation of pilot points in certain areas, which would leave other areas void, the distance between any two pilot points is not allowed to be smaller than λ. A value of $z = Z_i$ is assigned to the ith pilot point. The Z_i, $1 \leqslant i \leqslant N$ are random numbers, following a normal distribution with zero mean and standard deviation σ. Each of these pilot points defines an area in which the log permeability is 'close' to Z_i. The size (surface) of this area must be of order λ^2 and, since the whole domain contains N such areas, N must be given by:

$$N = \frac{\beta}{\lambda^2} \tag{2.19}$$

where β is an adjustable parameter, which depends on the way the correlation length is defined and must be determined by trial and error. In this work the value of β is taken equal to 0.5. (Lake, 1987; Moissis, 1988).

Let $\mathbf{x}$ be the location of any point in the domain. The value of z at $\mathbf{x}$ will be the sum of a long-scale contribution z_l and a short-scale contribution z_s

$$z = z_l + z_s \tag{2.20}$$

The long scale contribution z_l is a weighted average of the values of z at the pilot points:

$$z_l = \sum_{i=1}^{N} w_i Z_i \tag{2.21}$$

The weights are defined as follows: Let d_i be the distance between $\mathbf{x}$ and the ith pilot point, let $d_{\min}$ be the minimum of these distances and let $D_i = d_i/d_{\min}$. Then:

$$w_i = \frac{\exp(-\alpha D_i)}{\sum_{i=1}^{N} \exp(-\alpha D_i)} \tag{2.22}$$

where $\alpha > 0$ is an adjustable parameter. As defined in Eq. 2.22 the weights w_i are decreasing functions of D_i. The parameter α controls the value of the largest weight (corresponding to the closest pilot point) relative to the other weights. For large values of α ($\alpha > 4$) the largest weight becomes practically equal to one and the domain is divided into N regions, in each of which z is almost exactly equal to the log permeability of the corresponding pilot point Z_i. The regions are defined by sharp boundaries. Smaller values of α allow for a smooth transition between the regions. For very small values of α ($\alpha < 1$) the boundaries disappear and the effects of distant pilot points become too significant. It was found by trial and error that $\alpha = 2$ represents a reasonable choice between the two extremes.

The short-scale contribution z_s is essentially a random noise superimposed on z_l:

$$z_s = f\sigma(2R - 1)\exp\left(1 - \frac{\lambda}{d_{\min}}\right) \tag{2.23}$$

where R is a random number uniformly distributed in $[0, 1]$ and f is an adjustable parameter. Equation 2.23 is chosen in such a way that the maximum possible amplitude of z_s decreases with $d_{\min}$, approaching zero as $d_{\min} \to 0$ and becoming equal to $f\sigma$ for $d_{\min} = \lambda$. The expected value of z_s is, of course, zero. Thus f is a measure of the magnitude of the short-range heterogeneity relative to the magnitude of the long-range heterogeneity and, as such, it should normally be less than one. A value of $f = 0.5$ was used in most of the performed simulations, but values of $f = 0$ and $f = 1$ were also tried.

Finally, the permeability field is given by:

$$k = \exp(z) \tag{2.24}$$

The coefficient of permeability variation CV is related to the variance σ^2 of the log permeability distribution by the following equation:

$$\sigma^2 = \ln[1 + (CV)^2] \tag{2.25}$$

In the present study the permeability fields are generated on an 80×80 spatial grid. The permeability of the (i, j) grid block will be denoted by k_{ij}. The correlation length can be expressed more conveniently in grid block units, by setting $\lambda_D = 80\lambda$. The values of λ_D tried were 0, 2, 4 and 8 grid blocks (respectively $\lambda = 0$, 1/40, 1/20 and 1/10 of the porous medium). $\lambda = 0$ corresponds to uncorrelated random permeability fields, for which the autocorrelation function is the delta function. Several realizations were generated for each of the above correlation lengths. For each realization the coefficient of permeability variation CV was allowed to take the values 0.1, 0.2, 0.5 and (in a few cases) 1.0. Figure 1 shows autocorrelograms of some of the generated permeability fields. In all cases the correlation length obtained from the autocorrelogram agrees with the input value of the parameter λ, used in the synthesization algorithm. Figure 2 shows the permeability field for $\lambda_D = 0$, realization No. 1 and Fig. 3 for $\lambda_D = 8$, realization No. 1.

3. Results

3.1 General observations

Simulations of horizontal miscible flooding were conducted for each realization for viscosity ratios of 10 and 41. The porous medium was assumed to be the

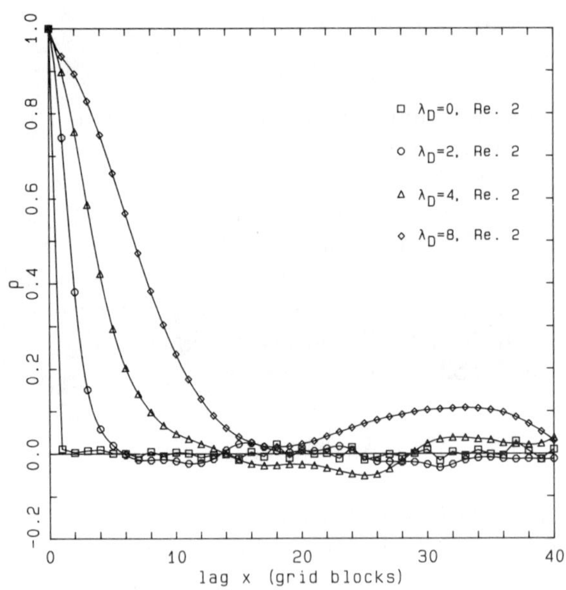

Figure 1. Autocorrelograms of generated permeability fields.

Figure 2. Permeability map of generated medium, $\lambda_D = 0$, Realization 1. Bright areas correspond to high permeability.

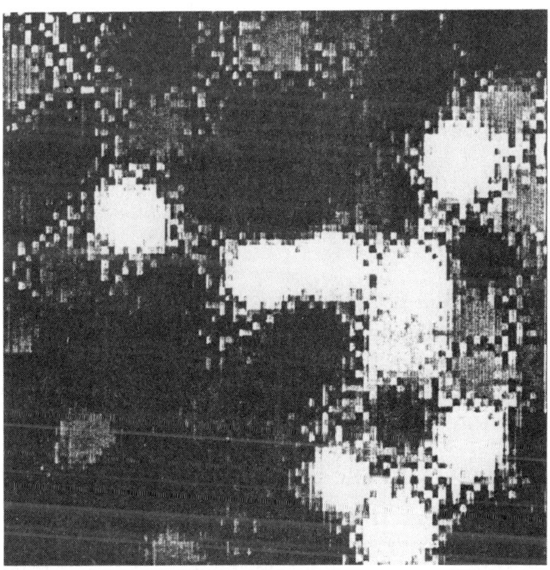

Figure 3. Permeability map of generated medium, $\lambda_D = 8$, Realization 1. Bright areas correspond to high permeability.

unit square. A constant porosity $\phi = 0.20$ was assumed. The longitudinal and transverse dispersion coefficients (see Eq. 2.5), normalized by the length of the medium, had the values $\alpha_l = 2 \times 10^{-3}$ and $\alpha_t = 5 \times 10^{-5}$. An 80×80 uniform grid (the same as the permeability grid) and a time step of 0.1 PVI were used for all simulations (time will be reported in PVI, pore volumes injected, throughout this chapter).

Concentration profiles (contours of constant concentration) at $t = 0.1$ and $t = 0.3$ are plotted for some of the simulations in Figs 7–10. The following qualitative characteristics are exhibited by all runs: At $t = 0.1$, the profile is quite irregular, with a certain number of small fingers. At later times some of the initial fingers grow faster than the rest and eventually dominate the displacement. These long fingers suppress the growth of smaller ones; this phenomenon is known as shielding. Later, smaller fingers merge with larger ones, leaving unswept areas, which may be fairly extensive. This process results in a number of large fingers, which grow quite independently from each other, at least until breakthrough. We shall refer to these fingers as 'active' fingers. The number of initial fingers, the number of active fingers and the extent of shielding and merging depend on CV, the correlation length and the viscosity ratio, as is discussed below.

A quantitative measure of the instability of the displacement before breakthrough is the root-mean-square length of the $c = 0.5$ profile, defined as follows:

$$\mathrm{RMS} = \left[\int_0^1 [x_{0.5} - \bar{x}]^2 \, \mathrm{d}y \right]^{\frac{1}{2}} \tag{3.1}$$

where $(x_{0.5}, y_{0.5})$ are all the points (x, y) of the domain for which $c(x, y) = 0.5$ and $\bar{x}$ is the mean advance of the profile:

$$\bar{x} = \int_0^1 x_{0.5} \, \mathrm{d}y \tag{3.2}$$

In this paper RMS is often plotted against time for a particular simulation. All such plots stop at breakthrough time, which we shall define conventionally as the time when the average outflow concentration surpasses 0.01.

3.2 Generation of fingers

To evaluate the effect of the permeability distribution near the inflow end ($x = 0$) on finger generation, the concentration profile at $t = 0.1$ is related to the permeability map near the inflow end. More precisely the quantities correlated are the average of $x_{0.5}(y)$ (distance from the inlet of location where $c = 0.5$) in each grid block j in the y direction (denoted by X_j) and the harmonic mean permeability $\bar{K}_j^m$ over the first m blocks in the x direction, i.e.

$$\frac{m}{\bar{K}_j^m} = \sum_{i=1}^{m} \frac{1}{k_{ij}} \tag{3.3}$$

Table 1. Correlation of the concentration profile at $t = 0.1$ with the permeability distribution near the inflow end.

λ_D	Realization	CV	M	m_{max}	r_{max}
0	1	0.1	10	5	0.796
0	1	0.1	41	5	0.709
0	1	0.2	10	4	0.777
0	1	0.2	41	4	0.676
0	1	0.5	10	6	0.614
0	1	0.5	41	10	0.508
0	2	0.1	10	5	0.722
2	1	0.1	10	9	0.837
2	1	0.1	41	8	0.753
2	1	0.2	10	10	0.827
2	1	0.2	41	11	0.773
2	1	0.5	10	14	0.804
2	1	0.5	41	15	0.721
4	1	0.1	10	12	0.844
4	1	0.1	41	13	0.732
4	1	0.2	10	12	0.807
4	1	0.2	41	13	0.755
4	1	0.5	10	25	0.754
4	1	0.5	41	25	0.704
4	2	0.2	10	12	0.873
4	2	0.2	41	13	0.840
8	1	0.1	10	14	0.792
8	1	0.1	41	15	0.598
8	1	0.2	10	15	0.783
8	1	0.2	41	16	0.670
8	1	0.5	10	52	0.815
8	1	0.5	41	52	0.784
8	2	0.2	10	14	0.764
8	2	0.2	41	14	0.624
8	3	0.2	10	12	0.820
8	3	0.2	41	12	0.621

Harmonic averages of permeability are used, because harmonic averaging gives the effective permeability in the direction of flow.

Correlation coefficients are calculated for different values of m. Let m_{max} be the value of m that gives the highest correlation coefficient for a particular simulation and let r_{max} be that coefficient. Values of m_{max} and r_{max} for some realizations are shown in Table 1. The coefficient r_{max} represents the range near the entrance of the porous medium, over which permeability variations are important in finger generation. It is found that m_{max} is independent of the viscosity ratio, increases with the correlation length and is roughly the same for all realizations having the same correlation length.

The fact that m_{max} increases with λ_D indicates that large-scale permeability variations are sensed over a longer range than small-scale variations, which is expected. The relatively high values of r_{max} suggest that the permeability distribution near the inflow end dictates the fingering pattern at early times. To illustrate this effect, the concentration profiles at $t = 0.1$ for $M = 10$ and $M = 41$ are juxtaposed to a sketch of $\bar{K}_j^m$ (plotted as a function of j, for $m = m_{max}$) for three typical runs in Figs 4–6. From these plots it is clear that the fingering

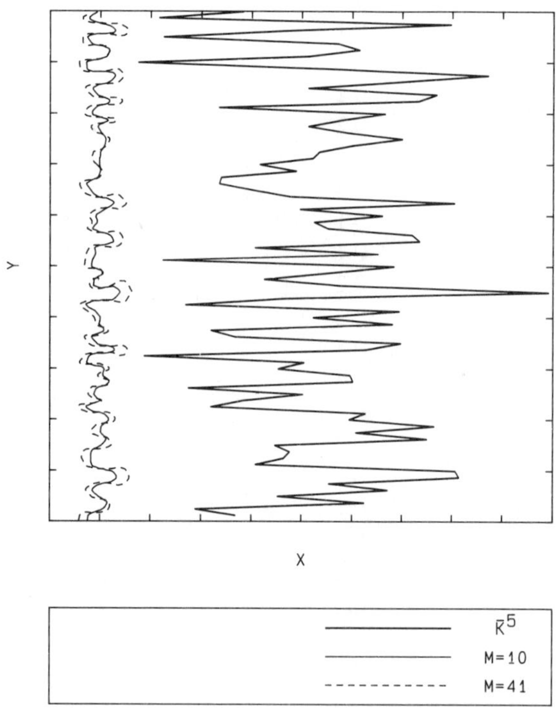

Figure 4. Effect of permeability distribution near the inflow end on viscous fingering. $c = 0.5$ profiles at $t = 0.1$; $\lambda_D = 0$, Realization 1, $CV = 0.1$.

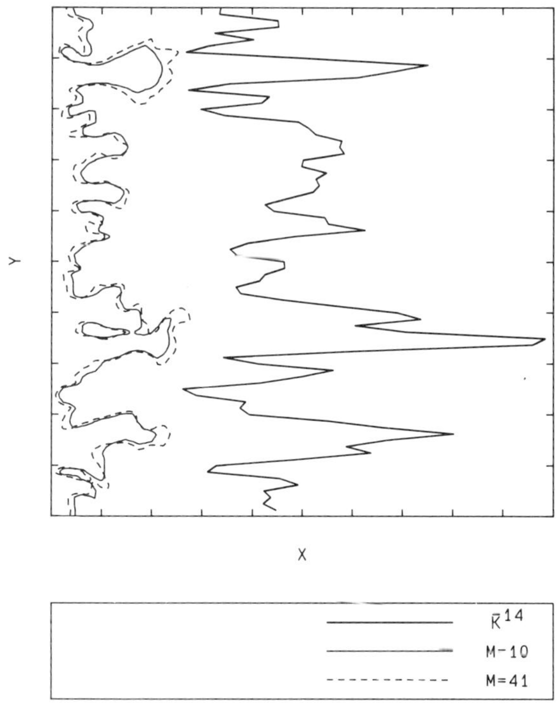

Figure 5. Effect of permeability distribution near the inflow end on viscous fingering. $c = 0.5$ profiles at $t = 0.1$; $\lambda_D = 2$, Realization 1, $CV = 0.5$.

pattern is independent of the viscosity ratio, in the sense that, for $t = 0.1$, the number of fingers, their locations and their relative lengths are the same for both viscosity ratios studied (for the same realization); the actual length of the fingers is, of course, larger for $M = 41$ than for $M = 10$. Furthermore, in all cases, the locations of the fingers correspond consistently to (local) maxima of $\bar{K}_j^m(y)$. Since the number of maxima in $\bar{K}_j^m$ decreases as the correlation length increases, the number of fingers initially formed is a decreasing function of the correlation length. This fact can be verified from Figs 7–10. The above-mentioned results have been confirmed for a wider range of viscosity ratios.

As time advances the correlation between X_j and $\bar{K}_j^m$ deteriorates. This correlation is good only during the initial stage of the displacement, whose duration is a function of CV and the viscosity ratio. Non-linear phenomena (mainly shielding, as discussed above) are already present at $t = 0.1$ and this is why the correlation is not perfect. Non-linear phenomena occur faster and to a larger extent at higher viscosity ratios; this explains the smaller value of r_{max} that was observed in all cases for $M = 41$ than for $M = 10$.

The dependence of the correlation coefficient r_{max} on the heterogeneity of the

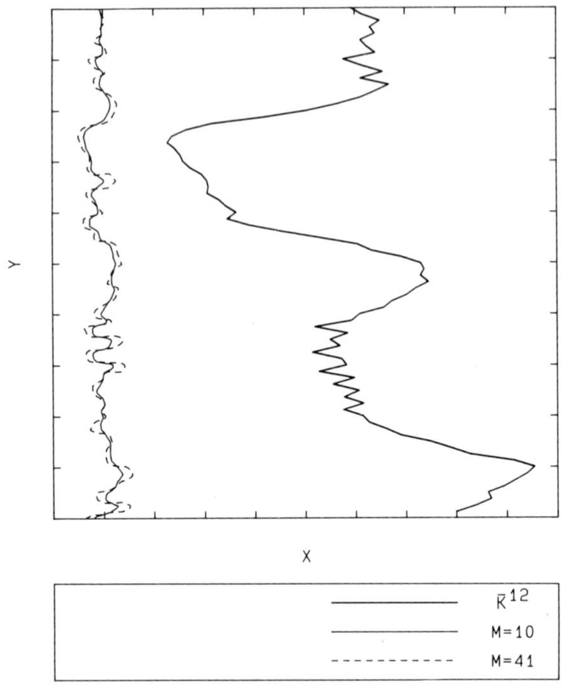

Figure 6. Effect of permeability distribution near the inflow end on viscous fingering. $c = 0.5$ profiles at $t = 0.1$; $\lambda_D = 8$, Realization 2, $CV = 0.2$.

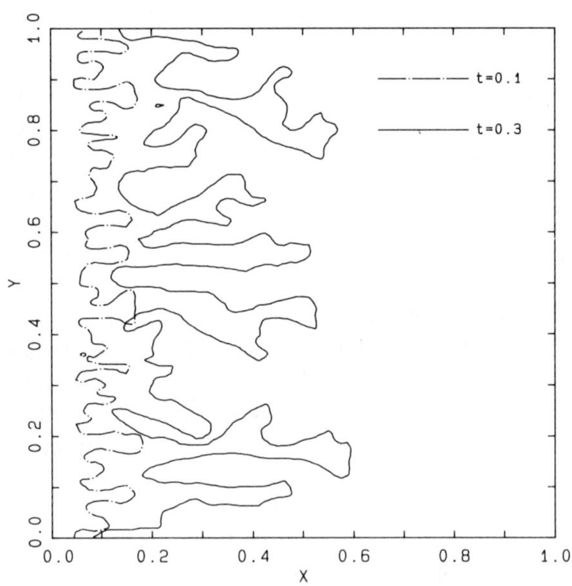

Figure 7. $c = 0.5$ profiles; $\lambda_D = 0$, Realization 1, $CV = 0.2$, $M = 41$.

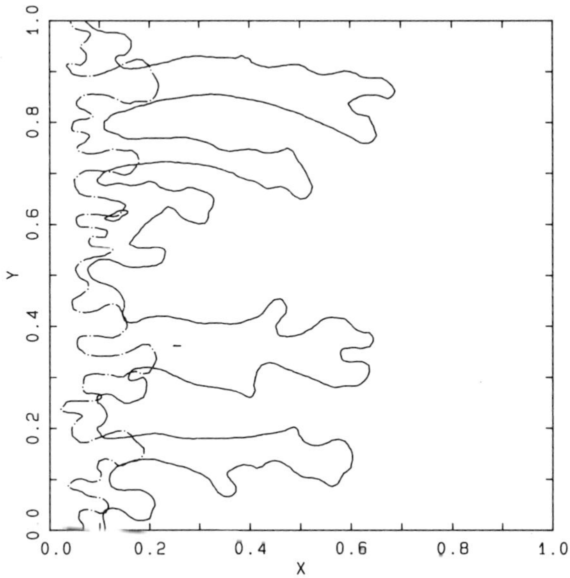

Figure 8. $c = 0.5$ profiles; $\lambda_D = 2$, Realization 1, $CV = 0.2$, $M = 41$.

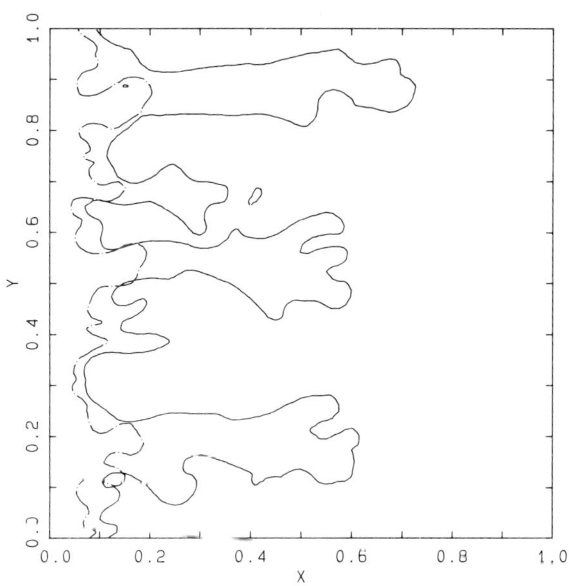

Figure 9. $c = 0.5$ profiles, $\lambda_D = 4$, Realization 1, $CV = 0.2$, $M = 41$.

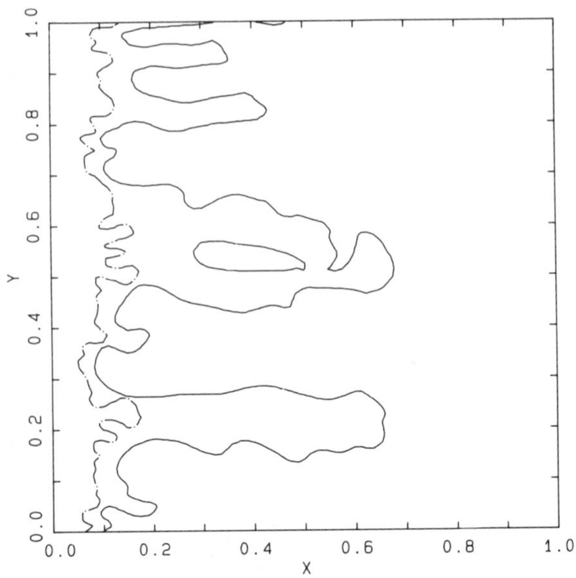

Figure 10. $c = 0.5$ profiles; $\lambda_D = 8$, Realization 1, $CV = 0.2$, $M = 41$.

medium (CV) is not clear; from Table 1 it can be observed that, as CV increases, r_{max} sometimes decreases and sometimes increases. The former tends to happen for small correlation lengths, the latter for larger correlation lengths. In fact the dependence of r_{max} on CV is determined by the effects of two competing mechanisms. On the one hand, a more heterogeneous permeability distribution tends to have a stronger effect on the initial concentration profile, hence to increase r_{max}. On the other hand, in displacement through a more heterogeneous medium non-linear phenomena are stronger and occur faster, tending to reduce r_{max}. This suggests that, as CV increases, r_{max} passes through a maximum. Our results seem to confirm this hypothesis; they also indicate that the value of CV for which the maximum r_{max} occurs depends on the correlation length.

Finally it is interesting to note that simulations of a stable displacement ($M = 1$) show an $r_{max} > 0.9$ at all times before breakthrough, with m_{max} increasing continuously. This suggests that for $M = 1$ the displacement is governed by permeability variations at all times, which is expected, since the displacement is stable and the heterogeneity of the porous medium is the only source of perturbation in the system.

3.3 Effect of permeability variations downstream

To study the importance of the permeability distribution downstream, a set of runs was performed, in which the downstream 80% of the permeability field was 'homogenized', i.e. the permeabilities of the downstream 64 × 80 blocks were replaced by their (arithmetic) average; in this way any effect of permeability variations downstream was removed. Figure 11 shows the RMS versus time curves for $M = 10$, $CV = 0.2$ and $\lambda_D = 0$ and $\lambda_D = 8$, for both the 'homogenized' and the 'base' case. As seen, the effect of the permeability distribution downstream, unimportant in uncorrelated media ($\lambda_D = 0$), becomes more significant as the correlation length increases.

The cause of this different behaviour is linked with the scale of the permeability variations. For a finger to be significantly affected by a permeability variation, the range of the variation has to be of the same order as the finger width. This is true for media with non-zero correlation length, as can be seen in Figs 8–10. In these media, the existence of extensive regions of high or low permeability causes channelling of the flow and thus affects finger growth mainly by causing additional merging. The fact that finger width increases with the viscosity ratio (Moissis, 1988) may be the reason why the effect of the permeability distribution downstream seems to be stronger for $M = 10$ than for $M = 41$. On the other hand, for $\lambda_D = 0$, the width of the fingers is always larger than the range of the permeability variations. Even though in the simulations with $\lambda_D = 0$ the range of permeability variations is actually of the order of a grid block, the width of the fingers is substantially larger (Fig. 7). When averaged over several blocks,

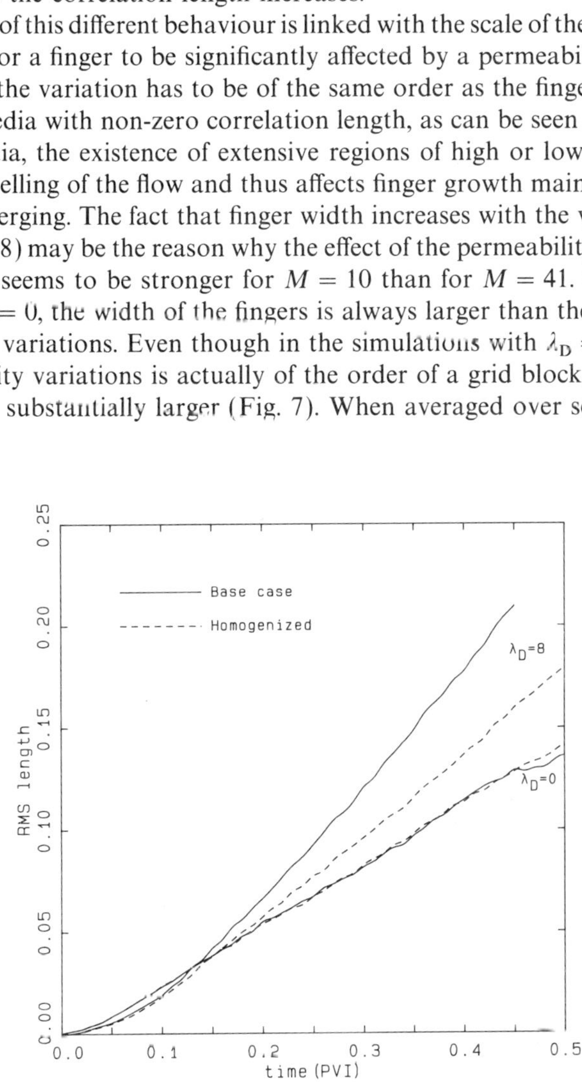

Figure 11. Effect of permeability variations downstream for different correlation lengths; $CV = 0.2$, $M = 10$.

the small-scale heterogeneities are offset and their net result is only short wavelength perturbations.

3.4 Effect of small-scale heterogeneities

As explained previously, the field $z = \ln k$ is a superposition of a long-scale contribution z_l and a short-scale contribution z_s (Eqs 2.20–2.23). A series of simulations was conducted to investigate the effect of small-scale permeability variations in media where long-scale variations exist as well. All permeability fields considered in these simulations had the same CV and exactly the same z_l field as the base ($\lambda_D = 8$, Realization 1, $CV = 0.2$; termed here Realization 1.1), but their z_s fields differed, as explained below.

For the second medium (Realization 1.0), z_s was identically zero, i.e. no short-scale random variations existed. This was equivalent to setting $f = 0$ in Eq. 2.23. To generate z_s for the third medium (Realization 1.2), the same sequence of random numbers was used as for the base case but f was taken equal to 1. This resulted in short-scale permeability variations of greater magnitude than in the base case. Finally, for the fourth medium (Realization 1.3), $f = 0.5$, as in the base case, but a different sequence of random numbers was chosen, resulting in a different z_s field. All simulations were conducted with $M = 41$.

Concentration profiles for realizations 1.0, 1.2 and 1.3 are plotted in Figs 12,

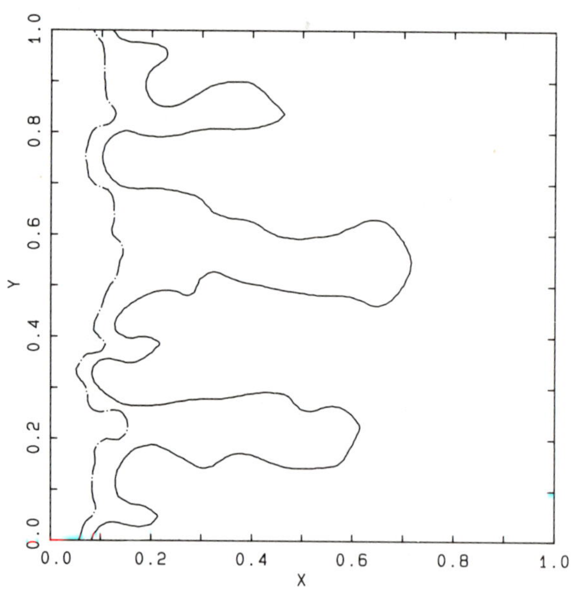

Figure 12. Effect of small-scale heterogeneities. $c = 0.5$ profiles; $\lambda_D = 8$, Realization 1.0, $CV = 0.2$, $M = 41$.

13 and 14, respectively. Profiles for the base case (Realization 1.1) are shown in Fig. 10. The profiles at $t = 0.1$ will be compared first. Since for Realization 1.0 there are no small-scale heterogeneities which would generate individual fingers, the profile at $t = 0.1$ (Fig. 12) shows three areas of potential finger growth, but there is only one well-defined finger. On the other hand, for Realization 1.2 (Fig. 13), the profile is essentially the same as in the base case, except that the fingers are slightly longer, as expected. For Realization 1.3 (Fig. 14), the profile at $t = 0.1$ is different from the base case in its details but it also shows finger growth in the same three areas. At $t = 0.3$, however, after active fingers have developed, all four profiles look essentially the same. Each profile clearly shows two active fingers in the same locations (about $y = 0.20$ and $y = 0.55$), of which the latter is longer and has clearly formed by merging (except in Realization 1.0), and a number of shielded fingers between $y = 0.8$ and $y = 1$. The location where the merging occurs is a high permeability area. Of course the profile for Realization 1.0 is smoother and for Realization 1.2 more irregular than for the base case, because of the effect of small-scale heterogeneities. More important is the fact that there is some difference in the length of the longest finger. However, this difference is rather small and it becomes even smaller later in the displacement, except for Realization 1.2, which has a slightly earlier breakthrough (0.39 as opposed to 0.41 for the other three realizations). The RMS curves for all four realizations are about the same (Fig. 15). In conclusion,

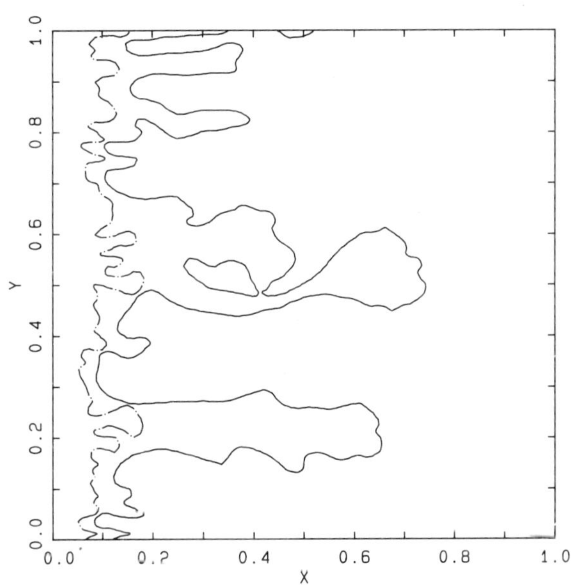

Figure 13. Effect of small-scale heterogeneities. $c = 0.5$ profiles; $\lambda_D = 8$, Realization 1.2, $CV = 0.2$, $M = 41$.

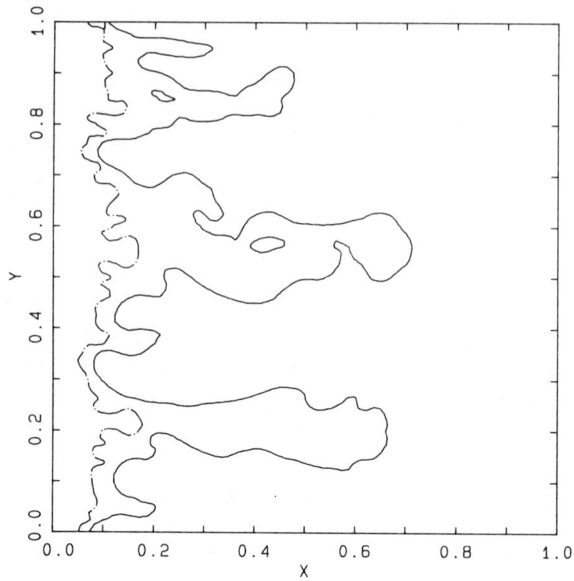

Figure 14. Effect of small-scale heterogeneities. $c = 0.5$ profiles; $\lambda_D = 8$, Realization 1.3, $CV = 0.2$, $M = 41$.

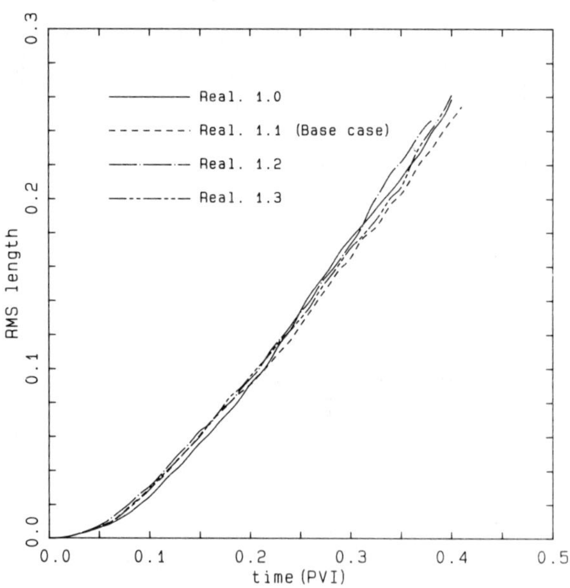

Figure 15. Effect of small-scale heterogeneities. $\lambda_D = 8$, Realization 1, $CV = 0.2$, $M = 41$.

it seems that small-scale heterogeneities in media with large correlation lengths do not significantly affect the displacement.

3.5 Effect of the correlation length

As seen from Figs 7–10, the number of active fingers is a decreasing function of the correlation length. There are two reasons for this. First, as discussed earlier, the initial number of fingers (e.g. at $t = 0.1$) is smaller for larger correlation lengths. Second, the effect of the downstream permeability distribution is more significant for large λ_D, resulting in more merging, which reduces the number of fingers. From the simulations performed it was found that the number of active fingers was independent of the specific permeability distribution, in the sense that different realizations having the same values of the parameters λ_D, CV and M produced the same number of active fingers. Thus, for example for $CV = 0.2$ and $M = 41$, five or six active fingers form for $\lambda_D = 0$ (Fig. 7), four for $\lambda_D = 2$ (Fig. 8), three for $\lambda_D = 4$ (Fig. 9), two for $\lambda_D = 8$ (Fig. 10).

RMS length for each of the cases simulated was calculated by averaging RMS values for different realizations. The agreement between realizations was quite good for $\lambda_D = 0$, 2 and 4 (typical example in Fig. 16), but not for $\lambda_D = 8$, where large differences between realizations were found (Fig. 17). These

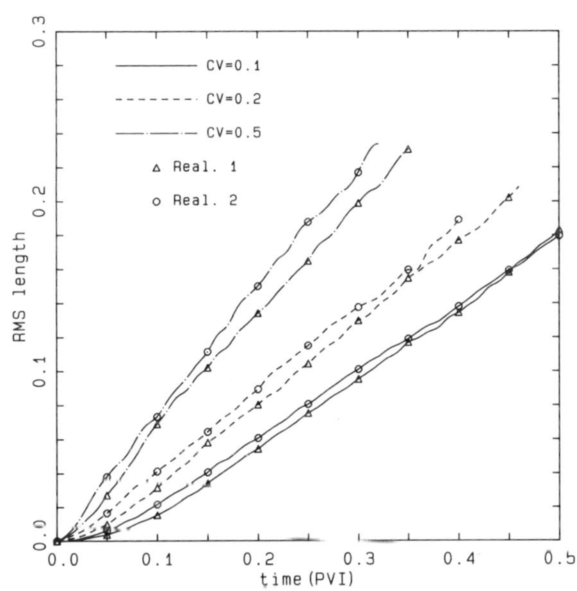

Figure 16. RMS length for different realizations; $\lambda_D = 4$, $M = 10$.

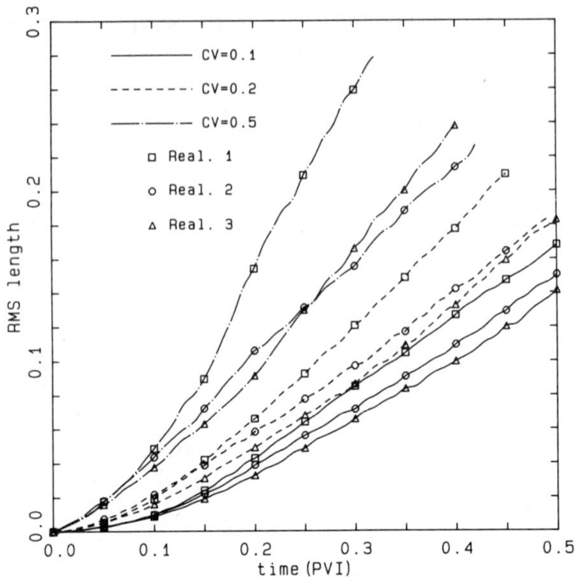

Figure 17. RMS length for different realizations; $\lambda_D = 8$, $M = 10$.

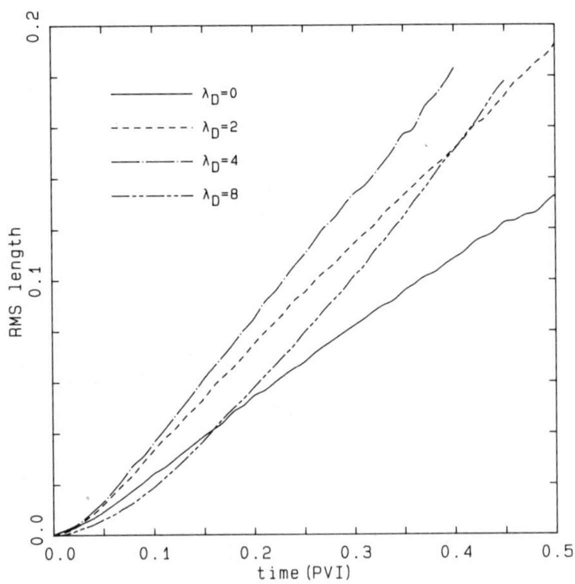

Figure 18. Average RMS length for different λ_D; $CV = 0.2$, $M = 10$.

differences become more pronounced as the coefficient of permeability variation increases. The reason for this behaviour is that for $\lambda_D = 8$ the correlation length is one-tenth of the size of the domain, hence comparable to it. The domain contains only about fifty 'permeability regions', roughly equal to the number of pilot points. This low number is not sufficient to average out the permeability variations from region to region and flow channelling may occur to varying degrees, resulting in large differences between realizations. In order to obtain statistically meaningful results for large correlation lengths, simulations must be performed for a large number of realizations.

Figure 18 shows RMS versus time curves, with λ_D as a parameter, for $CV = 0.2$ and $M = 10$. Average root-mean-square growth rates RMS′ for some of the cases simulated are listed in Table 2. RMS′ is the average slope of the RMS versus time curve; it was calculated by a linear least-square fit between $t = 0.1$

Table 2. Average RMS′ for the cases simulated.

λ_D	CV	M	t_{BT}	RMS′	r
0	0.1	10	>0.5	0.2770	0.9993
0	0.1	41	0.46 0.48	0.5054	0.9987
0	0.2	10	>0.5	0.2729	0.9991
0	0.2	41	0.43–0.46	0.5032	0.9994
0	0.5	10	>0.5	0.3295	0.9980
0	0.5	41	0.38–0.42	0.6348	0.9984
2	0.1	10	>0.5	0.3432	0.9993
2	0.1	41	0.47	0.5460	0.9999
2	0.2	10	>0.5	0.3929	0.9998
2	0.2	41	0.42–0.43	0.6094	0.9999
2	0.5	10	0.41–0.45	0.5131	0.9994
2	0.5	41	0.34–0.35	0.6640	0.9913
4	0.1	10	>0.5	0.4023	0.9998
4	0.1	41	0.45–0.46	0.5565	0.9999
4	0.2	10	0.41–0.47	0.4797	0.9998
4	0.2	41	0.36–0.39	0.6800	0.9998
4	0.5	10	0.33–0.36	0.6750	0.9996
4	0.5	41	0.27–0.31	0.9523	0.9999
8	0.1	10	>0.5	0.3642	0.9989
8	0.1	41	0.43–0.49	0.5463	0.9988
8	0.2	10	0.46–0.50	0.4556	0.9984
8	0.2	41	0.41–0.42	0.6247	0.9984
8	0.5	10	0.33–0.43	0.7739	0.9995
8	0.5	41	0.29–0.36	0.9857	0.9995

and $t = 0.5$ or breakthrough (whichever occurs first). The result was averaged over all realizations with the same set of parameters (λ_D, CV, M). The correlation coefficient r of the fit is included in the table. It can be seen that, in all cases, the growth of RMS with time can be considered linear. The growth is not linear however during an initial stage of the displacement, whose duration seems to be an increasing function of the correlation length (Fig. 18). Table 2 also includes the range of breakthrough times t_{BT} for the various realizations.

From Fig. 18 it is apparent that, for $CV = 0.2$, the growth rate of fingers increases with the correlation length up to $\lambda_D = 4$, then decreases sharply for $\lambda_D = 8$. The smaller RMS length for $\lambda_D = 8$ is due to the very long time that is needed for fingers to develop initially for this high correlation length. (In fact the slope of the RMS curve near $t = 0$ is smaller for $\lambda_D = 8$ than for any other correlation length.) This slow initial growth was observed in all realizations. Results for $M = 41$ are qualitatively similar.

This behaviour is consistent with linear stability analysis (Jerault *et al.*, 1984), which predicts that the growth rate of perturbations decreases with increasing wavelength. For large correlation lengths the permeability distribution tends to generate fingers of large wavelengths, which grow relatively slowly. However, once these fingers grow past the zone in which their growth is approximately described by linear stability analysis, it is easier for the invading fluid to develop flow channels and the root-mean-square growth rate (slope of the RMS curve) is higher for large correlation lengths. (For example at $t = 0.35$ the slope for $\lambda_D = 8$ is about the same as for $\lambda_D = 4$.) These two competing mechanisms (i.e. slower initial growth and faster subsequent growth as the correlation length increases) determine the actual root-mean-square growth rates for different correlation lengths. In the cases simulated with $CV = 0.1$ and 0.2, a maximum average RMS′ growth rate was observed for $\lambda_D = 4$. For more heterogeneous media (e.g. $CV = 0.5$), the correlation length for which maximum fingering occurs may depend on the value of CV (Fig. 23), as discussed in the next section.

3.6 Effect of the coefficient of permeability variation

The number of active fingers is a decreasing function of the coefficient of permeability variation CV. As the value of CV increases, the difference in permeability between a high permeability region and a low permeability region increases. In the initial stage of the displacement this difference favours the growth of the longest fingers, which grow in the high permeability regions. Thus, for highly heterogeneous media, the longest fingers grow much faster than the rest and dominate the displacement relatively early, shielding the smaller fingers. Later in the displacement the large difference in permeability from region to region facilitates merging. The result of these two mechanisms is reduction of the number of active fingers. Since almost all the invading fluid flows through one or a few fingers, breakthrough occurs early and sweep efficiency is poor (Fig. 19).

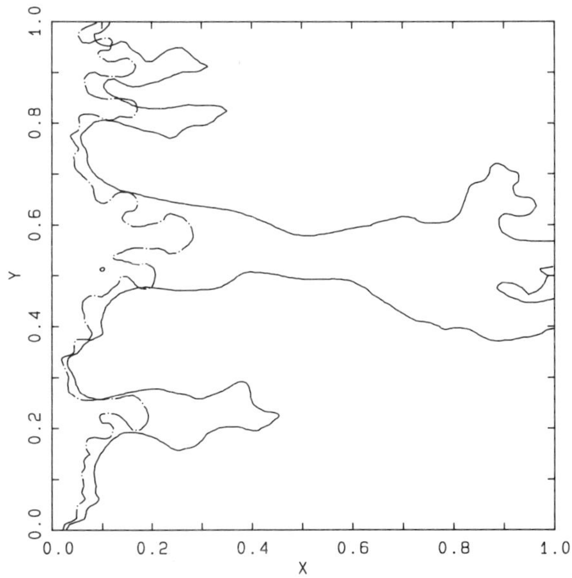

Figure 19. $c = 0.5$ profiles; $\lambda_D = 8$, Realization 1, $CV = 0.5$, $M = 41$.

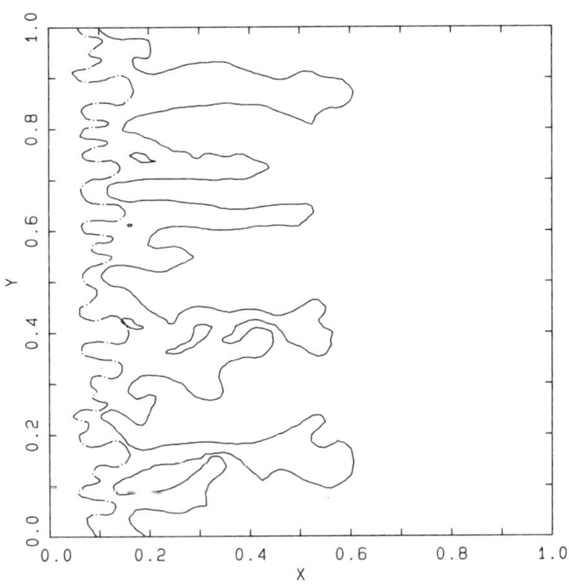

Figure 20. $c = 0.5$ profiles; $\lambda_D = 2$, Realization 1, $CV = 0.1$, $M = 41$.

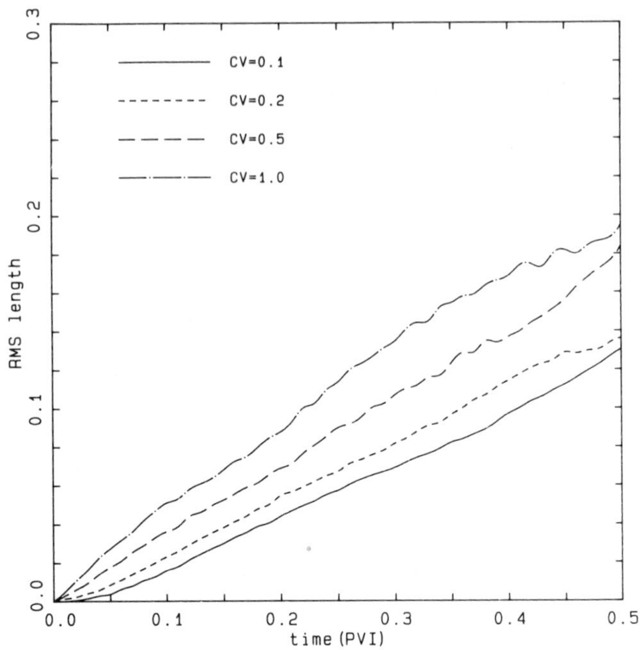

Figure 21. RMS length for different CV; $\lambda_D = 0$, Realization 1, $M = 10$.

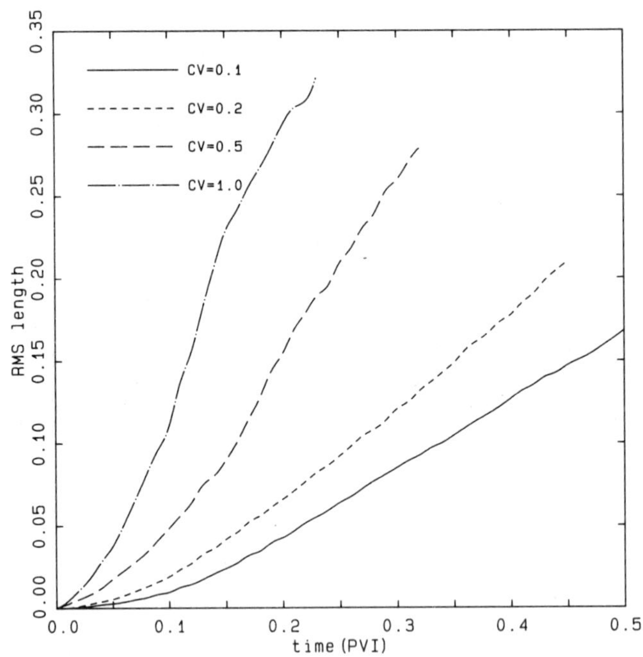

Figure 22. RMS length for different CV; $\lambda_D = 8$, Realization 1, $M = 10$.

Conversely, for low values of CV, the heterogeneity of the medium is small. In this case all fingers grow at comparable speeds, less shielding and merging occurs, the number of active fingers is larger and breakthrough is delayed (Fig. 20).

As expected, the root-mean-square growth rate increases with CV. However the effect of CV becomes much stronger as the correlation length increases. The difference in the effect of CV is especially great between the uncorrelated medium (Fig. 21) and the correlated media (e.g. Fig. 22). This quantitative difference is due to the large extent of flow channelling, which occurs in media with large correlation lengths. For large λ, increased heterogeneity favours flow of the invading fluid through the high permeability channels, causing high finger growth rates and early breakthrough. For uncorrelated media, or media with small correlation lengths, there is little or no channelling and the effect of increased heterogeneity is not as pronounced.

The dependence of RMS′ on both λ_D and CV is summarized (for $M = 10$) in Fig. 23.

4. Conclusions

The effects of the heterogeneity of the porous medium on unstable incompressible miscible displacement were studied by numerical simulation. Porous media were

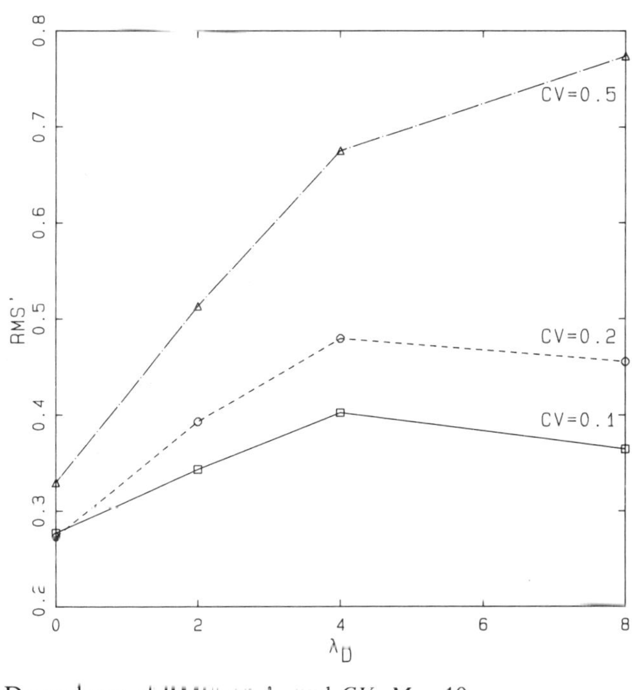

Figure 23. Dependence of RMS′ on λ_D and CV, $M = 10$.

generated by a statistical method and were characterized by two parameters, the coefficient of permeability variation CV and the correlation length λ. The third parameter considered in the study was the viscosity ratio M. The main conclusions are outlined below.

The permeability distribution near the inflow end of the porous medium determines the initial number, locations and relative growth rates of the fingers; none of these depends on the viscosity ratio. Locations where fingers form are dictated by maxima in the permeability near the inflow end. The initial number of fingers depends strongly on the correlation length of the permeability distribution. For highly correlated media permeability has fewer maxima and fewer fingers form. Subsequent merging reduces the initial number of fingers to a smaller number of 'active fingers', which dominate the displacement. Smaller fingers are shielded (their growth is suppressed) and eventually merge with longer fingers. The number of fingers is thus reduced as the displacement proceeds. The number of active fingers is a decreasing function of the correlation length, the heterogeneity of the medium (i.e. the coefficient of permeability variation) and the viscosity ratio but, with those parameters fixed, it does not depend on the specific permeability distribution. The effect of downstream permeability variations, negligible for uncorrelated media, becomes more important as the correlation length increases. The effect is more pronounced for small viscosity ratios.

After a short initiation period, fingers grow linearly in time until they start interacting with other fingers. The root-mean-square growth rate (RMS′) can be approximated by a constant, although the approximation deteriorates slightly at higher viscosity ratios. As the correlation length increases, RMS′ passes through a maximum. RMS′ also increases with increasing heterogeneity of the medium. This effect is much more pronounced in highly correlated media, since, in those, large heterogeneity facilitates channelling of flow.

Acknowledgements

This research was supported in part by the Department of Energy, under Grant No. 1553820. The authors wish to thank CRAY Research for providing the computer time that was necessary for the calculations.

References

Bartlett, M.S. (1975). 'The Statistical Analysis of Spatial Pattern'. Chapman and Hall, London.

Dawson, C.N., Russell, T.F. and Wheeler, M.F. (1989). Some improved error estimates for the modified method of characteristics, *SIAM J. Numer. Anal.*, in press.

Delhomme, J.P. (1979). Spatial variability and uncertainty in groundwater flow parameters: a geostatistical approach, *Water Resour. Res.* **15**, 269–280.

Douglas, J., Jr. and Russell, T.F. (1982). Numerical methods for convection-dominated diffusion problems based on combining the method of characteristics with finite difference or finite element procedures, *SIAM J. Numer. Anal.* **19**, 871–884.

Ewing, R.E., Russell, T.F. and Wheeler, M.F. (1983). Simulation of miscible displacement using mixed methods and a modified method of characteristics, SPE paper 12 241, *Reservoir Simulation Symposium*, San Francisco, November 1983.

Jerault, G.R., Nitsche, L.C., Teletzke, G.F., Davis, H.T. and Scriven, L.E. (1984). Frontal structure and stability in immiscible displacement, SPE paper 12 691, *59th Annual Technical Conference of SPE*, Houston, September 1984.

Journel, A.B. and Huijbregts, Ch.J. (1978). 'Mining Geostatistics'. Academic Press, New York.

Lake, L.W. (1987). A marriage of geology and reservoir engineering. *In* 'Numerical Simulation in Oil Recovery' (M.F. Wheeler, ed.), The IMA Volumes in Mathematics and its Applications, Vol. 11, pp. 177–198. Springer-Verlag, Berlin and New York.

Mantoglou, A. and Wilson, J.W. (1982). The turning bands method for simulation of random fields using line generation by a spectral method, *Water Resour. Res.* **18**, 1379–1394.

Matheron, G. (1973). The intrinsic random functions and their applications, *Adv. Appl. Prob.* **5**, 439–468.

Mejia, J. and Rodriguez-Iturbe, I. (1974). On the synthesis of random fields from the spectrum: an application to the generation of hydrologic spatial processes, *Water Resour. Res.* **10**, 705–711.

Moissis, D. (1988). Simulation of viscous fingering during miscible displacement in nonuniform porous media. Ph.D. Dissertation, Rice University, Houston.

Moissis, D., Miller, C.A. and Wheeler, M.F. (1987). A parametric study of viscous fingering. *In* 'Numerical Simulation in Oil Recovery' (M.F. Wheeler, ed.), The IMA Volumes in Mathematics and its Applications, Vol. 11, pp. 227–247. Springer-Verlag, Berlin and New York.

Nakata, M., Weiser, A. and Wheeler, M.F. (1985) Some superconvergence results for mixed finite element methods for elliptic problems on rectangular domains. *In* 'The Mathematics of Finite Elements and Applications V, MAFELAP 1984' (J.R. Whiteman, ed.), pp. 367–389. Academic Press, London.

Peaceman, D.W. (1966). Improved treatment of dispersion in numerical calculation of multi-dimensional miscible displacement, *Soc. Petrol. Eng. J.* **6**, 213–216.

Russell, T.F. (1982). Finite elements with characteristics for two-component incompressible miscible displacement, SPE paper 10 500, *Proc. Sixth SPE Symposium on Reservoir Simulation*, New Orleans, LA, January–February 1982.

Russell, T.F. and Wheeler, M.F. (1983). Finite element and finite difference methods for continuous flows in porous media. *In* 'The Mathematics of Reservoir Simulation' (R.E. Ewing, ed.), pp. 35–105. SIAM, Philadelphia.

Russell, T.F., Wheeler, M.F. and Chiang, C. (1986). Large-scale simulation of miscible displacement by mixed and characteristic finite element methods. *In* 'Mathematical Computational Methods in Seismic Exploration and Reservoir Modeling' (W.E. Fitzgibbons, ed.), pp. 85–107. SIAM, Philadelphia.

Wheeler, M.F. and Gonzalez, R. (1984). Mixed finite element methods for petroleum reservoir engineering problems. *In* 'Computing Methods in Applied Sciences and Engineering VI' (R. Glowinski and J.L. Lions, eds), pp. 639–658. North-Holland, New York.

Chapter X

Multigrid Methods for Solving Multiscale Transport Problems

C.R. Cole and H.P. Foote

Pacific Northwest Laboratory, Richland, WA 99352, USA

1. Introduction

Numerical simulation of groundwater systems with high permeability contrasts and multiple heterogeneity scales is necessary to clarify the flow and dispersive transport characteristics of these systems. These kinds of numerical studies frequently require the domain of interest to be discretized over two or three orders of magnitude of scale. Resulting systems of equations may have several million unknowns, necessitating a very efficient solution technique. It is in this context that we have investigated multigrid methods. The multigrid approach (Brandt, 1977) is probably the most efficient approach for solving large-elliptic problems because computational work and memory requirements scale linearly with the number of unknowns (Brandt, 1977; Van der Wees, 1988).

A multigrid approach was chosen to compute two- and three-dimensional, steady-state, groundwater flow because the approach can resolve the hierarchy of different heterogeneity scales important to multiscale transport simulations. The robustness of the multigrid approach coupled with its computational and memory efficiency allowed solution of problems with high permeability contrasts (in excess of 10 000 to 1) and a large number of unknowns (e.g. a three-dimensional problem with 7.5 million nodes) on a moderately sized SUN-4/280 computer with 80 megabytes of memory.

Multigrid is an iterative approach for solving algebraic equations arising from boundary-value problems on discrete spatial domains. A unique characteristic of multigrid is the use of a series of coarser grids to accelerate convergence. Multigrid's hierarchy of grid scales provides an appropriate framework for representing and solving problems with a range of heterogeneity scales. The various steps in the multigrid approach, as will be demonstrated, demand an

DYNAMICS OF FLUIDS IN HIERARCHICAL POROUS MEDIA
ISBN 0-12-200260-1

understanding of the parameter averaging process in groundwater systems, especially where complex spatial distributions of parameters involving several length scales and orientations are found to occur.

Basic flow and transport issues (e.g. scale-dependent dispersion) related to parameter heterogeneity (Cole and Foote, 1987; Cole and Rice, 1988; Codell, 1988) are being studied in 'numerical groundwater laboratories' (Cole *et al.*, 1985) using high-resolution simulations of complex groundwater systems. In a numerical laboratory approach, as discussed in Buneman *et al.* (1980), computer implementations of basic physical laws are used to understand the suitability and/or range of applicability of higher level approaches. The groundwater flow and transport issues being studied in our numerical groundwater laboratory are discussed later.

This chapter provides the reader with a basic understanding of the multigrid solution approach. It does not provide a theoretical analysis of multigrid methods using Fourier analysis techniques such as presented in Brandt (1989). However, the qualitative basis of the method is presented through a series of figures that illustrate how successively coarser grids systematically reduce solution error through classical iterative methods. As will be seen, the combination of fine and coarse grids provides optimal resolution of all spatial frequency components in the solution.

The reader should have a basic understanding of: (1) finite difference discretization; (2) techniques for solving partial differential equations and boundary-value problems; and (3) classical iterative (relaxation) methods for solving the resulting set of linear algebraic equations.

Discussions in this chapter are limited to the application of finite difference methods to the linear elliptic groundwater flow equation. Multigrid methods, it should be noted, enjoy a much wider range of application. Section 2 contains a short history of the multigrid and multilevel problem-solving approach and the types of problems to which it is being applied. Section 2 also provides an outline of the multigrid approach. Section 3 implements the multigrid method for the groundwater problem, discusses the steps in more detail, and illustrates the steps with one-dimensional examples. Section 4 discusses briefly how multigrid codes are being used to address various parameter heterogeneity issues and the approaches used to display complex three-dimensional results. Finally, Section 5 contains a suggested reading list on multigrid methods.

2. The multigrid method

2.1 Brief history of multigrid

The publication of the now classic paper 'Multi-level adaptive solutions to boundary-value problems' (Brandt, 1977) was for many, a first introduction to multigrid solution techniques. Brandt had published papers on this topic several years earlier; however, the 1977 paper is recommended as a starting point for anyone desiring to learn more about multigrid methods. In the introduction to *Multigrid Methods for Integral and Differential Equations* (Paddon and Holstein, 1985), Brandt provides a very readable historical development of multilevel processing in many contexts and partial differential equations (PDEs) in particular. He then describes briefly generalizations of multilevel techniques to such problems as image processing, pattern recognition, statistical physics, and network design. This is clearly a very active and expanding area of research. We have provided a short reading list at the end of the chapter that in turn contains references to hundreds of publications. Stephen McCormick's 1987 text on *Multigrid Methods* contains a multigrid bibliography with 607 entries. In the sections that follow we describe the basic features of multigrid that were established by the early 1970s. For a more comprehensive introduction to multigrid methods, we recommend the short monograph *A Multigrid Tutorial* (Briggs, 1987).

2.2 Outline of the multigrid approach

The most important elements of the multigrid approach are outlined here to provide an overall understanding and motivation for the expanded description in later sections. This discussion is limited to the solution of linear elliptic problems. The equation to be solved can be written in the following form:

$$Lu = f \tag{2.1}$$

where L is a linear differential operator, u is the dependent variable and exact solution to Eq. 2.1, and f is the source term. Let v be an approximate solution to Eq. 2.1. The error e is defined as the required correction to v, or

$$e = u - v \tag{2.2}$$

Because the approximate solution v does not satisfy Eq. 2.1, its substitution into Eq. 2.1 will leave a residual r, defined by

$$r = f - Lv \tag{2.3}$$

The error e satisfies an equation that is found by applying the linear operator to the error e as follows:

$$L(u - v) = Lu - Lv = f - Lv = r; \quad \text{or } Le = r \tag{2.4}$$

Equation 2.4 is referred to as the residual equation and provides a way of computing e. Given the correction e for the approximate solution v, the exact solution u is then estimated via Eq. 2.2 as $u = v + e$. Note that Eq. 2.4, the residual equation, has the same form as the original Eq. 2.1, but its source term is now the residual given by Eq. 2.3. This means that the method chosen to solve Eq. 2.1 will also work on the residual equation.

Defect correction (Jacobs, 1981; Brandt, 1977) is central to the multigrid approach and is described by the following three steps:

(1) Given an approximate solution v compute the residual r using Eq. 2.3.
(2) Solve Eq. 2.4, the residual equation, for the correction e.
(3) Add the correction e to the approximate solution v to obtain the exact solution u.

In practice the continuous problem is approximated on a discrete grid having a characteristic discretization scale h appropriate to represent the spatial features of interest. In this step of the multigrid approach any standard discretization method (e.g. finite difference or finite element) can be used. At the discrete level the operator L will be approximated by the matrix $\mathbf{A}^h$, and vectors $\mathbf{f}^h$, $\mathbf{u}^h$, $\mathbf{r}^h$, and $\mathbf{e}^h$ will be the discrete counterparts of the continuous variables f, u, r and e. The discrete form of Eq. 2.1 in matrix and vector notation is

$$\mathbf{A}^h\mathbf{u}^h = \mathbf{f}^h \tag{2.5}$$

In a standard approach this equation could be solved by an iterative method such as Gauss–Seidel or Jacobi relaxation. However, as illustrated in detail in the later sections the convergence rate of these methods deteriorate as iterations continue on grid h. Brandt (1977) has shown that the cause of this slow convergence is the ineffectiveness of the relaxation method in smoothing the low-frequency components of the error. The smoothing process has strong spatial frequency dependence. It rapidly reduces errors with a period of a few h (high-frequency error), but errors with periods greater than $10h$ (low-frequency error) are smoothed very slowly. A key observation is that the *relaxation process is very effective on error components whose periods are small relative to the grid scale h.* This is the reason for introducing the multiple grid scales used in the multigrid approach. Error components with relatively long periods on the fine grid h, will have relatively short periods on a sufficiently coarse grid; the relative period of the error will be reduced by the ratio of the grid scales. Therefore, the strategy of the multigrid approach is to move the low-frequency components of error to relatively coarse grids where they can be smoothed efficiently.

This multigrid approach is implemented by applying the defect correction procedure previously described across a hierarchy of grids. Given some initial approximation $\mathbf{v}^h$ for Eq. 2.5 on grid h, the steps in the multigrid approach for a system with two grids, h and $2h$, are as follows:

(1) Smooth the approximate solution $\mathbf{v}^h$ on grid h, removing high-frequency error components.

(2) Compute the residual vector $\mathbf{r}^h$ from $\mathbf{r}^h = \mathbf{f}^h - \mathbf{A}^h\mathbf{v}^h$.
(3) Transfer the grid h residuals to grid $2h$ to form $\mathbf{f}^{2h}$. (This intergrid transfer process is known as restriction or projection.)
(4) Solve the discrete residual equation, $\mathbf{A}^{2h}\mathbf{e}^{2h} = \mathbf{f}^{2h}$, on grid $2h$ to obtain a coarse grid approximation for the correction to the fine grid vector $\mathbf{v}^h$. This coarse grid correction vector $\mathbf{e}^{2h}$ contains the lower spatial frequencies that remained in $\mathbf{v}^h$ after Step 1.
(5) Transfer the correction $\mathbf{e}^{2h}$ to grid h and add it to $\mathbf{v}^h$, thus reducing the low-frequency error in $\mathbf{v}^h$. (This intergrid transfer process is known as interpolation or prolongation.)
(6) Repeat the cycle from Step 1 until $\mathbf{v}^h$ has reached the desired accuracy of approximation to $\mathbf{u}^h$.

Although we have described only the basic steps of the multigrid approach for two grid levels, these steps can be extended recursively as deeply as required to solve the problem by noting that Eq. 2.5 and the residual equation at any grid level have the same form. Thus at Step 4 the solution of the residual equation on grid $2h$ can be viewed as a new two grid process involving $2h$ and $4h$ grids. Additionally, the similarity of the form of the residual equation implies that the same relaxation procedure can be applied to all grid levels.

The preceding outline of the basic steps in the multigrid approach for a linear elliptic equation completes this section. Details regarding coarse grid operator construction and how the various intergrid transfers (e.g. restriction of residuals and interpolation of coarse grid corrections) are carried out are explained in succeeding sections.

3. Multigrid applied to groundwater flow equation

This section discusses the application of the multigrid method to the partial differential equation for groundwater flow. It covers the equation, its discretization by finite difference methods on the finest or desired grid as well as the various coarser grids required by the multigrid method, and the specific details regarding each of these basic steps.

3.1 Partial differential equation

The equation being solved is the steady-state, three-dimensional, continuum, mass balance equation for saturated, confined groundwater flow. The linear, elliptic, partial differential equation given below is known as the Poisson equation.

$$\nabla \cdot (K \nabla u) = f$$

or

$$\frac{\partial}{\partial x}\left(K_{xx}\frac{\partial u}{\partial x}\right)+\frac{\partial}{\partial y}\left(K_{yy}\frac{\partial u}{\partial y}\right)+\frac{\partial}{\partial z}\left(K_{zz}\frac{\partial u}{\partial z}\right)=f \tag{3.1}$$

where:

∇ is the gradient operator $\quad L^{-1}$

x, y, z are the Cartesian coordinates aligned along the major hydraulic conductivity axes (K_{xx}, K_{yy}, K_{zz}) {also $(x_1, \ldots, x_\alpha)$ where dimension, $\alpha = 1, 2$, or 3} $\quad L$

u is the potential at spatial location (x, y, z) $\quad L$

K is the saturated hydraulic conductivity tensor at location (x, y, z) $\quad LT^{-1}$

K_{xx}, K_{yy}, K_{zz} is saturated hydraulic conductivity in x, y and z directions, respectively, at location (x, y, z) $\quad LT^{-1}$

f is the source/sink term at location (x, y, z) as volumetric flux per unit volume $\quad L^3T^{-1}L^{-3}$

Algorithms for the various multigrid steps required to solve the one-, two- and three-dimensional form of this equation are discussed in the next section. Domains for the one-, two- and three-dimensional problems are a unit length line, a unit square and a unit cube. For this particular problem, Dirichlet boundary conditions of $u = 0$ on the left point, line, and plane respectively; a boundary condition of $u = 100$ on the right point, line, and plane respectively; and no-flow boundaries, $\partial u/\partial v = 0$, along the other exterior boundaries of the two- and three-dimensional domains (i.e. the two lines and four planes) are prescribed as illustrated in Fig. 1 (where n is the unit vector normal to the no-flow boundary).

3.2 Equation discretization

While multigrid can be used in conjunction with either finite element or finite difference discretization schemes, only the uniform finite difference scheme is discussed in this chapter. This simpler formulation allows discussions to concentrate on various steps of the multigrid approach rather than on nuances of implementation.

In both finite element and finite difference schemes, an approximation to the continuous domain, represented by Eq. 3.1, is determined at a finite set of n discrete points in space (x_i, y_i, z_i). Partial derivatives are approximated with differences between the functional values (e.g. u) at these discrete points and boundary condition information is specified at these discrete points along domain boundaries.

With a finite difference approach the continuous problem domain (Fig. 1) is approximated with a grid consisting of a finite number of discrete volume

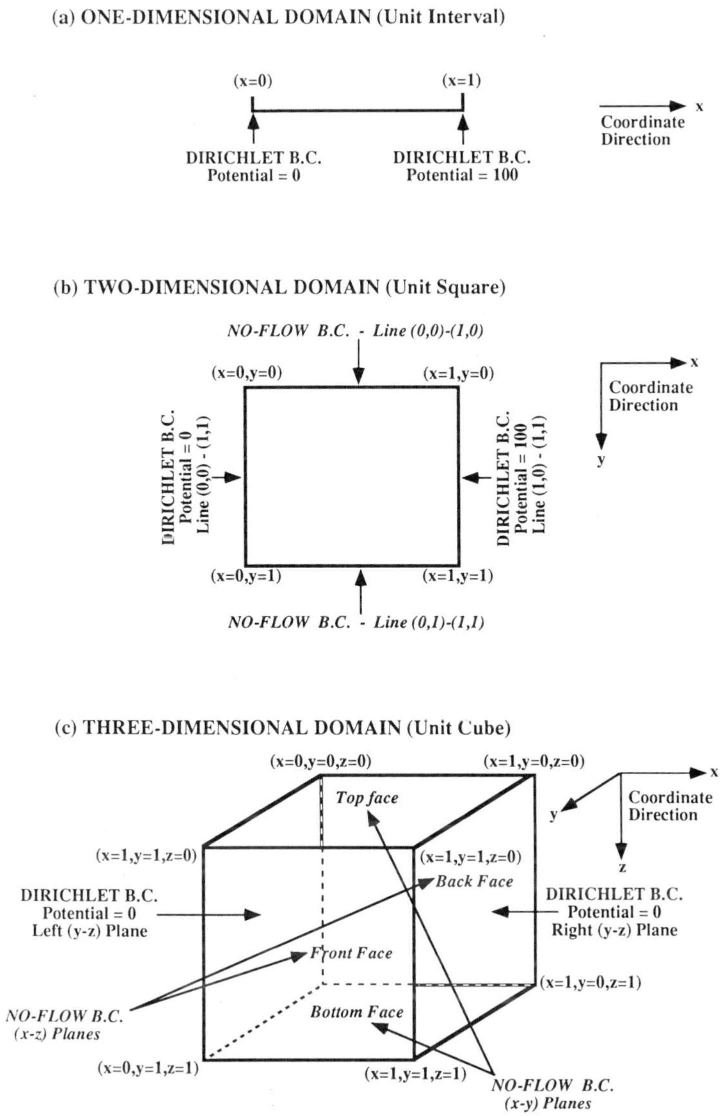

Figure 1. Coordinate directions and boundary conditions (B.C.) for: (a) one-dimensional; (b) two-dimensional; and (c) three-dimensional domains for Eq. 3.1.

elements by either of two standard conventions: (1) block-centred (or cell-centred); and (2) point-centred (sometimes called mesh-, node-, or vertex-centred), as shown in Fig. 2 for one-, two- and three-dimensional domains. Major differences occur at boundaries, where for point-centred schemes, block or cell volumes and cross-sectional areas vary with point location (i.e. corner,

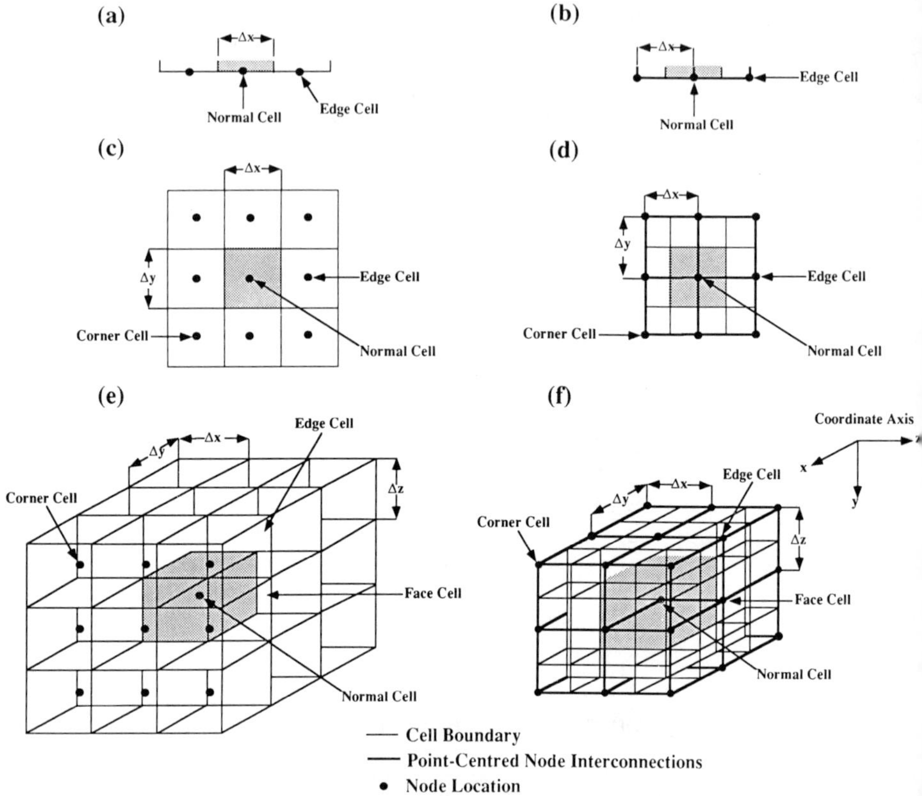

Figure 2. Finite difference discretization conventions: (a) 1-D block-centred; (b) 1-D point-centred; (c) 2-D block-centred; (d) 2-D point-centred; (e) 3-D block-centred; (f) 3-D point-centred.

edge, face or interior) as shown in Fig. 2. These two finite difference discretization schemes lead to the two standard multigrid grid coarsening schemes with the same names. Figure 3 illustrates both point-centred and block-centred coarsening for a two-dimensional problem where the next coarser grid is constructed from the previous grid by doubling the grid spacing. Point-centred discretization (Fig. 2b, d and f) with uniform spacing (e.g. Δx, Δy, Δz is proportional to h at the finest or desired grid level) and standard point-centred grid coarsening (Fig. 3a and b) are schemes described in this chapter.

Notation for grids on two levels (i.e. fine and coarse), matrices and vectors on these grids, and the various grid and intergrid operators are defined in Table 1. As mentioned earlier it is sufficient to discuss multigrid in terms of only two grids because the expansion to more grids is achieved by recursion of a two-grid process since all intergrid operations are between a fine and coarse grid.

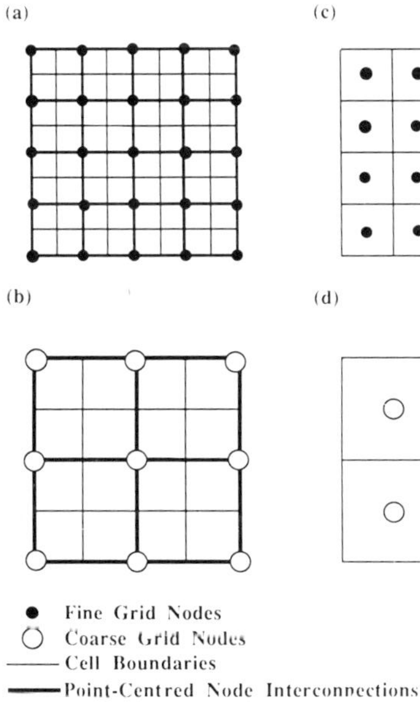

Figure 3. Point-centred (a, b) and block centred (c, d) coarsening for a two-dimensional grid formed by doubling the grid spacing of the finer grid.

3.3 Finite difference equations and stencils

Equation 3.2 is the unsimplified form of the finite difference form for Eq. 3.1 for a node at location (i, j, k) as shown in Fig. 4. It was developed by an integrated finite difference approach (Reisenauer *et al.*, 1970) that involves integrating around the nodal volume and using Green's theorem to complete the transformation.

$$\{[C_{i+1}(K_{xx})_{i,j,k}(u_{i+1,j,k} - u_{i,j,k})/\Delta x] - [C_{i-1}(K_{xx})_{i-1,j,k}(u_{i,j,k} - u_{i-1,j,k})/\Delta x]\}$$

$$+ \{[C_{j+1}(K_{yy})_{i,j,k}(u_{i,j+1,k} - u_{i,j,k})/\Delta y] - [C_{j-1}(K_{yy})_{i,j-1,k}(u_{i,j,k} - u_{i,j-1,k})/\Delta y]\}$$

$$+ \{[C_{k+1}(K_{zz})_{i,j,k}(u_{i,j,k+1} - u_{i,j,k})/\Delta z] - [C_{k-1}(K_{zz})_{i,j,k-1}(u_{i,j,k} - u_{i,j,k-1})/\Delta z]\}$$

$$= (\text{Vol}_{i,j,k}) f_{i,j,k} \qquad (3.2)$$

where:

C_l is the common transfer area for flux between node (i, j, k) and the six adjacent nodes along the x, y, and z directions (e.g. C_{i-1} is the common transfer area between node (i, j, k) and node $(i-1, j, k)$.

$\text{Vol}_{i,j,k}$ is the volume of node at location (i, j, k).

Table 1. Notation for grids (fine and coarse), matrices and vectors on grids, various grid and intergrid operators and stencils.

Notation	Description
$(x_1, \ldots, x_\alpha)$	Cartesian coordinates in α-dimensional space ($\alpha = 1$, 2 or 3). It may also be written in vector form as $\mathbf{x}$.
h and $2h$	Relative or generic fine and coarse grid spacing in a two-grid system. Actual fine and coarse grid spacing $(\Delta x_1, \ldots, \Delta x_\alpha)^{h \text{ or } 2h}$ is proportional to h and $2h$, respectively. Numbering grid levels sequentially with 0 denoting the finest or desired grid level would result in grid level l having a spacing proportional to $2^l h$.
$(\ )^h$ and $(\ )^{2h}$	Superscripts indicate relative grid spacing. In a two-grid system superscript h indicates fine grid and $2h$ coarse grid.
G^h and G^{2h}	Representation for fine and coarse grid function spaces respectively. Fine grid coordinates $(x_1, \ldots, x_\alpha)$ for point-centred computational grids for any dimensionality, $\alpha = 1$, 2 or 3, for the fine grid are: $$G^h = \{(x_1, \ldots, x_\alpha)\text{:}\quad x_\alpha = j_\alpha h_\alpha;\quad j_\alpha = 0, 1, \ldots, n_\alpha;\quad h_\alpha = L_\alpha/n_\alpha;\quad \alpha = 1, 2 \text{ or } 3\}$$ where n_α is number of discretization points in α-coordinate direction and L_α is α-coordinate domain length. The coarse grid is similarly defined as: $$G^{2h} = \{(x_1, \ldots, x_\alpha)\text{:}\quad x_\alpha = j_\alpha h_\alpha;\quad j_\alpha = 0, 1, \ldots, n_\alpha/2;\quad h_\alpha = L_\alpha/(2n_\alpha);\quad \alpha = 1, 2, 3\}$$
$\mathbf{A}^h$	Bold uppercase letters represent linear matrix operators defined on a given grid. In this case superscript h indicates this linear matrix operator is defined on the fine grid.
$a^h_{i,j}$	Double superscripted lowercase matrix operator letters represent matrix components. First subscript identifies matrix row and second subscript denotes column. In this case, $a^h_{i,j}$ is the component at row i and column j of linear matrix operator $\mathbf{A}^h$ defined on the fine grid.
$\mathbf{u}^h$	Bold lowercase letters represent vectors.

Table 1. *Continued*

Notation	Description
u_i^h	Single subscripted lowercase vector letters denote vector components. Subscript refers to vector component number and superscript denotes grid the vector is defined on. Thus component i of vector $\mathbf{u}^h$ on the fine grid is u_i^h.
$\mathbf{I}_h^{2h}$ and $\mathbf{I}_{2h}^h$	These are intergrid operators for transferring information between the different grids. $\mathbf{I}_h^{2h}$ is the restriction or projection operator that provides for the fine grid to coarse grid transfer of residual and other vectors. $\mathbf{I}_{2h}^h$ is the interpolation or prolongation operator for coarse to fine grid transfer of correction vectors. Operator subscript denotes the spacing for the source grid, and superscript denotes the spacing of the destination grid. Some of the many ways to define these operators are discussed in detail in the text.

Stencils

$$[s_{-1} \quad s_1 \quad s_{+1}]$$

1-D stencil

$$\begin{bmatrix} 0 & s_{0,-1} & 0 \\ s_{-1,0} & s_{0,0} & s_{1,0} \\ 0 & s_{0,1} & 0 \end{bmatrix}$$

2-D
5-point operator
stencil

$$\left]\begin{matrix} s_{-1,-1} & s_{0,-1} & s_{1,-1} \\ s_{-1,0} & s_{0,0} & s_{1,0} \\ s_{-1,1} & s_{0,1} & s_{1,1} \end{matrix}\right[$$

Full
2-D prolongation
stencil

$$\left[\begin{matrix} \begin{bmatrix} 0 & 0 & 0 \\ 0 & s_{0,0,-1} & 0 \\ 0 & 0 & 0 \end{bmatrix} \\ \begin{bmatrix} 0 & s_{0,-1,0} & 0 \\ s_{-1,0,0} & s_{0,0,0} & s_{1,0,0} \\ 0 & s_{0,1,0} & 0 \end{bmatrix} \\ \begin{bmatrix} 0 & 0 & 0 \\ 0 & s_{0,0,1} & 0 \\ 0 & 0 & 0 \end{bmatrix} \end{matrix}\right.$$

3-D
7-point operator
stencil

Stencils provide a convenient way to discuss the various grid operators. For a one-dimensional Laplace problem, $[1, 2, 1]$ would be a standard 3-point discretization stencil for the Laplace operator $\mathbf{A}^h$. The stencil for the full-weighting one-dimensional restriction operator $\mathbf{I}_h^{2h}$ would be $[0.25, 0.5, 0.25]$.

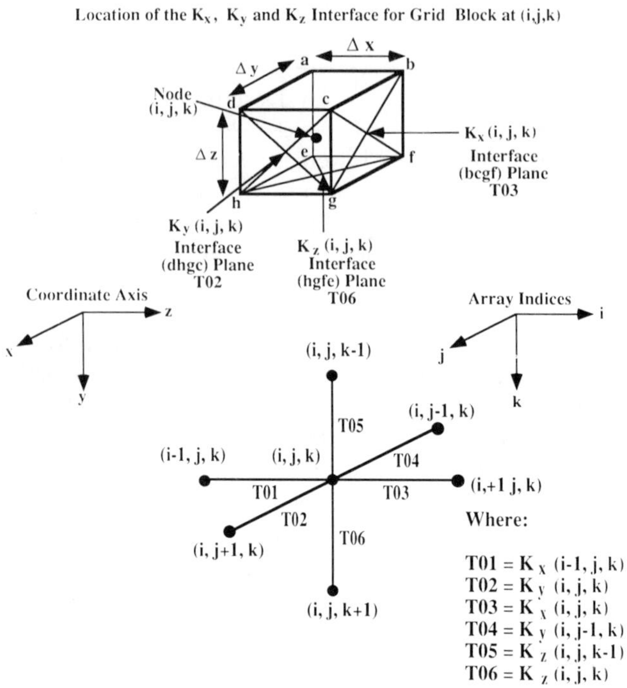

Figure 4. Control volume, K_x, K_y, and K_z, interface locations and areas, coordinate axes, and conventions for array indices for a point-centred, interior, finite difference node.

The six values of C_l for any node and the value of $\mathrm{Vol}_{i,j,k}$ are dependent on the node type (interior, face, edge or corner), and C_l also varies with orientation. For an interior node:

$$C_{i+1} = C_{i-1} = \Delta y \Delta z \qquad \mathrm{L}^2$$

$$C_{j+1} = C_{j-1} = \Delta x \Delta z \qquad \mathrm{L}^2$$

$$C_{k+1} = C_{k-1} = \Delta x \Delta y \qquad \mathrm{L}^2$$

$$\mathrm{Vol}_{i,j,k} = \Delta x \Delta y \Delta z \qquad \mathrm{L}^3$$

Collecting terms, multiplying through by $\mathrm{Vol}_{i,j,k}$, and rewriting Eq. 3.2 for an interior node, yields Eq. 3.3

$$[(K_{xx})_{i-1,j,k}/\Delta x^2]u_{i-1,j,k} + [(K_{xx})_{i,j,k}/\Delta x^2]u_{i+1,j,k} + [(K_{yy})_{i,j-1,k}/\Delta y^2]$$

$$+ [(K_{yy})_{i,j,k}/\Delta y^2]u_{i,j+1,k} + [(K_{zz})_{i,j,k-1}/\Delta z^2]u_{i,j,k-1} + [(K_{zz})_{i,j,k}/\Delta z^2]u_{i,j,k+1}$$

$$- \{[(K_{xx})_{i-1,j,k} + (K_{xx})_{i,j,k}]/\Delta x^2 + [(K_{yy})_{i,j-1,k} + (K_{yy})_{i,j,k}]/\Delta y^2$$

$$+ [(K_{zz})_{i,j,k-1} + (K_{zz})_{i,j,k}]/\Delta z^2\} u_{i,j,k} = f_{i,j,k} \tag{3.3}$$

Finally the familiar seven-point finite difference stencil is developed from Eq. 3.3 by setting $K_{xx} = K_{yy} = K_{zz} = 1$ and $\Delta x^2 = \Delta y^2 = \Delta y^2 = h^2$ at grid level 0 and multiplying through by h^2 yielding Eq. 3.4 for each interior node in three dimensions, and Eqs 3.5 and 3.6 illustrate the formulations for two and one dimensions, respectively.

$$u_{i,j,k-1} + u_{i,j-1,k} + u_{i-1,j,k} - 6u_{i,j,k} + u_{i+1,j,k} + u_{i,j+1,k} + u_{i,j,k+1} = f_{i,j,k}h^2 \tag{3.4}$$

$$u_{i,j-1} + u_{i-1,j} - 4u_{i,j} + u_{i+1,j} + u_{i,j+1} = f_{i,j}h^2 \tag{3.5}$$

$$u_{i-1} - 2u_i + u_{i+1} = f_i h^2 \tag{3.6}$$

3.4 Algebraic and residual equations

Finite difference or finite element discretization of Eq. 3.1 leads to a linear system of algebraic equations.

$$\mathbf{Au} = \mathbf{f} \tag{3.7}$$

where $\mathbf{A}$ is an $n \times n$ matrix of coefficients $a_{i,j}$ dependent on the approximations used and system discretization, $\mathbf{u}$ is a vector of n components u_i for the exact solution of Eq. 3.1 at the n discrete grid points (x_i, y_i, z_i), and $\mathbf{f}$ is a vector of n right-hand-side values f_i dependent on the values of the source term f at the n discrete grid points and boundary conditions imposed (see Fig. 1).

The exact solution $\mathbf{u}$ to Eq. 3.7 is not known, but an approximate solution $\mathbf{v}$ on the fine grid can be estimated. The error $\mathbf{e}$ in this approximation is given by:

$$\mathbf{e} = \mathbf{u} - \mathbf{v} \tag{3.8}$$

and the residual equation

$$\mathbf{Ae} = \mathbf{r} \tag{3.9}$$

is used to calculate corrections $\mathbf{e}$ to the approximate solution $\mathbf{v}$ as a function of the residual $\mathbf{r}$, which is a computable measure of approximate solution quality $\mathbf{v}$. The residual $\mathbf{r}$ is calculated and defined by:

$$\mathbf{r} = \mathbf{f} - \mathbf{Av} \tag{3.10}$$

The residual Eq. 3.9 is derived by subtracting the rearranged form of the residual computation (i.e. $\mathbf{Av} = \mathbf{f} - \mathbf{r}$) from Eq. 3.1 to obtain:

$$\mathbf{A}(\mathbf{u} - \mathbf{v}) = \mathbf{r} \tag{3.11}$$

which, when rewritten in terms of $\mathbf{e}$ via Eq. 3.8, forms the residual Eq. 3.9. The last approximate solution $\mathbf{v}$ can be updated with an estimate for the correction term $\mathbf{e}$ to obtain a new approximation $\mathbf{u}_{\text{new}}$, for the solution to Eq. 3.1 as follows:

$$\mathbf{u}_{\text{new}} = \mathbf{v} + \mathbf{e} \tag{3.12}$$

In the multigrid approach the residual equation is actually solved on coarser grids to reduce computational work and to provide optimal smoothing of solution errors with different spatial frequencies.

3.5 Detailed discussion of multigrid steps

Thus far we have outlined the basic steps in the multigrid approach and developed the equations. In this subsection we give more detailed descriptions of these steps, provide some implementation details, and illustrate these steps and the utility of the multiple discretizations used in the multigrid approach through a one-dimensional example of a multigrid V-cycle.

Grid scale and relaxation relationships

The relaxation operation is the basic method used at each grid level for error reduction. At any grid level, however, the effectiveness of the error reduction is dependent on the spatial frequency of the error component in question. In fact, the spatial frequency response of a relaxation technique is a key factor in evaluating its effectiveness for the multigrid approach. Figure 5 illustrates this point with examples of Gauss–Seidel and Jacobi iteration on a one-dimensional grid. The grid has 129 node points and 128 intervals. Basic Fourier theory shows that such a grid allows 64 independent frequency components (Bracewell, 1965). For these examples the input function Y_i was a sum of frequency components of equal amplitude on a unit interval.

$$Y_i = \sum_{j=1}^{N/2} \sin(2\pi j x_i) \quad i = 0, 1, 2, \ldots, N \tag{3.13}$$

where N is the number of intervals and $x_i = i/N$.

Figure 5a is a frequency magnitude plot of the input function used in the succeeding examples (frequency is in cycles per unit interval). These plots

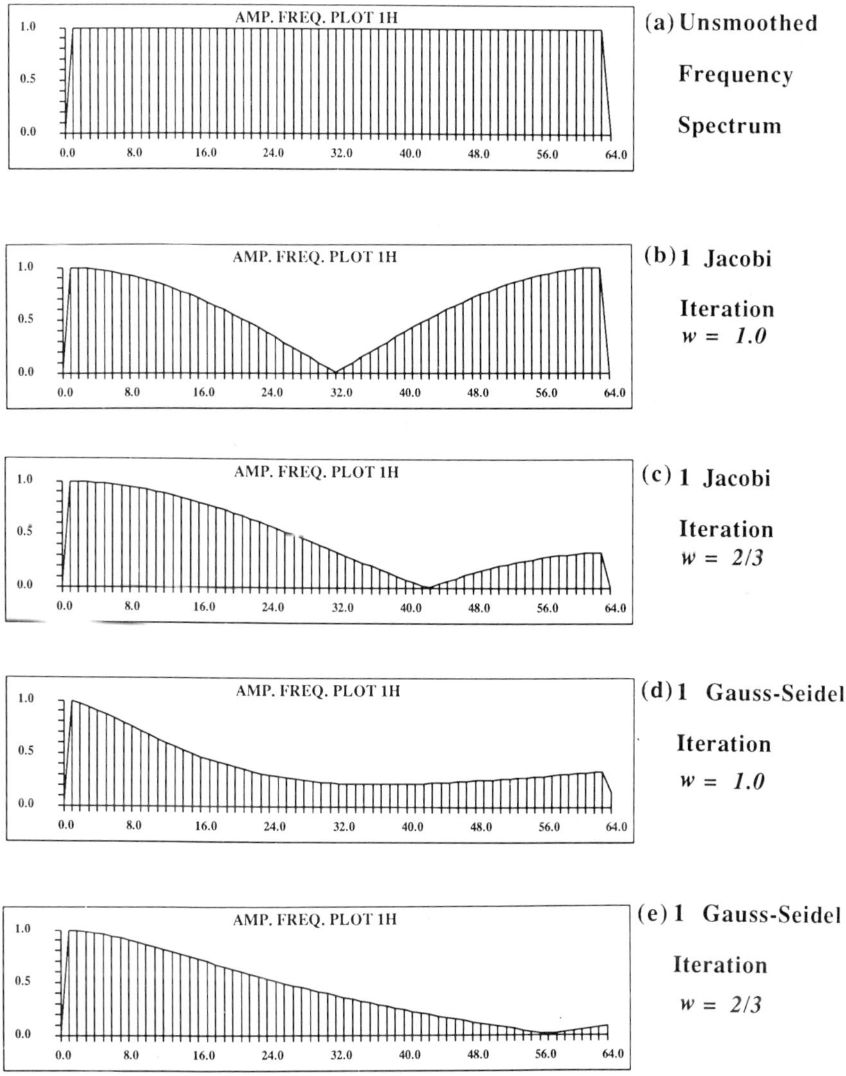

Figure 5. Spectral response curves for representative cases of Jacobi and Gauss–Seidel relaxation iterations.

resemble frequency response functions for filters. They clearly show the filtering or smoothing action resulting from the relaxation iterations. For example, in comparing Fig. 5b with 5c one observes that an unweighted Jacobi iteration ($w = 1$) does not smooth the higher frequency errors, whereas a weighted Jacobi with $w = 2/3$ does reduce the upper frequencies. Figure 5d shows that one iteration with Gauss–Seidel relaxation ($w = 1$) attenuates a larger portion of the frequency spectrum than Jacobi ($w = 2/3$), and illustrates why it has faster convergence than Jacobi. The frequency response curve provides a straightforward way to determine the effects of multiple relaxation sweeps; one simply multiplies the frequency response function by itself for each iteration.

Another important characteristic of the relaxation process is illustrated in Fig. 6. Here one Gauss–Seidel iteration has been applied to a flat input spectrum at three grid scales h, $2h$ and $4h$. One notes immediately that the three response curves have the same shape. This characteristic shows that the attenuation of a frequency of 64 cycles on a $1h$ grid is the same as that of 32 cycles on $2h$ grid and 16 cycles on $4h$ grid. Thus, when the ratio of wavelength to grid scale is the same, the smoothing factor is constant. This relationship is the basis for transferring low-frequency error components to coarser grids for more efficient smoothing. For example, observe the effect of transferring a frequency of 16 cycles from curve 6a to 6b to 6c. The attenuation increases from 0.27 to 0.52 and finally to 0.76 where it becomes an effective smoothing value. From the

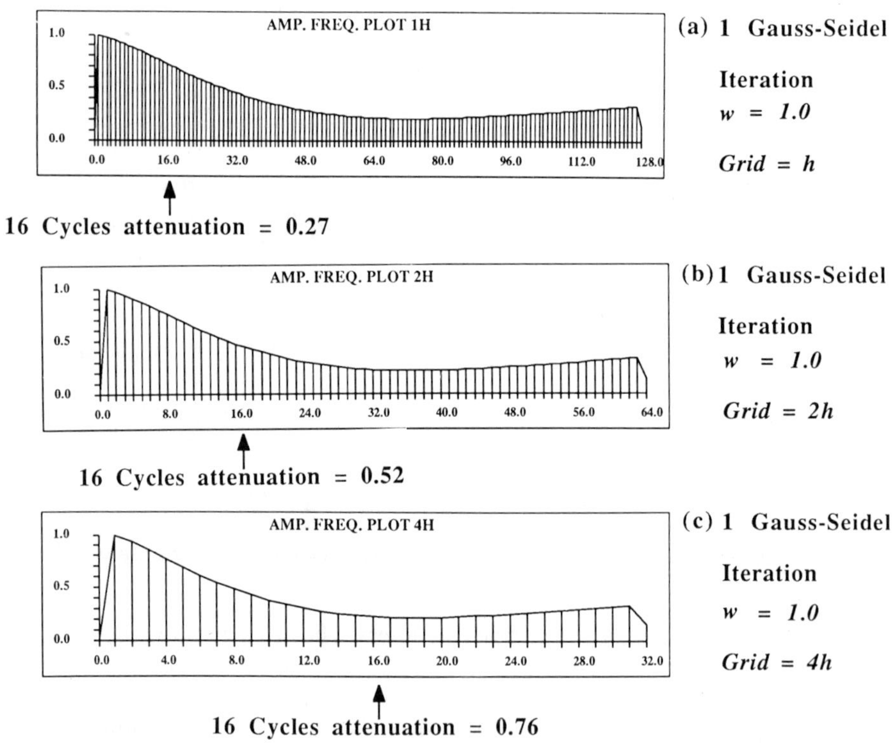

Figure 6. Spectral response curves for one Gauss–Seidel relaxation iteration at grid scales h, $2h$, $4h$.

preceding analysis the basic error smoothing strategy emerges:

(1) apply one or two relaxation sweeps on the fine grid h to smooth the high-frequency error;
(2) transfer the remaining smooth error to grid $2h$ to remove the next lower band of error frequencies;
(3) continue moving to successively coarser grids until the lowest frequency errors are efficiently removed.

The additional computational work performed on the coarser grids is not substantial. Standard grid-coarsening practice is to use a sequence of grids whose grid spacing is successively doubled. Thus for an n-dimensional problem the relaxation work and memory requirements decreases by a factor of $1/2^n$ each time the grid spacing is doubled. Memory requirements for the fine grid and all possible doubled grids for a three-dimensional problem is therefore $< 1 + 1/8 + 1/64 + 1/512 + \ldots \approx 1.143$, or about a 15% increase.

This additional work and memory is a small price to pay for a linear proportion between total work and the number of unknowns.

Intergrid transfer operations

The case for intergrid transfers has been established by the earlier discussion. There are two types of transfers required:

(1) interpolation (also called prolongation) of correction vectors $\mathbf{e}^{2h}$ from relative grid level $2h$ to level h. This linear interpolation operator is denoted $\mathbf{I}_{2h}^{h}$ and functions according to the equation:

$$\mathbf{I}_{2h}^{h}\mathbf{e}^{2h} = \mathbf{e}^{h} \tag{3.14}$$

A convenient way to represent this interpolation operator is with a stencil of interpolation weights:

1-D stencil: $$\left]\begin{matrix} 1/2 & 1 & 1/2 \end{matrix}\right[$$

2-D stencil: $$\left]\begin{matrix} 1/4 & 1/2 & 1/4 \\ 1/2 & 1 & 1/2 \\ 1/4 & 1/2 & 1/4 \end{matrix}\right[$$

The stencil is centred on the $2h$ grid node, and the node value is distributed to neighbouring fine ($1h$) grid nodes according to the weights. The interpolation operator can also be presented as a matrix that converts the full coarse grid error correction vector $\mathbf{e}^{2h}$ into $\mathbf{e}^{h}$, its fine grid representation.

(2) restriction of residuals from relative grid level h to grid $2h$. This linear operator will be denoted $\mathbf{I}_h^{2h}$ and is frequently chosen to be the transpose of $\mathbf{I}_{2h}^h$ within a constant factor. Thus,

$$\mathbf{I}_h^{2h}\mathbf{r}^h = \mathbf{f}^{2h} \tag{3.15}$$

and

$$\mathbf{I}_h^{2h} = c(\mathbf{I}_{2h}^h)^{\mathrm{T}} \tag{3.16}$$

where c is a constant.

The restriction operator also has a stencil representation:

1-D stencil: $1/4[1 \quad 2 \quad 1]$

2-D stencil: $1/16\begin{bmatrix} 1 & 2 & 1 \\ 2 & 4 & 2 \\ 1 & 2 & 1 \end{bmatrix}$

We now have the basic set of tools needed for multigrid operation.

One-dimensional multigrid examples

Figure 7 shows the steps in a two-level multigrid iteration. This is a one-dimensional problem with linearly increasing conductivity, prescribed potentials of 0 and 100 on the left and right sides respectively, and no source vector **f**. The initial solution estimate consists of a linear ramp function superposed on a 0.25 period sine function. The solution steps are as follows:

(1) one relaxation iteration on $1h$ grid (the period 0.25 error vector changed very little);

(2) compute the residual vector $\mathbf{r}^h$ on the $1h$ grid

$$(\mathbf{r}^h = \mathbf{f}^h - \mathbf{A}^h\mathbf{v}^h)$$

(3) apply restriction operator to residual $\mathbf{r}^h$ to form the source term $\mathbf{f}^{2h}$

$$(\mathbf{f}^{2h} = \mathbf{I}_h^{2h}\mathbf{r}^h)$$

(4) solve residual equation on $2h$ grid

$$(\mathbf{A}^{2h}\mathbf{e}^{2h} = \mathbf{f}^{2h})$$

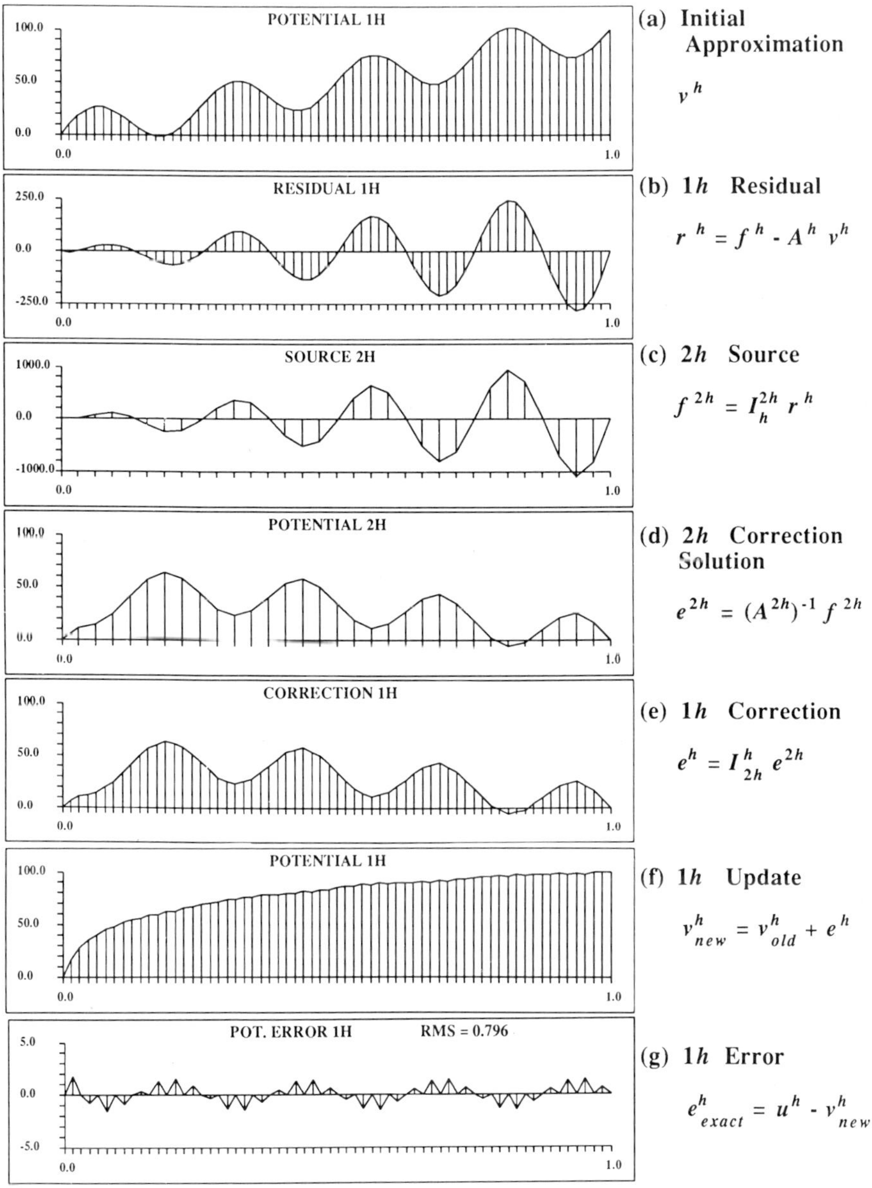

Figure 7. Steps in a two-level multigrid solution of a one-dimensional problem.

(5) interpolate correction from $2h$ to h

$$(\mathbf{e}^h = \mathbf{I}_{2h}^h \mathbf{e}^{2h})$$

(6) update $\mathbf{v}_{\text{new}}^h$ with interpolated

$$(\mathbf{v}_{\text{new}}^h = \mathbf{v}^h + \mathbf{e}^h)$$

Figure 7g shows the remaining error in $\mathbf{v}^h$ computed by a direct solution of the $1h$ equation. Most of the low-frequency error has been removed and the remaining high-frequency error can be effectively smoothed.

Figure 8 was designed to show how the components of the error spectrum are resolved by the hierarchy of grid scales. In this case the boundary conditions were both zero and the exact solution was therefore zero everywhere. The initial

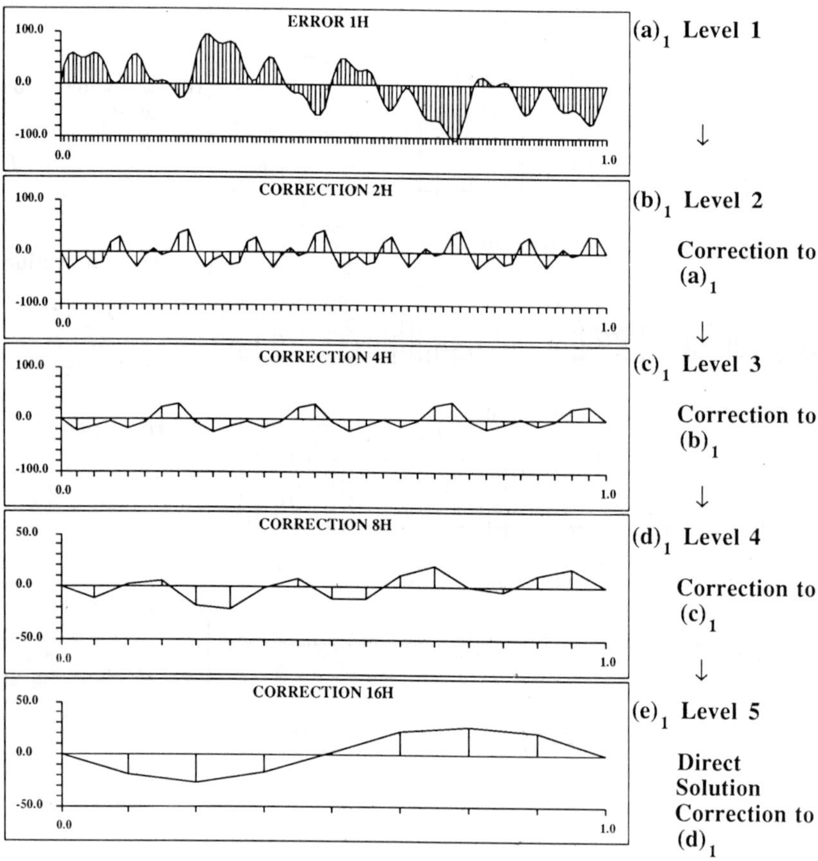

Figure 8. Descending leg of a five-level multigrid V-cycle with two Gauss–Seidel sweeps per level.

approximation consisted of the sum of five error components: equal amplitude sine waves with frequencies of 1, 4, 5, 16, and 32 cycles per interval. The unit interval was discretized into 129 nodes. Figure 8 shows how these error components were separated on the descending leg of a standard multigrid V-cycle. In Fig. 9 the successive corrections are accumulated during the upward leg of the V-cycle. At the top level we can see the large reduction achieved in all error frequencies.

Derivation of coarse grid operator

Up to this point we have said little about the form of the coarse grid matrix operator $\mathbf{A}^{2h}$. The simplest choice is to use the same operator on the coarse grids as on the fine grids. However, in the case of variable conductivities or heterogeneity, which is of particular interest here, an equivalent conductivity

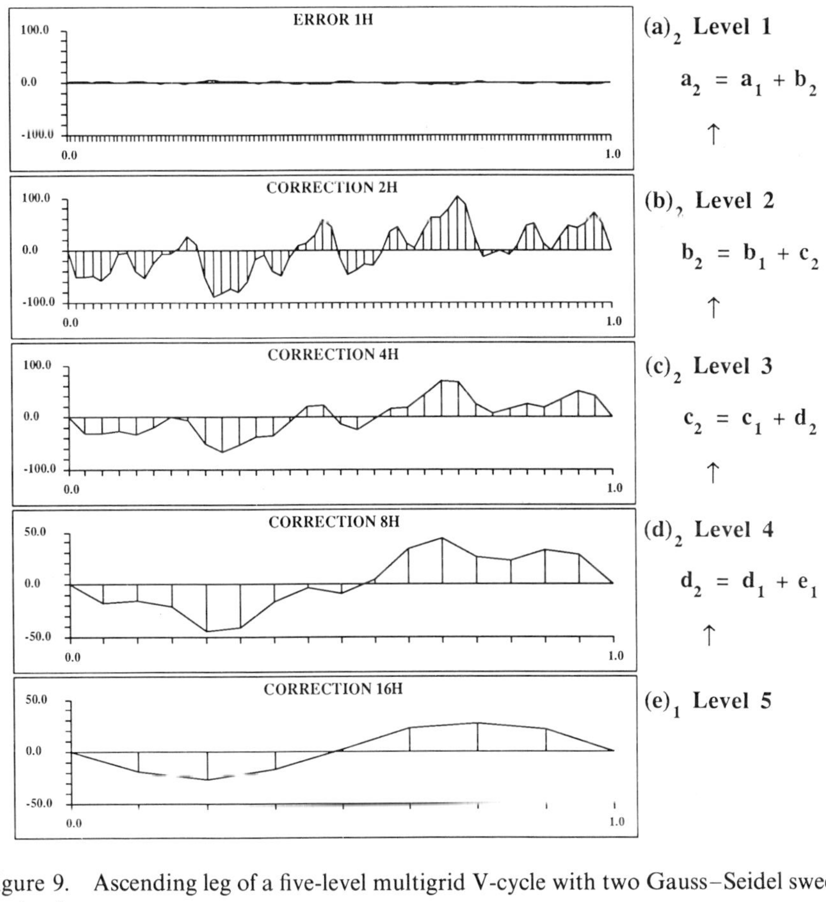

Figure 9. Ascending leg of a five-level multigrid V-cycle with two Gauss–Seidel sweeps per level.

must be calculated on the coarser grids. Consider first a one-dimensional example; we write the equation $\mathbf{A}^h\mathbf{u}^h = \mathbf{f}^h$ in stencil form

$$(1/h^2)[-K_{i-1}^h \quad K_{i-1}^h + K_i^h \quad -K_i^h]u_i^h = f_i^h \tag{3.17}$$

On grid $2h$ we have the same form:

$$(1/(2h)^2)[-K_{i-1}^{2h} \quad K_{i-1}^{2h} + K_i^h \quad -K_i^{2h}]u_i^{2h} = f_i^{2h} \tag{3.18}$$

where:

$$K_i^{2h} = 2(K_{2(i-1)+1}^h K_{2(i-1)+2}^h)/(K_{2(i-1)+1}^h + K_{2(i-1)+2}^h)$$

that is, the harmonic average of the fine grid conductivities connecting the i to $i+1$ nodes on the $2h$ grid. This averaging is carried out recursively to obtain effective conductivities for each of the successively coarser grids. Figure 10 shows an example of averaging a one-dimensional conductivity array over five grid levels. One should note that the low conductivities dominate the coarse grid average, indicating the importance of conductivity extremes in two- and three-dimensional distributions.

A more general way of defining the coarse grid operator is the Galerkin or variational approach. Starting from the residual equation on grid h we have $\mathbf{A}^h\mathbf{e}^h = \mathbf{r}^h$ and then let $\mathbf{e}^h$ be approximated by $\mathbf{I}_{2h}^h\mathbf{e}^{2h}$ or

$$\mathbf{A}^h\mathbf{I}_{2h}^h\mathbf{e}^{2h} = \mathbf{r}^h \tag{3.19}$$

Now transfer the residual to the $2h$ grid. Apply the $\mathbf{I}_h^{2h}$ operator to both sides of Eq. 3.19 and we have:

$$\mathbf{I}_h^{2h}\mathbf{A}^h\mathbf{I}_{2h}^h\mathbf{e}^{2h} = \mathbf{I}_h^{2h}\mathbf{r}^h = \mathbf{f}^{2h} \tag{3.20}$$

We can now write

$$\mathbf{A}^{2h}\mathbf{e}^{2h} = \mathbf{f}^{2h}$$

where we have taken $\mathbf{I}_h^{2h}\mathbf{A}^h\mathbf{I}_{2h}^h$ as our definition of the operator $\mathbf{A}^{2h}$. This is called the Galerkin condition:

$$\mathbf{A}^{2h} = \mathbf{I}_h^{2h}\mathbf{A}^h\mathbf{I}_{2h}^h \tag{3.21}$$

and

$$\mathbf{I}_h^{2h} = c(\mathbf{I}_{2h}^h)^{\mathrm{T}} \tag{3.22}$$

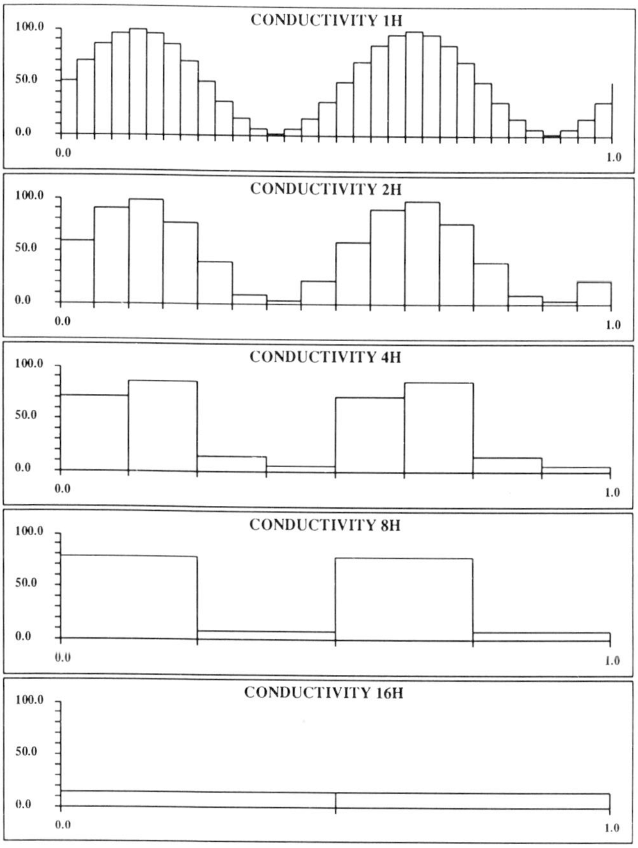

Figure 10. Illustration of harmonic averaging of conductivity over five grid scales.

Figure 11 illustrates this definition of $\mathbf{A}^{2h}$ with a numerical example. We can compute the coefficients of the $\mathbf{A}^{2h}$ stencil by performing the required operations on a unit vector $\mathbf{e}^{2h}$. In this example the conductivities, $\mathbf{K}_i$s, are all equal and the result is just the nine-point Laplacian stencil. However, when there is a high conductivity path along the diagonal as illustrated in Fig. 12, the result is more interesting. Here one can see that the strong diagonal coupling has been reflected in the computed operator weights. The harmonic averaging technique described earlier would not have produced this diagonal dominance of corner weights. Additionally, the interpolation and restriction operators have been modified in this example. The K-weighted designation in Fig. 12 refers to operator-weighted interpolation (Alcouffe *et al.*, 1981). This modification to the fixed interpolation and restriction is required when there are large discontinuities in the K_is. It is based on the requirement for preserving the continuity of flux ($K\nabla u$) across the step change in hydraulic conductivity. One can generally expect a more

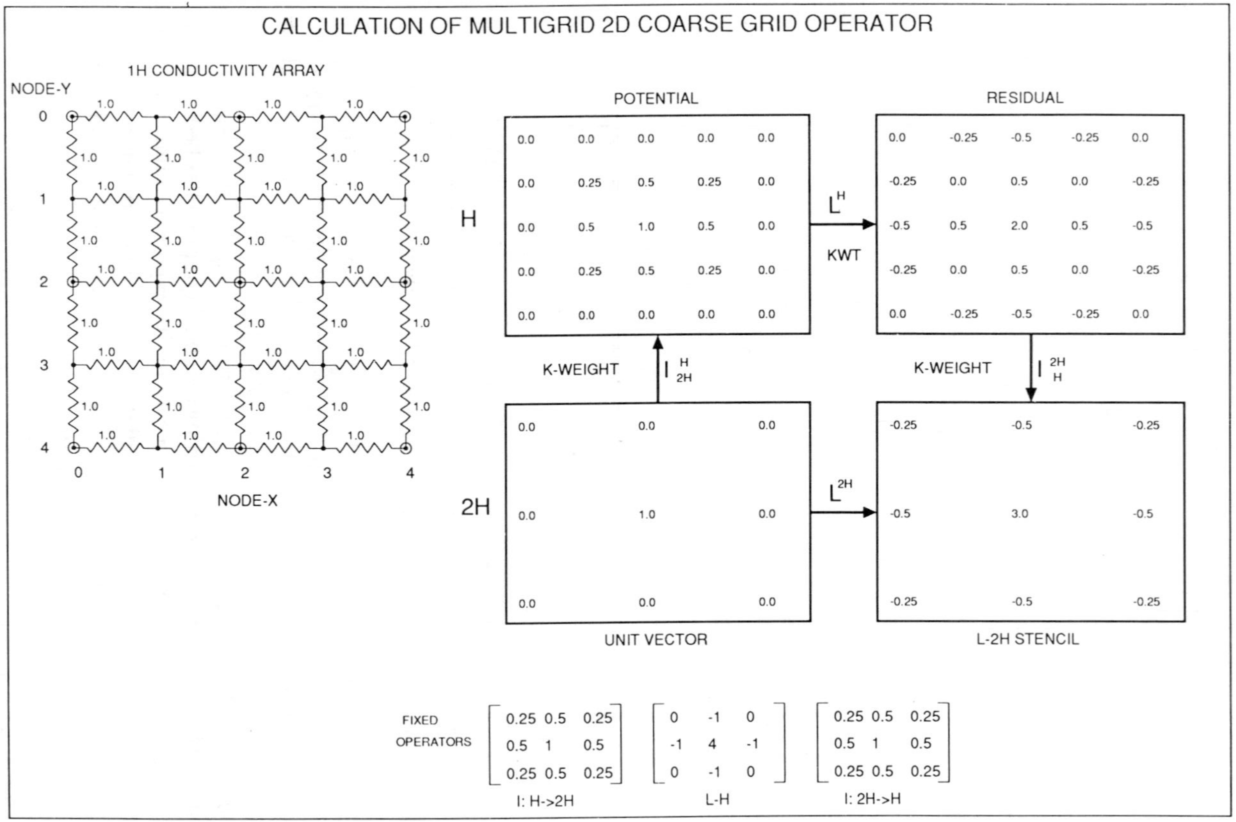

Figure 11. Calculation of two-dimensional Galerkin operator for uniform K grid.

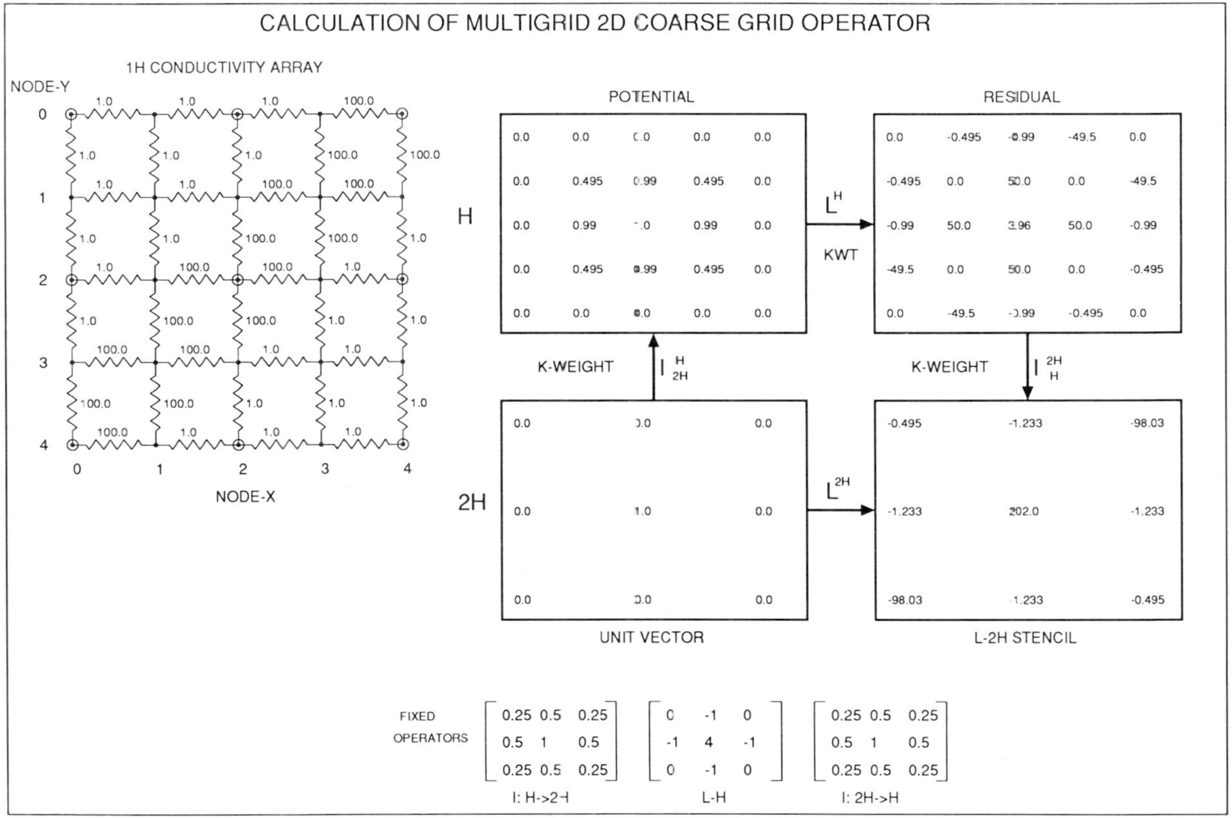

Figure 12. Calculation of a two-dimensional Galerkin operator with strong diagonal coupling.

accurate correction using the Galerkin form of $\mathbf{A}^{2h}$. In some cases this choice can make the difference between a solution converging or diverging. The cost of this robustness is the additional memory needed to store the coefficients of $\mathbf{A}^{2h}$ ($\mathbf{A}^{2h}$ is a nine point operator in two dimensions and a 27 point operator in three dimensions) and the associated increased computational work per unknown. A good compromise for large problems might be to initially identify nodes requiring the more robust operator and only store the weights for those nodes.

4. Application of multigrid codes

Improved understanding of spatial variability is required to predict contaminant migration and groundwater flow in heterogeneous and fractured rock masses. Spatial variability issues such as:

(1) interpretation of parameters from field measurements,
(2) interpolation of parameters to obtain estimates away from measurement points, and
(3) averaging of parameters from small-scale tests to obtain appropriate large scale values required in predictive flow and transport models

can be partially addressed through a numerical laboratory approach (Cole *et al.*, 1985).

Using large-scale computer simulations for conducting numerical experiments became feasible with the advent of inexpensive, fast computers with large memories and equation-solving approaches like multigrid. The approach uses computer solutions of basic equations to develop increased understanding of real systems and to test new hypotheses. In a numerical laboratory approach, large, complex, synthetic data sets such as the 1025×257 node hydraulic conductivity data set used to develop the solution shown in Fig. 13 (Cole *et al.*, 1985) are studied with high-resolution computer codes. In subsequent studies (Cole and Foote, 1987; Cole and Rice, 1988) we investigated parameter interpolation problems and the associated transport prediction errors that arise in groundwater systems of reasonable size with a high degree of spatial variability. Figure 14 shows a multimode arrival distribution resulting from multiple transport pathways. In this literature we were looking for arrival distribution signatures that might be used to identify various classes of heterogeneity.

Our two- and three-dimensional multigrid computer codes solve saturated zone groundwater problems that have parameter distributions with a much wider range of spatial frequency components than could be realistically measured with field tests. This freedom and capability provides the means for evaluating the effect of neglecting various components of a parameter's spatial frequency

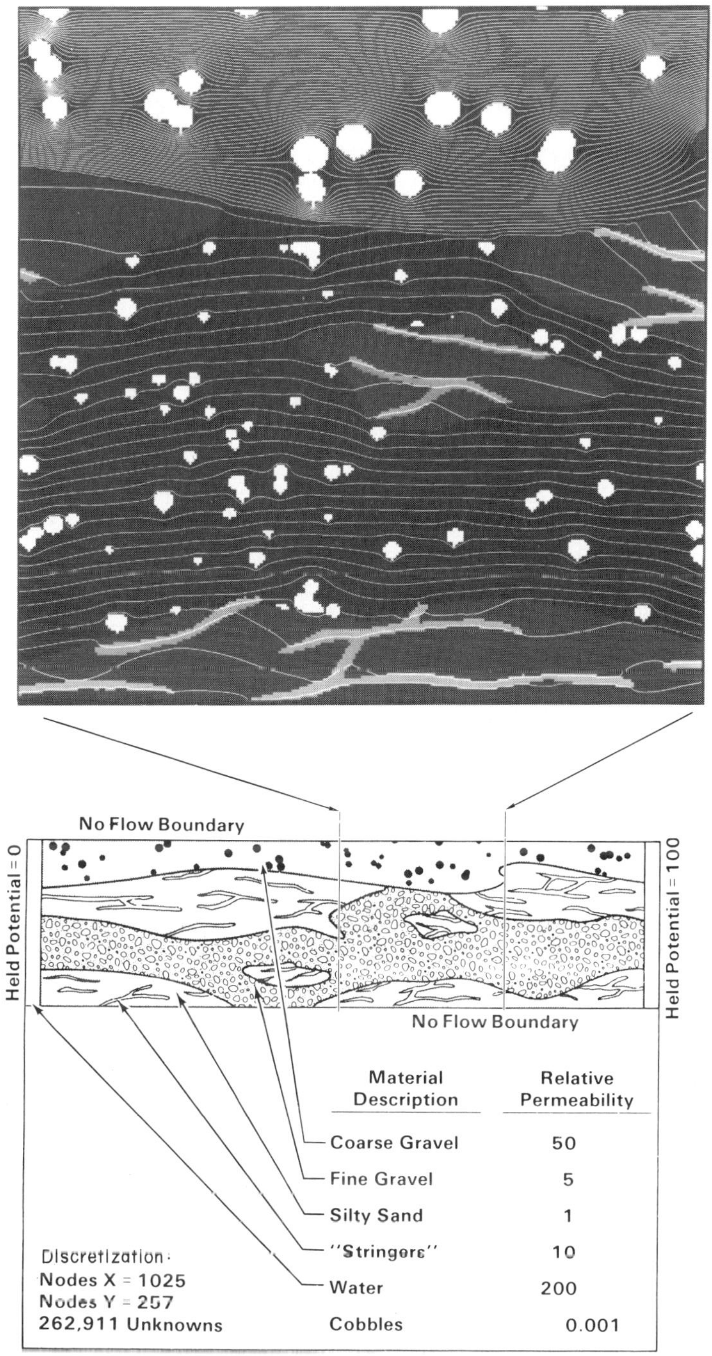

Figure 13. Streamline plot of flow through one-quarter of a 1025 × 257 node simulation of a layered heterogeneous aquifer cross-section.

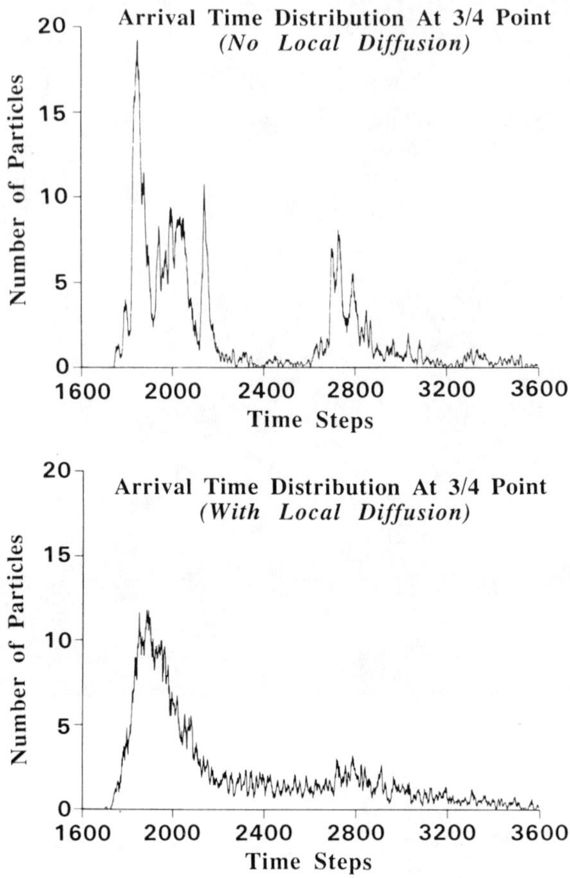

Figure 14. Arrival distributions for numerical tracer experiment performed with flow solutions from Fig. 13.

domain. The purpose of these efforts is to determine the sensitivity of groundwater transport predictions to the various spatial frequencies present in complex spatially correlated hydraulic conductivity distributions. We wished to determine how much of the variability in hydraulic conductivity must be resolved and incorporated into our predictive models to obtain a satisfactory level of detail in our predictions of fluid potential and contaminant arrival distributions.

4.1 Three-dimensional visualization

The multigrid methods described here allow efficient solution of very large problems on workstation class computers. For example, on a SUN-4/280 computer with 80 megabytes of physical memory and 350 megabytes of virtual memory we have solved a three-dimensional block with node dimensions $129 \times 129 \times 449$ (7.44×10^6 unknowns) in about 10 central processor unit (CPU) hours (speed of 1 megaflop). With this solution capability one can investigate scale-dependent dispersion over three orders of magnitude of scale in three dimensions. However, attached to the potential for greater complexity is the problem of visualization. The methods used for two-dimensional visualization (contours, vector fields, streamlines) require considerable modification when extended to large three-dimensional data sets. We are currently experimenting with a variety of volume-rendering techniques to address this problem. Figure 15 shows an example application of volume rendering. In this example, streamlines have been combined with colour-coded, equal potential surfaces to depict the solution to a $65 \times 65 \times 65$ node cube containing a high

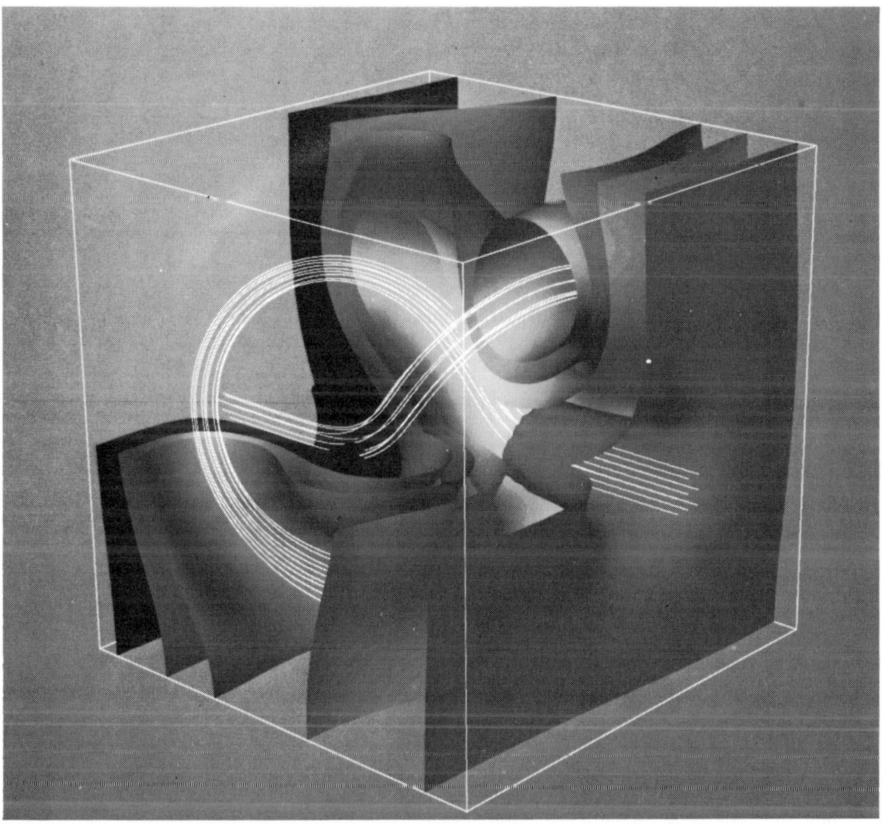

Figure 15. Cutaway view of combined equal potential surfaces and streamlines for solution to a $65 \times 65 \times 65$ node test problem. (Full colour reproduction of this figure appears as a frontispiece.)

conductivity channel. Cutting planes, hidden surface removal and colour shading techniques were used to generate this plot, which was reproduced on a high-resolution film recorder. We have found stereo, colour and motion are effective visualization techniques when used separately or in combination for improving understanding and analysis of complex three-dimensional data sets. However, we have also learned that it is even easier to create total confusion. Therefore, we are still in the early stages of gaining the skill and experience necessary to fully exploit three-dimensional visualization.

5. Suggested reading

Brandt, A. (1977). Multi-level adaptive solutions to boundary value problems, *Math. Comput.* **31**, 333–390.

Briggs, William L. (1987). 'A Multigrid Tutorial'. SIAM, Philadelphia.

Hackenbusch, Wolfgang (1985). 'Multi-Grid Methods and Applications'. Springer-Verlag, Berlin.

Hackenbusch, W. and Trottenberg, U. (1982). 'Multigrid Methods'. Springer-Verlag, Berlin.

Hackenbusch, W. and Trottenberg, U. (eds) (1985). 'Multigrid Methods II', Vol. 1228 of Lecture Notes in Mathematics (A. Dold and B. Eckman, eds). Springer-Verlag, Berlin.

Jesperson, D. (1984). Multigrid Methods for Partial Differential Equations. *In* 'Studies in Numerical Analysis', Vol. 24 of Studies in Mathematics. Mathematical Association of America.

McCormick, S. (1984). 'Multigrid Methods', Vol. 3 of the SIAM Frontiers Series. SIAM, Philadelphia.

Paddon, D.J. and Holstein, H. (eds) (1985). 'Multigrid Methods for Integral and Differential Equations'. Clarendon Press, Oxford.

Varga, R. (1962). 'Matrix Iterative Analysis'. Prentice-Hall, Englewood Cliffs, NJ.

Young, D. (1971). 'Iterative Solution of Large Linear Systems'. Academic Press, New York.

Acknowledgement

This work was supported by the US Department of Energy under Contract DE-AC06-76RLO 1830.

References

Alcouffe, R.E., Brandt, A., Dendy, J.E., Jr. and Painter, J.W. (1981). The multi-grid method for the diffusion equation with strongly discontinuous coefficients, *SIAM J. Sci. Stat. Comput.* **2**.

Bracewell, R. (1965). 'The Fourier Transform and Its Application'. McGraw-Hill, New York.

Brandt, A. (1977). Multi-level adaptive solutions to boundary value problems, *Math. Comput.* **31**, 333–390.

Brandt, A. (1989). Rigorous local mode analysis of multigrid. *In* 'Proceedings of the Fourth Copper Mountain Conference on Multigrid Methods', University of Colorado at Denver.

Buneman, O., Barnes, C.W., Green, J.C. and Nielsen, D.E. (1980). Principles and capabilities of 3-D particle simulations. *J. Comput. Phys.* **38**, 1–44.

Codell, R. (1988). Testing site characterization strategies with synthetic data bases. Paper presented at the Symposium on Hydrogeologic Aspects of Radioactive Waste Disposal, sponsored by the Northeastern Section of the Geological Society of America, Portland, Maine, 1988.

Cole, C.R. and Foote, H.P. (1987). Use of a multigrid technique to study effects of limited sampling of heterogeneity on transport prediction. *In* 'Proceedings of the Solving Ground-Water Problems with Models Conference and Exposition', Vol. 1, pp. 355–380. National Water Well Association, Dublin, Ohio.

Cole, C.R. and Rice, W.A. (1988). Use of geostatistical methods to study the effect of limited sampling of spatial heterogeneity on transport predictions. Paper presented at the Symposium on Hydrogeologic Aspects of Radioactive Waste Disposal, sponsored by the Northeastern Section of the Geological Society of America, Portland, Maine, 1988.

Cole, C.R., Foote, H.P., Zimmermann, D.A. and Simmons, C.S. (1985). Understanding, testing, and development of stochastic approaches to hydrologic flow and transport through the use of the multigrid method and synthetic data sets. *In* 'Preprints of the Symposium on the Stochastic Approach to Subsurface Flow', Greco 35 Hydrogeologie Ecole des Mines de Paris, Fontainbleau, pp. 364–378.

Dendy, J.E. (1983). Black box multigrid for nonsymmetric problems, *Appl. Math. Comput.* **13**, 261–283.

Jacobs, D.A.H. (1981). The exploitation of sparsity by iterative methods. *In* 'Sparse Matrices and their Uses' (Iain S. Duff, ed.), pp. 191–222. Academic Press, London.

McCormick, S.F. (1987). 'Multigrid Methods'. SIAM, Philadelphia.

Paddon, D.J. and Holstein, H. (1985). 'Multigrid Methods for Integral and Differential Equations'. Clarendon Press, Oxford.

Reisenauer, A.E., Cearlock, D.B. and Bryan, C.A. (1970). Mathematical Model for Approximating Transient Three-Dimensional Groundwater Systems, Pacific Northwest Laboratory, Richland Washington, BNWL-SA-3539.

Van der Wees, A.J. (1988). A nonlinear multigrid method for three-dimensional transonic flow. Dissertation, Technische Universiteit Delft, Amsterdam.

Chapter XI

Fluid Flow and Solute Transport in Fractal Heterogeneous Porous Media

Stephen W. Wheatcraft, Gregory A. Sharp and Scott W. Tyler

Desert Research Institute, University of Nevada System, P.O. Box 60220, Reno, Nevada 89506, USA

1. Introduction

Prediction of transport in heterogeneous porous media is one of the most important problems currently being addressed by groundwater hydrologists, chemical engineers and mechanical engineers, just to name a few. The specific goals differ somewhat, depending on the discipline, but the basic questions and problems are universal. Traditional porous media continuum theory averages microscopic balance equations to a Representative Elementary Volume (REV). This theory is based on the assumption that there is a natural scale of the porous medium below which the microscopic equations apply, and above which the geometry of the solids and voids can be averaged in a meaningful way to be replaced by a fictitious continuum at the macroscopic (REV) scale (Bachmat and Bear, 1986). The price for obtaining this fictitious continuum is that the unknown details of geometry at the microscopic scale are replaced with average parameters at the macroscopic scale that can be measured and used in the macroscopic equations to provide a predictive model.

Just as it is not possible to know the geometry of the pore space on the microscopic scale, the aquifer can be heterogeneous enough on a macroscopic scale that it is not possible (or at least not practical) to know all the geometry of the heterogeneity on a macroscopic, or field scale. A common approach to take here is to use stochastic theories, characterized by ensemble averages of the macroscopic variables (Gelhar, 1986). These theories result in scale-dependent parameters (such as macrodispersivity) which are functions of the statistical nature of the hydraulic conductivity distribution. Most of these theories also rely on the assumption that there is a natural scale to the heterogeneity, e.g. the correlation scale.

DYNAMICS OF FLUIDS IN HIERARCHICAL POROUS MEDIA
ISBN 0-12-200260-1

In many field situations, the existence of a natural scale may be a reasonable assumption, especially if some limit is placed as to how far into the future we require the model to be able to extrapolate. However, these assumptions of the existence of a natural scale are artificial separations that may not exist in many real situations. There may be multiple scales of heterogeneity, or there may be a continuous variation over space and time with no characteristic scale. The latter situation may be described as possessing fractal characteristics. One highly idealized conceptual model for a fractal porous medium is a Sierpinski carpet (or a Menger sponge, in three dimensions) shown in Fig. 1. A more realistic conceptual model for a fractal porous medium, based on geological depositional concepts, is presented and discussed as Fig. 1 in Wheatcraft and Tyler (1988). As in the Sierpinski carpet, this type of porous medium will have no characteristic scale, and the concepts of volume, porosity and other distributed parameters that should be characteristic of the medium are essentially meaningless. One cannot define an REV, and the correlation scale is infinite. Hence 'traditional' properties of a fractal porous medium are non-stationary, non-ergodic, and they are a function of the scale of measurement (Cushman, 1984; Wheatcraft and Tyler, 1988).

2. Fractal characteristics of porous media

Mandelbrot (1983) defined fractal geometry and recognized that many, perhaps even most, natural phenomena could be described using the concepts of fractal geometry. Fractional Gaussian noise (FGN) (Mandelbrot and Van Ness, 1968) has been used very successfully to generate artificial landscapes that are strikingly realistic (Voss, 1985). One can extend this concept by assuming that the geological materials underlying the fractal landscapes have fractal heterogeneous features that are reflected in the process of landform generation. Hence porous media comprised of natural geological materials may display fractal characteristics. Katz and Thompson (1985) have found sandstone porosity to display fractal characteristics. Burrough (1983a,b) studied the fractal properties of soils. The studies of pore-scale fractal processes (Adler, 1985), fracture networks (Barton and Larson, 1985; Ross, 1986), reservoir porosity (Hewett, 1986), and reservoir extent (Agterberg, 1982) all lead to the conclusion that fractal scaling concepts are operative over many scales of observation. Tyler and Wheatcraft (1989) develop a method to predict water retention curves in unsaturated soils based on a fractal model of particle size distribution in soils.

In addition to evidence that porous media display fractal scaling properties, it has been shown that some scale-dependent transport processes in porous media are fractal. Sapoval *et al.* (1985, 1986), and Rosso *et al.* (1985, 1986) examine the fractal nature of diffusional processes in porous media. Arya *et al.* (1985), Hewett (1986) and Wheatcraft and Tyler (1988) examine scale-dependent

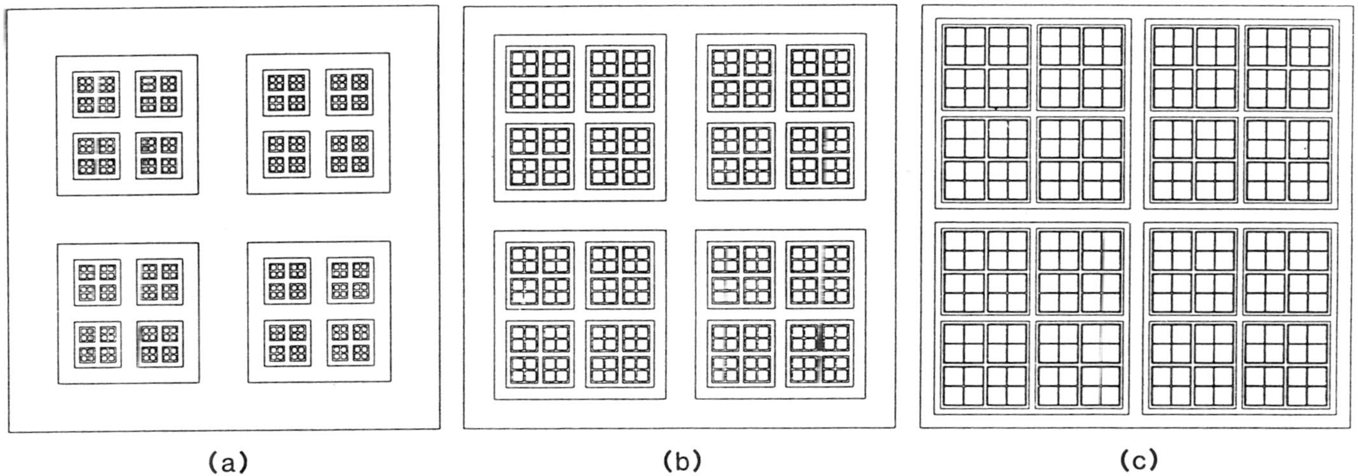

Figure 1. Varying fractal dimension of Sierpinski carpet: (a) $D = 1.25$; (b) $D = 1.5$; (c) $D = 1.75$.

dispersion processes in oil reservoirs and heterogeneous aquifers using fractal concepts. Wheatcraft and Tyler (1988) conclude that a fractal model is able to explain the data from field dispersion experiments better than current stochastic theories. It should be emphasized that their conclusion is tentative because the comparison is based on field data combined from many different sites. A valid comparison can be done only for an experiment conducted over a large scale at one site.

The simple fractal model developed by Wheatcraft and Tyler (1988) is based on the premise that a fractal porous medium will produce 'streamtubes', or particle tracks that are fractal in nature. The scale-dependent tortuosity, dispersivity, and variance relationships that are developed are based on this Lagrangian approach. In this chapter, we take a more physically based approach to looking at fractal behaviour of transport processes. We make the fundamental assumption that Darcy's law and the advection-dispersion equation are valid on the local scale and that field-scale dispersion is primarily caused by differential advection driven by field-scale variability of the hydraulic conductivity. This approach is common to most modern stochastic theories. We use a deterministic fractal model for the geometry of the hydraulic conductivity field which is based on modifications of the Sierpinski carpet. This fractal model is used to generate hydraulic conductivity distributions in a topologically two-dimensional domain. The flow equation is then solved for a case with simple boundary conditions of fixed potentials on two boundaries and no-flow on the other two boundaries. Once the velocity field is obtained from the steady-state solution to the flow equation, the advection-dispersion equation is solved through a particle-tracking scheme and transport statistics are collected and examined. The primary purpose of this study is to determine if a fractal hydraulic conductivity distribution will cause fractal scaling behaviour in the solution to the standard flow and transport equations and to compare such behaviour (if it exists) to the simple fractal model developed in Wheatcraft and Tyler (1988).

3. Basic fractal scaling concepts

In this section, we briefly review some fundamental ideas of fractal geometry. For more detailed review of these concepts, see Mandelbrot (1983), Wheatcraft and Tyler (1988) and Feder (1988). A fundamental property of fractals that is of particular importance for developing a theory of transport in fractal porous media is that quantities such as mass, length, area and volume (and other quantities such as density that are derived from these basic quantities) do not have intrinsic values. The length of a fractal line is a function of the measurement unit used to evaluate it, and it is scale-dependent. This relationship is given by

$$L(\varepsilon) = H_D \varepsilon^{1-D} \tag{3.1}$$

where ε is the size of the measurement unit (often referred to as the ruler size), H_D is an approximation to the Hausdorff measure of the fractal line corresponding to the Hausdorff (fractal) dimension D. This approximation is given by

$$H_D = N\varepsilon^D \tag{3.2}$$

where N is the number of measurement units of size ε needed to cover the fractal line. It should be emphasized that N is a function of ε. This relationship was formalized by Mandelbrot (1983). A very clear and lengthy discussion can be found in Chapter 2 of Feder (1988). The fractal dimension D is a fractional number $1 < D < 2$. If D is replaced with a number $d < D$, then the measure (length) of the line becomes infinite. If D is replaced with a number $d > D$, then the measure of the line becomes zero. Hence the fractal (or Hausdorff) dimension of the line is the critical dimension at which the measure of the line changes from zero to infinity, and this measure is referred to as the Hausdorff measure. Equation 3.1 can be thought of as an approximate transformation relationship between the constant Hausdorff measure of the fractal line, and the variable measure of the fractal line, as evaluated in the Euclidean norm. Equation 3.1 can be generalized to n-dimensional vector space

$$\mathbf{x}(\varepsilon) = \mathbf{H}_D\varepsilon^{n-D} \tag{3.3}$$

In Eq. 3.3, n is an integer: $n = 1$ refers to lines; $n = 2$ refers to surfaces; $n = 3$ refers to volumes; and $n = 4$ may refer to state variables (e.g. temperature) in three-dimensional space, and $n < D < n + 1$.

It is a relatively simple matter to develop fractal scaling relationships from the above equations (Wheatcraft and Tyler, 1988) that relate distance along a Cartesian coordinate system (straight-line distance x_s) to the approximate distance along a fractal line, or path (x_f). For a fractal line, this relationship is (Equation 7 in Wheatcraft and Tyler, 1988)

$$x_f = \varepsilon_c^{1-D} x_s^D \tag{3.4}$$

where ε_c is the fractal cut-off, as discussed in Wheatcraft and Tyler (1988). The scaling relationship between fractal velocity V_f and straight-line velocity, V_s, can be found by differentiating Eq. 3.4:

$$V_f = \varepsilon_c^{1-D} D x_s^{D-1} V_s \tag{3.5}$$

Equation 3.4 provides the source for scale-dependent dispersivity in the Wheatcraft and Tyler (1988) 'set of fractal streamtubes model' in which the dispersivity (α_m) is shown to behave according to

$$\alpha_m = \tfrac{1}{2}\sigma_c^2 x_s^{2D-1} \tag{3.6}$$

where σ_c^2 is the variance of ε_c^{1-D}, considered to be a constant. Equation 3.6 is the same as Eq. 16 in Wheatcraft and Tyler (1988), but with simplified notation. For the same model, the arrival time variance is shown to be

$$\sigma_{x_f}^2 = \sigma_c^2 x_s^{2D} \tag{3.7}$$

These fractal scaling relationships for travel distance, velocity, dispersivity and variance (Eqs 3.4–3.7) are of considerable interest in terms of developing more rigorous, physically based fractal transport models, primarily because their implications for transport behaviour are somewhat counter-intuitive, especially Eq. 3.5. A solute particle which behaves according to Eq. 3.4 will travel a greater distance (x_f) than indicated by field measured 'straight-line values' (x_s) alone. More importantly, once the projection of the particle has moved twice the original field-measured distance x_s, it will have travelled along the fractal path more than twice the original fractal length x_f. Equation 3.4 is not so difficult to accept, because it is exactly analogous to measuring the length of a fractal line with different measuring units, as discussed with Eqs 3.1–3.3. Equation 3.5 predicts that the relationship between actual velocity and field-measured (straight-line) velocity is scale-dependent and non-linear. If we assume that the velocity along the fractal path is constant (V_f = constant), then we would observe an apparent slow down in the field-measured velocity (V_s). Conversely, if we observed (in the field) that V_s is constant, Eq. 3.5 would insist that the actual velocity was increasing with scale, due to the scale-dependent tortuous nature of the fractal path (see discussion on fractal tortuosity in Wheatcraft and Tyler, 1988, p. 570).

Two things need to be emphasized: the equations are not valid for distances smaller than ε_c; and the observed relationships between these variables are all referenced to a starting point (the point where $x_s = 0$). If we start our measurements over in a new coordinate system, the discrepancy between V_s and V_f disappears. While this seems counter-intuitive, it should be interpreted as merely an artifact of the measurement process. More importantly, we are already familiar with the notion that scale-dependent dispersivity is referenced to a starting point. If we measure the dispersivity in a field tracer experiment at a time near the start of the experiment, and then measure it repeatedly during the course of the experiment, the value will grow (assuming that the aquifer is heterogeneous). Yet if we started a new experiment down-gradient, where we had just measured relatively large values of dispersivity in the original plume, we would again measure relatively small dispersivity values at early times during the new experiment.

The main objective of this chapter is to perform numerical experiments in which flow and transport equations are solved numerically using fractal hydraulic conductivity distributions and to see if the results show fractal scaling behaviour similar to that shown in the simple numerical experiments of Wheatcraft and Tyler (1988), and quantified by Eqs 3.4–3.7. Such behaviour in the solution of physically based flow and transport equations would be evidence that the fractal scaling relationships are valid for transport behaviour in fractal porous media, and not just an artifact of the way we constructed the simple fractal streamtube model.

4. Numerical experiments of transport in Sierpinski carpet fractal media

4.1 Sierpinski carpet construction

In this section, we use a topologically one-dimensional ($1\text{-}D_t$) form of Sierpinski carpet to represent a fractal heterogeneous media. As seen in Fig. 1, a Sierpinski carpet can be generated with different fractal dimensions. Also, many different choices can be made regarding the generating algorithm for the fractal media, each of which produces a distinct pattern and unique fractal dimension. Figure 1 shows a pattern that is an idealized version of the fractal facies depositional model discussed in Wheatcraft and Tyler (1988). For these simulations, we chose the model shown in Figs 2a–c. This particular Sierpinski carpet is created in such a way that the number of blocks increases by a factor of four with each recursive level of added detail. The first step results in the square of Fig. 2a, where $D = \log(N)/\log(1/\varepsilon) = \log(16-4)/\log(4) \approx 1.7925$, which would be the fractal dimension of the carpet shown in Fig. 2 after an infinite number of recursions. Subsequent divisions and removals of the remaining squares reduce the amount of area the carpet occupies, and gives it the more empty appearance of Fig. 2c. This fractal is very similar to the classic triadic Sierpinski carpet, which is divided into thirds (instead of fourths) with the centre square removed.

4.2 Application of the Sierpinski carpet to a hydraulic conductivity distribution

The Sierpinski carpet itself is merely a fractal pattern which can be thought of as area-removing. The original block (at the first recursion level) is said to be 'solid', and blocks are removed during application of the recursive algorithm. If we are to use this carpet for a hydraulic conductivity distribution, we must make some assumptions regarding how values of hydraulic conductivity are to be assigned to a binary pattern, in which the white areas are voids, and the black areas are considered solid. The only way to do this and retain the original fractal dimension of the carpet is to assign a binary hydraulic conductivity

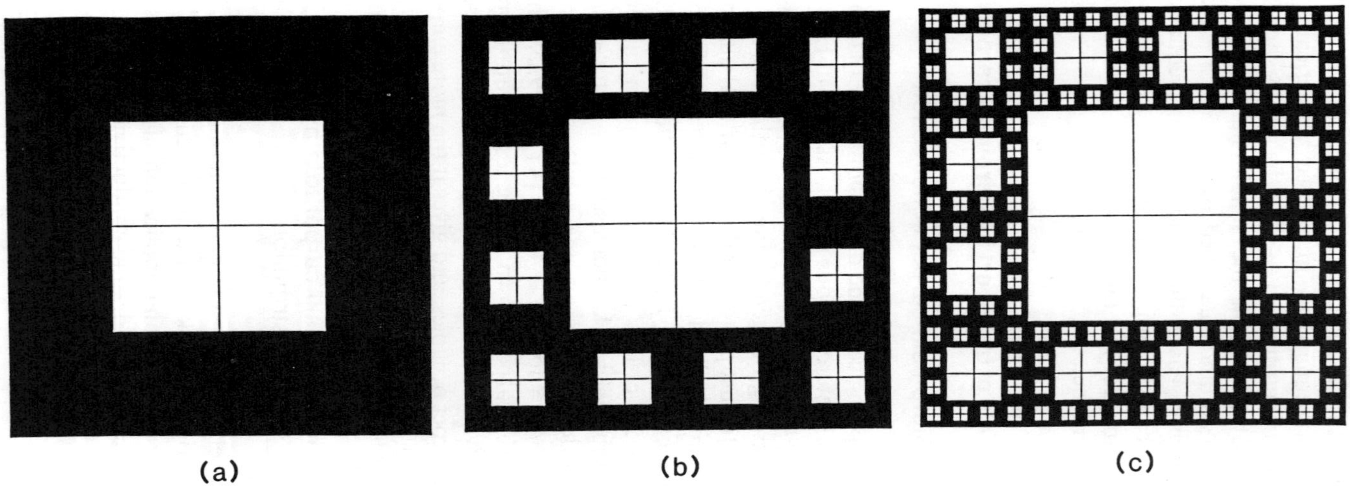

Figure 2. Four-in-four Sierpinski carpet used as a hydraulic conductivity distribution in the numerical simulations.

distribution, in which the white blocks have zero hydraulic conductivity, and the black blocks have a finite value for hydraulic conductivity. We do not consider this to be a physically realistic concept; moreover blocks of zero hydraulic conductivity complicate the problem of finding an efficient numerical scheme to solve the flow equation. As a result, we have chosen a compromise in which the white blocks are considered to be homogeneous and isotropic blocks of porous media with a hydraulic conductivity of one arbitrary hydraulic conductivity unit (cu). The black blocks are assigned a hydraulic conductivity of 100 cu's. This permeability contrast is large enough that the white blocks can be thought of as nearly zero hydraulic conductivity compared to the black blocks. Figure 3 shows the discretization of the Sierpinski carpet into a hydraulic conductivity distribution.

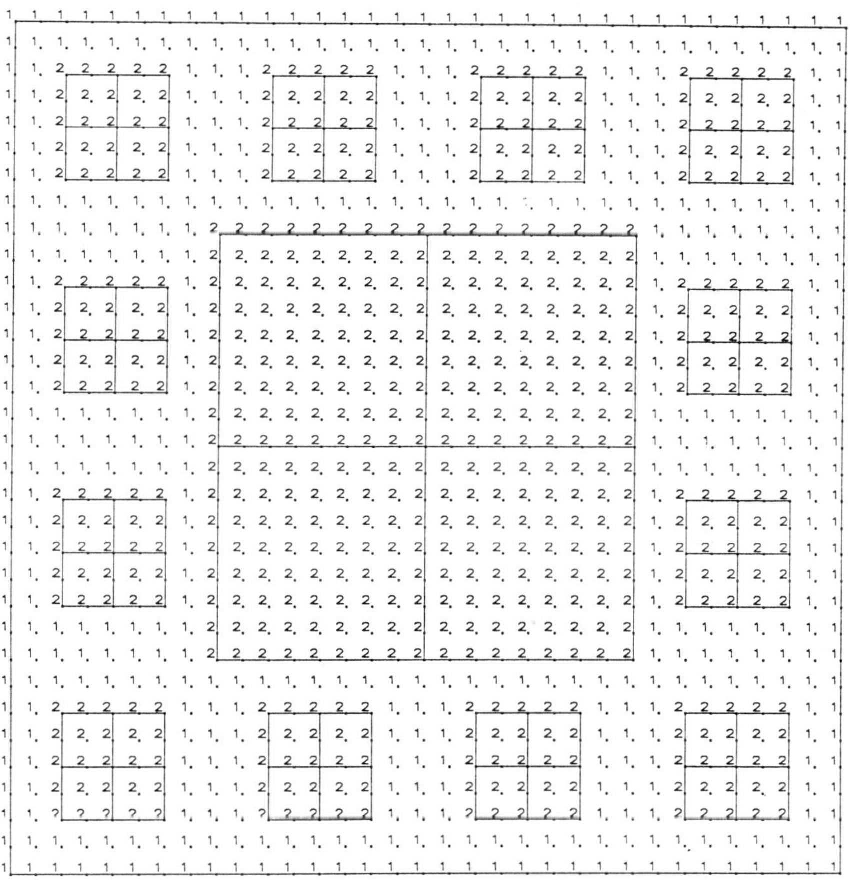

Figure 3. Discretization into nodes of different hydraulic conductivity: (1) $K - 100$; (2) $K = 1$.

It is important to realize that the hydraulic conductivity pattern has the same fractal dimension as the Sierpinski carpet only in the case where we assign zero hydraulic conductivity to the white blocks. This can readily be seen in the limiting case where we assign the same hydraulic conductivity to both the white and black blocks. Then the resulting porous medium is fractal only in its visible pattern; the hydraulic conductivity distribution is constant, homogeneous and non-fractal. Hence the fractal dimension (≈ 1.7925, discussed in the above paragraph) of the Sierpinski carpet is different from the fractal dimension of the hydraulic conductivity distribution. In fact, a 2-D_t hydraulic conductivity is thought of as a surface, in which elevation represents the hydraulic conductivity value, and its fractal dimension is $2 < D < 3$. In these experiments, our primary interest is to determine if fractal scaling behaviour occurs in the solution of fluid flow and transport equations for a self-similar fractal hydraulic conductivity distribution. The actual fractal dimension of such a distribution is unimportant because we would not find real aquifers which have such an idealized hydraulic conductivity pattern.

Each block is considered to be one length unit (lu) square, in which the units of length and area are arbitrary, but consistent with the units of hydraulic conductivity. An advantage of using arbitrary units is that no assumption is made regarding the lower limit of fractal behaviour, which we refer to as the fractal cutoff. For lengths below the fractal cutoff, length measures are non-fractal, i.e. measure becomes independent of ruler size. In this study, the fractal cutoff is given by the size of the smallest block at any scale. The fractal cutoff is therefore the same value (1 lu) at the 4, 16, 64 and 256 length scales. This results in an implicit assumption that the ratio of observation (length) scale to fractal cutoff grows proportionately to the length scale. Using a size of one square lu for each block, the numerical value of the length scale is identical to the ratio of length scale to fractal cutoff. In this set of experiments, the important difference between length scales is that each larger scale contains additional heterogeneity, and therefore a larger ratio of length scale to fractal cutoff.

We chose to discretize the domain into blocks that are one-half the length (one-quarter the area) of the fractual cutoff. Hence, at the 4 lu scale, there will be 8 unknown head values on each row, and 8^2, or 64 total unknowns. Similarly, at the 16, 64 and 256 lu scales, there will be 1024, 16 384 and 262 144 unknowns, respectively. The next larger length scale, 1024, contains over 1 million unknowns in the solution of the flow equation, and was too large for the computer resources that were available. Note that the 4 length unit (lu) scale of the fractal fits exactly in any of the remaining 64 blocks of the 16 lu scale of the fractal; thus it follows that the 4, 16 and 64 lu scales of the fractal all fit in that of the 256 lu scale, displaying exact self-similarity. This generating algorithm has the characteristic that the number of blocks are evenly divisible by two at any length scale. This algorithm is a necessity for the numerical scheme that we use to solve the flow equation, and it represents a compromise between the 'more realistic' Fig. 1, and the computational limitations imposed by the numerical scheme used to solve the flow equation. These limitations will be discussed in more detail below and are explained more fully in Sharp (1988) and Cole *et al.* (1985).

4.3 Numerical scheme

Simulation of fluid flow in the fractal porous medium is accomplished numerically using a finite difference multilevel (multigrid) adaptive solution method (Brandt, 1977) to solve the 2-D_t steady-state, continuity equation

$$\frac{\partial}{\partial x}\left(K(x,y)\frac{\partial h}{\partial x}\right)+\frac{\partial}{\partial y}\left(K(x,y)\frac{\partial h}{\partial y}\right)=0 \qquad (4.1)$$

where h is the hydraulic head, and $K(x, y)$ is the hydraulic conductivity. The boundary conditions for the left and right boundaries are determined based on the criterion that we want to have the same effective driving force for all simulations, regardless of the scale of the simulation. The effective driving force, Δh_f, is arbitrarily set to 0.1; the head on the left, h_l, is set to zero; and the head on the right, h_r, is determined by the simple linear relationship

$$h_r = \Delta h_f L_0 + h_l \qquad (4.2)$$

where L_0 is the scale of the simulation (4, 16, 64 or 256 lu's). The top and bottom boundaries are zero flux, as demonstrated by the streamlines for the solution at the 16 lu scale, shown in Fig. 4.

The particular multigrid solver employed is MGRID, developed at Pacific Northwest Laboratory (Cole *et al.*, 1985; Cole and Foote, 1987). The amount of work that MGRID has to perform for a particular solution is linearly proportional to the number of unknowns, thus it can solve a problem with over 250 000 unknowns very quickly (less than half an hour on a Sun 3/280 with a fast floating point accelerator). The MGRID implementation of the multigrid method divides the grid into quarters. Therefore, a major constraint of this implementation is that the problem must be discretizable into a grid containing 2^n grid blocks (n is a positive integer); thus the need for a fractal displaying self-similarity at scales of a power of four (two squared). Hence we are restricted to using recursive algorithms to generate our fractal distribution that go in powers of two, four, eight, etc. Power-of-two fractals are extremely simplistic, and power-of-eight fractals require more nodal points than power-of-four fractals to describe an equal number of levels, assuming the same grid spacing is used. The multigrid algorithm could be modified so that it divides the grid

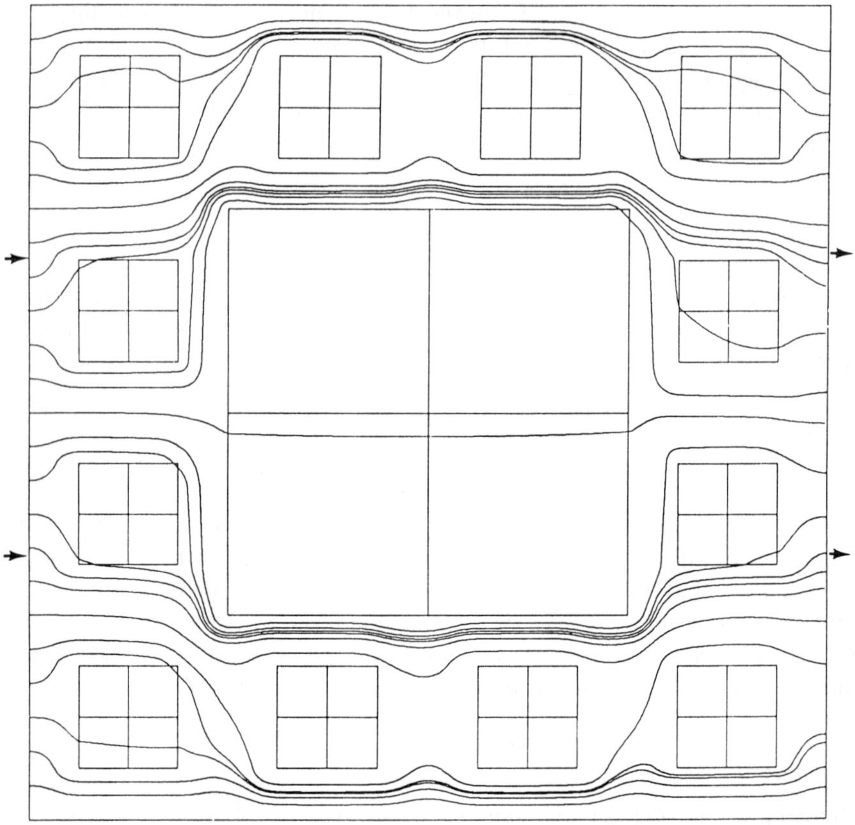

Figure 4. Streamlines for the solution at the 16 lu scale.

into thirds, or fifths, but for this study, it was easier to generate a power-of-four fractal carpet (Figs 2a–c).

The steady-state flow field calculated using MGRID is used in conjunction with a particle-tracking scheme to simulate solute particle transport. The computer code employed is a modified version of the Pacific Northwest Laboratory's particle-tracker GRADXYS (Cole *et al.*, 1985). This particle-tracking approach is advantageous in that only the validity of Darcy's law need be assumed, and, in addition, mechanical (Fickian) dispersion can be completely neglected so that dispersion is due only to differential advection. While it is possible to include statistical molecular diffusion in particle-tracking simulations, it is not done here because these experiments are meant to isolate the dispersion process due to differential advection. Input required for GRADXYS is the conductivity and potential field (MGRID solution), from which a velocity vector at each nodal point is calculated using a central finite difference representation for the potential gradient. Particle movement is determined from the four nodal point velocity vectors surrounding it. Breakthrough curves and other statistics were collected using simulations with 1000 particles; simulations were repeated using 200 particles for graphical visualization purposes in the figures.

4.4 Results

Figures 5a–c show the steady-state potential distribution, obtained from the solution to Eq. 4.1 by MGRID, for the 16, 64 and 256 lu scales, respectively. The mesh diagrams show the effect on the potential field of the additional heterogeneity that appears at the 64 and 256 lu scale, due to the recursive nature of the fractal algorithm that generates the hydraulic conductivity distribution. The self-similar nature of the solution is also obvious in comparing Figs 5a–c. The solution at the 16 lu scale appears to be a subset of the potential solution at the 64 lu scale, which in turn appears to be a subset of the 256 lu scale, although it is not exactly self-similar, due to the complex interactions of fluid flow at the larger scale.

Figures 6a–d show the synoptic positions of the particle swarms at various points in time as the simulation progresses through the 16 lu scale. It is evident that the particles become quite spread out as they are transported by differential advection primarily around the low hydraulic conductivity blocks and through the high hydraulic conductivity matrix. Figure 7 shows the particle tracks for the 64 lu simulation. The time steps have been chosen sufficiently close that they appear to be streamlines. This figure can be compared to Fig. 4, which shows the streamlines for the 16 lu simulation. Qualitatively, one can see that the additional heterogeneity present in the 64 lu simulation causes the streamlines to take a much more tortuous path through the system, in keeping with the concepts of the fractal scaling relationships.

The breakthrough curves, or arrival time distributions, for the four length scales are shown in Fig. 8. The only dispersion process operating in these simulations is differential advection, yet Fig. 8 shows that the dispersion increases greatly as a function of the scale of the experiment. This is quantified in Fig. 9, which is a log–log plot of the variance of the arrival time distribution for each length scale simulation. The slope of the log–log plot in Fig. 9 shows that arrival time variance is proportional to a power of ≈ 2.12. For traditional advection-dispersion, in which the dispersion is caused by mechanical dispersion, as defined in Bear (1972), the variance is proportional to total travel distance. For pure differential advection in a simple layered system such as that described by Mercado (1967) (referred to hereafter as Mercado dispersion), the arrival time variance grows proportional to the square of the travel distance. In the simple fractal model of Wheatcraft and Tyler (1988), variance grows proportional to the straight-line travel distance raised to the $2D$ power (x_s^{2D}), as in Eq. 3.7.

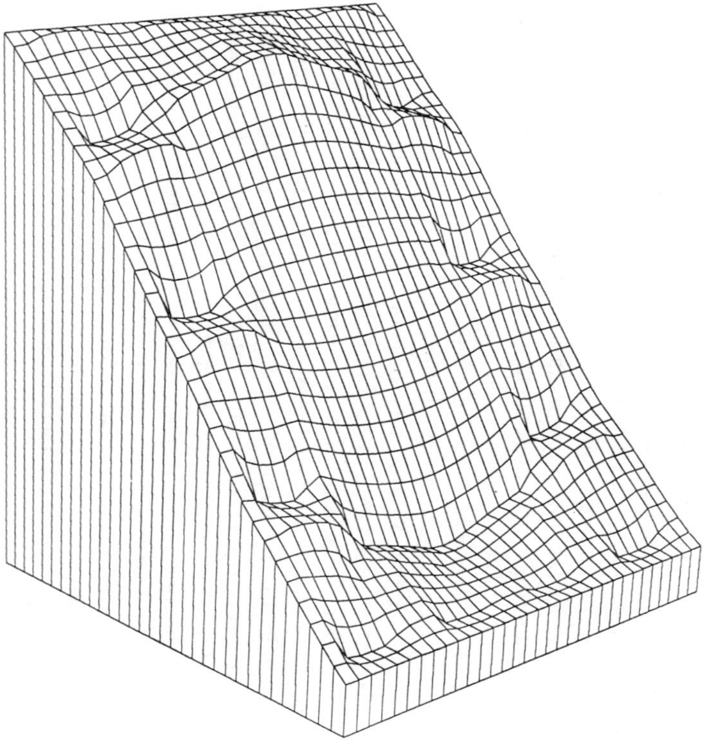

(a)

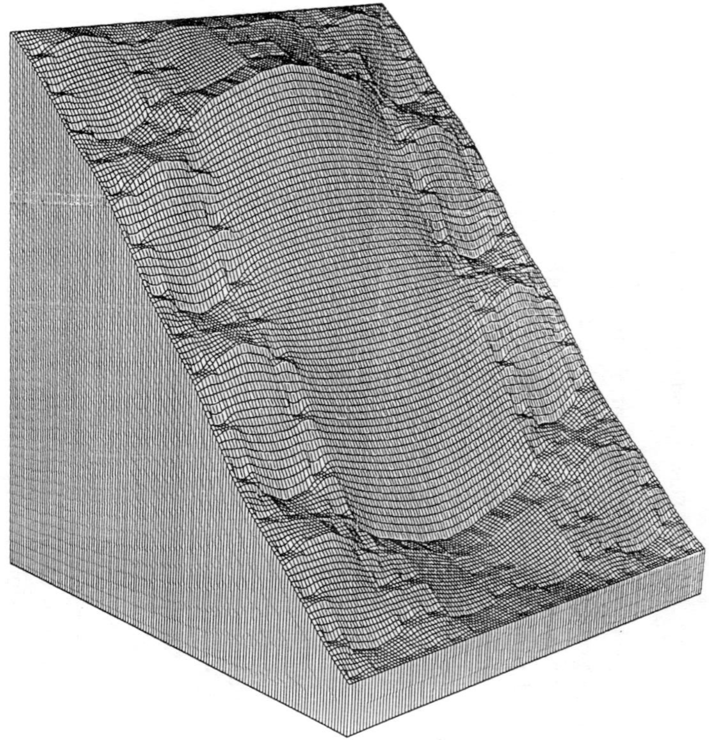

(b)

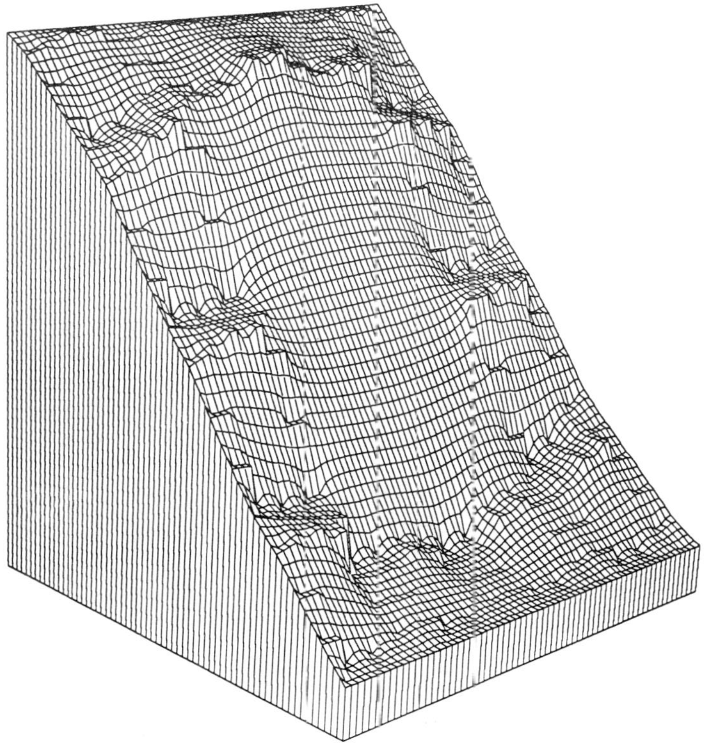

(c)

Figure 5. Multigrid solution of the potential field: (a) 16 lu scale; (b) 64 lu scale; (c) 256 lu scale.

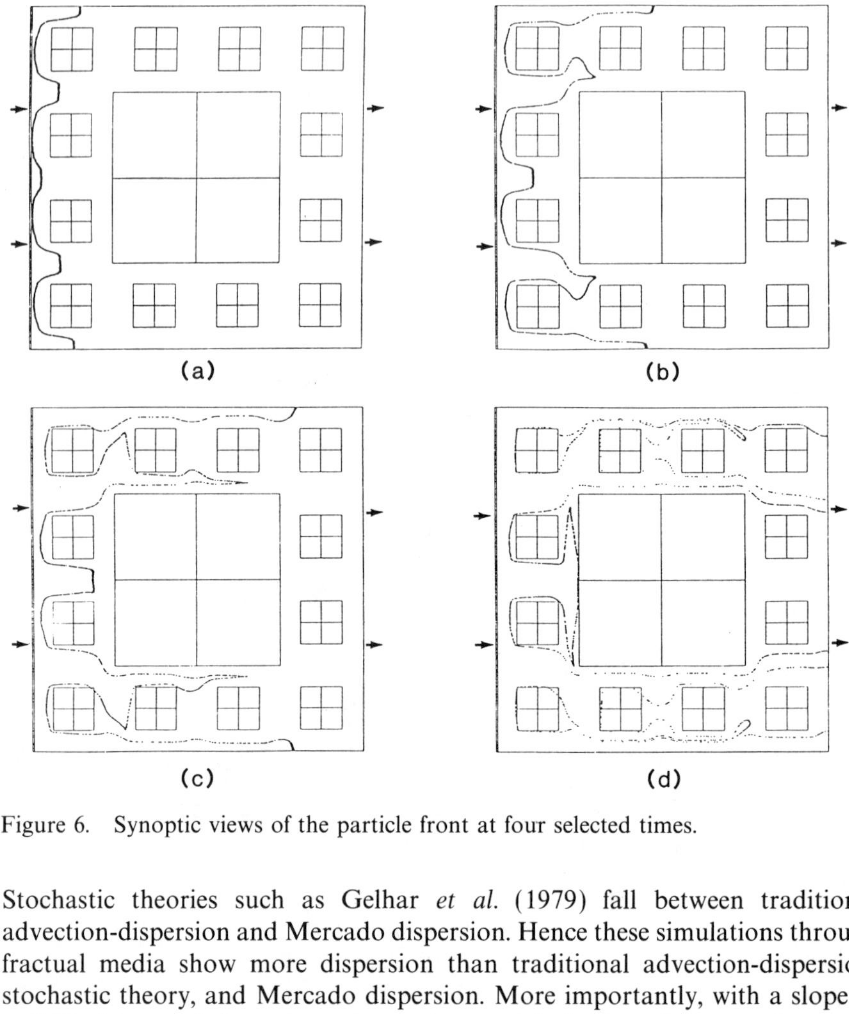

Figure 6. Synoptic views of the particle front at four selected times.

Stochastic theories such as Gelhar *et al.* (1979) fall between traditional advection-dispersion and Mercado dispersion. Hence these simulations through fractual media show more dispersion than traditional advection-dispersion, stochastic theory, and Mercado dispersion. More importantly, with a slope of ≈ 2.12, the growth in variance is consistent with the fractal streamtube model of Wheatcraft and Tyler (1988).

A few particle traces were saved for each simulation so that we could determine their fractal dimensions. The number of traces saved for a particular simulation was limited to about ten, at the time of these simulations, by available disc space. Ten of these particle traces are shown, along with their fractal dimensions, in Fig. 10. The fractal dimensions were determined by a Fortran computer program that automates the ruler method and box-counting method for computing fractal dimension (Sharp, 1988). The fractal dimensions for these traces vary, depending on the path that the particle is forced to take through the flow field; the smallest is ≈ 1.008, and the largest is ≈ 1.06, the average

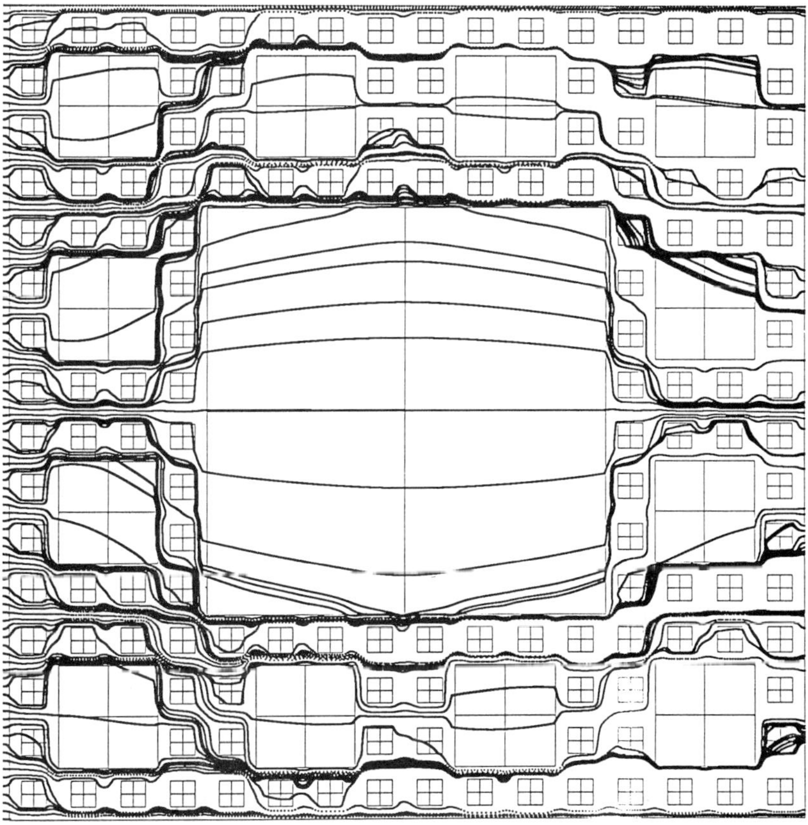

Figure 7. Particle tracks (same as streamlines) for the 64 lu scale.

being about 1.03. Another type of average fractal dimension was determined by calculating the total travel distance for each particle L_f, taking the average, and doing this for each length scale simulation L_s. The results are plotted in Fig. 11. By application of Eq. 3.4, the slope of the curve in Fig. 11 is the average fractal dimension of the particle traces, $D \approx 1.025$. This average agrees quite well with the independent box-counting method of determining the fractal dimensions of the ten particle traces in Fig. 10.

5. Conclusions

Through Fig. 9, it has been shown that the arrival time variances grow with the scale of the experiment in a manner that is consistent with the fractal streamtube model of Wheatcraft and Tyler (1988). Dispersion increases at a

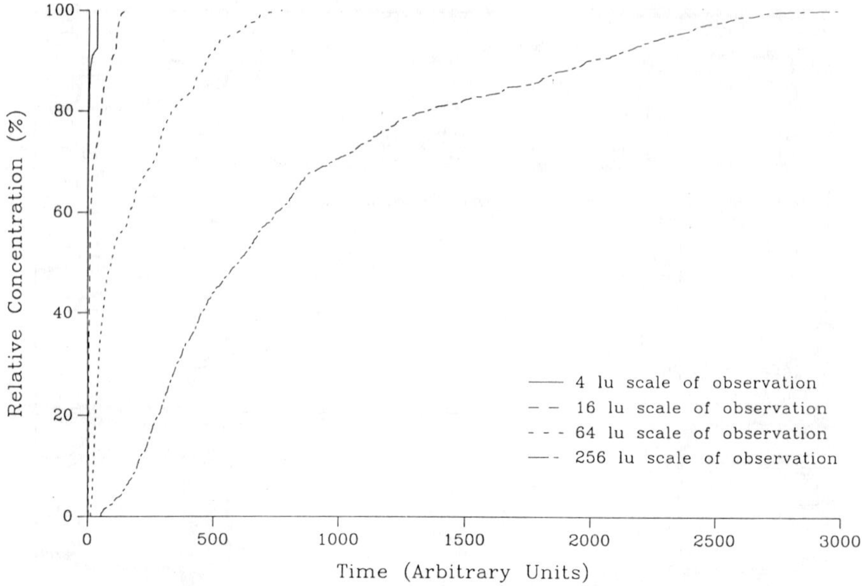

Figure 8. Breakthrough curves for the four simulations at four different scales.

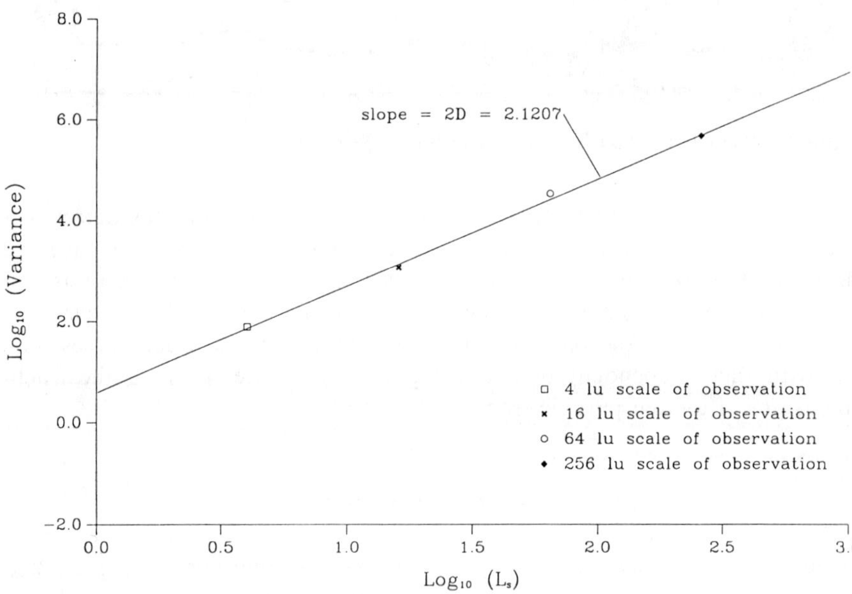

Figure 9. Variance of the arrival time distribution as a function of scale.

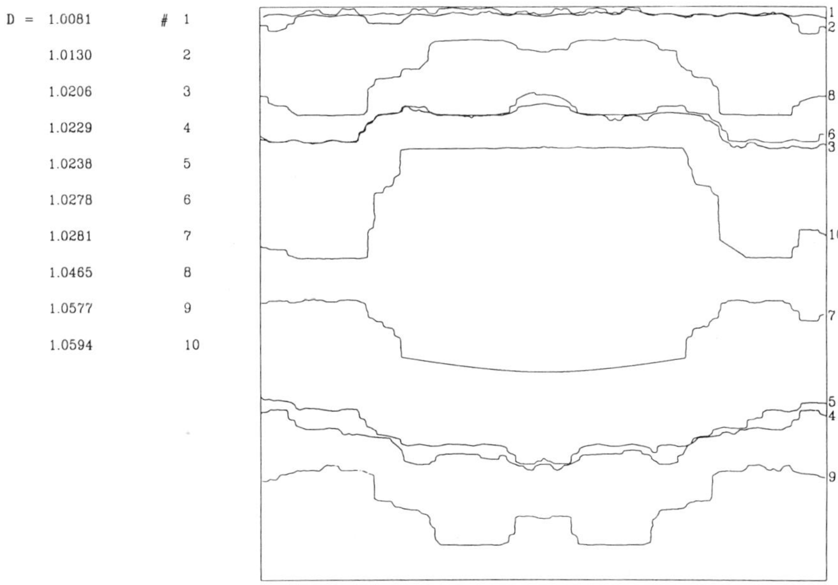

Figure 10. Ten particle tracks and their fractal dimensions, as calculated by the ruler (box-counting) method.

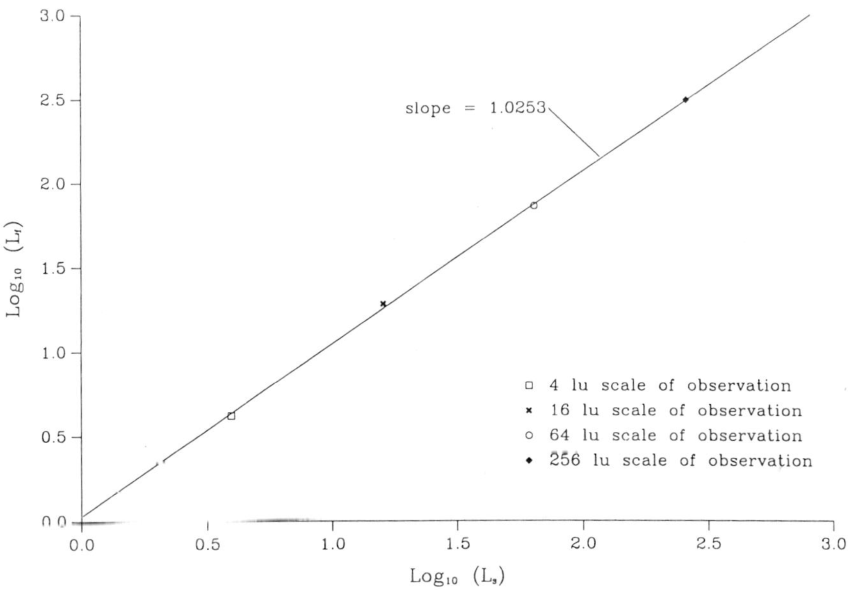

Figure 11. Increase in actual travel distance (L_f) as a function of Euclidean distance (L_s).

rate which is faster than that which can be predicted by any other model. Individual particle traces were examined (Fig. 10) and determined to have a fractal dimension greater than one. In Fig. 11, the particle traces are shown to follow a fractal scaling relationship similar to Eq. 3.4, which says that the actual travel distance is related to the straight-line distance by a power law. It appears that the exponent in this power law is numerically equal to the fractal dimension, as measured by the box-counting method. If the arrival time variances agreed exactly with the streamtube model, the slope, according to Eq. 3.7, would be $2D$. The arrival time variances show a power law growth with a slope of 2.12, which would indicate a fractal dimension of 1.06, which is at the high end of the fractal dimensions of the particle traces. However, we would not expect the arrival time variances in this model to agree exactly with Eq. 3.7, which is generated from the simple fractal streamtube model.

We have shown that a fractal hydraulic conductivity distribution leads to fractal scaling in the numerical solution of flow and transport equations through such media. The Sierpinski carpet distribution is highly idealized with respect to what might be expected in the field, but it is consistent with the notion of evolving heterogeneity, as discussed by Sposito *et al.* (1986) and Wheatcraft and Tyler (1988). The consequences of such fractal scaling are of particular importance with respect to measuring aquifer properties for input into predictive models. Field sampling schemes designed to obtain the statistical properties of the aquifer will fail because the average properties and their variances will be dependent on the scale at which they are measured. Stochastic methods rely on the correlation scale to be small compared to the scale of the experiment, but a fractal hydraulic conductivity distribution will be correlated over the entire scale of observation. Hence for aquifers that display fractal heterogeneity, we need sampling schemes that obtain information on the fractal nature of the hydraulic conductivity distribution. New theoretical techniques will be needed to predict transport in fractal media that take into account the fractal dimension of the media, the scale over which an experiment will be made, and the size of instruments that are used in the measurement of aquifer properties, fluid potential, and solute concentration.

Acknowledgements

This work was funded by the Nevada Agency for Nuclear Projects/Nuclear Waste Projects Office under Department of Energy Grant No. DE-FG08-85-NV10461. The opinions expressed in this paper do not necessarily represent those of the State of Nevada or the US Department of Energy.

References

Adler, P.M. (1985). Transport processes in fractals, III, Taylor dispersion in two examples of fractal capillary networks, *Int. J. Multiphase Flow*, **11**, 241–254.

Agterberg, F.P. (1982). Recent developments in geomathematics, *Geo Processing* **2**, 1–32.

Arya, A., Hewett, T.A., Larson, R. and Lake, L.W. (1985). Dispersion and reservoir heterogeneity. Paper SPE 14364 presented at the 60th Annual Technical Conference, Society of Petroleum Engineers, Las Vegas, Nevada, 22–25 September.

Bachmat, Y. and Bear, J. (1986). Macroscopic modelling of transport phenomena in porous media. 1: The continuum approach, *Transport in Porous Media* **1**, 213–240.

Barton, C.C. and Larson, E. (1985). Fractal geometry of two-dimensional fracture networks at Yucca Mountain, Southwest Nevada. Paper presented at the International Symposium on Fundamentals of Rock Joints, Swedish National Group of the International Society of Rock Mechanics, Bjorkliden, Sweden, 15–20 September.

Bear, J. (1972). 'Dynamics of Fluids in Porous Media', 764 pp. Elsevier Science, New York.

Brandt, A. (1977). Multi-level adaptive solutions to boundary value problems, *Math Comput.* **31**, 333–390.

Burrough, P.A. (1983a). Multiscale sources of spatial variation in soil. I: The application of fractal concepts to nested levels of soil variation, *J. Soil Sci.* **34**, 577–597.

Burrough, P.A. (1983b). Multiscale sources of spatial variation in soil. II: A non-Brownian fractal model and its applications in soil survey, *J. Soil Sci.* **34**, 599–620.

Cole, C.R. and Foote, H.P. (1987). Use of a multigrid technique to study effects of limited sampling and heterogeneity on transport prediction. Paper presented to the National Water Well Association Conference, Solving Ground Water Problems with Models, Denver, Colorado, 10–12 February.

Cole, C.R., Foote, H.P., Zimmermann, D.A. and Simmons, C.S. (1985). Understanding, testing, and development of stochastic approaches to hydrologic flow and transport through the use of the multigrid method and synthetic data sets. Preprints of symposium on Stochastic Approach to Subsurface Flow, Greco 35 Hydrogeologic Ecole des Mines de Paris, pp. 364–378.

Cushman, J.H. (1984). On unifying the concepts of scale, instrumentation, and stochastics in the development of multiphase transport theory. *Water Resour. Res.* **20**, 1668–1678.

Feder, J. (1988). 'Fractals', 283 pp. Plenum Press, New York.

Gelhar, L.W. (1988). Stochastic subsurface hydrology from theory to applications, *Water Resour. Res.* **22**, 135S–145S.

Gelhar, L.W., Gutjahr, A.L. and Naff, R.L. (1979). Stochastic analysis of macrodispersion in a stratified aquifer, *Water Resour. Res.* **15**, 1387–1397.

Hewett, T.A. (1986). Fractal distributions of reservoir heterogeneity and their influence on fluid transport. Paper SPE 15386 presented at the 61st Annual Technical Conference, Society of Petroleum Engineers, New Orleans, LA, 5–8 October.

Katz, A.J. and Thompson, A.H. (1985). Fractal sandstone pores: Implications for conductivity and pore formation, *Phys. Rev. Lett.* **54**, 1325–1328.

Mandelbrot, B.B. (1983). 'The Fractal Geometry of Nature', 468 pp. W.H. Freeman, New York.

Mandelbrot, B.B. and Van Ness, J.W. (1968). Fractional Brownian motions, fractional noises and applications, *SIAM Rev.* **10**, 422.

Mercado, A. (1967). The spreading pattern of injected water in a permeability-stratified aquifer, *IAHS AISH Publ.* **72**, 23–36

Ross, B. (1986). Dispersion in fractured fractal networks, *Water Resour. Res.* **22**, 823–827.

Rosso, M., Gouyet, J.F. and Sapoval, B. (1985). Determination of percolation probability from the use of a concentration gradient, *Phys. Rev. B* **32**, 6053–6054.

Rosso, M., Gouet, J.F. and Sapoval, B. (1986). Gradient percolation in three dimensions and relation to diffusion fronts, *Phys. Rev. Lett.* **57**, 3195–3198.

Sapoval, B., Rosso, M. and Gouyet, J.F. (1985). The fractal nature of a diffusing front and the relation to percolation. *J. Phys. Lett.* **46**, L149–L156.

Sapoval, B., Rosso, M., Gouyet, J.F. and Colonna, J.F. (1986). Dynamics of the creation of fractal objects by diffusion and $1/f$ noise, *Solid State Ionics* **18**, 21–30.

Sharp, G.A. (1988). Solute transport in fractal heterogeneous porous media. M.S. Thesis, University of Nevada–Reno, Reno, 71 pp.

Sposito, G., Jury, W.A. and Gupta, V.K. (1986). Fundamental problems in the stochastic convection–dispersion model of solute transport in aquifers and field soils, *Water Resour. Res.* **22**, 77–88.

Tyler, S.W. and Wheatcraft, S.W. (1989). Application of fractal mathematics to soil water retention estimation, *Soil Sci. Soc. Am. J.* **53**, 987–996.

Voss, R.F. (1985). Random fractal forgeries. In 'Fundamental Algorithms for Computer Graphics' (R.A. Earnshaw, ed.), pp. 805–835. Springer-Verlag, New York.

Wheatcraft, S.W. and Tyler, S.W. (1988). An explanation of scale-dependent dispersivity in heterogeneous aquifers using concepts of fractal geometry, *Water Resour. Res.* **24**, 566–578.

Chapter XII

Lie Group Invariance of the Richards Equation

Garrison Sposito
Department of Soil Science, University of California, Berkeley, CA 94720, USA

1. Introduction

The physics of water movement in partially saturated soil became a predictive science with the advent of the Richards equation. This non-linear partial differential equation describes the time evolution of the volumetric water content of a porous medium, as governed by a potential with components arising from the nature of the medium itself and from the terrestrial gravity field (Richards, 1931). In its conventional 'mass picture' form, the Richards equation can be expressed (Hillel, 1980, Ch. 9; Sposito, 1986):

$$\frac{\partial \theta}{\partial t} = \nabla \cdot [D(\theta) \nabla \theta] - \frac{\partial K}{\partial \theta} \frac{\partial \theta}{\partial z} \tag{1.1}$$

where θ $[L^3 L^{-3}]$ is volumetric water content, z is a vertical position coordinate measured negative upward, $D(\theta)$ $[L^2 T^{-1}]$ is soil water diffusivity, and $K(\theta)$ $[LT^{-1}]$ is hydraulic conductivity. The last two functions are 'transport coefficients' that appear in the Buckingham flux law, which traditionally is combined with a differential law of mass balance to derive Eq. 1.1 (Sposito, 1986). These transport coefficients depend strongly on θ, leading to the non-linearity of the Richards equation (Hillel, 1980, Ch. 9).

Implicit in Eq. 1.1 is the assumption of uniform, constant temperature, applied pressure, and dissolved solute concentration, as well as a restriction to homogeneous, isotropic, non-deformable porous media (Sposito, 1981, Ch. 7). It is possible, however, to relax these conditions by rewriting Eq. 1.1 in a covariant form through the use of a generalized definition of the water potential and a transformation to material position coordinates (Gupta *et al.*, 1977; Sposito, 1981, 1986; Sposito and Chu, 1982). This covariant form of the Richards equation, however, is not essential to the present discussion, nor is the

DYNAMICS OF FLUIDS IN HIERARCHICAL POROUS MEDIA
ISBN 0-12-200260-1

three-dimensional character of the first term on the right-hand side of Eq. 1.1. For isotropic porous media, no loss of generality ensues from consideration of the basic symmetry properties of Eq. 1.1 when restricted to scalar transport coefficients and the vertical spatial dimension only.

The solution of Eq. 1.1, whether analytical or numerical and under either weak conditions or explicit assumptions about the dependence of the two transport coefficients on θ, continues to be a complex topic of investigation in theoretical soil physics (Philip, 1969; Parlange, 1980; Parlange *et al.*, 1980; Parlange and Fleming, 1984; Broadbridge and White, 1988; Sander *et al.*, 1988a,b). The complexity of Eq. 1.1 is made even greater if it is generalized to describe isothermal water movement through inhomogeneous porous media. In this case, the transport coefficients and the water potential become explicit functions of the position coordinates and water content, and the term in brackets on the right-hand side of Eq. 1.1 must be replaced by gradient expressions involving both scalar and vector flux potentials (see, e.g. Gupta *et al.*, 1977). An alternative to confronting this added mathematical complexity directly is to assume that Eq. 1.1 applies locally at all points in an inhomogeneous porous medium (e.g. a field soil), with local values of the transport coefficients at different places connected through simple scaling transformations (Warrick *et al.*, 1977; Warrick and Nielsen, 1980).

For example, if $\psi(\theta)$ represents the soil water matric potential in a neighbourhood of some point and $\psi'(\theta')$ does the same at some other point, then the physical scaling relationship (Miller, 1980; Warrick and Nielsen, 1980),

$$\lambda'\psi'(\mu'\theta') = \lambda\psi(\mu\theta) \tag{1.2}$$

where λ and μ are positive scaling factors, might be used to relate the two local matric potentials if the functional dependence of each on θ has the same general mathematical form. The scaling factor μ is usually taken to be proportional to the inverse of the volumetric water content at saturation, θ_s (Sposito and Jury, 1985). This choice can be justified by the physical postulate that the pore-size distribution in a field soil be invariant under an arbitrary scaling of pore dimensions. Since a conventional graph of the cumulative pore-size distribution function versus pore diameter is equivalent to a graph of θ/θ_s versus ψ (see, for example, Danielson and Sutherland, 1986), invariance of the pore-size distribution implies invariance of $\tilde{\psi}(S) \equiv \alpha\psi(\theta/\theta_s)$, with $S \equiv \theta/\theta_s$ taken as a reduced water content variable and α interpreted as a relative scaling factor for pore size. Strict invariance of the pore-size distribution in an inhomogeneous soil is not likely in nature, but the use of S instead of θ is a convenient starting point for consideration of the symmetry properties of Eq. 1.1. Doing this, one writes $K(S)$ and $D(S)$, such that (in one spatial dimension) the Richards equation becomes:

$$\frac{\partial S}{\partial t_R} = \frac{\partial}{\partial z}\left[D(S)\frac{\partial S}{\partial z}\right] - \frac{\partial K}{\partial S}\frac{\partial S}{\partial z} \tag{1.3}$$

where $t_R \equiv t/\theta_S$ is a reduced time variable. Equation 1.3 is assumed to describe infiltration locally, with reduced water content and time variables, and to be generalizable to inhomogeneous porous media if $D(S)$ and $K(S)$ have a uniform functional dependence on S (Warrick and Amoozegar-Fard, 1979).

The use of scaling operations (technically called *stretching transformations*) to simulate inhomogeneity can be effected through the algorithm termed *Warrick similitude* (Sposito and Jury, 1985; Sposito, 1986). Warrick similitude involves stretching transformations of t_R and z that leave Eq. 1.3 invariant in form, thus generating a scaled Richards equation that is to apply globally to an inhomogeneous soil (Warrick and Amoozegar-Fard, 1979; Warrick *et al.*, 1985; Sposito and Jury, 1985). Solutions of the global Richards equation are then specialized to individual sites in a field soil by operating on them with the inverse of the stretching transformation used to create the global equation (Warrick and Nielsen, 1980). Details of these procedures as applied to experimental data on field soils are given by Simmons *et al.* (1979) and Jury *et al.* (1987).

The stretching transformations that underlie the scaling of the Richards equation can be studied formally with methods in the theory of Lie groups as applied to partial differential equations (Hill, 1982; Olver, 1986). Lie group analysis of the Richards equation for the purpose of obtaining a deeper understanding of Warrick similitude does not appear to have been carried out previously, although the more general Lie–Bäcklund group analysis has been applied to find solutions of the Richards equation for special classes of inhomogeneous porous media (Bluman and Kumei, 1980; Rosen, 1982; Broadbridge, 1988). In the present chapter, attention will be restricted to continuous point groups of transformations of the Richards equation, with the objective of developing conditions necessary for the invariance of Eq. 1.3 under Warrick similitude. Because of the second term on the right-hand side of Eq. 1.3, it can be expected that the symmetry properties of the Richards equation will be more restricted than those of the non-linear diffusion equation (i.e. Eq. 1.3 without the last term). The Lie group analysis of this latter partial differential equation is well known (Bluman and Cole, 1974, Section 2.11; Hill, 1982, Ch. 7; Ovsiannikov, 1982, pp. 68–73). From this analysis one learns that the sets of transformations of dependent and independent variables which leave the non-linear diffusion equation invariant are connected intimately with the functional form of $D(\theta)$ (see, e.g. Bluman and Cole, 1974, p. 297). Stretching transformations and simple translations turn out to be consistent with an arbitrary $D(\theta)$ in this case. One of the objectives of the present study is to determine whether the same result holds for the Richards equation in one spatial dimension.

In the following section the use of Eq. 1.3 to interpret field data on water movement through soil is summarized in a way that introduces the need to consider formal stretching transformations of the Richards equation. The next two sections give a general Lie group analysis and an explicit calculation of the point transformation group of Eq. 1.3. The latter result leads directly to conditions on the mathematical forms of $D(\theta)$ and $K(\theta)$, which are discussed in the last section of the paper.

2. The phenomenological picture

Prototypical field experiments for which a mathematical description by the one-dimensional Richards equation is appropriate have been discussed in detail by Nielsen *et al.* (1973), Simmons *et al.* (1979) and Rao *et al.* (1983). In outline, these experiments involve the establishment of instrumented rectangular plots at a number of sites over a relatively large land area. The field plots are managed as individual irrigated units through which the infiltration of water is monitored by measurements of the volumetric water content, θ, and the matric potential, ψ. Water movement in soil profiles beneath the field plots is assumed to be described physically by use of the Buckingham flux law (Sposito, 1986) and an area-averaged Richards equation in which only the vertical spatial coordinate appears, as in Eq. 1.3.

The physical interpretation of the field plot data is based on the assumption that water movement as a whole can be represented by a suitable average of the behaviour observed in the individual plots. The averaging process is designed to produce a mean Richards equation of the form (Warrick and Amoozegar-Fard, 1979):

$$\frac{\partial S}{\partial t} = \frac{\partial}{\partial z}\left[D_{\mathrm{m}}(S)\frac{\partial S}{\partial z}\right] - \frac{\partial K_{\mathrm{m}}}{\partial S}\frac{\partial S}{\partial z} \tag{2.1}$$

where $K_{\mathrm{m}}(S)$ is a field-wide mean hydraulic conductivity,

$$D_{\mathrm{m}}(S) = K_{\mathrm{m}}(S)\frac{\partial \psi_{\mathrm{m}}}{\partial S} \tag{2.2}$$

is a mean soil water diffusivity, and $\psi_{\mathrm{m}}(S)$ is a mean matric potential.

In the scaling approach, Eq. 2.1 is related to the Richards equation describing water movement below a specific field plot (denoted by the subscript P) through a Warrick similitude transformation (Warrick and Amoozegar-Fard, 1979;

Sposito and Jury, 1985):

$$t_P = \alpha_P \omega_P^2 t_R, \qquad z_P = \alpha_P z \tag{2.3}$$

where α_P and ω_P are positive scale factors. The application of Eq. 2.3 to Eq. 2.1 produces a plot-specific Richards equation:

$$\frac{\partial S}{\partial t_P} = \frac{\partial}{\partial z_P}\left[D_P(S)\frac{\partial S}{\partial z_P}\right] - \frac{\partial K_P}{\partial S}\frac{\partial S}{\partial z_P} \tag{2.4}$$

where

$$D_P(S) = \alpha_P D_m(S)/\omega_P \tag{2.5}$$

$$K_P(S) = K_m(S)/\omega_P^2 \tag{2.6}$$

are the plot-specific soil water diffusivity and hydraulic conductivity, respectively. Equations 2.2, 2.5, and 2.6 combine to yield the additional transformation:

$$\psi_P(S) = \alpha_P \psi_m(S) \tag{2.7}$$

relating a plot-specific matric potential to the field-average value. Equations 2.6 and 2.7 can be used to define α and ω^2 as scaling factors for the matric potential and hydraulic conductivity, respectively (Miller, 1980; Sposito and Jury, 1985).

The estimation of ψ_m, K_m, and the sets of α_P and ω_P in terms of measured values of $\psi_P(S)$ and $K_P(S)$ has been discussed by Simmons *et al.* (1979) and Jury *et al.* (1987). Given sets of ψ_P and K_P data, one may invoke model expressions to describe their S-dependence and then fit these expressions to a field-wide set of data subject to Eqs 2.6 and 2.7 along with a normalization constraint on the α_P and ω_P^2 (Simmons *et al.*, 1979). Alternatively, one can use measured values of ψ_P and K_P at fixed values of S to estimate parameters in an assumed field-wide probability distribution for the two soil water variables, then use these empirical distributions to calculate ψ_m and K_m as statistical averages subject to Eqs 2.6 and 2.7 along with a chosen normalization of the α_P and ω_P^2 (Jury *et al.*, 1987). This latter method requires a larger data set than the former, but it trades the need to postulate explicit model forms of $\psi_P(S)$ and $K_P(S)$ for the physically less stringent condition of assuming that parametric probability distributions exist for the two soil water variables.

Sposito and Jury (1985) have shown that the scaling relationships developed in Warrick similitude can be considered special cases of 'inspectional analysis', which involves scaling the Richards equation through stretching transformations of the independent variables and both stretching and simple translation of the dependent variable. The general set of transformations is (from unprimed to primed variables):

$$t' = \beta t, \quad z' = \gamma z, \quad S' = \mu S + \delta \tag{2.8}$$

where the Greek letter symbols are constant parameters. For the purpose of investigating the invariance of the Richards equation under general stretching transformations, it is useful to retain the flexibility implied by Eq. 2.8 and to define *generalized Warrick similitude* with the equations:

$$t' = \alpha\omega^2 t_{\mathrm{R}}, \quad z' = \alpha z, \quad S' = \mu S + \delta \tag{2.9}$$

such that Eq. 2.4 would feature S' instead of S, with the prime replacing the subscript P, and Eq. 2.6 would become:

$$K(S') = \mu K(S)/\omega^2 \tag{2.10}$$

Equations 2.5 and 2.7 are not altered under generalized Warrick similitude (except for dropping the subscript P in favour of the prime).

The scaling approach is based on the fundamental physical assumption that Eq. 1.3 describes any one-dimensional infiltration process within the entire volume occupied by a vadose zone at all times of observation after an initial input of water through the land surface. This physical hypothesis is given a mathematical form by invoking the stretching transformations in Eqs 2.9, 2.5, 2.7 and 2.10, which take time and space coordinates and transport coefficients from their field-wide mean values to a set of plot-specific values. It is evident, after consideration of two such transformations applied to two different plots, that the time and space coordinates and transport parameters in any two field plots can also be related by stretching transformations. The physical condition all these transformations must meet is to leave the mathematical form of Eq. 1.3 unchanged.

Therefore, if the transformation of dependent and independent variables in Eq. 1.3 from one field plot to another is described by the equations:

$$S' = f(t,z,S) \tag{2.11a}$$

$$z' = g(t,z,S) \tag{2.11b}$$

$$t' = h(t,z,S) \tag{2.11c}$$

where (dropping now the R subscript on the time variable) t,z,S refers to one plot and t',z',S' to another, then these equations must transform Eq. 1.3 into the primed Richards' equation:

$$\frac{\partial S'}{\partial t'} = \frac{\partial}{\partial z'}\left[D(S')\frac{\partial S'}{\partial z'}\right] - \frac{\partial K(S')}{\partial S'}\frac{\partial S'}{\partial z'} \tag{2.12}$$

where $D(S')$ and $K(S')$ are the transformed $D(S)$ and $K(S)$. Clearly, not all continuous transformations having the general form of Eq. 2.11 will be consistent with Eq. 2.12. Therefore, the central problem to address in a study of the foundations of the scaling approach is to determine the properties of Eq. 2.11 such that stretching transformations will leave the Richards equation invariant in form. The resolution of this problem should lead to conditions on $D(S)$ and $K(S)$ that make their transformation, induced by Eq. 2.11, consistent with the invariance of the Richards equation, as illustrated in Eq. 2.12.

3. Formal Lie group invariance of the Richards equation

The explicit mathematical form of Eq. 2.11 can be calculated with methods in the theory of one-parameter Lie groups (Hill, 1982; Ovsiannikov, 1982; Olver, 1986). Group properties are imposed on Eq. 2.11 by introducing the real-valued parameter ε:

$$S' = f(t,z,S;\,\varepsilon) \tag{3.1a}$$

$$z' = g(t,z,S;\,\varepsilon) \tag{3.1b}$$

$$t' = h(t,z,S;\,\varepsilon) \tag{3.1c}$$

Equations 3.1 determine a *local Lie group* parametrized by ε if smooth identity and inverse transformations can be defined and the group closure property exists under any associative transformation of S, z, and t (Hill, 1982, Ch. 2; Olver, 1986, Ch. 1). These three characteristics can be illustrated with Eq. 3.1b:

$$z = g(t,z,S;0) \quad \textit{identity} \tag{3.2a}$$

$$z = g(t',z',S';-\varepsilon) \quad \textit{inverse} \tag{3.2b}$$

$$z'' = g(t,z,S;\varepsilon+\delta) \quad \text{if} \quad z'' = g(t',z',S';\delta) \quad \text{and}$$

$$z' = g(t,z,S;\varepsilon) \quad \textit{closure} \tag{3.2c}$$

The closure property in Eq. 3.2c can be used to demonstrate the associativity of the transformations in Eq. 3.1. Expressions like Eq. 3.2 apply to Eqs 3.1a and 3.1c as well. The study of the invariance of Eq. 1.3 is tantamount to deducing a realization of Eq. 3.1 subject to Eq. 3.2, such that $S'(t',z')$ solves Eq. 2.12 whenever $S(t,z)$ solves Eq. 1.3 (Bluman and Cole, 1974, Ch. 2; Hill, 1982, Ch. 2; Olver, 1986, Ch. 2).

The algorithm for establishing the Lie group that leaves the Richards equation invariant in form is based on the fact that Eq. 3.1 can be calculated by letting ε become an infinitesimal quantity, so as to expand the right-hand side of Eq. 3.1 in a MacLaurin series to first order in ε; then determining the series coefficients such that Eq. 1.3 transforms into Eq. 2.12; and finally solving a set of first-order differential equations based on the series coefficients and the condition in Eq. 3.2a (Bluman and Cole, 1974, Ch. 2; Hill, 1982, Ch. 2). Following this algorithm, one expands Eq. 3.1 to first order in ε about $\varepsilon = 0$:

$$S' = S + \varepsilon\zeta(t,z,S) \tag{3.3a}$$

$$z' = z + \varepsilon\xi(t,z,S) \tag{3.3b}$$

$$t' = t + \varepsilon\eta(z,t,S) \tag{3.3c}$$

where, for example,

$$\zeta(t,z,S) = (\partial S'/\partial\varepsilon)_{\varepsilon=0} \tag{3.4}$$

and similarly for ξ and η. Expansions like Eq. 3.3 also must be developed for the transport coefficients in Eq. 2.12. To this end Eq. 2.12 is rewritten in the form:

$$\frac{\partial S'}{\partial t'} = D(S')\frac{\partial^2 S}{\partial z'^2} + \frac{\partial D}{\partial S'}\left(\frac{\partial S'}{\partial z'}\right)^2 - \frac{\partial K}{\partial S'}\frac{\partial S'}{\partial z'} \tag{3.5}$$

which makes clear the need to consider both $D(S)$ and its first derivative:

$$D(S') = D(S) + \varepsilon\zeta(t,z,S)\frac{\partial D}{\partial S} \tag{3.6a}$$

$$\frac{\partial D(S')}{\partial S'} = \frac{\partial D}{\partial S} + \varepsilon\zeta(t,z,S)\frac{\partial^2 D}{\partial S^2} \tag{3.6b}$$

$$\frac{\partial K(S')}{\partial S'} = \frac{\partial K}{\partial S} + \varepsilon\zeta(t,z,S)\frac{\partial^2 K}{\partial S^2} \tag{3.6c}$$

The second term on the right-hand side of Eq. 3.6 is calculated with the chain rule and 3.3a.

The functions ζ, ξ, and η are determined by the condition that Eq. 2.12 result from Eq. 1.3, to first order in ε, through the application of Eqs 3.3 and 3.6. Then the transformations in Eq. 3.1 are calculated by solving the differential equations (Hill, 1982, Ch. 2):

$$\frac{dS'}{d\varepsilon} = \zeta(t',z',S') \tag{3.7a}$$

$$\frac{dz'}{d\varepsilon} = \xi(t',z',S') \tag{3.7b}$$

$$\frac{dt'}{d\varepsilon} = \eta(t',z',S') \tag{3.7c}$$

subject to the initial condition implied by the identity transformation (e.g. in Eq. 3.2a for z). More elegant techniques for generating the transformations in Eq. 3.1 with Lie series (Hill, 1982, Ch. 2; Olver, 1986, Ch. 1) lead to the same results as the simple differential equation approach in Eq. 3.7 (Dickson, 1924).

The calculations through which Eqs 3.3 are determined for a non-linear partial differential equation have been discussed by Bluman and Cole (1974, Section 2.14), Hill (1982, Ch. 7), and Olver (1986, Sections 2.4 and 3.2). The procedure is to express each term in Eq. 3.5 as the sum of a corresponding term in Eq. 1.3 plus a term in ε that results from the application of the chain rule and the transformations in Eqs 3.3 and 3.6. The resulting equations are given in the Appendix (see also Hill, 1982, Section 7.1). These equations are combined according to Eq. 3.5 to produce an expression involving a primed Richards equation, an unprimed Richards equation, and a set of terms multiplied by the same factor ε. The terms in ε are also given in the Appendix. They contain partial derivatives of S whose coefficients must vanish identically to make Eq. 3.5 equal Eq. 1.3 to lowest order in ε. This condition leads to a series of constraints on ζ, ξ and η in the form of coupled partial differential equations. In the present case, these constraint equations are:

$$\frac{\partial \eta}{\partial S} \equiv 0 \quad \frac{\partial \eta}{\partial z} \equiv 0 \quad \frac{\partial \xi}{\partial S} \equiv 0 \tag{3.8}$$

$$\frac{\partial \xi}{\partial t} - D(S)\frac{\partial^2 \xi}{\partial z^2} + 2\frac{\partial}{\partial S}\left[D(S)\frac{\partial \zeta}{\partial z}\right] = \frac{\partial K}{\partial S}\left[\frac{\partial \xi}{\partial z} - \zeta\frac{\partial D/\partial S}{D(S)}\right] + \zeta\frac{\partial^2 K}{\partial S^2} \tag{3.9}$$

$$D(S)\frac{\partial \eta}{\partial t} + \zeta\frac{\partial D}{\partial S} - 2D(S)\frac{\partial \xi}{\partial z} = 0 \tag{3.10}$$

$$D(S)\frac{\partial^2 \zeta}{\partial S^2} + \frac{\partial^2 D}{\partial S^2}\zeta + \frac{\partial D}{\partial S}\frac{\partial \zeta}{\partial S} + \frac{\partial D}{\partial S}\frac{\partial \eta}{\partial t} - 2\frac{\partial D}{\partial S}\frac{\partial \xi}{\partial z} = 0 \tag{3.11}$$

$$\frac{\partial \zeta}{\partial t} - D(S)\frac{\partial^2 \zeta}{\partial z^2} + \frac{\partial K}{\partial S}\frac{\partial \zeta}{\partial z} = 0 \tag{3.12}$$

Equations 3.8 to 3.12 are derived in the Appendix.

For the case of the linear diffusion equation, in which $D(S)$ is a constant and $K(S) \equiv 0$, these expressions reduce to conditions found by Bluman and Cole (1974, p. 209). If $K(S) \equiv 0$, Eq. 1.3 becomes a non-linear diffusion equation, and the expressions above reduce to conditions given by Hill (1982, Section 7.2) for determining the Lie group of that equation. The presence of $\partial K/\partial S$ in the Richards equation creates a non-linear Fokker–Planck equation for which the more general constraints in Eqs 3.8 to 3.12 apply. These constraints must be solved for $\zeta(t,z,S)$, $\xi(t,z)$ and $\eta(t)$ consistently with the mathematical forms of $D(S)$ and $K(S)$.

4. The Lie group of the Richards equation

Equation 3.10 can be rearranged to be an expression for ζ:

$$\zeta(t,z,S) = \frac{D(S)}{\partial D/\partial S}\left(2\frac{\partial \xi}{\partial z} - \frac{\mathrm{d}\eta}{\mathrm{d}t}\right) \tag{4.1}$$

This result, in turn, can be substituted into Eq. 3.12 to yield the equation:

$$2\frac{\partial^2 \xi}{\partial z \partial t} - \frac{\mathrm{d}^2 \eta}{\mathrm{d}t^2} - 2D(S)\frac{\partial^3 \xi}{\partial z^3} + 2\frac{\partial K}{\partial S}\frac{\partial^2 \xi}{\partial z^2} = 0 \tag{4.2}$$

Since ξ and η do not depend on S (cf. Eq. 3.8), this equation cannot be satisfied in general unless *each* of the first two terms on the left-hand side, the coefficient of $D(S)$, and the coefficient of $\partial K/\partial S$ vanishes identically:

$$\frac{\mathrm{d}^2 \eta}{\mathrm{d}t^2} = 2\frac{\partial^2 \xi}{\partial z \partial t}, \quad \frac{\partial^3 \xi}{\partial z^3} \equiv 0, \quad \frac{\partial^2 \xi}{\partial z^2} \equiv 0 \tag{4.3}$$

Therefore, $\xi(t,z)$ has the form:

$$\xi(t,z) = \tfrac{1}{2} z \frac{\mathrm{d}\eta}{\mathrm{d}t} + bz + c \tag{4.4}$$

where $\eta(t)$ is as yet arbitrary, and b and c are constants. Given Eqs 4.1 and 4.4, it follows that:

$$\zeta(S) = \frac{D(S)}{\partial D/\partial S} 2b \tag{4.5}$$

The substitution of Eqs 4.4 and 4.5 into the constraint Eq. 3.11 produces the condition:

$$2bD(S)\partial^2\left[\frac{D(S)}{\partial D/\partial S}\right]\bigg/\partial S^2 \equiv 0 \tag{4.6}$$

Given that $D(S)$ is not zero identically, this condition can be met only if $b \equiv 0$, or if

$$\frac{\partial^2}{\partial S^2}\left[\frac{D(S)}{\partial D/\partial S}\right] \equiv 0 \tag{4.7}$$

The substitution of Eqs 4.4 and 4.5 into Eq. 3.9 brings this latter constraint into the form:

$$\begin{aligned}\tfrac{1}{2}z\frac{\mathrm{d}^2\eta}{\mathrm{d}t^2} &= \frac{\partial K}{\partial S}\left(\frac{1}{2}\frac{\mathrm{d}\eta}{\mathrm{d}t} - b\right) + 2b\frac{D(S)}{\partial D/\partial S}\frac{\partial^2 K}{\partial S^2} \\ &= \frac{\partial K}{\partial S}\left[\frac{1}{2}\frac{\mathrm{d}\eta}{\mathrm{d}t} + (2\beta - 1)b\right] + 2b\left[\frac{D(S)}{\partial D/\partial S}\frac{\partial^2 K}{\partial S^2} - \beta\frac{\partial K}{\partial S}\right]\end{aligned} \tag{4.8}$$

where β is an arbitrary scale parameter. Since η is independent of S, Eq. 4.8 cannot be true in general unless $\mathrm{d}^2\eta/\mathrm{d}t^2$ vanishes identically and the conditions

$$\frac{1}{2}\frac{\mathrm{d}\eta}{\mathrm{d}t} + (2\beta - 1)b \equiv 0 \tag{4.9}$$

$$\frac{D(S)}{\partial D/\partial S} \equiv \beta\frac{\partial K/\partial S}{\partial^2 K/\partial S^2} \tag{4.10}$$

are met as well. It follows from Eqs 4.4, 4.5 and 4.9 that the Lie group of the Richards equation is defined through the functions:

$$\zeta(S) = 2b\frac{D(S)}{\partial D/\partial S} \tag{4.11a}$$

$$\xi(z) = -2b(\beta - 1)z + c \tag{4.11b}$$

$$\eta(t) = -2b(2\beta - 1)t + d \tag{4.11c}$$

where β, b, c and d are constants, with $D(S)$ and $K(S)$ subject to Eqs 4.6 and 4.10.

The combination of Eqs 3.7 and 4.11 leads to five transformations of z and t that will leave the Richards equation invariant in form. These transformations are listed in Table 1. For each row in the table, all parameters except that listed

Table 1. The independent space–time transformations in the Lie group of the one-dimensional Richards equation

Parameter	Equation for z'	Equation for t'
c	$z' = z + c\varepsilon$	$t' = t$
$b(\beta \neq 1/2 \text{ or } 1)$	$z' = \exp(-2b(\beta - 1)\varepsilon)z$	$t' = \exp(-2b(2\beta - 1)\varepsilon)t$
$b(\beta = 1/2)$	$z' = \exp(b\varepsilon)z$	$t' = t$
$b(\beta = 1)$	$z' = z$	$t' = \exp(-2b\varepsilon)t$
d	$z' = z$	$t' = t + d\varepsilon$

in the first column are set equal to zero, and the corresponding form of Eqs 4.11b and 4.11c are introduced into Eqs 3.7b and 3.7c to calculate z' and t'. It is evident from Table 1 that the parameters c and d correspond respectively to space and time translations. The parameter b corresponds to a scaling transformation of z and t whose specific form depends on the value of β. If $\beta \neq \frac{1}{2}$ or 1, the scaling transformation acts on z and t simultaneously.

The transformation of S, mediated by $\zeta(S)$ in Eq. 3.3a, depends on the constraints in Eqs 4.5 and 4.6. If $b \equiv 0$, then $\zeta \equiv 0$ and $S' = S$ (identity transformation). This case also corresponds to simple translations of z and t. If $b \neq 0$, then Eq. 4.7 must be met by $D(S)$. This expression is equivalent to the non-linear second-order differential equation:

$$D(S)\frac{\mathrm{d}^2 D}{\mathrm{d}S^2} - \frac{(m-1)}{m}\left(\frac{\mathrm{d}D}{\mathrm{d}S}\right)^2 = 0 \tag{4.12}$$

where $1/m$ is the constant value of $\mathrm{d}[D(S)/(\mathrm{d}D/\mathrm{d}S)]/\mathrm{d}S$. The solutions of Eq. 4.12 are then (Ince, 1956, pp. 335–337; Ovsiannikov, 1982, pp. 68–73):

$$D(S) = (pS + q)^m \tag{4.13}$$

$$D(S) = A\exp(kS) \tag{4.14}$$

where p, q, m, A and k are real-valued parameters. For these two possibilities, Eq. 4.5 leads to the expressions:

$$\zeta(S) = 2b(pS + q)/mp \tag{4.15}$$

$$\zeta(S) = 2b/k \tag{4.16}$$

which correspond, respectively, to scaling/translation or simple translation of S, as can be seen after the substitution of Eq. 4.15 or Eq. 4.16 into Eq. 3.7a. Therefore, it may be concluded that scaling transformations of z and t are connected with $D(S)$ as given in Eqs 4.15 and 4.16 along with stretching and/or translation of S.

The introduction of Eq. 4.13 or Eq. 4.14 into Eq. 4.10 produces a first-order differential equation for $\mathrm{d}K/\mathrm{d}S$ whose solution is (Ince, 1956, Section 2.13):

$$\frac{\mathrm{d}K}{\mathrm{d}S} = c_1(pS + q)^{m\beta} \tag{4.17}$$

in the case of Eq. 4.13 and

$$\frac{\mathrm{d}K}{\mathrm{d}S} = c_2 \exp(\beta k S) \tag{4.18}$$

in the case of Eq. 4.14, where c_1 and c_2 are arbitrary constants. These two expressions can be integrated to reveal that $K(S)$ shows either a power-law or exponential dependence on S (if $m\beta$ and $\beta k \neq 0$):

$$K(S) = [c_1/p(1 + m\beta)](pS + q)^{1+m\beta} + c_0 \tag{4.19}$$

if $m\beta \neq -2$,

$$K(S) = c_1(K_1 + K_2 S)[(qK_2 - pK_1)(pS + q)]^{-1} + c_0 \tag{4.20}$$

if $m\beta = -2$ (Petit Bois, 1961, p. 2), and

$$K(S) = (c_2/\beta k)\exp(\beta k S) + c_0 \tag{4.21}$$

where K_1, K_2 and c_0 are arbitrary constants.

5. Discussion

Equation 2.2, combined with Eqs 4.13 or 4.14, and 4.19, 4.20 or 4.21, leads to a first-order differential equation for $\psi(S)$ that can be solved under chosen boundary conditions. For Eqs 4.19 and 4.21, the resulting general forms of the matric potential are:

$$\psi(S) = \psi_1(pS + q)^{m(1-\beta)} + k_1 \quad (m\beta \neq -2) \tag{5.1}$$

$$\psi(S) = \psi_2 \exp(k(1 - \beta)S) + k_2 \tag{5.2}$$

where

$$\psi_1 = (1 + m\beta)/c_1 m(1 - \beta) \tag{5.3}$$

$$\psi_2 = A/c_2 k(1 - \beta) \tag{5.4}$$

and k_1, k_2 are constants to be determined by boundary conditions. The expressions in Eqs 5.1 and 5.2 are consistent with the Lie group of the Richards equation. (Note that $m \neq 0$, $k \neq 0$ and $\beta \neq 1$ in Eqs 5.1 and 5.2). If $\beta = 1$, then the resulting $\psi(S)$ will show a logarithmic dependence on S instead of the power-law dependence in Eq. 5.1, or it will be proportional to S instead of showing the exponential dependence in Eq. 5.2.

The $K(S)$ given by Eq. 4.20 has been studied by Sander *et al.* (1988) in conjunction with the choice $m = -2$, $\beta = 1$. This choice of the exponents leads to a matric potential of the form (Petit Bois, 1961, p. 2):

$$\psi(S) = (1/c_1)\ln\left(\frac{K_1 + K_2 S}{pS + q}\right) + c_3 \tag{5.5}$$

where c_3 is an arbitrary constant and $c_0 = 0$. The choice $\beta = 1$, however, corresponds to a stretching transformation of time only, as shown in Table 1.

Equations 4.13, 4.19 and 5.1 have a power-law form that has often been applied and even recommended to model the dependence of the transport coefficients or the matric potential on the water content of unsaturated soil (Russo and Bresler, 1980; Parlange *et al.*, 1980; Rao *et al.*, 1983; Mualem, 1986; Gregson *et al.*, 1987; Sivapalan *et al.*, 1987; Broadbridge and White, 1988; Russo, 1988). Perhaps the best known special case of these expressions is the Brooks–Corey–Bresler–Russo (BCBR) model (Russo and Bresler, 1980):

$$D(S) = -(K_0\psi_0/\beta_{\mathrm{RB}})S^{(\eta_{\mathrm{RB}} - \beta_{\mathrm{RB}} - 1)/\beta_{\mathrm{RB}}} \tag{5.6a}$$

$$K(S) = K_0 S^{\eta_{\mathrm{RB}}/\beta_{\mathrm{RB}}} \tag{5.6b}$$

$$\psi(S) = \psi_0 S^{-1/\beta_{\mathrm{RB}}} \tag{5.6c}$$

where

$$\beta_{\mathrm{RB}} = -1/m(1 - \beta) \tag{5.7a}$$

$$\eta_{\mathrm{RB}} = -(1 + m\beta)/m(1 - \beta) \tag{5.7b}$$

$$\psi_0 = \psi_1 p^{m(1-\beta)} \tag{5.7c}$$

$$K_0 = c_1 p^{1+\beta m} \tag{5.7d}$$

and $q = k_1 \equiv 0$. Since the exponents β_{RB} and η_{RB} must be positive in the BCBR model, $m > 0$, $\beta > 1$, or $m < 0$, $\beta < 1$. Russo (1988) has discussed parameter estimation in the BCBR model and compared Eq. 5.6 with other model expressions for soil water properties.

The exponential dependence on S featured in Eqs 4.14, 4.21, and 5.2 has not been applied as extensively as the power-law dependence, but Nielsen *et al.* (1973) and Simmons *et al.* (1979) have shown that an exponential model gave a good description of soil water parameters in field plots established at a 150 ha site on soil of variable texture and infiltration rate.

Sander *et al.* (1988a,b) and Kühnel *et al.* (1989) have discussed and applied the expressions for $D(S)$, $K(S)$, and $\psi(S)$ in Eqs 4.13, 4.20 and 5.5 in the special case:

$$D(S) = D_0(1 - vS)^{-2} \tag{5.8a}$$

$$K(S) = \frac{K_s(1 - v)S}{(1 - vS)} \tag{5.8b}$$

$$\psi(S) = \psi_e + [D_0/K_s(1 - v)] \ln [(1 - v)S/(1 - vS)] \tag{5.8c}$$

where $m = -2$, $\beta = 1$, $c_0 = 0$, $K_1 = 0$,

$$D_0 = 1/q^2 \tag{5.9a}$$

$$v = -p/q \tag{5.9b}$$

$$K_s = c_1/q(q + p) = c_1/qK_2 \tag{5.9c}$$

$$\psi_e = c_3 \tag{5.9d}$$

The model expressions in Eqs 5.8 lead to an exactly solvable Richards equation (under flux boundary conditions) with physically reasonable results for the moisture profile as well as the S-dependence of D, K and ψ.

The Lie group of the Richards' equation as epitomized in Table 1 exhibits only stretching transformations and translations as symmetry operations, and these only if the transport coefficients show either power-law or exponential dependence on the degree of saturation. This result contrasts with the Lie group of the non-linear diffusion equation (Bluman and Cole, 1974, p. 297; Ovsiannikov, 1982, pp. 68–73), which admits more symmetry operations and permits stretching transformations and translations for *arbitrary* $D(S)$. Evidently the presence in the Richards equation of the 'gravitational term' involving $\partial K/\partial S$ reduces the number of symmetry operations and mathematical forms of $D(S)$ consistent with Lie group invariance. The only symmetry operation permitted for arbitrary $D(S)$ is simple translation (the case $b \equiv 0$ in Eq. 4.6 and Table 1).

A comparison of Eq. 2.9 with the second row of transformations in Table 1 yields the relationships:

$$\alpha = \exp[-2b(\beta - 1)\varepsilon] \tag{5.10a}$$

$$\omega = \exp(-b\beta\varepsilon) \tag{5.10b}$$

which connect the Warrick similarity transformation with the Lie group of the Richards equation. The corresponding relationships for $D(S)$ and $\psi(S)$ based on Eqs 2.5 and 2.7 are:

$$D(S') = \exp(2b\varepsilon)D(S) \tag{5.11a}$$

$$\psi(S') = \exp(-2b(\beta - 1)\varepsilon)\psi(S) \tag{5.11b}$$

The transformation of $K(S)$ depends on that of S, which, in turn, depends on the explicit mathematical form of $D(S)$. Using Eqs 3.7a and 4.15, for example, one derives the result:

$$\begin{aligned} S' &= \exp(2b\varepsilon/m)S + (q/p)[\exp(2b\varepsilon/m) - 1] \\ &= (\alpha/\omega^2)^{1/m}S + (q/p)[(\alpha/\omega^2) - 1]^{1/m} \end{aligned} \tag{5.12}$$

for a power-law $D(S)$. Similarly, using Eqs 3.7a and 4.16, one finds the transformation:

$$S' = S + (2b\varepsilon/k) \tag{5.13}$$

It follows from Eq. 2.10 that:

$$K(S') = \exp\left[2b\left(\beta + \frac{1}{m}\right)\varepsilon\right]K(S) \tag{5.14a}$$

for a power-law $K(S)$ and

$$K(S') = \exp(2b\beta\varepsilon)K(S) \tag{5.14b}$$

for an exponential $K(S)$.

Equations 5.12 and 5.13 ensure that Eqs 5.11 and 5.14 are satisfied when Eq. 1.3 is transformed into Eq. 2.12. Taking Eq. 5.12 as an example, one may use the results in Table 1 to write Eq. 1.3 in the form:

$$\frac{\exp[-2b(2\beta-1)\varepsilon]}{\exp(2b\varepsilon/m)}\frac{\partial S'}{\partial t'} = \frac{\exp[-4b(\beta-1)\varepsilon]}{\exp(2b\varepsilon/m)}\frac{\partial}{\partial z'}\left[D(S)\frac{\partial S'}{\partial z'}\right] - \frac{\exp(-2b(\beta-1)\varepsilon)}{\exp(2b\varepsilon/m)}\cdot\frac{\partial K}{\partial S}\frac{\partial S'}{\partial z'} \tag{5.15}$$

which reduces to Eq. 2.12 after division by the ratio of exponentials on the left-hand side, reference to Eq. 5.10, and incorporation of the result

$$\frac{\partial K(S')}{\partial S'} = \exp(2b\beta\varepsilon)\frac{\partial K(S)}{\partial S} \tag{5.16}$$

which follows from Eqs 4.17 and 5.12. The invariance of Eq. 1.3 under the stretching transformations in Table 1 and the transformation of S in Eq. 5.13 is demonstrated in the same fashion.

Generalized Warrick similitude as epitomized in Eq. 2.9 features three scaling parameters, α, ω and μ. Equations 5.8 lead to a relationship between the first two scaling parameters:

$$\alpha = \omega^{2(1-1/\beta)} \tag{5.17}$$

with the parameter β not equal to 1 or 1/2, since stretching transformations of both t and z are considered (see Table 1). Equation 5.17 implies that equality between α and ω does not exist except in the special case $\beta = 2$, corresponding to rather particular examples of Eqs 4.19–4.21 that cannot be expected to hold for unsaturated soils in general. The stretching transformations under which the Richards equation is invariant require two parameters, b (or $b\varepsilon$) and β, and these parameters do not permit a global equality between the scale factors for the hydraulic conductivity and matric potential. Similar conclusions, but from a different perspective, have been drawn by Sposito and Jury (1985) and Jury *et al.* (1987).

6. Conclusions

The Lie group of the Richards equation comprises only stretching transformations and translations. This relative lack of symmetry contrasts with the larger set of transformations under which the non-linear and linear diffusion equations are invariant in form. The cause of the reduction in symmetry for the Richards equation is the 'gravitational' term in $\partial K/\partial S$. Indeed, if this term were not present in Eq. 3.9, then the right-hand side of Eq. 4.8 would vanish identically, the conditions in Eqs 4.9 and 4.10 would not be imposed, and the Richards' equation would be invariant under stretching transformations for arbitrary $D(S)$ and $K(S)$ with $\zeta(S) \equiv 0$. This invariance would correspond to Warrick similitude in its original form (Sposito and Jury, 1985).

The present study has also shown that the Richards equation will be invariant under arbitrary stretching transformations only if the dependence of the soil water diffusivity and hydraulic conductivity on water content is of a power-law form or an exponential form, as in Eqs 4.13 and 4.19 or Eqs 4.14 and 4.21. Therefore, generalized Warrick similitude for the Richards equation is possible, not for an arbitrary $D(S)$ and $K(S)$, but only for two special mathematical forms. One of them—the exponential—is the basis for a successful scaling of the Richards equation known as Nielsen similitude (Sposito and Jury, 1985).

Experimental tests of Warrick similitude with power-law transport coefficients have been summarized by Sposito and Jury (1985) and Jury *et al.* (1987). In some cases the scaling parameters, α and ω, have been found to be equal or highly correlated in field studies, and in other cases not. Equation 5.17 makes clear the point that both results are consistent with the scaling invariance of the Richards equation. A lack of invariance under scaling is expected only if $D(S)$ and $K(S)$ show neither a power-law nor an exponential dependence on S. If the transport coefficients do not have one of these mathematical forms, scaling of the Richards equation in heterogeneous porous media is not possible.

Acknowledgements

The research reported in this chapter was supported in part by NSF grant no. ECE-8513726. Gratitude is expressed to Ms Terri DeLuca for her excellent typing of the manuscript.

Appendix

Hill (1982, Section 7.1) has derived equations relating partial derivatives of S' and S under the infinitesimal transformations prescribed by Eq. 3.3. For the three partial derivatives in Eq. 3.5, these equations are:

$$\frac{\partial S'}{\partial t'} = \frac{\partial S}{\partial t} + \varepsilon\left[\frac{\partial \zeta}{\partial t} + \left(\frac{\partial \zeta}{\partial S} - \frac{\partial \eta}{\partial t}\right)\frac{\partial S}{\partial t} - \frac{\partial \xi}{\partial t}\frac{\partial S}{\partial z} - \frac{\partial \eta}{\partial S}\left(\frac{\partial S}{\partial t}\right)^2 - \frac{\partial \xi}{\partial S}\frac{\partial S}{\partial z}\frac{\partial S}{\partial t}\right] \quad \text{(A1)}$$

$$\frac{\partial S'}{\partial z'} = \frac{\partial S}{\partial z} + \varepsilon\left[\frac{\partial \zeta}{\partial z} + \left(\frac{\partial \zeta}{\partial S} - \frac{\partial \xi}{\partial z}\right)\frac{\partial S}{\partial z} - \frac{\partial \eta}{\partial z}\frac{\partial S}{\partial t} - \frac{\partial \xi}{\partial S}\left(\frac{\partial S}{\partial z}\right)^2 - \frac{\partial \eta}{\partial S}\frac{\partial S}{\partial t}\frac{\partial S}{\partial z}\right] \tag{A2}$$

$$\begin{aligned}\frac{\partial^2 S'}{\partial z'^2} = \frac{\partial^2 S}{\partial z^2} + \varepsilon\Bigg[& \frac{\partial^2 \zeta}{\partial z^2} + \left(2\frac{\partial^2 \xi}{\partial z \partial S} - \frac{\partial^2 \xi}{\partial z^2}\right)\frac{\partial S}{\partial z} - \frac{\partial^2 \eta}{\partial z^2}\frac{\partial S}{\partial t} \\ & + \left(\frac{\partial^2 \zeta}{\partial S^2} - 2\frac{\partial^2 \xi}{\partial z \partial S}\right)\left(\frac{\partial S}{\partial z}\right)^2 - 2\frac{\partial^2 \eta}{\partial z \partial S}\frac{\partial S}{\partial t}\frac{\partial S}{\partial z} \\ & - \frac{\partial^2 \xi}{\partial S^2}\left(\frac{\partial S}{\partial z}\right)^3 - \frac{\partial^2 \eta}{\partial S^2}\frac{\partial S}{\partial t}\left(\frac{\partial S}{\partial z}\right)^2 + \Bigg[\left(\frac{\partial \zeta}{\partial S} - 2\frac{\partial \xi}{\partial z}\right) \\ & - 3\frac{\partial \xi}{\partial S}\frac{\partial S}{\partial z} - \frac{\partial \eta}{\partial S}\frac{\partial S}{\partial t}\Bigg]\frac{\partial^2 S}{\partial z^2} - 2\left(\frac{\partial \eta}{\partial z} + \frac{\partial \eta}{\partial S}\frac{\partial S}{\partial z}\right)\frac{\partial^2 S}{\partial z \partial t}\Bigg]\end{aligned} \tag{A3}$$

Equation A3 is simplified by using the Richards equation (Eq. 3.5 in unprimed form) to eliminate $\partial^2 S/\partial z^2$ in favour of $\partial S/\partial t$ and $\partial S/\partial z$.

The combination of Eqs 3.5 and 3.6 and A1–A3 yields the primed Richards equation as equal to the sum of the unprimed Richards equation and a complicated set of terms whose coefficient is ε. (Terms having coefficients in ε^2 are dropped.) This set of terms must vanish if Lie group invariance is to hold:

$$\begin{aligned}&\left[-\frac{\partial \zeta}{\partial t} + D(S)\frac{\partial^2 \zeta}{\partial z^2} - \frac{\partial K}{\partial S}\frac{\partial \zeta}{\partial z}\right] + \frac{\partial S}{\partial t}\left[\frac{\partial \eta}{\partial t} - D(S)\frac{\partial^2 \eta}{\partial z^2} + \frac{\partial K}{\partial S}\frac{\partial \eta}{\partial z}\right. \\ &\left. - 2\frac{\partial \xi}{\partial z} + \zeta\frac{\partial D/\partial S}{D(S)}\right] + \frac{\partial S}{\partial z}\left[\frac{\partial \xi}{\partial t} - D(S)\frac{\partial^2 \xi}{\partial z^2} - \frac{\partial K}{\partial S}\frac{\partial \xi}{\partial z}\right. \\ &\left. + \zeta\left(\frac{\partial K}{\partial S}\frac{\partial D/\partial S}{D(S)} - \frac{\partial^2 K}{\partial S^2}\right) + 2\frac{\partial D}{\partial S}\frac{\partial \zeta}{\partial z} + 2D(S)\frac{\partial^2 \zeta}{\partial z \partial S}\right] \\ &+ \left(\frac{\partial S}{\partial z}\right)^2\left\{D(S)\frac{\partial^2 \zeta}{\partial S^2} + \frac{\partial D}{\partial S}\frac{\partial \zeta}{\partial S} + \left[\frac{\partial^2 D}{\partial S^2} - \frac{(\partial D/\partial S)^2}{D(S)}\right]\zeta - 2\frac{\partial K}{\partial S}\frac{\partial \xi}{\partial S}\right\} \\ &+ 2\frac{\partial S}{\partial z}\frac{\partial S}{\partial t}\left[\frac{\partial \xi}{\partial S} - \frac{\partial D}{\partial S}\frac{\partial \eta}{\partial z} - D(S)\frac{\partial^2 \eta}{\partial z \partial S}\right] \\ &- \left(\frac{\partial S}{\partial z}\right)^2\frac{\partial S}{\partial t}\left[D(S)\frac{\partial^2 \eta}{\partial S^2} - 2\frac{\partial D}{\partial S}\frac{\partial \eta}{\partial S}\right] \\ &+ \left(\frac{\partial S}{\partial z}\right)^3\left[-D(S)\frac{\partial^2 \xi}{\partial S^2} + \frac{\partial D}{\partial S}\frac{\partial \xi}{\partial S}\right] + 2\left(\frac{\partial^2 S}{\partial z \partial t}\frac{\partial S}{\partial z}\frac{\partial \eta}{\partial S} - \frac{\partial^2 S}{\partial z \partial t}\frac{\partial \eta}{\partial z}\right) \equiv 0\end{aligned} \tag{A4}$$

The identity above cannot be satisfied in general unless each of the terms in brackets, as well as the last two terms on the left-hand side, individually vanish. When this requirement is imposed on the terms in $\partial^2 S/\partial z \partial t$ and $(\partial S/\partial z)(\partial S/\partial t)$, Eq. 3.8 results. Equation 3.9 follows from the vanishing of the term in $\partial S/\partial z$; Eq. 3.10 follows from Eq. 3.8 and the vanishing of the term in $\partial S/\partial t$; Eq. 3.11 follows from Eqs 3.8, 3.10, and the vanishing of the term in $(\partial S/\partial z)^2$, and Eq. 3.12 follows from the vanishing of the first term on the left-hand side of Eq. A4. All other terms in Eq. A4 are redundant.

References

Bluman, G.W. and Cole, J.D. (1974). 'Similarity Methods for Differential Equations'. Springer-Verlag, New York.

Bluman, G.W. and Kumei, S. (1980). On the remarkable nonlinear diffusion equation $(\partial/\partial x)(a(u+b)^{-2}(\partial u/\partial x)) - (\partial u/\partial t) = 0$, *J. Math. Phys.* **21**, 1019–1023.

Broadbridge, P. (1988). Integrable forms of the one-dimensional flow equation for unsaturated heterogeneous porous media, *J. Math. Phys.* **29**, 622–627.

Broadbridge, P. and White, I. (1988). Constant rate rainfall infiltration: A versatile nonlinear model. 1. Analytic solution, *Water Resour. Res.* **24**, 145–154.

Danielson, R.E. and Sutherland, P.L. (1986). Porosity, *Agronomy* **9**, 443–461.

Dickson, L.E. (1924). Differential equations from the group standpoint, *Ann. Math.* **25**, 287–378.

Gregson, K., Hector, D.J. and McGowan, M. (1987). A one-parameter model for the soil water characteristic, *J. Soil Sci.* **38**, 483–486.

Gupta, V.K., Sposito, G. and Bhattacharya, R.N. (1977). Toward an analytical theory of water flow through inhomogeneous porous media, *Water Resour. Res.* **13**, 208–210.

Hill, J.M. (1982). 'Solution of Differential Equations by Means of One-Parameter Groups'. Pitman, Boston.

Hillel, D. (1980). 'Fundamentals of Soil Physics'. Academic Press, New York.

Ince, E.L. (1956). 'Ordinary Differential Equations'. Dover, New York.

Jury, W.A., Russo, D. and Sposito, G. (1987). The spatial variability of water and solute transport properties in unsaturated soil. II. Scaling models of water transport, *Hilgardia* **55**, 33–56.

Kühnel, V., Dooge, J.C.I., Sander, G.C. and O'Kane, J.P.J. (1989). Duration of atmosphere-controlled and of soil-controlled phases of infiltration for constant rainfall at a soil surface. *Ann. Geophys.* (in press).

Miller, E.E. (1980). Similitude and scaling of soil-water phenomena. *In* 'Applications of Soil Physics (D. Hillel, ed.), pp. 300–318. Academic Press, New York.

Mualem, Y. (1986). Hydraulic conductivity of unsaturated soils: prediction and formulas. *In* 'Methods of Soil Analysis', Part One. *Agronomy* **9**, 799–823.

Nielsen, D.R., Biggar, J.W. and Erh, K.T. (1973). Spatial variability of field-measured soil-water properties, *Hilgardia* **42**, 215–260.

Olver, P.J. (1986). 'Applications of Lie Groups to Differential Equations'. Springer-Verlag, New York.

Ovsiannikov, L.V. (1982). 'Group Analysis of Differential Equations'. Academic Press, New York.

Parlange, J.-Y. (1980). Water transport in soils, *Ann. Rev. Fluid Mech.* **12**, 77–102.

Parlange, J.-Y. and Fleming, J.F. (1984). First integrals of the infiltration equation: 1, *Theory. Soil Sci.* **137**, 391–394.

Parlange, J.-Y., Braddock, R.D. and Chu, B.T. (1980). First integrals of the diffusion equation: An extension of the Fujita solutions, *Soil Sci. Soc. Am. J.* **44**, 908–911.

Petit Bois, G. (1969). 'Tables of Indefinite Integrals'. Dover, New York.

Philip, J.R. (1969). Theory of Infiltration, *Adv. Hydrosci.* **5**, 215–290.

Rao, P.S.C., Jessup, R.E., Hornsby, A.C., Cassel, D.K. and Pollans, W.A. (1983). Scaling soil microhydrologic properties of Lakeland and Konowa soils using similar media concepts. *Agric. Water Manag.* **6**, 277–290.

Richards, L.A. (1931). Capillary conduction of liquids through porous mediums, *Physics* **1**, 318–333.

Rosen, G. (1982). Method for the exact solution of a nonlinear diffusion-convection equation, *Phys. Rev. Lett.* **49**, 1844–1847.

Russo, D. (1988). Determining soil hydraulic properties by parameter estimation: On the solution of a model for the hydraulic properties, *Water Resour. Res.* **24**, 453–459.

Russo, D. and Bresler, E. (1980). Scaling soil hydraulic properties of a heterogeneous field, *Soil Sci. Soc. Am. J.* **44**, 681–684.

Sander, G.C., Parlange, J.-Y., Kühnel, V., Hogarth, W.L., Lockington, D.L. and O'Kane, J.P.J. (1988a). Exact nonlinear solution for constant flux infiltration, *J. Hydrol.* **97**, 341–346.

Sander, G.C., Parlange, J.-Y. and Hogarth, W.L. (1988b). Extension of the Fujita solution to air and water movement in soils, *Water Resour. Res.* **24**, 1187–1191.

Simmons, C.S., Nielsen, D.R. and Biggar, J.W. (1979). Scaling of field-measured soil-water properties, *Hilgardia* **47**, 77–174.

Sivapalan, M., Beven, K. and Wood, E.F. (1987). On hydrologic similarity. 2. A scaled model of storm runoff production, *Water Resour. Res.* **23**, 2266–2278.

Sposito, G. (1981). 'The Thermodynamics of Soil Solutions'. Clarendon Press, Oxford.

Sposito, G. (1986). The 'physics' of soil water physics, *Water Resour. Res.* **22S**, 83–88. Reprinted in 1987 with corrections in 'History of Geophysics' (E.R. Landa and S. Ince, eds), Vol. 3, pp. 93–98. American Geophysical Union, Washington, D.C.

Sposito, G. and Chu, S.-Y. (1982). Internal energy balance and the Richards equation, *Soil. Sci. Soc. Am. J.* **46**, 889–893.

Sposito, G. and Jury, W.A. (1985). Inspectional analysis in the theory of water flow through unsaturated soil, *Soil Sci. Soc. Am. J.* **49**, 791–798.

Warrick, A.W. and Amoozegar-Fard, A. (1979). Infiltration and drainage calculations using spatially scaled hydraulic properties, *Water Resour. Res.* **15**, 1116–1120.

Warrick, A.W. and Nielsen, D.R. (1980). Spatial variability of soil physical properties in the field. *In* 'Applications of Soil Physics' (D. Hillel, ed.), pp. 319–344. Academic Press, New York.

Warrick, A.W., Mullen, G.W. and Nielsen, D.R. (1977). Scaling field-measured soil hydraulic properties using a similar media concept, *Water Resour. Res.* **13**, 355–362.

Warrick, A.W., Lomen, D.O. and Yates, S.R. (1985). A generalized solution to infiltration, *Soil Sci. Soc. Am. J.* **49**, 34–38.

Chapter XIII

Quasilinear Analysis, Universal Scaling, and Lagrangian Simulation of Dispersion in Complex Geological Media

Shlomo P. Neuman, You-Kuan Zhang and Oscar Levin
Department of Hydrology and Water Resources, University of Arizona, Tucson, AZ 85721, USA

1. Introduction

During the last two decades there has been a growing recognition that chemical transport in geological media is strongly influenced by spatial variations in hydraulic conductivity. Such variations produce fluctuations in the subsurface fluid velocity which in turn cause dissolved chemicals to disperse over distances considerably greater than those normally observed in laboratory column experiments. A plausible explanation is that, contrary to traditional Fickian theory (e.g. Bear and Verruijt, 1987), the rates at which dissolved chemicals spread are not constant but vary with time or travel distance. Evidence that an individual plume may spread at variable rates for a period of over 380 hours is provided by the tracer experiment of Peaudecerf and Sauty (1978) in France. That the same can happen over a much longer period of at least three years in a relatively homogeneous geological medium is vividly demonstrated by the celebrated tracer experiment at Borden, Ontario, Canada (cf. Freyberg, 1986). In both experiments the plumes appear to have reached asymptotically a constant rate of longitudinal spread with time, but the same does not seem to have happened with respect to transverse spread. In fact, we do not expect the rate of spread in any direction to remain constant indefinitely. As a plume grows, it inevitably encounters heterogeneities on scales larger than those already experienced and thus enters a new phase of non-Fickian, or time-dependent, dispersion. One manifestation of this non-Fickian behaviour is the apparent increase in longitudinal dispersivity with the scale of the study which has been pointed out by Lallemand-Barres and Peaudecerf (1978), Anderson (1979),

DYNAMICS OF FLUIDS IN HIERARCHICAL POROUS MEDIA
ISBN 0-12-200260-1

Pickens and Grisak (1981), Gelhar (1986), Arya *et al.* (1988), and Lake (1988, 1989). This scale effect has been interpreted by Arya *et al.* (1988) and Wheatcraft and Tyler (1988) to imply that field dispersivities have the fractal dimensions 1.12 and 1.3, respectively. A different interpretation of the same effect will be described below.

The issue of time-dependent dispersion is addressed in a number of recent articles by Dagan (1984, 1987, 1988), Güven and Molz (1986), Gelhar (1987), Koch and Brady (1987, 1988), Sposito and Barry (1987), Yeh (1987), Barry *et al.* (1988), Kitanidis (1988), Naff *et al.* (1988), Graham and McLaughlin (1989a, b), Rinaldo *et al.* (1989), Rubin (1989) and others. In this chapter we analyse the manner in which the ensemble mean concentration evolves during the pre-asymptotic or non-Fickian stage toward an asymptotic or Fickian regime in three-dimensional geological media. The ensemble mean is of interest as the most easily obtained estimate of the actual concentration, which, of course, cannot be predicted without knowing the hydraulic conductivity at each point in space. To estimate concentration within a finite sampling volume, the point ensemble concentration must be convolved over this volume with the characteristic function of the sampling instrument (e.g. Cushman, 1987).

Dagan followed Taylor's (1921) theory of diffusion by continuous motions to derive a non-linear integrodifferential equation for the spatial covariance of ensemble mean concentrations in a statistically homogeneous random field of mildly fluctuating log hydraulic conductivities. He then linearized this expression by disregarding deviations from the mean trajectory of plume 'particles' under the integral sign, a step that corresponds to the first iteration in a procedure suggested by Phythian (1975). Dagan further neglected local dispersion and thereby limited his analysis to an infinite Peclet number.

Neuman and Zhang (1989) and Zhang and Neuman (1989) recovered an important aspect of the non-linearity effect lost in Dagan's (1984, 1987, 1988) analysis by adopting Corrsin's conjecture in the manner of Lundgren and Pointin (1976), and then quasilinearizing the result. Corrsin's conjecture introduces the plume covariance under the integral sign and their quasi-linearization consists of replacing it by an 'effective Fickian covariance tensor' that grows linearly with time. This latter covariance is associated with an 'effective Fickian dispersion tensor' which is generally much larger than the local dispersion tensor. It is therefore associated with an 'effective Peclet number P' which is much smaller than that arising from local dispersion alone. A similar effective Peclet number can be used to recover higher-order effects in existing Fickian theories such as those of Gelhar and Axness (1983) and Neuman *et al.* (1987). Thus, while Dagan dealt with an infinite Peclet number, the theory of Neuman and Zhang includes P values that are quite small. This is doubly important when one considers that even traditional Peclet numbers associated with local dispersion are in reality finite and may be low where groundwater seepage velocity is very slow due to mild hydraulic gradients and/or low hydraulic conductivities.

Many geological media are structured and therefore anisotropic. In stochastic models of subsurface flow and transport it is common to postulate that log hydraulic conductivity is locally a scalar but possesses a covariance structure, the length scale of which depends on direction. There are indications that this postulate 'works' in some sedimentary materials (cf. Byers and Stephens, 1983; Hufschmied, 1985; Sudicky, 1986) as well as in some fractured crystalline rocks (Neuman, 1987; Neuman and Depner, 1988). In sedimentary environments the principal directions of statistical anisotropy tend to be parallel and normal to the bedding with a relatively short correlation scale in the normal direction. In fractured rocks the principal axes are controlled by the orientations of fracture planes and channels that develop along discontinuities or their intersections. The principal axes are therefore difficult to predict on the basis of fracture mapping alone yet may sometimes be determined by other means as described by Neuman and Depner (1988). These authors have used data from an experimental field site near Oracle, Arizona, to predict Fickian dispersivities in anisotropic fractured granites on the basis of a theory developed by Neuman *et al.* (1987). This theory, as well as earlier work by Gelhar and Axness (1983), suggests that anisotropy has an important effect on the spread of dissolved chemicals under Fickian conditions. Dagan (1988) has shown that the same is true under non-Fickian conditions for an infinite Peclet number when the medium is isotropic in a plane containing the mean velocity vector.

The quasilinear theory of Neuman and Zhang (1989) and Zhang and Neuman (1989) accounts for dispersion in geological media that have arbitrary degrees of statistical anisotropy. In this paper, we briefly introduce their theory and use it to illustrate how the dispersivity tensor $\boldsymbol{\alpha}$ and spatial covariance tensor $\mathbf{X}$ of ensemble mean concentrations in a plume having zero initial spread vary with dimensionless time (mean travel distance relative to correlation scale) t_μ and effective Peclet number P under various anisotropic conditions. The theory assumes that log hydraulic conductivities form a mildly fluctuating, statistically homogeneous field characterized by an anisotropic exponential covariance. This is sufficient to explain how a plume evolves in a non-Fickian fashion until it has covered all scales of heterogeneity present in the medium, and in a Fickian fashion thereafter. We use the Borden experiment in Ontario, Canada, as an example of such behaviour and as a means to validate the theory.

According to the quasilinear theory, non-Fickian conditions prevail as long as a plume keeps encountering new scales of heterogeneity while it spreads. Evidence that this actually happens is provided by the spatial scale effect according to which longitudinal dispersivities from laboratory and field tracer studies in porous and fractured media throughout the world grow with the scale of the study. We show how the observed scale effect can be interpreted to imply that: (a) the media have a self-similar log hydraulic conductivity structure with a mean fractal dimension $D \simeq 1.75$ and a universal semivariogram $\gamma(s) \simeq 0.017\sqrt{s}$; (b) fractured rocks and porous earth materials are self-similar world-wide; (c) local deviations from global self-similarity occur within discrete, statistically homogeneous geological units on a hierarchy of scales; (d) there is no unique representative elementary volume (REV) in any locale, the notion being limited at best to such homogeneous units on a succession of scales; (e) Fickian dispersion regimes develop at best intermittently as a plume spreads through similar homogeneous units on a hierarchy of scales; and (f) the scale parameter which controls dispersivities in a given transport model diminishes as the density of information about groundwater velocities increases.

We conclude the paper by pointing out that Eulerian grid methods such as finite-difference and finite-element are not well suited to the numerical simulation of non-Fickian plumes. We provide the blueprint for an alternative Lagrangian method based on a semidiscrete superposition of Gaussian miniplumes that grow according to the quasilinear theory described below.

2. Quasilinear theory

2.1 Field-scale dispersivity and spatial covariance tensors

Following Zhang and Neuman (1989) we define a system of Cartesian coordinates $\tilde{\mathbf{x}}$ in which $\tilde{x}_1$, $\tilde{x}_2$ and $\tilde{x}_3$ point in the principal directions of log hydraulic conductivity covariance. These coordinates are related to $\mathbf{x}$, a coordinate system in which x_1 is parallel to the ensemble mean seepage velocity $\boldsymbol{\mu}$, through

$$\tilde{\mathbf{x}} = \boldsymbol{\beta}\mathbf{x} \tag{2.1}$$

where $\boldsymbol{\beta}$ is a unitary matrix (of direction cosines), β_{ij} being the cosine of the angle formed by $\tilde{x}_i$ and x_j. The authors assume that the log hydraulic conductivities $Y = \ln K$ have an anisotropic exponential covariance given in the domain of $\tilde{\mathbf{x}}$ by

$$\rho_Y(\tilde{\mathbf{x}}) = \sigma_Y^2 \exp\left[-\left(\frac{\tilde{x}_1^2}{L_1^2} + \frac{\tilde{x}_2^2}{L_2^2} + \frac{\tilde{x}_3^2}{L_3^2}\right)^{\frac{1}{2}}\right] \tag{2.2}$$

where σ_Y^2 is the variance of Y and L_1, L_2 and L_3 are its integral scales in the principal directions of anisotropy. The covariance becomes isotropic when $\tilde{\mathbf{x}}$ is transformed according to

$$\mathbf{x}^* = \mathbf{L}^{-1}\tilde{\mathbf{x}} \tag{2.3}$$

where $\mathbf{L}$ is a diagonal matrix such that $\mathbf{L} = \operatorname{diag}(L_1, L_2, L_3)$.

Let **d** be an 'effective dispersion tensor' given by the sum of an 'effective non-local dispersion tensor' $\mathbf{d}_0$ and a local dispersion tensor $\mathbf{d}_l$. To avoid conceptual difficulties it is easiest to think of $\mathbf{d}_l$ as molecular diffusion coefficient though other definitions are possible. Both $\mathbf{d}_0$ and $\mathbf{d}_l$ are taken to be time-independent with eigenvectors parallel and normal to the velocity vector $\boldsymbol{\mu}$. This means that in the **x** coordinates **d** is diagonal with principal components (eigenvalues) d_1 parallel to x_1 and $\boldsymbol{\mu}$, d_2 parallel to x_2, and d_3 parallel to x_3. While $\mathbf{d}_l$ is assumed known, $\mathbf{d}_0$ is chosen so that $2\mathbf{d}_0 t$ constitutes a first-order approximation for the spatial covariance tensor **X** of ensemble mean concentrations in a plume having zero initial spread. The latter depends in a non-linear fashion on **d** and is therefore obtained by iteration (the theory is said to be quasilinear because **d** is a constant and hence the integrodifferential equation (Eq. 7) in Neuman and Zhang (1989) can be solved in closed form for **X**).

Define the directional effective Peclet number P and dimensionless time t_μ as

$$P = \frac{\mu L_\mu}{d_1}, \quad t_\mu = \frac{\mu t}{L_\mu} \tag{2.4}$$

where μ is the magnitude of $\boldsymbol{\mu}$,

$$L_\mu^{-2} = \boldsymbol{\beta}_1^{\mathrm{T}} \mathbf{L}^{-2} \boldsymbol{\beta}_1 \tag{2.5}$$

and $\boldsymbol{\beta}_1$ is a unit vector representing the direction cosines of $\boldsymbol{\mu}$ with respect to $\tilde{\mathbf{x}}$, superscript T indicating transpose. The scalar L_μ is the length of a radius vector parallel to $\boldsymbol{\mu}$ of an ellipsoid having semiaxes L_1, L_2 and L_3 oriented along $\tilde{x}_1$, $\tilde{x}_2$ and $\tilde{x}_3$, respectively. It represents the 'directional integral scale' of the covariance function $\rho_Y(\tilde{\mathbf{x}})$ parallel to the mean seepage velocity vector. The dispersivity tensor in the principal $\tilde{\mathbf{x}}$ coordinates is given by (Zhang and Neuman, 1989)

$$\tilde{\boldsymbol{\alpha}}(P, t_\mu) = \tilde{\boldsymbol{\alpha}}_\infty(P) - \tilde{\boldsymbol{\gamma}}(P, t_\mu) \tag{2.6}$$

where

$$\tilde{\boldsymbol{\alpha}}_\infty(P) = \frac{P q_2 q_3 L_\mu \sigma_Y^2}{4\pi \lambda_1 \lambda_2 \lambda_3} \int_0^{2\pi} \int_{-1}^{1} \frac{1}{R(1 + w)^2} \mathbf{G} \, \mathrm{d}u \, \mathrm{d}\psi \tag{2.7}$$

$$\tilde{\boldsymbol{\gamma}}(P, t_\mu) = \frac{P q_2 q_3 L_\mu \sigma_Y^2}{4\pi \lambda_1 \lambda_2 \lambda_3} \int_0^{2\pi} \int_{-1}^{1} (M_0 + M_1 + M_2) \mathbf{G} \, \mathrm{d}u \, \mathrm{d}\psi \tag{2.8}$$

$$M_0 = \frac{2}{R^2(1-w^2)}\sqrt{\left(\frac{t_\mu}{\pi P}\right)}\exp\left(-\frac{u^2 P t_\mu}{4}\right) - \frac{2w}{R(1-w^2)^2}\,\mathrm{erfc}\left[\frac{u\sqrt{(Pt_\mu)}}{2}\right] \tag{2.9a}$$

$$M_1 = \frac{[PR^2-(2+w)(1+w)t_\mu]}{2PR^3(1+w)^2}\exp\left[\frac{t_\mu}{PR^2}(1+w)\right]\mathrm{erfc}\left[\frac{2+w}{2R}\sqrt{\left(\frac{t_\mu}{P}\right)}\right] \tag{2.9b}$$

$$M_2 = \frac{[PR^2-(2-w)(1-w)t_\mu]}{2PR^3(1-w)^2}\exp\left[\frac{t_\mu}{PR^2}(1-w)\right]\mathrm{erfc}\left[\frac{2-w}{2R}\sqrt{\left(\frac{t_\mu}{P}\right)}\right] \tag{2.9c}$$

Included in these expressions are the diagonal tensor

$$\mathbf{q}^2 = d_1\mathbf{d}^{-1} \tag{2.10}$$

with elements $q_1 = 1, q_2 = \sqrt{(d_1/d_2)}$, and $q_3 = \sqrt{(d_1/d_3)}$, the coordinate vector

$$\boldsymbol{\sigma} = (u, \sqrt{(1-u^2)}\cos\psi, \sqrt{(1-u^2)}\sin\psi)^{\mathrm{T}} \tag{2.11}$$

and

$$\mathbf{Q} = \boldsymbol{\beta}\mathbf{q}\boldsymbol{\sigma}, \quad S^2 = \mathbf{Q}^{\mathrm{T}}\mathbf{Q} = \boldsymbol{\sigma}^{\mathrm{T}}\mathbf{q}^2\boldsymbol{\sigma} \tag{2.12a}$$

$$w = uPR, \quad R^2 = \mathbf{Q}^{\mathrm{T}}\boldsymbol{\lambda}^{-2}\mathbf{Q} = \boldsymbol{\sigma}^{\mathrm{T}}\mathbf{q}\boldsymbol{\beta}^{\mathrm{T}}\boldsymbol{\lambda}^{-2}\boldsymbol{\beta}\mathbf{q}\boldsymbol{\sigma} \tag{2.12b}$$

$$\boldsymbol{\lambda} = L_\mu\mathbf{L}^{-1} \tag{2.12c}$$

$$\mathbf{G} = \left(\beta_1 - \frac{\mathbf{Q}\sigma_1}{S^2}\right)\left(\beta_1^{\mathrm{T}} - \frac{\mathbf{Q}^{\mathrm{T}}\sigma_1}{S^2}\right) \tag{2.12d}$$

Note that if the medium is isotropic, one can always choose $\tilde{\mathbf{x}}$ to coincide with $\mathbf{x}$ such that $\boldsymbol{\beta} = \boldsymbol{\lambda} = \boldsymbol{\delta}$ (identity tensor) and $R^2 = S^2$.

It is obvious that $\tilde{\boldsymbol{\alpha}}$ is a symmetric second-rank tensor which is generally non-diagonal but becomes diagonal when the mean velocity is parallel to any of the three principal coordinates $\tilde{x}_1$, $\tilde{x}_2$ and $\tilde{x}_3$.

The spatial covariance $\tilde{\mathbf{X}}$ of a plume having initially zero spread is given by

$$\tilde{\mathbf{X}}(P, t_\mu) = \frac{Pq_2q_3L_\mu^2\sigma_Y^2}{2\pi\lambda_1\lambda_2\lambda_3}\int_0^{2\pi}\int_{-1}^{1}\left[\frac{t_\mu}{R(1+w)^2} + N_0 + N_1 + N_2\right]\mathbf{G}\,\mathrm{d}u\,\mathrm{d}\psi \tag{2.13}$$

where

$$N_0 = \frac{3w - 8w^2 + 6w^3 - w^5}{u(1 - w^2)^3} - \frac{2(3 + w^2)}{(1 - w^2)^2}\sqrt{\left(\frac{Pt_\mu}{\pi}\right)}\exp\left(-\frac{u^2 Pt_\mu}{4}\right)$$

$$+\left[\frac{2wt_\mu}{R(1 - w^2)^2} + \frac{8w^2}{u(1 - w^2)^3}\right]\operatorname{erfc}\left[\frac{u\sqrt{(Pt_\mu)}}{2}\right] \tag{2.14a}$$

$$N_1 = \frac{[u(1 + w)(2 + w)t_\mu - Rw(3 + w)]}{2Ru(1 + w)^3}\exp\left[\frac{t_\mu(1 + w)}{PR^2}\right]\operatorname{erfc}\left[\frac{2 + w}{2R}\sqrt{\left(\frac{t_\mu}{P}\right)}\right] \tag{2.14b}$$

$$N_2 = \frac{[u(1 - w)(2 - w)t_\mu - Rw(3 - w)]}{2Ru(1 - w)^3}\exp\left[\frac{t_\mu(1 - w)}{PR^2}\right]\operatorname{erfc}\left[\frac{2 - w}{2R}\sqrt{\left(\frac{t_\mu}{P}\right)}\right] \tag{2.14c}$$

Like the dispersivity $\tilde{\boldsymbol{\alpha}}$, the covariance $\tilde{\mathbf{X}}$ is a second-rank tensor that is generally non-diagonal.

2.2 Asymptotic expressions for field-scale dispersivity tensor

Zhang and Neuman (1989) developed asymptotic expressions for the dispersivity and spatial covariance tensors $\tilde{\boldsymbol{\alpha}}$ and $\tilde{\mathbf{X}}$. We present below only results for $\tilde{\boldsymbol{\alpha}}$ as corresponding expressions for $\tilde{\mathbf{X}}$ are easily obtained from

$$\mathbf{X}(P, t_\mu) = 2L_\mu \int_0^{t_\mu} \boldsymbol{\alpha}(P, t_\mu)\,\mathrm{d}t_\mu \tag{2.15}$$

During the early stages of plume development ($t_\mu \ll 1$ and $t_\mu \ll P$)

$$\tilde{\boldsymbol{\alpha}}(t_\mu) = \frac{t_\mu q_2 q_3 L_\mu \sigma_Y^2}{4\pi\lambda_1\lambda_2\lambda_3}\int_0^{2\pi}\int_{-1}^{1}\frac{1}{R^3}\mathbf{G}\,\mathrm{d}u\,\mathrm{d}\psi \tag{2.16}$$

which shows that dispersivity is linearly proportional to t_μ and independent of P.

For small effective Peclet numbers ($P \ll 1$ and $P \ll t_\mu$) one obtains

$$\tilde{\boldsymbol{\alpha}}(P) = \frac{P q_2 q_3 L_\mu \sigma_Y^2}{4\pi\lambda_1\lambda_2\lambda_3}\int_0^{2\pi}\int_{-1}^{1}\frac{1}{R}\mathbf{G}\,\mathrm{d}u\,\mathrm{d}\psi \tag{2.17}$$

which means that $\tilde{\boldsymbol{\alpha}}$ is linearly proportional to P and does not vary with time.

This agrees with an earlier finding by Neuman *et al.* (1987, for $\mathbf{d}_0 = 0$) that in anisotropic media $\tilde{\boldsymbol{\alpha}}$ is proportional to P at small Peclet numbers during asymptotic (i.e. Fickian dispersion). Here we see that the same holds true during the early stages of plume development whenever $P \ll 1$ and $P \ll t_\mu$.

The asymptotic dispersivity tensor for $t_\mu \to \infty$ is given by Eq. 2.7.

When $P \to \infty$ and t_μ is finite,

$$\tilde{\boldsymbol{\alpha}}(t_\mu) = \frac{t_\mu L_\mu \sigma_Y^2}{4\pi\lambda_1\lambda_2\lambda_3} \int_0^{2\pi} \int_{-1}^{1} \frac{1}{B^3} \exp\left(-\frac{ut_\mu}{B}\right) \mathbf{G}\, du\, d\psi \tag{2.18}$$

where $B^2 = \boldsymbol{\sigma}^{\mathrm{T}} \boldsymbol{\lambda}^{-2} \boldsymbol{\sigma}$. This is a generalization of Dagan's (1984, 1987, 1988) linearized solutions for isotropic media and a restricted class of anisotropic media at infinite Peclet number. When the medium is statistically isotropic $B = 1$.

As $t_\mu \to \infty$ and $P \to \infty$ the dispersivity tensor $\tilde{\boldsymbol{\alpha}}$ degenerates to a single non-zero component representing longitudinal dispersivity parallel to $\boldsymbol{\mu}$. The latter is given by

$$\alpha_L = L_\mu \sigma_Y^2 \tag{2.19}$$

3. Influence of effective Peclet number and anisotropy on dispersion and spatial covariance

In the remainder of this paper we simplify the notation by writing $\boldsymbol{\alpha}$ and $\mathbf{X}$ instead of $\tilde{\boldsymbol{\alpha}}$ and $\tilde{\mathbf{X}}$. Zhang and Neuman (1989) investigated the manner in which $\boldsymbol{\alpha}$ and $\mathbf{X}$ vary with dimensionless time t_μ, effective directional Peclet number P, and statistical anisotropy by evaluating Eqs 2.6 and 2.13 numerically with the aid of Gaussian quadrature. We present below some of their results in terms of dimensionless quantities $\boldsymbol{\alpha}_D$ and $\mathbf{X}_D$ defined, respectively, as

$$\boldsymbol{\alpha}_D = \frac{\boldsymbol{\alpha}}{L_\mu \sigma_Y^2} \tag{3.1}$$

$$\mathbf{X}_D = \frac{\mathbf{X}}{L_\mu^2 \sigma_Y^2} \tag{3.2}$$

We characterize the two tensors by their principal components in the $\tilde{\mathbf{x}}$ system of coordinates where $\tilde{x}_1$, $\tilde{x}_2$ and $\tilde{x}_3$ point in the principal directions of log hydraulic conductivity covariance. We shall see that when the mean seepage velocity $\boldsymbol{\mu}$ is inclined to the $\tilde{\mathbf{x}}$ axes the principal directions of $\boldsymbol{\alpha}$ and $\mathbf{X}$ generally differ from those of $\tilde{\mathbf{x}}$ and $\boldsymbol{\mu}$.

Since Eqs 2.6 and 2.13 are quasilinear, they are solved for given P by modifying q_2 and q_3 iteratively so as to satisfy Eq. 2.10, while insuring that $d_{01}/d_{02} = X_1/X_2$ and $d_{01}/d_{03} = X_1/X_3$. For simplicity, the results below correspond to $\mathbf{d}_1 = 0$ unless stated otherwise. In this case, there is a simple non-linear relationship between P and the log hydraulic conductivity variance σ_Y^2. We take d_1 to represent asymptotic longitudinal dispersion so that $d_1 = \alpha_{\infty L}\mu$. Hence $P = L_\mu/\alpha_{\infty L}$ by virtue of Eq. 2.4 and since $\alpha_{DL} = \alpha_L/L_\mu\sigma_Y^2$, it follows that

$$\alpha_{\infty DL}P = \frac{1}{\sigma_Y^2} \tag{3.3}$$

Neuman and Zhang (1989) show that in statistically isotropic media $P \gg 1$ implies $\alpha_{\infty DL} = 1$ so that $P = 1/\sigma_Y^2$; the condition $P \gg 1$ is thus equivalent to $\sigma_Y^2 \ll 1$. When $P \ll 1$ in isotropic media, $\alpha_{\infty DL}$ is proportional to P so that P is inversely proportional to σ_Y, implying that the condition $P \ll 1$ is equivalent to $\sigma_Y^2 \gg 1$. The quasilinear theory is nominally restricted to $\sigma_Y^2 < 1$ and one, therefore, cannot guarantee the validity of results corresponding to $P < 1$. However, arguments presented by the above authors and by Zhang and Neuman (1989) suggest that this restriction is less severe in the quasilinear than in the linear case and we, therefore, present results for arbitrary values of P.

It follows from the above that when $P \ll 1$ ($\sigma_Y^2 \gg 1$) in statistically isotropic media, the asymptotic dispersivity and plume covariance tensors $\boldsymbol{\alpha}_\infty$ and $\mathbf{X}_\infty$ are both proportional to σ_Y. On the other hand when $P \gg 1$ ($\sigma_Y^2 \ll 1$) then $\alpha_{\infty DL} = 1$ and hence P is inversely proportional to σ_Y^2. Therefore, the asymptotic longitudinal quantities $\alpha_{\infty L}$ and $X_{\infty L}$ are proportional to σ_Y^2 while the asymptotic transverse quantities $\alpha_{\infty T}$ and $X_{\infty T}$ are proportional to σ_Y^4 when $\sigma_Y^2 \ll 1$.

We start by examining the behavior of $\boldsymbol{\alpha}_D$ when the principal integral scales of log hydraulic conductivity covariance are related through $L_1/L_3 = L_2/L_3 = 10$. This corresponds to an imperfectly layered medium that is statistically isotropic parallel to the layers in the plane of $\tilde{x}_1$ and $\tilde{x}_2$. Figures 1, 2 and 3 show how the principal dimensionless dispersivities α_{Di} for $i = 1$, 2 and 3, respectively, vary with t_μ at various effective Peclet numbers P when the mean seepage velocity vector $\boldsymbol{\mu}$ is collinear with $\tilde{x}_1$ (i.e. parallel to the layers). Figure 1 shows that $\alpha_{\infty DL} = 1$ for $P \geqslant 10$ and $\alpha_{\infty DL} < 1$ for $P < 10$ so that $P \geqslant 10$ corresponds to $\sigma_Y^2 \leqslant 0.1$ and $P < 10$ to $\sigma_Y^2 > 0.1$. Analogously, $\alpha_{\infty DL} = 1$ when $\sigma_Y^2 \leqslant 0.1$ and $\alpha_{\infty DL} < 1$ when $\sigma_Y^2 > 0.1$. We mentioned earlier that in isotropic media $\alpha_{\infty DL}$ does not fall below 1 as long as $\sigma_Y^2 < 1$; in the linear theory of Dagan (1988) $\alpha_{\infty DL} = 1$ under both isotropic and anisotropic conditions. This means that non-linearity, which in isotropic media causes a reduction in dimensionless longitudinal dispersivity when $\sigma_Y^2 > 1$, does the same in an anisotropic medium when $\sigma_Y^2 > 0.1$.

Returning to Figs 1–3, we see that the dispersivities are linearly proportional to t_μ and independent of P when $t_\mu < 1$ and $t_\mu \ll P$ (equivalently, they are

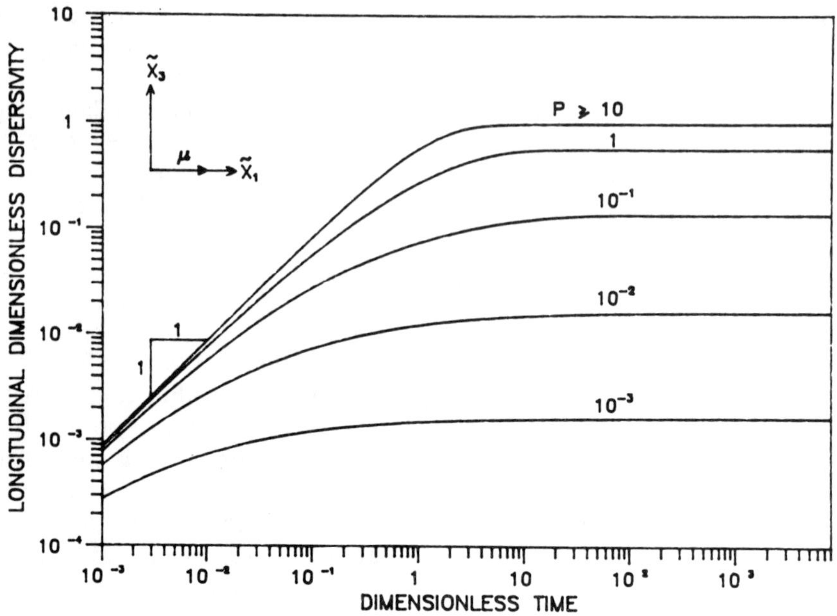

Figure 1. Longitudinal dimensionless dispersivity versus dimensionless time as a function of Peclet number when $L_1/L_3 = L_2/L_3 = 10$ and mean velocity is parallel to $\tilde{x}_1$.

independent of σ_Y^2 when $t_\mu < 1$ and $t_\mu \ll 1/\alpha_{\infty DL}\sigma_Y^2$). As t_μ increases the longitudinal dispersivities at all P (all σ_Y^2) and the transverse dispersivities at $P \ll 1$ ($\sigma_Y^2 \gg 1.7$ because $\alpha_{\infty DL} \cong 0.58$ when $P = 1$) increase monotonically toward their asymptotic Fickian limits which are attained no later than at about $t_\mu = 10$. At larger Peclet numbers ($P \geqslant 1$ or $\sigma_Y^2 \leqslant 1.7$) the transverse dispersivities climb to a peak before eventually descending to their asymptotic limits. These limits are attained at t_μ values that increase at a rate proportional to $\sqrt{P}$. If one adopts the linearized approximation of Dagan (1988), where $P \to \infty$ ($\sigma_Y^2 \to 0$), one finds that the transverse dispersivities decrease forever at a rate proportional to t_μ^{-2}.

The dimensionless transverse dispersivity along $\tilde{x}_3$ at any P (any σ_Y^2) and t_μ (Fig. 3) is larger than that along $\tilde{x}_2$ at the same P (or σ_Y^2) and t_μ (Fig. 2). As the integral scale of log hydraulic conductivity is smaller by a factor of 10 parallel to $\tilde{x}_3$ than it is parallel to $\tilde{x}_2$, the theory implies that spatial correlation has a negative effect on transverse dispersivity at all dimensionless times. A similar negative effect is implied for Fickian conditions by Gelhar and Axness (1983) and Neuman *et al.* (1987).

Next, we examine the effect of varying P and the anisotropy ratio $L_1/L_3 = L_2/L_3$ on the dimensionless covariance tensor $\mathbf{X}_D$ as a function of t_μ. Figure 4 shows how the dimensionless longitudinal variance X_{D1} increases with

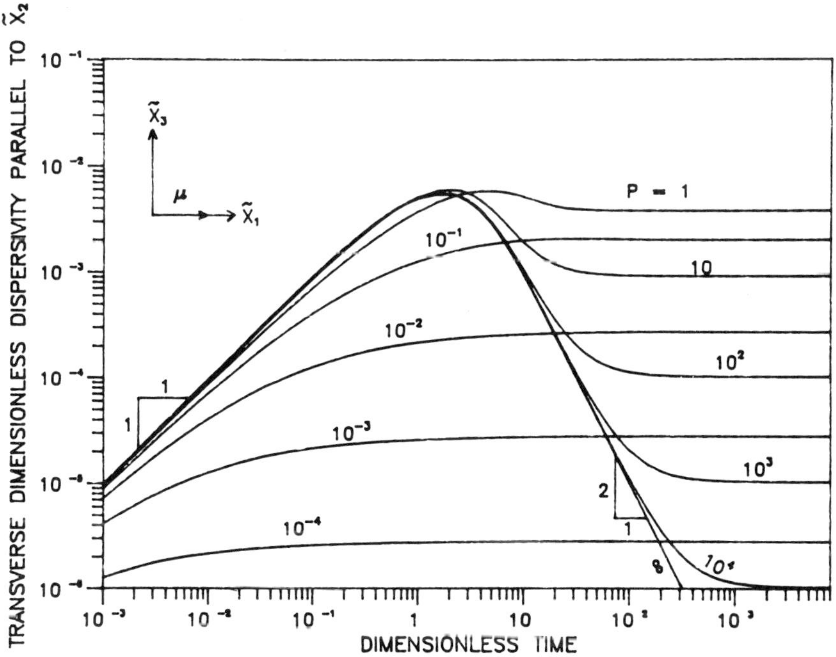

Figure 2. Transverse dimensionless dispersivity parallel to $\tilde{x}_2$ versus dimensionless time as a function of Peclet number when $L_1/L_3 = L_2/L_3 = 10$ and mean velocity is parallel to $\tilde{x}_1$.

t_μ when $P \geqslant 10$ and $L_1/L_3 = L_2/L_3 \leqslant 1$. The figure confirms an earlier finding by Dagan (1988) that in the linearized case where $P \to \infty$, longitudinal spread diminishes as $L_1/L_2 = L_2/L_3 \to 0$ and is therefore less in two-dimensional than in three-dimensional systems. We further learn from Fig. 4 that the same happens independently of Peclet number when $P \geqslant 10$ ($\sigma_Y^2 \leqslant 1$).

Figure 5 shows a similar graph for $L_1/L_3 = L_2/L_3 \geqslant 1$. The broken curves correspond to Dagan's (1988) linearized theory and suggest that a passage from isotropic to stratified conditions as $L_1/L_3 = L_2/L_3 \to \infty$ causes a slight increase in the longitudinal spread. The solid curves which represent a quasilinear solution for $P = 10$ show a very different behaviour: instead of increasing, X_{D1} decreases rapidly toward zero as $L_1/L_3 = L_2/L_3 \to \infty$. This means that as the medium becomes more stratified while mean flow takes place parallel to the bedding, the influence of field-scale velocity fluctuations on the longitudinal variance of a plume (having zero initial spread) gradually diminishes.

Zhang and Neuman (1988) showed that when $P \geqslant 10$ the dimensionless transverse variance X_{D2} in the isotropic plane of $\tilde{x}_1$ and $\tilde{x}_2$ goes up as $L_1/L_3 = L_2/L_3$ decreases below 1, and diminishes as this ratio increases

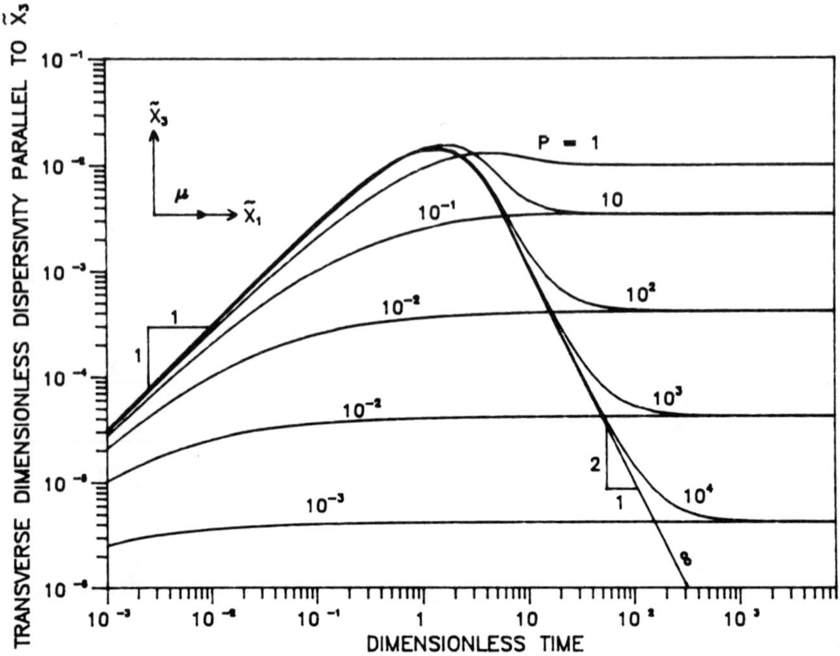

Figure 3. Transverse dimensionless dispersivity parallel to $\tilde{x}_3$ versus dimensionless time as a function of Peclet number when $L_1/L_3 = L_2/L_3 = 10$ and mean velocity is parallel to $\tilde{x}_1$.

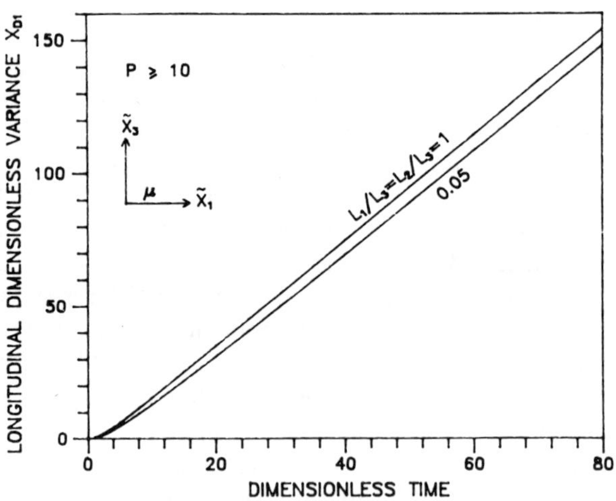

Figure 4. Longitudinal dimensionless variance X_{D1} versus dimensionless time for $P \geqslant 10$ when $L_1/L_3 = L_2/L_3 \leqslant 1$ and mean velocity is parallel to $\tilde{x}_1$.

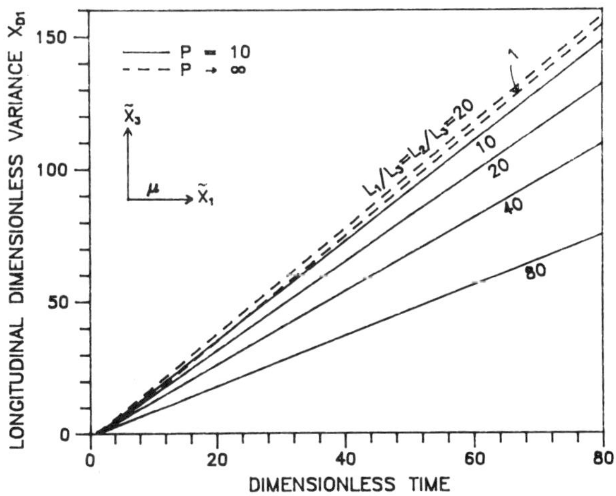

Figure 5. Longitudinal dimensionless variance X_{D1} versus dimensionless time for $P \geqslant 10$ when $L_1/L_3 = L_2/L_3 \geqslant 1$ and mean velocity is parallel to $\tilde{x}_1$.

above 1. This means that the more stratified a medium becomes, the less influence does field-scale differential flow exert on transverse spread parallel to the bedding when mean flow is in the plane of stratification and the plume initially has zero spread. Like in the case of X_{D1} the clue to understanding this phenomenon lies in the reduced ability of the plume to spread in a direction normal to the layers. Indeed, the dimensionless transverse variance X_{D3} in the plane of $\tilde{x}_1$ and $\tilde{x}_3$ decreases with anisotropy contrast when $P = 10$ regardless of whether $L_1/L_3 = L_2/L_3$ is less or greater than 1. In the limit as $L_1/L_3 = L_2/L_3$ tends either to zero (two-dimensional case) or infinity (perfect stratification), all field-scale influences on transverse spread in the direction of $\tilde{x}_3$ vanish.

Up to this point the mean velocity $\boldsymbol{\mu}$ has always been taken parallel to a principal axis of statistical anisotropy so that $\boldsymbol{\alpha}$ and $\mathbf{X}$ remained diagonal in the $\tilde{\mathbf{x}}$ coordinates. Under these conditions, field-scale longitudinal spread occurred parallel to $\boldsymbol{\mu}$ and transverse spread took place in the plane normal to $\boldsymbol{\mu}$. We now consider the more general situation where the mean seepage velocity vector $\boldsymbol{\mu}$ is inclined to the principal coordinates $\tilde{\mathbf{x}}$. We mentioned earlier that this generally causes the eigenvectors of $\boldsymbol{\alpha}_D$ and $\mathbf{X}_D$ to become non-collinear with both $\boldsymbol{\mu}$ and $\tilde{\mathbf{x}}$. Since we take the eigenvectors of the local dispersion tensor $\mathbf{d}_l$ to be parallel and normal to $\boldsymbol{\mu}$, we expect the eigenvectors of the combined (local and non-local) dispersion tensor

$$\mathbf{D} = \mathbf{d}_l + \alpha\mu \tag{3.3}$$

and the combined plume covariance tensor

$$\boldsymbol{\Sigma} = 2\mathbf{d}_l t + \mathbf{X} \tag{3.4}$$

to be generally non-collinear with $\boldsymbol{\mu}$ and $\tilde{\mathbf{x}}$ as well as with the eigenvectors of $\boldsymbol{\alpha}$ and $\mathbf{X}$. To study this effect, we examine the manner in which the eigenvector $\mathbf{e}_1$ of $\boldsymbol{\alpha}$ and $\mathbf{X}$, which corresponds to the largest (longitudinal) eigenvalues of these two tensors, and the eigenvector $\mathbf{E}_1$, which corresponds to the largest (longitudinal) eigenvalues of $\mathbf{D}$ and $\boldsymbol{\Sigma}$, vary with dimensionless time, directional effective Peclet number, and the ratios between L_1, L_2 and L_3. For simplicity we again set $L_1 = L_2$ and take $\boldsymbol{\mu}$ to lie in the $\tilde{x}_1$–$\tilde{x}_3$ plane so that it forms a 45° angle with both coordinates. We designate the angle between $\boldsymbol{\mu}$ and $\mathbf{e}_1$ by θ and that between $\boldsymbol{\mu}$ and $\mathbf{E}_1$ by Θ. Both angles are in the $\tilde{x}_1$–$\tilde{x}_3$ plane and we take them to be positive when between $\boldsymbol{\mu}$ and $\tilde{x}_3$, negative when between $\boldsymbol{\mu}$ and $\tilde{x}_1$.

Figure 6 shows how θ and Θ vary with t_μ for various anisotropy ratios $L_1/L_3 = L_2/L_3$ when $P = 10$. The angle θ is independent of local dispersion and remains constant at early t_μ as indicated by the broken curves. The angle Θ depends on the local directional Peclet number

$$P_1 = \frac{\mu L_\mu}{d_{1L}} \tag{3.5}$$

defined with respect to the local longitudinal dispersion coefficient d_{1L}. When P_1 is finite (a case illustrated here for a value of 100 and $\sigma_Y^2 = 0.5$), Θ is initially zero due to the dominance of local dispersion at early dimensionless time and our assumption that the eigenvectors of $\mathbf{d}_1$ are parallel and normal to $\boldsymbol{\mu}$. As dimensionless time progresses, Θ gradually increases toward θ as illustrated by the solid curves. This growth in Θ is equivalent to a rotation of the longitudinal plume axis away from its initial direction parallel to $\boldsymbol{\mu}$, toward the direction of largest correlation, due to the growing influence of non-local dispersion as t_μ increases. Zhang and Neuman (1989) have shown that the rate of this rotation and the dimensionless time at which it starts (as well as ends) increase in a manner inversely proportional to P_1. At later t_μ the two angles are identical and jointly follow the solid curves. As t_μ approaches 1, the two angles diminish toward their Fickian limits which are reached in the vicinity of $t_\mu = 10$ and depend on the effective Peclet number P. This is equivalent to a rotation of the longitudinal plume axis back toward $\boldsymbol{\mu}$.

The angles θ and Θ are seen to be positive when $L_1/L_3 = L_2/L_3 < 1$ so that both $\mathbf{e}_1$ and $\mathbf{E}_1$ are offset from $\boldsymbol{\mu}$ toward the direction of largest log hydraulic conductivity correlation parallel to $\tilde{x}_3$. When $L_1/L_3 = L_2/L_3 > 1$, the angles are negative which again means that $\mathbf{e}_1$ and $\mathbf{E}_1$ are offset toward the direction of largest correlation, this time parallel to $\tilde{x}_1$. This complements the finding of Neuman *et al.* (1987) that longitudinal dispersion is generally offset from $\boldsymbol{\mu}$ toward the direction of largest log hydraulic conductivity correlation under Fickian conditions. We now see, however, that the phenomenon is much more pronounced under non-Fickian than under Fickian conditions. Zhang and

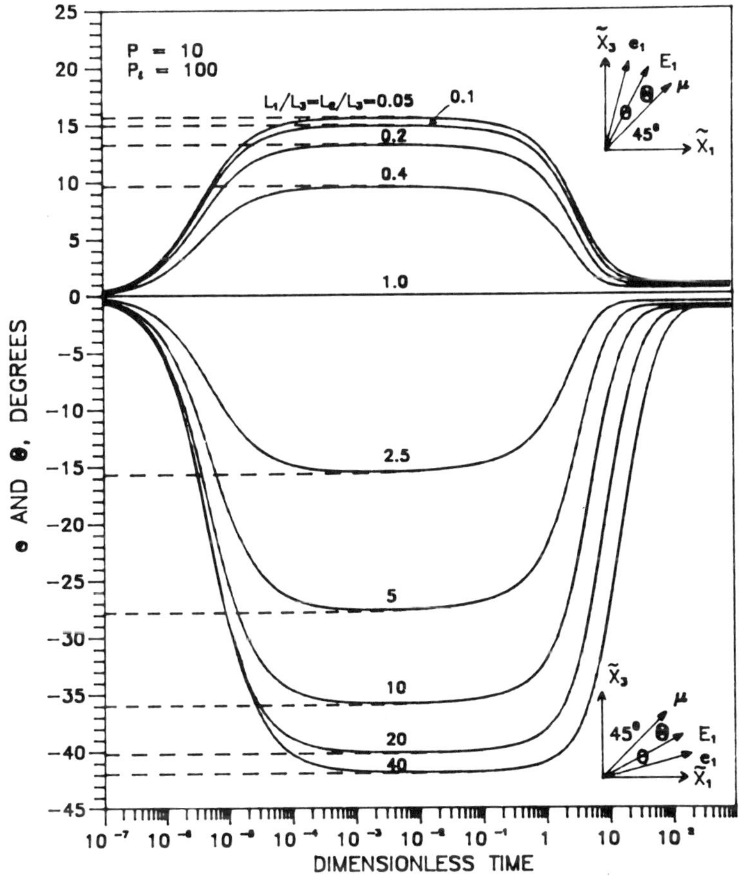

Figure 6. Deviation angles θ (dashed) and Θ (solid) versus dimensionless time for various anisotropy contrasts when $P = 10$, $P_1 = 100$, $\sigma_Y^2 = 0.5$, and mean velocity is at $45°$ to $\tilde{x}_1$ and $\tilde{x}_2$.

Neuman (1989) have shown that when $P \geqslant 10$, θ and Θ are both less than 2^0 at large t_μ. However, Fig. 6 shows that during the pre-asymptotic regime the two angles are generally much larger.

As the contrast between the integral scales increases, the maximum angle of deflection is seen to increase. However, the rate at which this angle increases goes down with the anisotropy contrast and appears to approach an asymptotic limit which depends on whether $L_1/L_3 = L_2/L_3$ tends to 0 or ∞. Large ratios of $L_1/L_3 = L_2/L_3$ correspond to stratified media in which field-scale longitudinal dispersion is seen to occur essentially parallel to the layers during much of the preasymptotic period even though the mean seepage velocity is inclined at 45^0 to the stratification.

4. Field validation of the quasilinear theory

Zhang and Neuman (1989) have validated their quasilinear dispersion theory against data from a plume of inorganic tracer extensively monitored during a natural gradient experiment conducted in a shallow anisotropic sand aquifer at the Canadian Forces Base, Borden, Ontario (Sudicky *et al.*, 1983; Mackay *et al.*, 1986; Roberts and Mackay, 1986; Roberts *et al.*, 1986). Mathematical analyses of the manner in which the plume, containing bromide and chloride, advanced and/or spread during the three-year experiment have been published by Freyberg (1986), Sudicky (1986), Sposito and Barry (1987), Barry *et al.* (1988) and Naff *et al.* (1988). The first two authors have made attempts to represent the horizontal spread of vertically averaged concentrations by means of a two-dimensional version of Dagan's (1984, 1987) theoretical model. Based on this model Freyberg and Sudicky estimated the asymptotic field-scale longitudinal dispersivity α_L at the Borden site to be 0.49 m and 0.45 m, respectively. Since the model they used is linearized and corresponds to an infinite Peclet number, it yields a zero asymptotic field-scale transverse dispersivity α_T. Sudicky also used three-dimensional Fickian formulae derived by Gelhar and Axness (1983) to calculate an asymptotic α_L for the site of 0.61 m and transverse dispersivities of zero.

Log hydraulic conductivity data from the site have been represented quite successfully (Sudicky, 1986) by an exponential covariance model according to which the aquifer is isotropic in the horizontal plane with corresponding integral scale $L_1 = L_2 = 2.8$ m but possesses a relatively short vertical integral scale $L_3 = 0.12$ m. This yields an anisotropy ratio of $L_1/L_3 = L_2/L_3 \cong 23$ which means that the aquifer is strongly (albeit imperfectly) stratified. For this reason, Naff *et al.* (1988) suggested that the vertically-averaged Borden concentration data should fit a three-dimensional model with the above anisotropy ratio better than a two-dimensional model. Yet according to Zhang and Neuman (1989) not only do two-dimensional models fit the original data better than three-dimensional models but, more importantly, the latter are contradicted by the observation that vertical spreading is much smaller than horizontal transverse spreading. We noted earlier that all stochastic theories discussed in this paper predict a greater transverse spread in the direction of lesser (in this case the vertical) than greater (in this case the horizontal) correlation, a prediction which is contrary to observations not only at Borden but also at other sites. Why is it that existing three-dimensional theories are unable to reproduce such observations, and why do two-dimensional theories provide a better match with experimental data?

We are not sure about the answers to these questions but agree with Zhang and Neuman (1989) that part of the problem may stem from the inability of existing stochastic theories to fully account for hydraulic anisotropy, due in part to their approximate nature but perhaps more so due to having so far disregarded local hydraulic anisotropy. Suppose that local and/or structural anisotropy at Borden is larger than has been deduced from the data in the past. In particular, suppose that the ratio between horizontal hydraulic conductivity K_H, and vertical hydraulic conductivity K_V, exceeds that estimated by Sudicky (1986) by a factor $\kappa^2 > 1$. Local phenomena, which appear to be only partially accounted for by the available hydraulic conductivity data and theory, include [*ibid*, p. 2073] 'the presence of numerous lenses of coarse to silty fine-grained sand embedded in a fine- to medium-grained sand' with a 'usually sharp and near-horizontal'... 'contact between zones having a large textural contrast,' and the fact that 'the thickness of individual beds generally varied from a few centimetres to a few tens of centimetres... although fine laminations on the order of a millimetre to a few millimetres thickness were sometimes encountered.' The net effect of such phenomena may be similar to that of the hypothetical horizontal boundaries invoked by Dagan (1987, 1989) and Barry *et al.* (1988) to justify their application of a two-dimensional theory to the Borden data. We can formally account for this excess anisotropy through a transformation of coordinates involving, say, a contraction of the horizontal axes by a factor $1/\kappa$ or expansion of the vertical axis by a factor κ. In the first case, a relatively small value of $\kappa^2 = K_H/K_V = 1.8$ would suffice to reduce L_1 and L_2 by 25% which, as shown by Naff *et al.* (1988), would bring about a reasonably good match between linear three-dimensional theory and longitudinal plume variance data from the Borden site. In the second case, a κ^2 greater than 530 would cause L_3 to exceed L_1 and L_2, while a much bigger value could render L_3 so large that a three-dimensional description of the Borden plume might become essentially identical to a two-dimensional description. Whether the latter happens or not at the Borden site remains an open question requiring further investigation.

By fitting straight lines to the variance of the depth-averaged Borden plume data Freyberg (1986, Fig. 10) determined the 'apparent' longitudinal and transverse dispersivities of this plume to be, respectively, 0.36 m and 0.039 m. Freyberg (1986) also determined that for a mean seepage velocity $\mu = 0.091$ m/day and horizontal integral scales $L_1 = L_2 = 2.8$ m, the apparent longitudinal dispersivity of 0.36 m yields an effective Peclet number $P \cong 7.8$. Zhang and Neuman (1989) modified Freyberg's data slightly to obtain $P \cong 7$. They then calculated the plume variance according to Eq. 2.13 by arbitrarily setting the apparent vertical dispersivity equal to 0.004 m and selecting a very large vertical integral scale to simulate two-dimensional behaviour. Local dispersion was taken to be zero so that $P_l \to \infty$. According to Sudicky (1986) the log hydraulic conductivity variance after nugget removal is $\sigma_Y^2 = 0.29$. Freyberg (1986) found that in order to fit Dagan's two-dimensional model to the Borden data he had to reduce σ_Y^2 from 0.29 to 0.24. Zhang and Neuman used Sudicky's original value of $\sigma_Y^2 = 0.29$ and their calculations are compared with both his and Freyberg's in Figs 7 and 8. All three sets of results involve a reduction of σ_Y^2

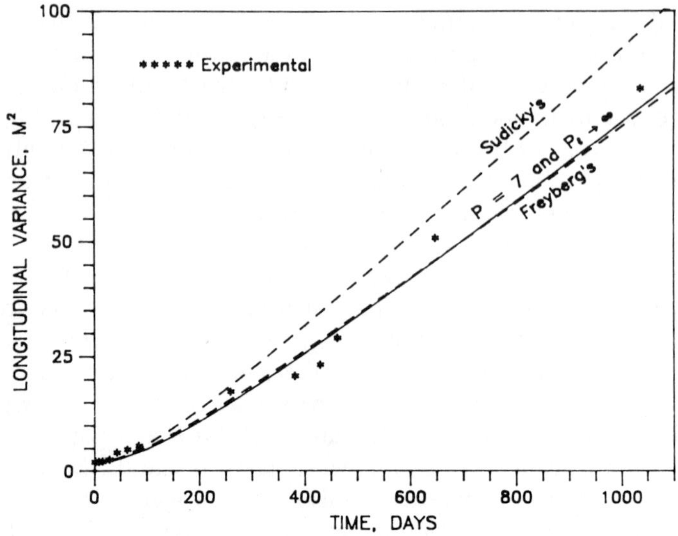

Figure 7. Longitudinal variance of Borden plume as a function of time. Solid curves are for $P = 7$ and $P_1 \to \infty$. Dashed curves correspond to results obtained by Sudicky (1986) and Freyberg (1986) with infinite Peclet number.

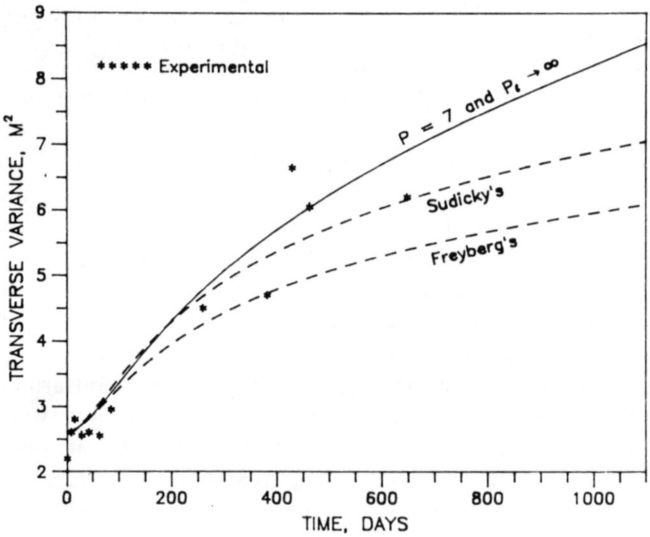

Figure 8. Transverse variance of Borden plume as a function of time. Solid curves are for $P = 7$ and $P_1 \to \infty$. Dashed curves correspond to results obtained by Sudicky (1986) and Freyberg (1986) with infinite Peclet number.

by a factor of 0.74 to account for the two-dimensional nature of the analysis as suggested by Dagan (1987) on the basis of integration over one vertical integral scale.

Figure 7 shows that by setting $P = 7$ one obtains a better fit between theory and longitudinal variance data than does Sudicky (1986) with Dagan's (1987) linearized model in which $P \to \infty$. The new fit, obtained without parameter modification, is as good as that achieved by Freyberg (1986) with Dagan's model after having 'calibrated' it by reducing σ_Y^2 from its measured value of 0.29 down to 0.24. This shows that in the Borden experiment non-linearity reduces longitudinal spread by a factor of 0.83.

Figure 8 shows how the quasilinear model with $P = 7$ fits transverse horizontal variance data from the Borden experiment. Also shown are (broken) curves fitted to the same data by Sudicky (1986) and Freyberg (1986) on the basis of Dagan's (1987) linearized model in which $P \to \infty$. The latter curves become asymptotically parallel to the time axis which reflects the fact that setting $P \to \infty$ is tantamount to disallowing transverse spread at late time. The (solid) curve of Zhang and Neuman (1989) fits the available data quite well despite their considerable scatter and, more importantly, continues to climb without ever becoming flat. One experimental point at 1038 days represents a transverse variance of 18.8 m^2 and is too large to include in the figure; it has been treated as an outlier by most authors. Although the curve of Zhang and Neuman does not climb sufficiently high to reach this point, it nevertheless comes much closer to it than do the other curves in Fig. 8.

We conclude this section by showing in Fig. 9 how the longitudinal and transverse field-scale dispersivities, α_L and α_T, vary with time during the Borden experiment when $P = 7$ and $P_1 \to \infty$. The figure suggests that, strictly speaking, longitudinal spread is non-Fickian until approximately day 1000 (2.8 years), at which time α_L attains its asymptotic value of 0.51 m (the experiment lasted 1038 days which corresponds to a dimensionless time $t_\mu \cong 34$). Transverse spread does not become strictly Fickian until approximately day 30 000 (82 years) at which time α_T attains an asymptotic value of 0.0095 m. A somewhat less strict definition of what constitutes Fickian behaviour might agree more closely with Sposito and Barry (1987), who have estimated the longitudinal limit to be reached after 1 year and the transverse after 60 years. The ratio between longitudinal and transverse dispersivities is initially 2.6. Following day 10, this ratio gradually increases until it stabilizes at the asymptotic limit of 53.7 after about 82 years. The peak transverse dispersivity of 0.056 m is attained on about day 100 and is almost six times as large as the asymptotic value.

5. Universal scaling of dispersivities and log hydraulic conductivities

In the Borden experiment the plume has reached a constant (i.e. Fickian) rate of longitudinal spread with time although the same has not happened with

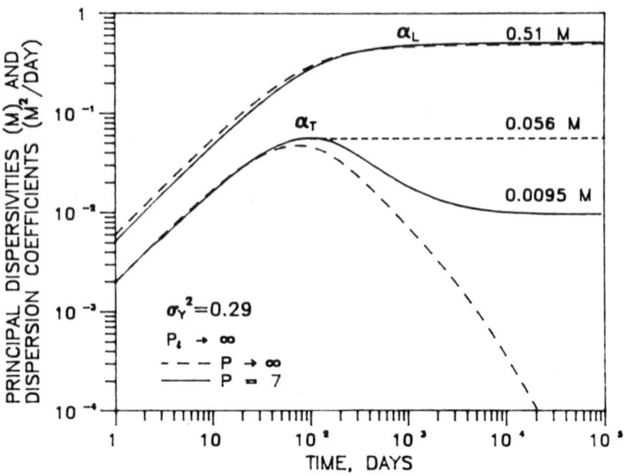

Figure 9. Longitudinal and transverse dispersivities of Borden plume as functions of time when $P_l \to \infty$. Solid curves are for $P = 7$, dashed for $P \to \infty$.

respect to transverse spread. This type of Fickian or quasi-Fickian behaviour has also been observed in numerous other field and laboratory studies. Yet the rate at which a plume spreads in any direction is not expected to remain constant forever. As the plume migrates and grows, it inevitably encounters heterogeneities on scales larger than those already sampled, and each such encounter brings about a new phase of non-Fickian, or time-dependent, dispersion. In the long run, a plume samples heterogeneities on ever-increasing scales and is therefore under a never-ending non-Fickian regime. Yet on discrete scales the plume may experience intermittent episodes of Fickian or quasi-Fickian mixing. This dual behaviour can be understood by recognizing that geological media tend to form discrete units on a hierarchy of scales (cf. Cushman, 1987, Fig. 1). A geological unit may exhibit second-order stationarity (homogeneity) with respect to log hydraulic conductivity on a given scale and to have a mean that is drastically different from that of all neighbouring units. As long as a plume evolves primarily within the confines of such a homogeneous unit, its spread is controlled almost entirely by the properties of the unit. If the plume travels a mean distance in excess of one integral scale while still remaining within the confines of the same unit, it may start showing Fickian or quasi-Fickian behaviour.

As the plume starts spreading into neighbouring units, larger-scale fluctuations in log hydraulic conductivity come into play. Their effect can be included in the analysis through the notion of nested structures as done for various soil properties by Burrough (1983a, b). We represent the covariance of log hydraulic conductivities over all scales and directions by a mean semivariogram $\gamma(s)$ in which s is a measure of distance. We assume in a manner similar to Journel and Huijbregts (1978, p. 150) that $\gamma(s)$ is the sum of nested semivariograms γ_j as illustrated in Fig. 10. Each of these nested semivariograms represents a second-order stationary process characterized by a finite variance σ_{Yj}^2 and a finite integral scale L_j such that $L_j < L_{j+1}$ for all $j \geqslant 0$. Their sum,

$$\gamma(s) = \sum_{j \geqslant 0} \gamma_j(s) \tag{5.1}$$

possesses neither a finite integral scale nor a finite variance. However, within a discrete homogeneous unit of size L such that $L_i \leqslant L \ll L_{i+1}$, the semivariogram is

$$\gamma_i(s) = \sum_{j=0}^{i} \gamma_j(s) \tag{5.2}$$

which has an integral scale of order L_i and a finite variance

$$\sigma_{Yi}^2 = \sum_{j=0}^{i} \sigma_{Yj}^2 \tag{5.3}$$

Consider a plume having initially zero spread whose centre of gravity has travelled a distance $s = \mu t$ within a homogeneous geological unit of size L such that $L_i \leqslant L \ll L_{i+1}$. In accord with Eqs 2.4 and 3.1 we can define the dimensionless time t_μ and the dimensionless dispersivity tensor $\boldsymbol{\alpha}$ with respect

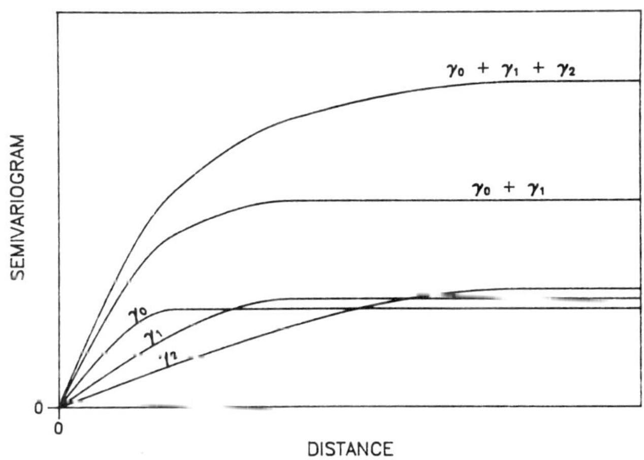

Figure 10. Schematic representation of nested semivariograms.

to this scale of plume migration as

$$t_\mu = \frac{\mu t}{L_i}, \quad \alpha_D = \frac{\alpha}{L_i \sigma_{Yi}^2} \tag{5.4}$$

If we use Fig. 1 as a guide, we find that according to the quasilinear theory the dimensionless longitudinal dispersivity α_{DL} rapidly approaches a constant value as dimensionless time increases past $t_\mu = 1$. This, together with Eq. 5.4, implies that longitudinal dispersivity is a constant or nearly so when $L_i < \mu t$ provided the plume is still essentially within the unit. In other words, longitudinal spread becomes Fickian or quasi-Fickian and stays so until the plume starts spreading across the geological boundaries of the homogeneous unit. At this time the semivariogram γ_{i+1} comes into play and a new non-Fickian regime develops which persists indefinitely unless another homogeneous geological unit is encountered on a scale larger than L.

Experimental evidence to support our view that dispersion in geological media is inherently non-Fickian, with at best intermittent episodes of Fickian or quasi-Fickian behaviour, exists in the form of over 130 longitudinal dispersivities from laboratory and field tracer studies in porous and fractured media throughout the world recorded by Lallemand-Barres and Peaudecerf (1978), Pickens and Grisak (1981), Gelhar *et al.* (1985) and Arya (1986). These dispersivities, which we consider 'apparent' because all were interpreted on the basis of traditional Fickian models, vary from less than 1 mm to more than 1 km and correspond to studies conducted on scales ranging from less than 10 cm to more than 100 km. We show that it is possible to interpret the spatial scale effect exhibited by these apparent dispersivities in a manner which is consistent with the quasilinear theory. Such an interpretation leads to the remarkable finding that the recorded dispersivities fluctuate about a single theoretical curve representing non-Fickian behaviour due to a self-similar log hydraulic conductivity field characterized by the universal semivariogram $\gamma(s) \cong 0.017\sqrt{s}$.

Figure 11 is a logarithmic plot of apparent longitudinal dispersivities α_{La} versus the scale of the study L_s as reported by the aforementioned authors. For reasons to be discussed later we have excluded from the figure dispersivities that had been obtained by calibrating numerical grid models against solute concentrations of large-scale plumes. A visual inspection of the figure suggests that α_{La} increases more rapidly with scale when $L_s \leqslant 100\,\text{m}$ than when $L_s \geqslant 100\,\text{m}$. This, together with theoretical considerations discussed below, has motivated us to fit by regression a separate straight line of the form

$$\log_{10} \alpha_{La} = b \log_{10} L_s + c \tag{5.5}$$

to the data in each domain. The best fit lines and their 95% confidence regions

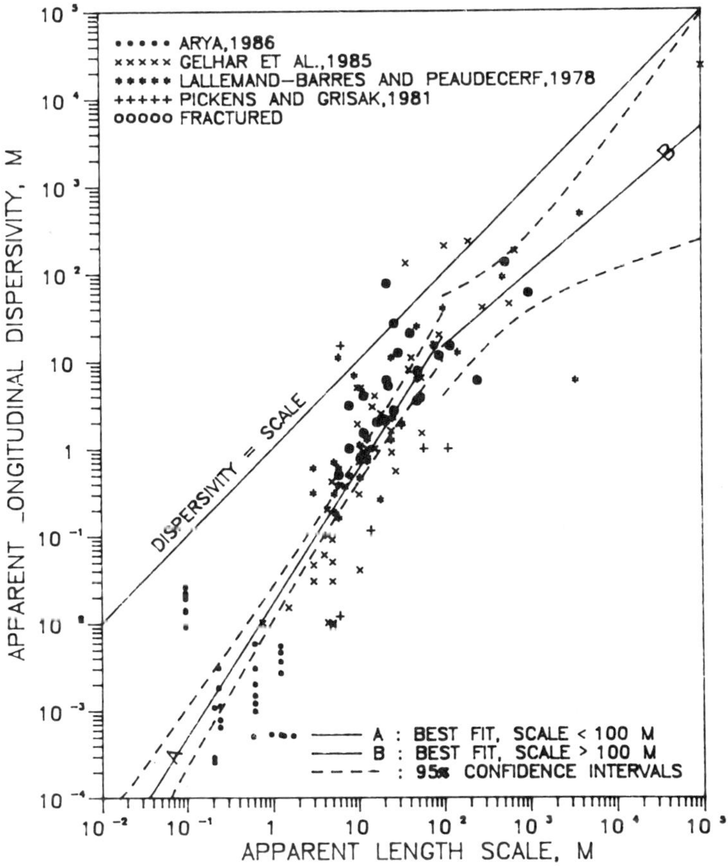

Figure 11. Apparent longitudinal dispersivities versus scale of study excluding numerical model calibration results.

are shown in Fig. 11. The line labelled A represents 119 data points in the domain $L_s \leqslant 100$ m, and line B corresponds to 16 points in the domain $L_s \geqslant 100$ m (one point lies at $L_s = 100$ m and belongs to both domains).

Line A can be expressed as

$$\alpha_{La} \cong 0.017 L_s^{1.5} \tag{5.6}$$

with a regression coefficient $R^2 = 0.71$ and 95% confidence intervals (0.011, 0.026) about the coefficient 0.017, (1.35, 1.70) about the exponent 1.5. We check the extent to which this is consistent with the quasilinear theory by considering a plume having initially zero spread whose centre of gravity has travelled a distance $s = \mu t = L_s$. This means that we treat the recorded experimental scales

L_s as mean travel distances. We select an arbitrary reference scale L_i that satisfies $L_i \gg L_s$ and use it to renormalize t_μ and $\boldsymbol{\alpha}_D$ according to Eq. 5.4. Then $t_\mu \ll 1$ and if we again use Fig. 1 as a theoretical guide, restricting our attention for the moment to $P \geqslant 10$, we conclude that the dimensionless longitudinal dispersivity α_{DL} must continue to increase with time.

The data in Fig. 11 were obtained from laboratory and field tracer studies interpreted on the basis of traditional Fickian models. These models ignore the time-dependent nature of α_{DL} and thus, according to Fig. 1, are equivalent to setting $\alpha_{DL} = 1$ when $P \geqslant 10$. This in turn implies that t_μ is of order 1 and since $\mu t = L_s$, that L_s is the effective integral scale controlling dispersion. The traditional Fickian interpretation is thus equivalent to computing an apparent longitudinal dispersivity α_{La} according to

$$\frac{\alpha_{La}}{L_s \sigma_Y^2} = 1 \tag{5.6}$$

The same interpretation is further equivalent to stating that

$$\sigma_Y^2 = \gamma(L_s) \tag{5.7}$$

Upon substituting Eq. 5.7 into Eq. 5.6 we obtain

$$\alpha_{La} = L_s \gamma(L_s) \tag{5.8}$$

This theoretical result is consistent with the regression formula (Eq. 5.6) provided only that the semivariogram of log hydraulic conductivities is of the form

$$\gamma(s) \cong 0.017 s^{2H} \tag{5.9}$$

where $H = 0.25$ with a 95% confidence interval (0.17, 0.35).

The above result was obtained from Fig. 1 for $P \geqslant 10$. The figure corresponds to a case where mean flow takes place parallel to the bedding in a weakly stratified medium. Reference to Fig. 6 of Neuman and Zhang (1989) reveals that in isotropic media the same result must hold for $P \geqslant 1$. The traditional Fickian interpretation is equivalent to redefining P in Eq. 2.4 according to $P = \mu L_s / d_1$. If we take $\mathbf{d}_1$ to represent molecular diffusion and consider it to be negligibly small, we can write $d_1 \cong \mu \alpha_{La}$ to obtain

$$P \cong \frac{L_s}{\alpha_{La}} \tag{5.10}$$

Our analysis then applies to $L_s \geqslant 10\alpha_{La}$ in weakly stratified media and $L_s \geqslant \alpha_{La}$ in isotropic media. Since most data points associated with line A in Fig. 11 lie below the line of unit slope corresponding to $\alpha_{La} = L_s$, they satisfy one or both of these conditions and are therefore consistent with our analysis.

Semivariograms of the form of Eq. 5.9 are known to represent self-similar stochastic processes with a fractal dimension $D = 2 - H$ (cf. Burrough, 1983a). We therefore conclude that all the soils and rocks represented by line A have, on the average, self-similar log hydraulic conductivity fields characterized by a fractal dimension $D \cong 1.75$ with a 95% confidence interval (1.65, 1.83). Since these materials include unconsolidated sediments as well as consolidated porous and fractured rocks of diverse origin from a variety of sites, the self-similarity appears to be a universal property of many geological media. The fact that data in Fig. 11 show a wide scatter about line A suggests that deviations from this mean property can be large and the self-similarity pattern is not perfect. While some of this scatter is clearly due to variations in the manner of sampling tracer concentrations, a lack of clarity about the relationship between mean travel distance and the reported length scale, variability in the mean flow pattern, and experimental as well as interpretive errors, much of it must be attributed to local deviations from mean behaviour due to approximate second-order stationarity of discrete geological units on a hierarchy of scales and associated anisotropy.

When $L_s \gg 100$ m, the apparent dispersivities in Fig. 11 vary more slowly with the experimental scale L_s than they do in the domain of line A. They are represented more closely by line B which can be expressed as

$$\alpha_{La} \cong 0.32 L_s^{0.83} \tag{5.11}$$

with a low regression coefficient $R^2 = 0.44$ and broad 95% confidence intervals (0.026, 3.95) about the coefficient 0.32, (0.30, 1.37) about the exponent 0.83.

To show that this is consistent with the quasilinear theory we again refer to Fig. 1. The figure implies that when mean flow is parallel to the bedding in mildly stratified media under a Fickian regime with $P < 1$, α_{DL} is proportional to P so that

$$\alpha_{DL} = cP \tag{5.12}$$

where $c \cong 1$. From Fig. 5 we deduce that as the stratification becomes more pronounced, α_{DL} (related to the slopes of the curves in the figure) for a given P decreases. Since Eq. 5.12 continues to hold, c becomes less than 1 and the expression becomes valid for $cP < 1$.

When data representing a non-Fickian regime are interpreted on the basis of a Fickian model, analogy to Eq. 5.6 implies that Eq. 5.12 should take the form

$$\frac{\alpha_{La}}{L_s \sigma_Y^2} = c \frac{L_s}{\sigma_{La}} \tag{5.13}$$

whenever $\alpha_{La} > cL_s$. This, together with Eq. 5.9, means that α_{La} should vary with L_s according to

$$\alpha_{La} \cong \sqrt{(0.017c)}L_s^{1.25} \tag{5.14}$$

and delineate a slope of 1.25:1 on logarithmic paper. However, upon testing the null hypothesis H_0 that such a slope is admissible under the regression leading to line B, we find that it must be rejected at the 80% confidence level. The reason is as follows.

Equation 5.14 derives from a quasilinear theory in which the effective Peclet number P, defined in Eq. 2.4, is a function of the longitudinal dispersion coefficient (and hence dispersivity α_L) which in turn depends on P. Standard Fickian interpretations are based on the traditional Peclet number P_l, defined in Eq. 3.5, which is independent of α_L. Since P_l is generally very large, these interpretations are equivalent to Eq. 5.6 rather than Eq. 5.13. As such, they are analogous to relying on the top curve in Fig. 1 rather than the lower curves regardless of how small is P or how stratified is the medium. This is the same as setting the left-hand side of Eq. 5.13 equal to 1 and thus calculating α_{La} not from Eq. 5.14, as one should according to the quasilinear theory, but from

$$\alpha_{La} = cL_s \tag{5.15}$$

where $c \leqslant 1$.

To show that Eq. 5.15 is consistent with line B in Fig. 11, we examine the null hypothesis H_0 that the exponent in Eq. 5.11 is not 0.83 but 1, i.e. that the latter equation can be replaced by

$$\alpha_{La} = bL_s \tag{5.16}$$

where b is a coefficient. The examination shows that H_0 cannot be rejected at the 50% confidence level and therefore Eq. 5.16 constitutes an acceptable alternative to line B. Another null hypothesis which states that $b \geqslant 1$ must be rejected at the 70% confidence level and we conclude that $b < 1$. Hence Eq. 5.15 is consistent with the data in Fig. 11.

From Eqs 5.8, 5.10 and 5.12 we infer that apparent longitudinal dispersivities obtained from laboratory and field studies on the basis of traditional Fickian models should, on the average, vary with the experimental scale according to Eq. 5.6 when $\alpha_{La} \leqslant cL_s$, and according to Eq. 5.15 when α_{La} becomes as large as cL_s. If we fit Eq. 5.15 by regression to the data scattered about line B in Fig. 11 we find that $c = 0.11$. This implies that most dispersivities within one log cycle of the line $\alpha_{La} = L_s$, shown in the figure, represent stratified media. It explains why the rate at which the apparent dispersivities increase diminishes as they come to within one log cycle of this line; why their mean position remains about one log cycle below this line as the scale continues to increase; and why lines A and B intersect approximately one log cycle below the same line.

It should remain clear that if the dispersivities in Fig. 11 would have been determined from the quasilinear theory rather than by means of traditional Fickian models, they would have varied according to Eq. 5.14 rather than Eq. 5.15. Therefore, their increase with scale would have been more rapid than it is in Fig. 11 when $L_s \geqslant 100$ m.

Figure 12 is the same as Fig. 11, but includes dispersivities obtained from the calibration of numerical models against data corresponding to large-scale plumes. Most of these dispersivities are associated with scales in excess of 100 m and lie below the lower confidence limit of line B. Applying regression to the calibrated dispersivities yields a straight line with a slope of 0.54. When the

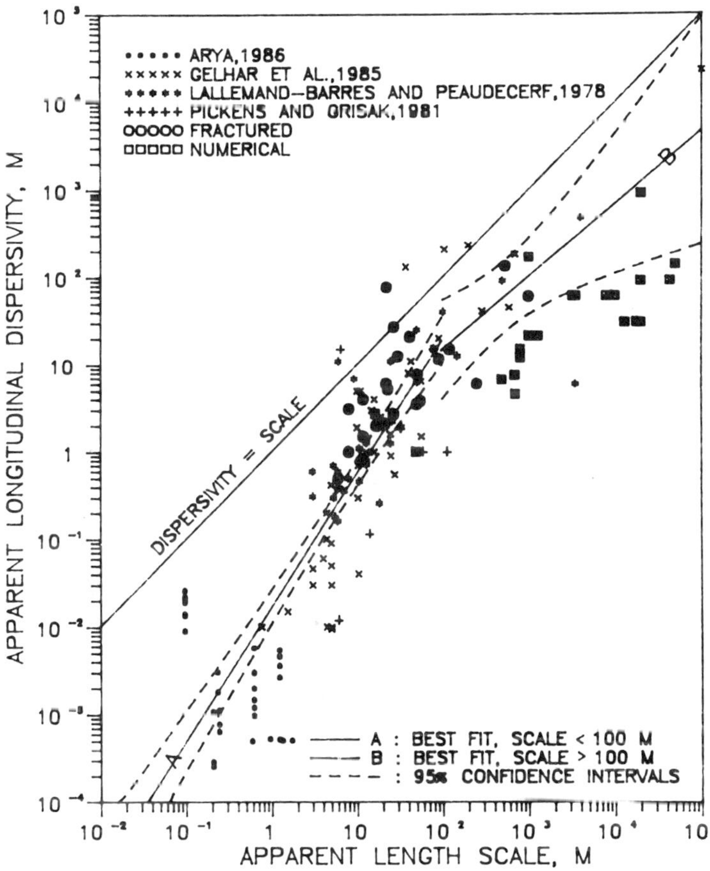

Figure 12. Apparent longitudinal dispersivities versus scale of study including numerical model calibration results.

regression includes all dispersivities corresponding to $L_s \geqslant 100$ m, the slope reduces to 0.48. In both cases the null hypothesis that the slope is 1 must be rejected at the 99% confidence level. Hence the calibrated dispersivities are consistent neither with our earlier theoretical interpretation nor with values determined by other means.

The calibrated dispersivities in Fig. 11 grow approximately in proportion to $\sqrt{L_s}$ as does the discrepancy between them and theoretical values predicted by Eq. 5.15. Hence they can be made consistent with theory and other values if one associates them with a reduced scale L_r proportional to $\sqrt{L_s}$. It has been suggested to us by J. Carrera (pers. commun.) that this reduction in scale, brought about by numerical model calibration, can be explained as follows: to calibrate a numerical advection-dispersion model against concentration data from an observed plume, it is common practice to start by calibrating a flow model against hydraulic head, flux, and hydraulic conductivity or transmissivity data. The latter is equivalent to producing information about the variation of groundwater velocities on scales equal to or larger than the spacing between nodes in a finite difference or finite element grid. Since this scale is generally smaller than the mean travel distance of a plume, flow model calibration generates information about velocity variations on scales smaller than L_s. In calibrating advection-dispersion models against plume data, it is common practice to treat the estimated velocity variations as deterministic information. This is equivalent to accounting deterministically for (i.e. filtering out) velocity variations on scales L_r that are smaller than L_s. Hence the dispersivities are associated with a length scale L_r which depends on the density of information about velocities and not on the mean travel distance L_s of the plume. The above regression analysis suggests that in past calibration exercises L_r increased with the mean travel distance at an average rate proportional to $\sqrt{L_s}$. This is the same as saying that the density of information about velocities decreased in a manner roughly proportional to the inverse of $\sqrt{L_s}$.

6. Numerical simulation by semidiscrete Gaussian superposition

6.1 Computational method

The quasilinear theory is Lagrangian in that dispersion depends on the residence time of plume 'particles' within the flow field. When all the particles which make up a plume enter a uniform mean flow field at time τ, it is possible to simulate the non-Fickian evolution of the plume with computational grid methods originally developed for Fickian transport. The only additional requirement is that the dispersivity be made to vary with residence time according to formulae such as Eqs 2.6 or 2.16–2.18. However, since the dispersivity is initially very small, a very fine spacing between nodes may be required to prevent the grid Peclet number ($\mu\Delta x/D_L$ where μ is mean velocity, Δx is nodal spacing parallel to μ, and D_L is time-dependent longitudinal component of the dispersion tensor $\mathbf{D}$ defined in Eq. 3.3) from becoming high and thereby causing large numerical errors. As most grid methods require that Courant numbers ($\mu\Delta t/\Delta x$ where Δt is time interval) be 1 or less, the time steps may also have to be severely curtailed. Applying Eulerian methods to non-uniform mean flow fields and/or cases where different solute molecules have entered the flow field at different times is not straightforward by any measure.

In this paper we describe a Lagrangian method of computation based on the concept of semidiscrete Gaussian superposition. For simplicity we restrict the analysis to uniform mean velocities in infinite or rectangular domains that are aligned with a system of coordinates, $\mathbf{x}$, which also form the principal axes of $\mathbf{D}(t-\tau)$. In statistically isotropic log hydraulic conductivity fields the latter requirement is achieved by orienting the mean velocity parallel to one of the coordinates. In statistically anisotropic media the same is guaranteed only if the coordinates are also principal directions for the log hydraulic conductivity covariance. Under these conditions, the semidiscrete Gaussian superposition method does not recognize limitations on Peclet and Courant numbers, yet is able to account for retardation and exponential decay. Since an extension of the method to more complex situations is still the subject of investigation, the description below should be considered merely a blueprint.

The method derives naturally from Taylor's (1921) theory of diffusion by continuous motions which is the basis for the quasilinear theory described in this paper. An excellent introduction to the theory from the standpoint of transport in geological media without decay or retardation can be found in the review article by Dagan (1987). Based on this theory we find that the simplest sources to treat computationally are point sources. Consider a point source of mass M_n introduced instantaneously into the flow field at time τ_n at point $\boldsymbol{\chi}_n$. We represent this mass by a labelled particle, n, that advances with a mean velocity $\boldsymbol{\mu} = \mathbf{v}/r$ where $\mathbf{v}$ is the mean groundwater velocity and $r \geqslant 1$ is a constant retardation coefficient (usually due to equilibrium sorption controlled by a linear isotherm). The contribution of the particle to the mean concentration at point $(\mathbf{x}, t)$ in space–time is

$$\langle c_n(\mathbf{x}, t)\rangle = \frac{M_n}{\phi r} \exp[-\lambda(t-\tau_n)] f_n(\mathbf{x}; t) \tag{6.1}$$

where ϕ is porosity, λ is a constant accounting for exponential decay (usually due to radioactivity or biodegradation), and f_n is the multivariate normal probability distribution function (pdf) of $\mathbf{x}$ at time t with mean $\boldsymbol{\chi}_n + \boldsymbol{\mu}(t-\tau_n)$ and covariance $\boldsymbol{\Sigma} = 2\mathbf{d}_1(t-\tau_n) + \mathbf{X}$ in accord with Eq. 3.4. The tensor $\mathbf{X}$ can be calculated as a function of the residence time $t-\tau_n$ with the aid of formulae such as Eqs 2.13 and 2.15.

A continuous point source can be represented by a sequence of particles released at successive time intervals Δt_n that are small compared to the Lagrangian time scale L_μ/μ (residence time corresponding to $t_\mu = 1$ beyond which longitudinal spread is Fickian). If a mass M_n is injected during Δt_n then the particle contribution is again given by Eq. 6.1. In the vicinity of no-flow boundaries the effect of each particle can be offset by an image of equal sign and strength, which comes into play only when one desires to compute the corresponding concentration.

To account for the instantaneous introduction of solute mass over a finite domain Ω_0 at time $t = \tau_0$, the domain is subdivided into a dense system of non-overlapping box-shaped parcels Ω_n, with sides $2\alpha_{ni}$ parallel to the coordinates x_i. We assume for simplicity that the mass initially included in Ω_n is evenly distributed so that the concentration $c_0(\mathbf{x})$ due to the instantaneous non-point source can be replaced locally by the equivalent uniform initial concentration

$$c_{pn} = \frac{M_n}{\phi r 2^m \prod_{i=1}^{m} \alpha_i} \tag{6.2}$$

where m is the number of space dimensions. Then the contribution of Ω_n to the mean concentration is obtained by integration over the box of contributions from point sources with infinitesimal mass,

$$\langle c_n(\mathbf{x}, t)\rangle = \frac{c_{pn}}{2^m} \exp[-\lambda(t-\tau_0)] \prod_{i=1}^{m} g_n(x_i, \mathbf{t}) \tag{6.3}$$

where

$$g_n(x_i, t) = \operatorname{erf}\left(\frac{\alpha_{ni} + \langle X_{ni}\rangle - x_i}{\sqrt{(2\Sigma_{0i})}}\right) + \operatorname{erf}\left(\frac{\alpha_{ni} - \langle X_{ni}\rangle + x_i}{\sqrt{(2\Sigma_{0i})}}\right) \tag{6.4}$$

Here $\langle \mathbf{X}_n\rangle$ is the mean position of the centroid of Ω_n and Σ_{0i} the ith associated variance at time t. This representation is exact if $c_0(\mathbf{x})$ is uniform in each Ω_n, otherwise its accuracy can be improved by decreasing Ω_n and increasing the number of parcels. Parcels that create measurable effects at no-flow boundaries can be counteracted by placing identical image parcels at equal distances on the other side. A single box-shaped parcel in an infinite medium with time-dependent dispersion coefficients was considered in earlier analyses by Güven and Molz (1986) and Sposito and Barry (1987).

Concentrations prescribed along inflow boundaries are handled similarly by defining box-shaped parcels Ω_n which have lengths equal to $\mu\Delta t_n$ parallel to the direction of flow. The value of c_{pn} assigned to each such parcel is the corresponding average boundary value during Δt_n. The time of origin τ_n occurs when half the box has crossed the boundary.

Continuous distributed sources emit densely-spaced non-overlapping box-shaped parcels Ω_n over time intervals Δt_n centred about times of origin τ_n. Taking the rate of mass generation in each parcel to be constant and uniform, the contribution of Ω_n to the mean concentration is obtained by integration over the box and Δt_n,

$$\langle c_n(\mathbf{x}, t) \rangle = \frac{\dot{c}_{pn} \Delta t_n}{2^m} \exp[-\lambda(t - \tau_0)] \prod_{i=1}^{m} g_n(x_i, t) \tag{6.5}$$

where $\dot{c}_{pn}$ is the time derivative of c_{pn}. The total mean concentration at any point $(\mathbf{x}, t)$ is the sum of contributions by all individual particles and parcels.

6.2 Examples

We present below five three-dimensional examples which illustrate some elementary aspects of pre-asymptotic behaviour as computed by semidiscrete Gaussian superposition. All five examples correspond to $P \gg 1$, $P_1 = \infty$, and $t_\mu \leqslant 2.5$ in a statistically isotropic medium with an integral scale L. Under these conditions $\mathbf{\Sigma} = \mathbf{X}$ and, from Figs 6 and 7 of Neuman and Zhang (1989), it is clear that $\mathbf{X}$ can be approximated by means of Dagan's (1984) formulae

$$X_1(t_\mu) = 2L^2\sigma_Y^2\left(t_\mu - \frac{8}{3} + \frac{4}{t_\mu} - \frac{8}{t_\mu^3} + \frac{8}{t_\mu^2}\left(1 + \frac{1}{t_\mu}\right)\exp(-t_\mu)\right) \tag{6.6}$$

$$X_2(t_\mu) = X_3(t_\mu) = 2L^2\sigma_Y^2\left(\frac{1}{3} - \frac{1}{t_\mu} + \frac{4}{t_\mu^3} - \left(\frac{4}{t_\mu^3} + \frac{4}{t_\mu^2} + \frac{1}{t_\mu}\right)\exp(-t_\mu)\right) \tag{6.7}$$

where the mean velocity is oriented parallel to the x_1 coordinate, all coordinates being normalized with respect to L. The magnitude of σ_Y^2 is arbitrarily set equal to unity and concentrations are given relative to a uniform background with reference zero.

Example 1 concerns the propagation in an unbounded domain of a plume initially confined to a unit cube. The initial concentration is uniform in the x_2 and x_3 directions but varies in a unimodal fashion along x_1 as shown in Fig. 13. The plume is discretized into parcels that form parallel slices of equal width normal to x_1 as exemplified in Fig. 14. Figures 14 and 15 illustrate the superposition effect at $t_\mu = 1.0$ with six and 50 slices, respectively. These and additional results, summarized in Fig. 13, demonstrate that six slices provide an adequate representation of plume shape in this case.

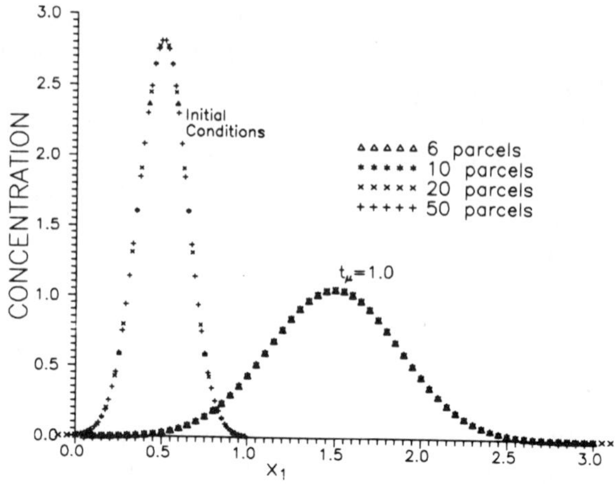

Figure 13. Concentration along x_1 at $t_\mu = 0$ and 0.5 in Example 1 as represented by a varying number of parcels.

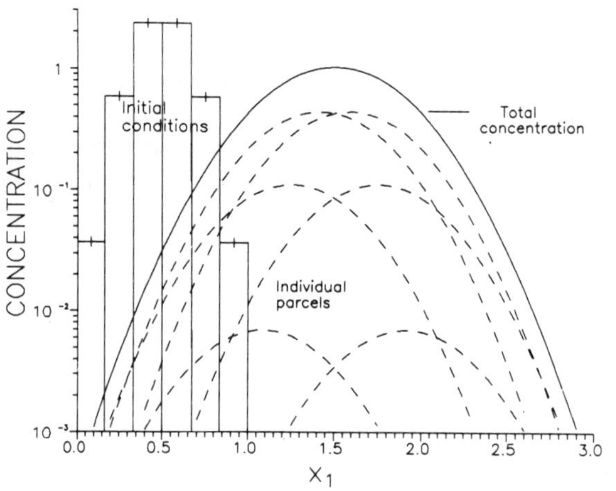

Figure 14. Concentration along x_1 at $t_\mu = 0$ and 0.5 in Example 1 as represented by six parcels individually and after superposition.

Example 2 deals with a plume of uniform initial concentration in the unit cube, represented by a single parcel of similar volume. Figure 16 shows what happens at three consecutive t_μ values (0.5, 1.0, 1.5) when the x_2–x_3 plane at $x_1 = 0$, coinciding with the left side of the initial plume, acts as an impermeable boundary. The spread of the plume is seen to be much less than that which an asymptotic Fickian rate of dispersion would predict.

Example 3 is similar to Example 2 except that the boundary is maintained at zero concentration, the plume is divided into two slices of equal width normal to x_1, and the results are shown at $t_\mu = 0.5$, 1.5 and 2.0 (Fig. 17). As pre-asymptotic dispersion causes some positive accumulation of mass along the boundary at $t_\mu = 0.5$, the effect must be offset by means of a parcel with negative mass introduced along this surface. Figure 17 illustrates the superposition of the two original slices and one boundary parcel at $t_\mu = 0.5$. Near-zero

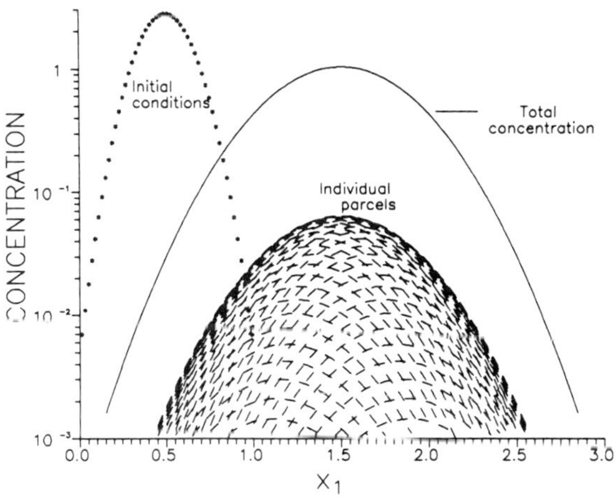

Figure 15. Concentration along x_1 at $t_\mu = 0$ and 0.5 in Example 1 as represented by 50 parcels individually and after superposition.

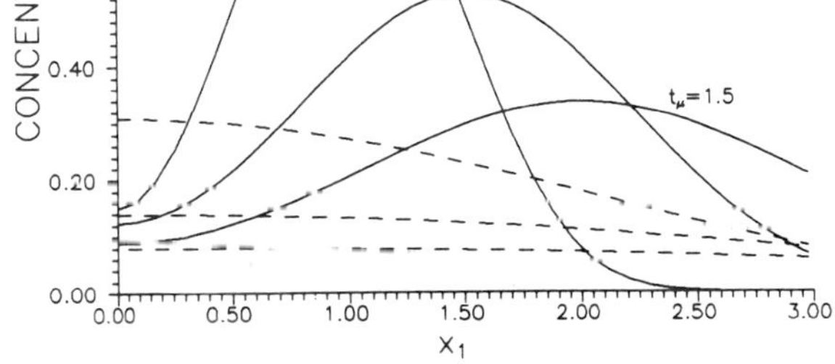

Figure 16. Effect of impermeable boundary (at $x_1 = 0$) on concentration along x_1 at $t_\mu = 0$, 0.5, 1.0 and 1.5 in Example 2 as represented by one parcel. Comparison between pre-asymptotic and asymptotic model results.

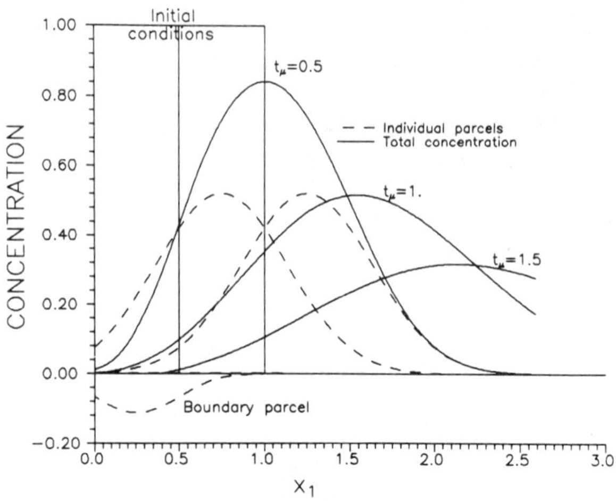

Figure 17. Effect of zero concentration boundary (at $x_1 = 0$) on concentration along x_1 at $t_\mu = 0$, 0.5, 1.5 and 2.0 in Example 3 as represented by two parcels. Also shown is the superposition of these two parcels and a boundary parcel with negative mass at $t_\mu = 0.5$.

concentration is seen to be achieved at the boundary with only one boundary parcel of size identical to each of the slices representing the initial plume.

Example 4 involves a plume having dimensions 8, 8, 1 parallel to the x_1, x_2, x_3 axes, repsectively, in a semi-infinite domain with an impermeable boundary located along the x_2–x_3 plane at $x_1 = 0$ (coinciding with the left side of the initial plume). Initial concentration varies in the x_1–x_2 plane according to the contours in Fig. 18 with a maximum of 4 at the centre. The plume is represented by a maximum of 16 × 16 equal size parcels (see grid in Fig. 18), each of which extends one unit in the x_3 direction. Figures 19 and 20 show how the number of parcels, which varies from 4 × 4 through 8 × 8 to 16 × 16, affects the calculated plume parallel to x_1 and x_2, respectively, along its centre in the x_3 direction, at two consecutive dimensionless times ($t_\mu = 0.5$ and 2.5). The effect of discretization into parcels on concentration contours at the centre of the plume in the x_3 direction is illustrated in Figs 21 and 22. It appears that 16 × 16 parcels capture adequately the essential features of the plume geometry. However, these features become grossly distorted if one relies on a Fickian instead of non-Fickian model as Figs 23 and 24 amply demonstrate.

Example 5 follows the evolution of the same plume at a retarded velocity ($r = 2$) with an exponential decay constant $\lambda = 0.5$. Figure 25 shows the important difference between pre-asymptotic and asymptotic behaviour at $t_\mu = 0.5$ and 2.5 along the x_1 direction in the plane of symmetry normal to x_3. Figure 26 shows a similar comparison between contours in the same plane at $t_\mu = 0.5$.

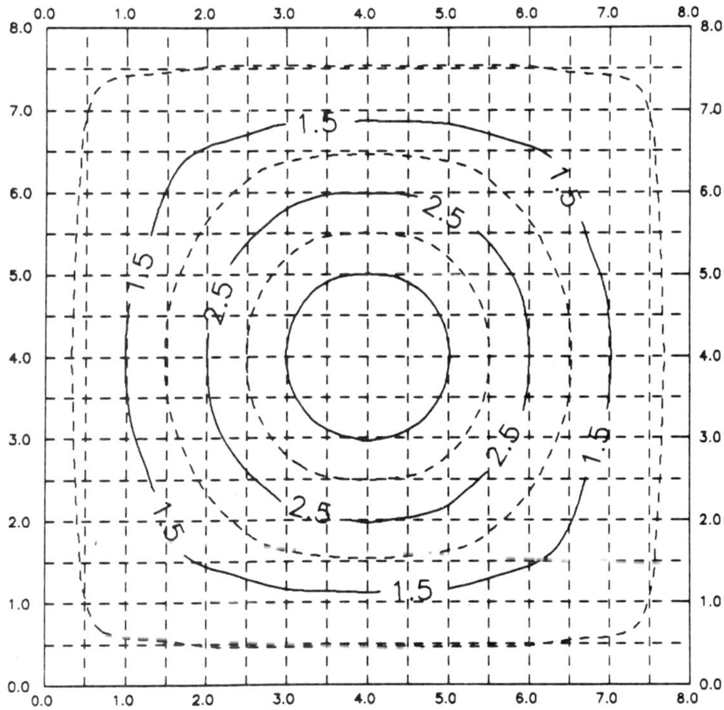

Figure 18. Contours of initial concentration along x_1–x_2 plane of symmetry normal to x_3 in Examples 4 and 5. The left side coincides with x_2 which acts as an impermeable boundary and mean flow is to the right. Also shown is discretization into a maximum of 16 × 16 parcels.

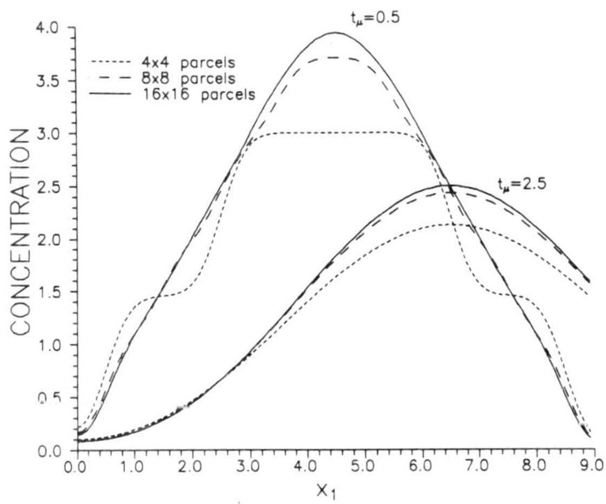

Figure 19. Concentration along centre line of plume parallel to x_1 at $t_\mu = 0.5$ and 2.5 in Example 4 as represented by a varying number of parcels.

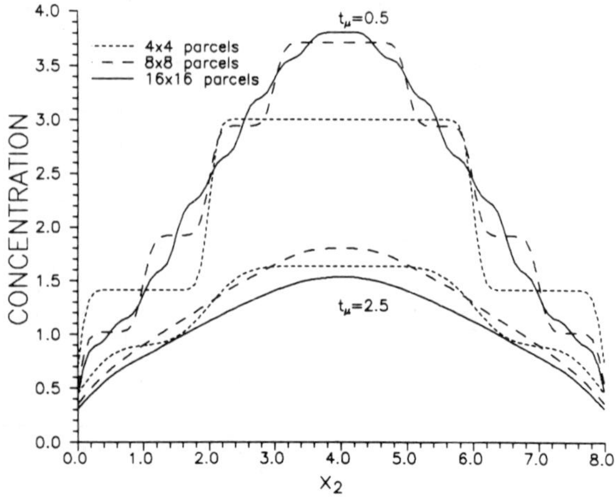

Figure 20. Concentration along centre line of plume parallel to y'_1 at $t_\mu = 0.5$ and 2.5 in Example 4 as represented by a varying number of parcels.

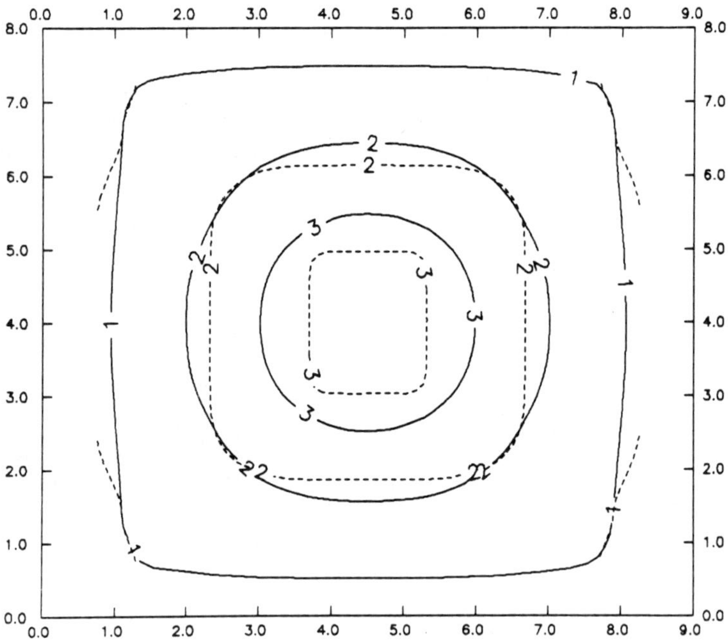

Figure 21. Concentration contours along x_1–x_2 plane of symmetry in Example 4 at $t_\mu = 0.5$ as represented by 4×4 (dashed) and 16×16 (solid) parcels.

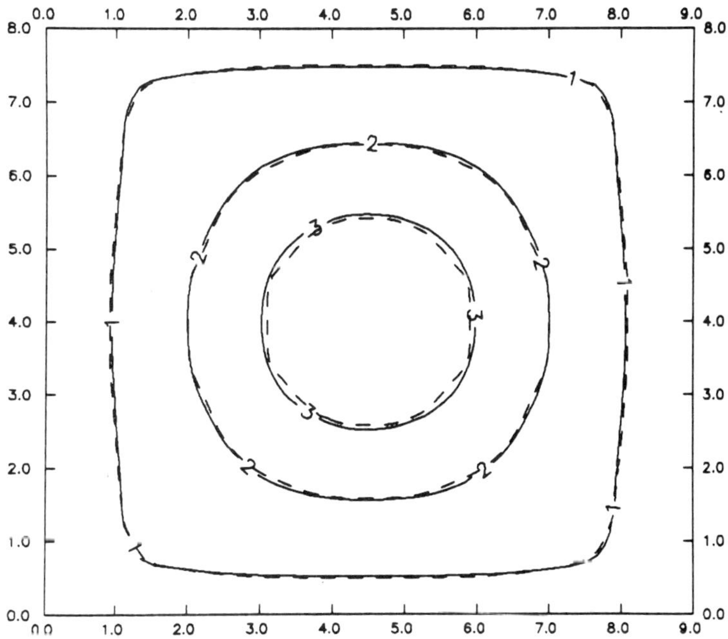

Figure 22. Concentration contours along x_1–x_2 plane of symmetry in Example 4 at $t_\mu = 0.5$ as represented by 8 × 8 (dashed) and 16 × 16 (solid) parcels.

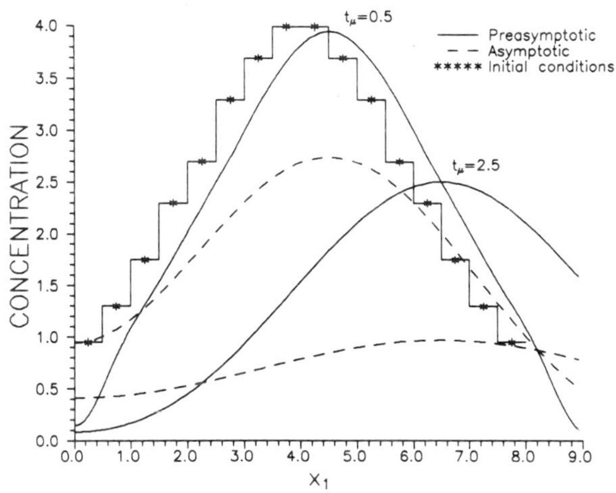

Figure 23. Concentration along centre line of plume parallel to x_1 at $t_\mu = 0, 0.5$ and 2.5 in Example 4 as represented by 16 × 16 parcels and obtained using pre-asymptotic (solid) and asymptotic (dashed) models.

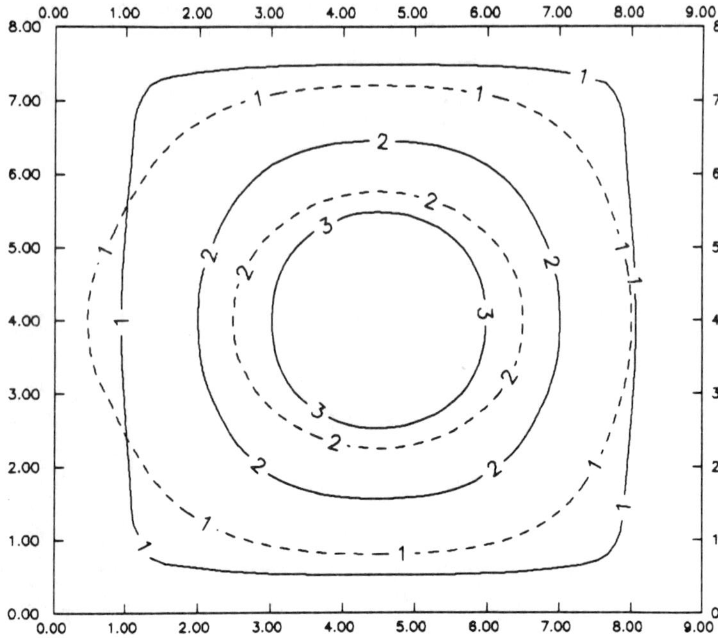

Figure 24. Concentration contours along x_1–x_2 plane of symmetry in Example 4 at $t_\mu = 0.5$ as represented by 16×16 parcels and obtained using pre-asymptotic (solid) and asymptotic (dashed) models.

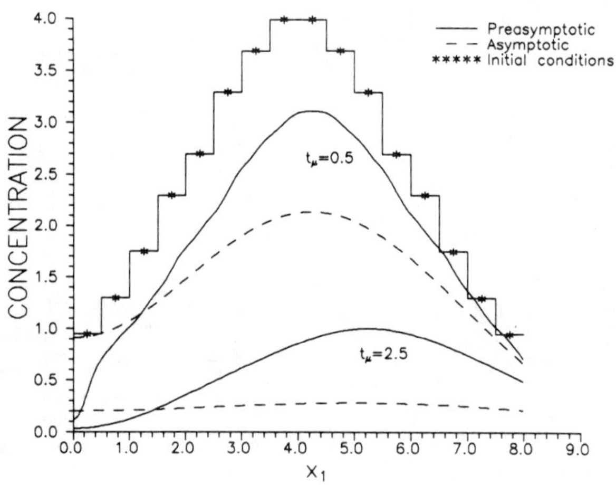

Figure 25. Concentration along centre line of plume parallel to x_1 at $t_\mu = 0, 0.5$ and 2.5 in Example 5 considering retardation ($r = 2$) and decay ($\lambda = 0.5$) as represented by 16×16 parcels and obtained using pre-asymptotic (solid) and asymptotic (dashed) models.

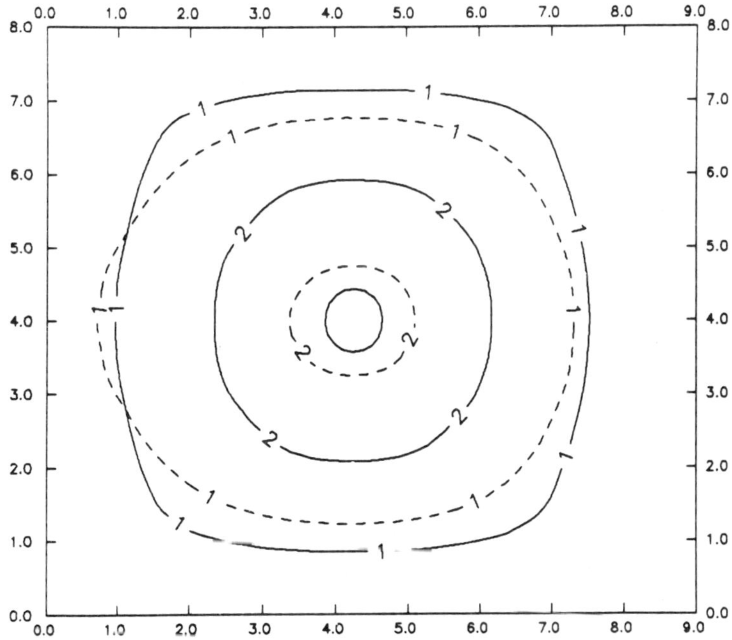

Figure 26. Concentration contours along x_1–x_2 plane of symmetry in Example 5 considering retardation ($r = 2$) and decay ($\lambda = 0.5$) at $t_\mu = 0.5$ as represented by 16 × 16 parcels and obtained using pre-asymptotic (solid) and asymptotic (dashed) models.

7. Summary

A quasilinear theory was described which explains how plumes of dissolved chemicals evolve with time when groundwater flows through multiscale geological media under non-Fickian and Fickian conditions. Expressions were given for the dispersivity tensor $\boldsymbol{\alpha}$ and plume covariance tensor $\mathbf{X}$ as functions of an effective directional Peclet number P and a dimensionless time t_μ (mean travel distance relative to log hydraulic conductivity correlation or integral scale) under various degrees of statistical anisotropy. These expressions show that in statistically homogeneous media in which the integral scales and variance of log hydraulic conductivity are finite, $\boldsymbol{\alpha}$ initially increases in direct proportion to t_μ but tends asymptotically to a constant as t_μ becomes large. While the longitudinal dispersivity increases monotonically toward its Fickian asymptote, the transverse dispersivities climb to a peak before descending towards their asymptotes when $P > 1$. When the mean seepage velocity vector $\boldsymbol{\mu}$ is parallel to a principal axis of statistical anisotropy, the longitudinal dispersivity acts parallel to $\boldsymbol{\mu}$. When the mean velocity is not collinear with any such axis, the direction of longitudinal spread may be offset significantly toward that of the largest integral scale and then rotate back to stabilize at a smaller angle of deviation as the regime becomes Fickian. The theory has been successful in reproducing the observed spread of depth-averaged concentrations in a three-year tracer experiment at Borden, Ontario, during which dispersivities varied with time in a non-Fickian fashion.

Many stochastic theories of flow and transport in the subsurface (refer to Cushman, 1987) are limited to media in which log hydraulic conductivity is a second-order stationary field with finite integral scales and variance. In such statistically homogeneous media it is sometimes possible to associate effective hydraulic conductivity values with suitably defined representative elementary volume's (REVs); a plume spreading through a homogeneous medium of infinite extent may eventually reach a Fickian mode of transport. However, real geological media often consist of discrete units which may exhibit statistical homogeneity on a hierarchy of scales. In such multiscale media there is no unique REV because units defined on different scales may have different effective hydraulic conductivities or none at all. As long as a plume is confined largely to a discrete, statistically homogeneous geological unit it may reach a Fickian or quasi-Fickian mode of transport on the corresponding scale. In the long run, the plume spreads through a hierarchy of such units and its mode of transport is therefore non-Fickian. In other words, dispersion in geological media is inherently non-Fickian with at best intermittent episodes of Fickian or quasi-Fickian behaviour.

The above conclusions appear to be supported by numerous longitudinal dispersivities from laboratory and field tracer studies in porous and fractured media throughout the world which grow consistently with the scale of the study. When this global scale effect is interpreted on the basis of a recent quasilinear dispersion theory due to Neuman and Zhang (1989) and Zhang and Neuman (1989) it leads to the remarkable conclusion that the media have a self-similar log hydraulic conductivity structure with a mean fractal dimension $D \cong 1.75$ and a universal semivariogram $\gamma(s) \cong 0.017\sqrt{s}$, regardless of whether they are porous or fractured. Since this semivariogram is associated with neither a finite integral scale nor a finite variance, it renders either the definition of a unique REV or the development of a long-term Fickian regime impossible.

Longitudinal dispersivities obtained from the calibration of numerical advection-dispersion models against concentration data from observed plumes were found to vary more slowly with the scale of the study than do dispersivities determined by other means. These models usually rely on groundwater velocities prescribed at the nodes of a grid on the basis of information about hydraulic heads, fluxes, and hydraulic conductivities or transmissivities. It has been suggested to us by J. Carrera (pers. commun.) that this is equivalent to accounting deterministically for (or filtering out) velocity variations on scales smaller than the mean travel distance of the plume so that the dispersivities become dependent on integral scales of similar order. This leads to the important operational conclusion that the scale parameter which controls dispersivities in a given transport model diminishes as the density of information about groundwater velocities goes up.

Finally, we have pointed out that Eulerian methods such as finite difference and finite element are not well suited for the numerical simulation of non-Fickian transport. We developed the blueprint for an alternative Lagrangian method based on the semidiscrete superposition of Gaussian miniplumes whose spread is calculated according to the quasilinear theory.

Acknowledgements

This work was supported in part by the U.S. Nuclear Regulatory Commission under contract number NRC-04-86-123. The second author (Y.-K.Z.) was a recipient of the John and Margaret Harshbarger Fellowship offered by the Department of Hydrology and Water Resources at the University of Arizona, and the third author (O.L.) was a recipient of the Rothschild Fellowship from Yad Hanadiv in Israel, while engaged in portions of this study. We wish to thank Professor Jesus Carrera of the Escuela de Caminos, Canales y Puertos, Universita Politecnica de Cataluña, Barcelona, Spain, for his valuable comments concerning the relationship between dispersivities and density of information about groundwater velocities.

References

Anderson, M.J. (1979). Modelling of groundwater flow systems as they relate to the movement of contaminants. *C.R.C. Crit. Rev. Environ. Control* **9**, 97–156.

Arya, A. (1986). Dispersion and reservoir heterogeneity. PhD dissertation, University of Texas, Austin, TX.

Arya, A., Hewett, T.A., Larson, R.G. and Lake, L.W. (1988). Dispersion and reservoir heterogeneity, *SPE Reservoir Engng* **3**, 139–148.

Barry, D.A., Coves, J. and Sposito, G. (1988). On the Dagan Model of Solute Transport in Groundwater: Application to the Borden Site, *Water Resour. Res.* **24**, 1805–1817.

Bear, J. and Verruijt, A. (1987). 'Modeling Groundwater Flow and Pollution'. D. Reidel Publishing Co., Dordrecht.

Burrough, P.A. (1983a). Multiscale sources of spatial variation in soil: I. The application of fractal concepts to nested levels of soil variation. *J. Soil Sci.* **34**, 577–597.

Burrough, P.A. (1983b). Multiscale sources of spatial variation in soil: II. A non-Brownian fractal model and its application in soil survey, *J. Soil Sci.* **34**, 599–620.

Byers, E. and Stephens, D.B. (1983). Statistical and stochastic analysis of hydraulic conductivity and particle size in a fluivial sand, *Soil Sci. Soc. Am. J.* **47**, 1072–1081.

Cushman, J.H. (1987). Development of stochastic partial differential equations for subsurface hydrology, *Stochastic Hydrol. Hydraul.* **1**, 241–262.

Dagan, G. (1984). Solute transport in heterogeneous porous formations, *J. Fluid Mech.* **145**, 155–177.

Dagan, G. (1987). Theory of solute transport by groundwater, *Ann. Rev. Fluid Mech.* **19**, 183–215.

Dagan, G. (1988). Time-dependent macrodispersion for solute transport in anisotropic heterogeneous aquifers, *Water Resour. Res.* **24**, 1491–1500.

Freyberg, D.L. (1986). A natural gradient experiment of solute transport in a sand aquifer, 2. Spatial moments and the advection and dispersion of nonreactive tracers, *Water Resour. Res.* **22**, 2031–2046.

Gelhar, L.W. (1988). Stochastic subsurface hydrology from theory to application, *Water Resour. Res.* **22**, 1355–1455.

Gelhar, L.W. and Axness, C.L. (1983). Three-dimensional stochastic analysis of macrodispersion in aquifers, *Water Resour. Res.* **19**, 161–180.

Gelhar, L.W., Mantoglou, A., Welty, C. and Rehfeldt, K.R. (1985). A review of field-scale physical solute transport processes in saturated and unsaturated porous media, EPRI EA-4190, Project 2485-5, August 1985.

Graham, W. and McLaughlin, D. (1989a). Stochastic analysis of nonstationary subsurface solute transport: 1. Unconditional moments. *Water Resour. Res.* **25**, 215–232.

Graham, W. and McLaughlin, D. (1989b). Stochastic analysis of nonstationary subsurface solute transport: 2. Conditional moments, *Water Resour. Res.* In press.

Güven, O. and Molz, F.J. (1986). Deterministic and stochastic analyses of dispersion in an unbounded stratified porous medium, *Water Resour. Res.* **22**, 1565–1574.

Hufschmied, P. (1985). Estimation of three-dimensional statistically anisotropic hydraulic conductivity field by means of single well pumping tests combined with flowmeter measurements. *In*: 'Proceedings of the International Symposium on the Stochastic Approach to Subsurface Flow', pp. 110–116. International Association for Hydraulic Research, Montvillargenne.

Journel, A.G. and Huijbregts, Ch.J. (1978). 'Mining Geostatistics'. Academic Press, London.

Kitanidis, P.K. (1988) Prediction by the method of moments of transport in a heterogeneous formation, *J. Hydrology* **102**, 453–473.

Koch, D.L. and Brady, J.F. (1987). A non-local description of advection-diffusion with application to dispersion in porous media, *J. Fluid Mech.* **180**, 387–403.

Koch, D.L. and Brady, J.F. (1988). Anomalous diffusion in heterogeneous porous media, *Phys. Fluids* **31**, 965–973.

Lake, L.W. (1988). A marriage of geology and reservoir engineering. *In*: 'Numerical Simulation of Oil Recovery' (M.F. Wheeler, ed.), pp. 177–198. Springer-Verlag, New York.

Lake, L.W. (1989). 'Enhanced Oil Recovery', Prentice Hall, Englewood Cliffs, NJ.

Lundgren, T.S. and Pointin, Y.B. (1976). Turbulent self-diffusion, *Phys. Fluids* **19**, 355–358.

Lallemand-Barres, A. and Peaudecerf, P. (1978). Recherche des relations entre la valeur de la dispersivite macroscopique d'un aquifere, ses autres caracteristiques et les conditions de mesure, *Bull. du BRGM, 2e Serie, Section III*, No. 4.

Mackay, D.M., Freyberg, D.L., Roberts, P.B. and Cherry, J.A. (1986). A natural gradient experiment on solute transport in a sand aquifer: 1. Approach and overview of plume movement, *Water Resour. Res.* **22**, 2017–2029.

Naff, R.L., Jim Yeh, T.-C. and Kemblowski, M.W. (1988). A note on the recent natural gradient tracer test at the Borden site, *Water Resour. Res.* **24**, 2099–2103.

Neuman, S.P. (1987) Stochastic continuum representation of fractured rock permeability as an alternative to the REV and fracture network concepts. *In*: 'Proceedings of the 28th U.S. Symposium on Rock Mechanics' (I.W. Farmer, J.J.K. Daemen, C.S. Desai, C.E. Glass and S.P. Neuman, eds), pp. 533–561. A.A. Balkema, Rotterdam.

Neuman, S.P. and Depner, J.S. (1988). Use of variable-scale pressure test data to estimate the log hydraulic conductivity covariance and dispersivity of fractured granites near Oracle, Arizona, *J. Hydrol.* **102**, 475–501.

Neuman, S.P. and Zhang, Y.-K. (1989). A quasilinear theory of non-Fickian and Fickian subsurface dispersion: 1. Theoretical analysis with application to isotropic media, *Water Resour. Res.* In press.

Neuman, S.P., Winter, C.L. and Newman, C.M. (1987). Stochastic theory of field-scale Fickian dispersion in anisotropic porous media, *Water Resour. Res.* **23**, 453–466.

Peaudecerf, G. and Sauty, J.-P. (1978). Application of a mathematical model to the characterization of dispersion effects on groundwater quality, *Progr. Water Technol.* **10**, 443–454.

Phytian, R. (1975). Dispersion by random fields, *J. Fluid Mech.* **67**, 145–153.

Pickens, J.F. and Grisak, G.E. (1981). Scale-dependent dispersion in a stratified granular aquifer, *Water Resour. Res.* **17**, 1191–1211.

Rinaldo, A., Marani, A. and Bellin, A. (1989). On mass response functions, *Water Resour. Res.* **25**, 1603–1617.

Roberts, P.V. and Mackay, D.M. (eds) (1986). A natural gradient experiment on solute transport in a sand aquifer, Technical Report 292, Department of Civil Engineering, Stanford University, Stanford, CA, September 1986.

Roberts, P.V., Goltz, M.N. and Mackay, D.M. (1986). A natural gradient experiment on solute transport in a sand aquifer, 3. Retardation estimates and mass balances for organic solutes, *Water Resour. Res.* **22**, 2047–2058.

Rubin, Y. (1989). Stochastic modeling of macrodispersion in heterogeneous porous media, *Water Resour. Res.* In press.

Sposito, G. and Barry, D.A. (1987). On the Dagan model of solute transport in groundwater: Foundational aspects, *Water Resour. Res.* **23**, 1867–1875.

Sudicky, E.A. (1986). A natural gradient experiment on solute transport in a sand aquifer: Spatial variability of hydraulic conductivity and its role in the dispersion process, *Water Resour. Res.* **22**, 2069–2082.

Sudicky, E.A., Cherry, J.A. and Frind, E.O. (1983). Migration of contaminants in groundwater at a landfill: A case study, 4. A natural gradient dispersion test, *J. Hydrol.* **63**, 81–108.

Taylor, G.J. (1921). Diffusion by continuous movements, *Proc. London Math. Soc.* **2**, 196–214.

Wheatcraft, S.W. and Tyler, S.W. (1988). An explanation of scale-dependent dispersivity in heterogeneous aquifers using concepts of fractal geometry, *Water Resour. Res.* **24**, 566–578.

Yeh, T.-C. (1987). Comment on 'Modeling of scale-dependent dispersion in hydrogeologic systems' by J.F. Pickens and G.E. Grisak, *Water Resour. Res.* **23**, 522.

Zhang, Y.-K. and Neuman, S.P. (1989). A quasilinear theory of non-Fickian and Fickian subsurface dispersion: 1. Application to anisotropic media and the Borden site, *Water Resour. Res.* In press.

Chapter XIV

Self-similar Randomness and Spectral Conditioning: Analysis of Scale Effects in Subsurface Hydrology

Rachid Ababou
Department of Civil Engineering and Operations Research, Princeton University, Princeton, NJ 08544, USA
(Present address: Southwest Research Institute (Div. 20), San Antonio, TX 78228, USA)

Lynn W. Gelhar
Department of Civil Engineering, Massachusetts Institute of Technology, Cambridge, MA 99352, USA

1. Introduction and overview

Stochastic approaches to groundwater flow and solute dispersion in heterogeneous aquifers have been extensively developed in the past decade (see Bakr *et al.*, 1978; Smith and Freeze, 1979a,b; Gutjhar and Gelhar, 1981; Dagan, 1981, 1987; Gelhar and Axness, 1983; Winter *et al.*, 1984; Naff and Vecchia, 1986; Gelhar, 1987; Ma *et al.*, 1987; Ababou *et al.*, 1988, 1990; McLaughlin and Wood, 1988a,b; and Graham and McLaughlin, 1989, among others). Essentially, these approaches seek approximate solutions of the stochastic partial differential equations of porous media hydrodynamics and transport, assuming that the hydraulic conductivity $K(x)$ is a random function of space, or 'random field'.

An important aspect, and perhaps the main ambiguity of stochastic approaches, resides in the elusive concept of randomness as applied to field problems. Broadly speaking, it seems that two main lines of thought have been pursued, although not always explicitly stated as such. In the 'ensemble approach', the theoretical probability space is viewed as the ensemble of all possible spatial realizations of the heterogeneous porous medium under certain constraints of smoothness, stationarity, etc. (this view is reflected at least in part in the work of Smith and Freeze, 1979a,b; Dagan, 1981; McLaughlin and Wood, 1988a,b, and a few others). In this framework, the ensemble variance $\sigma^2 = \mathrm{Var}(\ln K(x))$ may be interpreted as the uncertainty of log-conductivity at any given point x in space.

DYNAMICS OF FLUIDS IN HIERARCHICAL POROUS MEDIA
ISBN 0-12-200260-1

The other approach we have in mind may be qualified as the 'single world' approach of randomness. Here, the ergodic hypothesis (Yaglom, 1962) is used to justify a formal equivalence between the ensemble moments and the spatial averages of random fields over infinitely large domains. The variance σ^2 is viewed as an index of large-scale spatial variability, rather than uncertainty. In fact, ensemble moments are calculated as before using the formalism of probability theory, but they are only meaningful in the ergodic case where they can be identified to spatial moments. This spatial interpretation of randomness may lead to a simple methodology for characterizing the heterogeneity of real world phenomena, but it requires assuming that the spatial extent of these phenomena is somehow infinite.

In this chapter, we develop an alternative approach to spatial heterogeneity and uncertainty that provides a framework for resolving the apparent conflict between the different conceptualizations of randomness just described (see Cushman, 1988, for a different line of thought emphasizing the ensemble versus information viewpoints). Specifically, our objective is to develop a practical methodology for characterizing the variability and uncertainty of finite-scale flow and contaminant transport systems in naturally heterogeneous, random-like porous formations. The essence of the new approach may perhaps be grasped by considering briefly the problem of predicting solute concentrations and plume spreading in randomly heterogeneous flow systems. It has been shown that the hydrodynamic dispersion of passive tracers is determined, at large scales, by the spatial covariance structure or the spectral density tensor of groundwater velocity (Gelhar and Axness, 1983; Gelhar, 1987; Dagan, 1987). The validity of the velocity covariance tensor obtained from the spectral solution of Gelhar and Axness (1983), was recently confirmed numerically through single replicate simulations of stochastic groundwater flow (Ababou *et al.*, 1988).

The important role of spatial structure in transport suggests that the physical process of macrodispersion could be modelled like a deterministic one in a single replicate of a heterogeneous porous medium, instead of being modelled as a collection of such processes drawn from an ensemble space. On the other hand, specific engineering problems may require formulating transport models in the framework of risk analysis, e.g. with a view to evaluating the uncertainty of predicted concentration plumes. But this can only be achieved by considering the collection of all possible realizations of the porous medium—among other sources of uncertainty. In other words, it would seem that the ensemble approach of porous medium heterogeneity remains indispensable for purposes of risk analysis.

Based on these preliminary observations, we will develop in this chapter a slightly unorthodox interpretation of the notion of uncertainty. For instance, we argue that uncertainty about the predicted effective behaviour of contaminant plumes is essentially due to the large-scale heterogeneities of the flow system, perhaps of the order of a fraction of the plume size or larger. On the other hand, there are numerous smaller scale heterogeneities that interact in such a way that their global effects on contaminant spreading can be predicted with much more confidence. This view will be implemented in a more precise way through 'spectral conditioning' of stochastic equations, at least for the flow problem. This will lead to a unified treatment of spatial variability and uncertainty, relative to the specified scale of observation and/or to the characteristic scale of the physical process at hand.

A brief overview of the contents of this chapter follows. In Section 2, we present a summary of the spectral perturbation method, and apply it to the solution of the stochastic groundwater flow equation in the restrictive case of infinite domains. This serves as a background for subsequent developments.

In Section 3, the limitations of infinite domains, stationarity, and ergodicity hypotheses are discussed. We present a scenario to account for finite size effects in the case of a growing contaminant plume, discuss some available field data on the spatial distribution of hydraulic conductivity in aquifers, and finally argue in favour of a truncated self-similar random field model of conductivity variability. Some background material on random self-similarity is also provided.

In Section 4, we solve the multidimensional stochastic groundwater flow equations in finite domains using a spectral cut-off method. This only serves as a preliminary analysis of scale dependence when the flow system is either bounded or observed at some finite scale. Solutions of the groundwater flow equations are obtained in Fourier space, assuming that the log-conductivity field has a truncated self-similar spectrum. Conductivity data from boreholes are used to infer the exponent of the self-similar spectrum, and the domain size (L) is used to determine the low wavenumber spectral cut-off (L^{-1}). As domain size increases, the effective conductivity is found to decrease in one dimension, remains constant in two dimensions, and increases in three dimensions. Furthermore, an heuristic evaluation of three-dimensional macrodispersivity for an advected tracer suggests that longitudinal macrodispersivity linearly increases with respect to plume size or mean displacement.

Section 5 is devoted to the development and application of the 'spectral conditioning' approach. Many of the previously developed concepts and methods are used here, notably self-similarity, spectral cut-off, and the spectral perturbation method for solving stochastic equations. In addition, the method bears some resemblance to the conditional averaging technique used in Renormalization Group Theory. The spectral conditioning method is applied specifically to the case of one-dimensional saturated flow in a self-similar porous medium, given some scale of observation or plume size (L). The results of spectral conditioning clearly indicate the decreasing uncertainty of spatial statistics (head variance and effective conductivity) as the domain size increases.

A brief summary of the most significant concepts and results developed in this chapter will be found in the concluding Section 6.

2. Background on spectral method for infinite domains

Stochastic approaches to subsurface flow and transport are based usually on the assumption that the hydraulic coefficients of the porous medium—such as the natural logarithm of Darcy's conductivity ($\ln K$)—are normally distributed random functions of space with specified mean, variance, and covariance function. First- and second-order moments are often assumed invariant by translation, i.e. statistically homogeneous. For normally distributed random fields, this implies homogeneity for all other moments as well.

In what follows, we focus more specifically on the spectral approach to stochastic flow problems. As mentioned earlier, the infinite domain spectral theory is basically a 'single world' approach of randomness. Its aim is to characterize heterogeneous flow fields and solute concentration fields at the large scale, in particular through second-order moments and 'effective' coefficients of conductivity and dispersion. Approximate solutions of governing stochastic equations are obtained in Fourier space by using the spectral representation theorem of stationary random processes and statistically homogeneous random fields. The main approximations involved are neglecting high order fluctuation terms, and implicitly assuming that the random field solutions are normally distributed and statistically homogeneous over infinite domains. Under these assumptions, the spatial–statistical structure of a random field, e.g. the hydraulic head, is completely determined by its mean and its spectral density function.

Anticipating subsequent developments, we give below a concrete example of application of the spectral theory, namely for multidimensional groundwater flow. The results obtained in the literature are in fact much more general: they include closed form expressions for the variances and spectral densities of groundwater velocity, hydraulic head and concentration, and either closed form or quasi-analytical expressions for the effective conductivity and macrodispersivity tensors (Bakr *et al.*, 1978; Gelhar and Axness, 1983; Gelhar, 1986, 1987; Vomvoris, 1986; Ababou *et al.*, 1988). In treating the case of stochastic groundwater flow, we will follow in part the work of Bakr *et al.* (1978) and of Gelhar and Axness (1983), however with some differences. We explicitly use a regular perturbation method or 'σ-expansion' in the spectral solution procedure, as in Ababou *et al.* (1988) and Ababou and Gelhar (1990).

The steady-state partial differential equation governing hydraulic head in a multidimensional saturated porous medium with spatially variable conductivity $K(\mathbf{x})$ is obtained from the mass conservation equation and the local Darcy equation. These are, respectively:

$$\nabla \cdot \mathbf{Q} = 0 \tag{2.1}$$

$$\mathbf{Q} = -K(\mathbf{x}) \cdot \nabla H \tag{2.2}$$

where $\mathbf{Q}$ represents the specific discharge rate vector (Darcy velocity), and H the hydraulic head potential. Substituting Eq. 2.2 in Eq. 2.1 yields:

$$\nabla \cdot (\mathbf{K}(\mathbf{x}) \cdot \nabla H) = 0 \tag{2.3}$$

Let us now assume specifically that the conductivity $\mathbf{K}(\mathbf{x})$ in Eq. 2.3 is a second-order stationary random function of space, with log-normal probability distribution. The statistical properties of the log-conductivity, $F(\mathbf{x}) = \ln K(\mathbf{x})$, are entirely determined by its mean and its covariance function. These quantities are defined as follows:

$$F(\mathbf{x}) = \ln K(\mathbf{x}) \tag{2.4}$$

$$\langle F(\mathbf{x}) \rangle = \ln K_G \tag{2.5}$$

$$R_{ff}(\xi) = \langle (F(\mathbf{x}) - \langle F \rangle) \cdot (F(\mathbf{x} + \xi) - \langle F \rangle) \rangle \tag{2.6}$$

where $\langle\ \rangle$ stands for ensemble averaging (expectation operator). K_G is the geometric mean conductivity. R_{ff} is the covariance of log-conductivity between two points in space, and ξ is the separation vector between the two points.

We will now use a more convenient form of the groundwater flow equation. Equation 2.3 can be decomposed and expressed directly in terms of the log-conductivity, $F(\mathbf{x})$, as follows:

$$\nabla^2 H + \nabla F(\mathbf{x}) \cdot \nabla H = 0 \tag{2.7}$$

Let us now manipulate Eq. 2.7 by splitting each random field quantity into two parts, its ensemble average and its perturbation. The perturbations of the random log-conductivity and hydraulic head fields are defined as follows, using lower case symbols:

$$f(\mathbf{x}) = F(\mathbf{x}) - \langle F(\mathbf{x}) \rangle = \ln [K(\mathbf{x})/K_G] \tag{2.8}$$

$$h(\mathbf{x}) = H(\mathbf{x}) - \langle H(\mathbf{x}) \rangle \tag{2.9}$$

where $\langle f \rangle = 0$ and $\langle h \rangle = 0$ by construction. Taking the ensemble average of Eq. 2.7 leads to the mean flow equation:

$$\nabla^2 \langle H \rangle + \nabla \langle f \rangle \cdot \nabla \langle H \rangle = -\langle \nabla f \cdot \nabla H \rangle \tag{2.10}$$

and subtracting Eq. 2.10 from the original flow Eq. 2.7 yields the perturbation equations:

$$\nabla^2 h + \nabla \langle F \rangle \cdot \nabla h + \nabla f \cdot \nabla \langle H \rangle = -\{\nabla f \cdot \nabla h - \langle f \cdot \nabla h \rangle\}. \tag{2.11}$$

Finally, assuming that the mean log-conductivity $\langle F \rangle$ or the geometric mean K_G is constant, the mean and perturbation equations given above can be simplified as follows:

$$\nabla^2 \langle H \rangle = -\langle \nabla f \cdot \nabla h \rangle \tag{2.12a}$$

$$\nabla^2 h + \nabla f(\mathbf{x}) \cdot \nabla \langle H \rangle = -\{\nabla f \cdot \nabla h - \langle \nabla f \cdot \nabla h \rangle\} \tag{2.12b}$$

Observe that the mean and perturbation equations in Eq. 2.12 form a system of two stochastic equations with three unknowns: the mean head $\langle H(\mathbf{x}) \rangle$, the head perturbation $h(\mathbf{x})$, and the second-order term $\langle \nabla f \cdot \nabla h \rangle$. A complete solution of this problem would require solving an infinite hierarchy of equations governing higher order moments. This turns out to be an impossible task in the general case. However, various expansion methods may be used to arrive at approximate solutions that are asymptotically exact as some 'small parameter' goes to zero.

One way to obtain a first-order solution for the random hydraulic head field $H(\mathbf{x})$ is to expand H as a power series of the dimensionless parameter σ_f, the standard deviation of log-conductivity. By the same token, we now implicitly assume that σ_f, or σ for short, is a small parameter—*a priori* much smaller than unity. Accordingly, let:

$$H(\mathbf{x}) = H_0(\mathbf{x}) + \sigma H_1(\mathbf{x}) + \sigma^2 H_2(\mathbf{x}) + \cdots \tag{2.13}$$

and let $g(\mathbf{x})$ be the normalized log-conductivity random field having unit variance and zero mean, i.e.:

$$g(\mathbf{x}) = f(\mathbf{x})/\sigma \tag{2.14}$$

Note that the conductivity $K(\mathbf{x})$ is related to $g(\mathbf{x})$ via the relation:

$$K(\mathbf{x}) = K_G \exp(\sigma g(\mathbf{x})) \tag{2.15}$$

Substituting $g(\mathbf{x})$ in Eq. 2.7 with K_G constant gives:

$$\nabla^2 H + \sigma \nabla g \cdot \nabla H = 0 \tag{2.16}$$

Finally, substituting the former expansion of H in the latter equation yields the infinite hierarchy of equations (Ababou *et al.*, 1988; Ababou and Gelhar, 1990):

$$\left\{ \begin{array}{lll} \text{Order } \sigma^0: & \nabla H_0 & = 0 \\ \text{Order } \sigma^1: & \nabla^2 H_1 + \nabla g \cdot \nabla H_0 & = 0 \\ \vdots & \vdots & \end{array} \right. \tag{2.17}$$

Now, the zero-order equation given just above may be compared with the mean equation in Eq. 2.12. This shows that H_0 is just an approximation for the mean head $\langle H \rangle$, second-order accurate in σ. Likewise, the first-order term σH_1 in the series expansion appears to be an approximation for the head perturbation h, also second-order accurate in σ. Accordingly, we may now rewrite the mean and perturbation equations in Eq. 2.12 as follows:

$$\nabla^2 \langle H \rangle = O(\sigma^2) \tag{2.18a}$$

$$\nabla^2 h + \nabla f \cdot \nabla \langle H \rangle = O(\sigma^2) \tag{2.18b}$$

It turns out that these equations are tractable if the high order terms on the right-hand side are neglected, and if the mean hydraulic gradient $\langle \nabla H \rangle$ is assumed constant in the second equation. Note that $\langle \nabla H \rangle$ is indeed constant to second order in σ, since the first equation for the mean head satisfies the infinite domain Laplace equation to second order in σ. Accordingly, we assume from now on that the flow is uniform in the mean, and let the constant mean hydraulic gradient vector be denoted as:

$$\mathbf{J} = -\langle \nabla H \rangle \tag{2.19}$$

The perturbation equation can now be expressed, to second order in σ, in the simple form:

$$\nabla^2 h = \mathbf{J} \cdot \nabla f \tag{2.20}$$

Note that this is just a stochastic Poisson equation governing the head perturbation $h(\mathbf{x})$.

We now proceed to solve the stochastic Poisson equation (Eq. 2.20) in Fourier space by using the spectral representations of $f(\mathbf{x})$ and $h(\mathbf{x})$ in terms of Fourier–Stieltjes integrals (Yaglom, 1962; Loève, 1978; Priestley, 1981):

$$f(\mathbf{x}) = \int_{\mathbb{R}^m} e^{i\mathbf{kx}} \, dZ_f(\mathbf{k}) \tag{2.21a}$$

$$h(\mathbf{x}) = \int_{\mathbb{R}^m} e^{i\mathbf{kx}} \, dZ_h(\mathbf{k}) \tag{2.21b}$$

where $dZ_h(\mathbf{k})$ is a complex zero-mean random increment representing the Fourier amplitude of $h(\mathbf{x})$ at wavenumber $\mathbf{k}$. In a more general setting, $dZ_h(\mathbf{k})$ should be replaced by the random measure $Z_h(d\mathbf{k})$. The important point is that the spectral representation Eq. 2.21 exists and is unique for any zero-mean homogeneous random field that is sufficiently smooth. Homogeneity implies that the two point covariance function depends only on the separation vector between the two points; non-smooth conductivity fields having, say, discontinuous sample functions, may be represented in a similar way by using generalized random fields based on the Schwartz theory of distributions (Yaglom, 1962); an example of application is given in Cushman (1986). By writing Eq. 2.21 for $h(\mathbf{x})$, we assume in particular that $h(\mathbf{x})$ is a homogeneous random field, i.e. statistically invariant under translation.

The usefulness of the spectral representation, besides its uniqueness, lies in the fact that it is 'orthogonal'. This means that the random $\mathrm{d}Z(\mathbf{k})$ terms have the interesting property of being uncorrelated:

$$\langle \mathrm{d}Z(\mathbf{k})\mathrm{d}Z^*(\mathbf{k}')\rangle = 0 \quad \text{for} \quad \mathbf{k} \neq \mathbf{k}' \tag{2.22}$$

In addition, for each given $\mathbf{k}$, the real and imaginary parts of $\mathrm{d}Z(\mathbf{k})$ are zero-mean, uncorrelated, and identically distributed. Also, since the random fields at hand are real-valued, it is easily seen that the real part of $\mathrm{d}Z(\mathbf{k})$ must be even in $\mathbf{k}$ while the imaginary part must be odd:

$$\mathrm{d}Z(-\mathbf{k}) = \mathrm{d}Z^*(\mathbf{k}) \tag{2.23}$$

If the random field is Gaussian, as we assumed for f, then the complex $\mathrm{d}Z(\mathbf{k})$ terms are normally distributed, i.e. their real and imaginary parts are jointly Gaussian random variables for each fixed $\mathbf{k}$. Finally, the expectation of $|\mathrm{d}Z|^2$ is twice the variance of the real part of $\mathrm{d}Z$, and represents the spectral content of fluctuations of the random field in the wavenumber range $(\mathbf{k}, \mathbf{k} + \mathrm{d}\mathbf{k})$. More precisely, it can be shown that:

$$\langle |\mathrm{d}Z(\mathbf{k})|^2 \rangle = S(\mathbf{k})\,\mathrm{d}\mathbf{k} \tag{2.24}$$

where $S(\mathbf{k})$ is the spectral density function. Last but not least, the Wiener–Khinchine theorem states that the spectral density function is just the Fourier transform of the covariance function $R(\xi)$.

By the uniqueness of the above spectral representation, substituting Eq. 2.21 into Eq. 2.13 yields the following relation between the complex Fourier increments of $h(\mathbf{x})$ and $f(\mathbf{x})$:

$$\mathrm{d}Z_h(\mathbf{k}) \cong -i\frac{(J_1 k_1)}{k^2}\,\mathrm{d}Z_f(\mathbf{k}) \tag{2.25}$$

Finally, multiplying both sides by the complex conjugate $\mathrm{d}Z_h^*(\mathbf{k})$, and averaging, gives a simple relation between the spectral density functions of $h(\mathbf{x})$ and $f(\mathbf{x})$:

$$S_{hh}(\mathbf{k}) = \frac{(J_1 k_1)^2}{k^4}\,S_{ff}(\mathbf{k}) \tag{2.26}$$

where we have employed Einstein's summation convention on repeated indices. Observe that Eq. 2.26 is *a priori* valid for any number of spatial dimensions. The symbol k denotes the radial wavenumber, so we have $k^2 = k_1^2 + k_1^2 + k_3^2$ in the three-dimensional case. Finally, the head covariance function $R_{hh}(\xi)$ can be obtained by applying an inverse Fourier transform to its spectral density function, i.e.:

$$R_{hh}(\xi) = \int_{\mathbb{R}^m} e^{i\mathbf{k}\xi} S_{hh}(\mathbf{k})\, d\mathbf{k} \tag{2.27}$$

In particular, the head variance is given by $\sigma_h^2 = R_{hh}(0)$, which is obtained directly by integrating the head spectral density function.

The spectral method may also be used to obtain the moments of the velocity vector in addition to those of the hydraulic head. The procedure is analogous to the above and will not be developed here. Observe, however, that Gelhar and Axness (1983) obtained the velocity spectrum by inserting the previous head solution in the stochastic Darcy equation; an alternative method directly based on velocity equations was recently developed by Ababou *et al.* (1988, Section 4.3). The shape of the spectral density tensor of velocity turns out to be the same in both cases—only the velocity variances differ.

Finally, the effective conductivity $\bar{K}_{ij}$ is a constant symmetric tensor that satisfies a relation of the form:

$$\langle Q_i \rangle = -\bar{K}_{ij} \frac{\partial \langle H \rangle}{\partial x_j} \tag{2.28}$$

Briefly, $\bar{K}_{ij}$ is obtained by averaging the velocity or flux vector $Q(x)$ given by the local Darcy equation, using the spectrum of hydraulic head known from previous results (Gelhar and Axness, 1983). A particularly simple approximate expression for $\bar{K}_{ij}$ in the case of statistically anisotropic three-dimensional media was recently proposed (Ababou *et al.*, 1988, Section 4.2.6). This conjecture was based, in part, on spectral solutions, and was found to be exact for at least three special cases where exact solutions are available.

In closing, a few more comments on the nature of these spectral perturbation solutions may be in order. As we have seen, such solutions are approximate due to the regular perturbation expansion involved. However, the errors due to truncating the σ-expansion can be at least partially compensated by replacing terms like $(1 - \sigma_f^2/2)$, by their exponential equivalents like $\exp(-\sigma_f^2/2)$. In fact, this procedure gives the exact result in one dimension. Moreover, we have found that, in the one dimensional case, one obtains the exact solution by summing all the terms of the infinite series Eq. 2.13 (Ababou and Gelhar, 1990). This shows that, in one dimension at least, the σ-expansion is convergent, a fact that does not seem to have been noted previously.

3. Finite domain, spectral cut-off and self-similarity

One of the major difficulties encountered in field applications of the spectral method comes from the fact that groundwater flow and, most importantly solute concentration, cannot be truly ergodic and statistically homogeneous since the physical domains are necessarily bounded. There arises the problem of identifying a 'large scale' such that the infinite domain approximation holds.

Other stochastic methods have been developed in the literature in order to deal explicitly with non-ergodicity and non-stationarity. For instance, approximate non-stationary solutions of groundwater flow with prescribed boundary conditions and internal sources can be obtained by using regular perturbation methods similar to Eqs 2.13–2.18. One such method is the approximate Green's function method (Dagan, 1981). Another alternative is to solve numerically the approximate moment equations (McLaughlin and Wood, 1988a,b; Graham and McLaughlin, 1989). Provided mild variability of hydraulic conductivity, these methods may be particularly appropriate for studying irreducibly non-stationary phenomena—e.g. for risk analysis of drawdowns near a pumping well and other 'near source' problems. However, it seems desirable to retain the analytical simplicity of the spectral method for problems that exhibit less extreme forms of non-stationary behaviour. The case of groundwater flow in semi-infinite aquifers bounded by two planes was tackled by Naff and Vecchia (1986) using a combination of Green's function and infinite domain spectral representations. Here, we propose a direct extension of the spectral method to analyse the statistics of subsurface flow over finite domains of arbitrary size.

The perturbing effects due to finite domain size and to boundary conditions (if any) may be quite mild in certain situations. For instance, in the case of an ergodic log-conductivity field, comparisons with detailed numerical simulations of single realizations of three-dimensional groundwater flow have indicated that the infinite domain spectral theory is indeed quite accurate for domain size of the order of a few tens of log-conductivity correlation lengths or more (Ababou *et al.*, 1985, 1988, 1990). In the case of anisotropic correlation structure of the log-conductivity, the correlation length to be taken into account is essentially the one parallel to the mean flow. For a comprehensive account of results with both isotropic and anisotropic correlations, see the report by Ababou *et al.* (1988, Vol. 2, Sections 6.3, 6.4, 6.5).

In the simulations mentioned above, the overall flow pattern was chosen to be more or less 'uniform in the mean'. In particular, there were no sources or sinks, and the boundary conditions were such that the driving force for the flow was aligned with lateral domain boundaries. In addition, as mentioned above, the input log-conductivity fields were ergodic. The spectral method seems best adapted to the study of such flows that appear statistically uniform, at least at some sufficiently large scale. In other words, the ideal situation is one where the inhomogeneities that may exist have no individual influence on the overall mean flow pattern.

However, in field contamination problems, this may not be the case. For instance, the scale of analysis may be imposed by specific policy considerations, such as a target time or the distance between the source and compliance boundaries. In fact, many hydrological phenomena like saturated or unsaturated solute transport, originate from local sources and take place over finite regions whose scales increase with time. The spatial structure of the geological formation, and the associated flow field, need not appear statistically homogeneous at the scales of interest. Moreover, experimental studies published in the literature indicate that the statistical properties of conductivities or transmissivities measured at different scales—or by different methods—may vary greatly. In particular, the 'correlation scale' is quite frequently found to be either of the order of measurement spacing, or else of the order of domain size: see results of aquifer data analysis by Hoeksema and Kitanidis (1985). In addition, both the correlation and the variance may be strongly influenced by certain subjective choices like detrending (a striking example is given by Devary and Doctor, 1982). In our view, these observations indicate that, at least in some cases, the inferred correlation structure of the conductivity field depends strongly on the chosen scale of observation (see the discussion in the introductory chapter to this text), and on other artifacts like the removal of large-scale trends.

Scale-dependent effects of this nature may play an equally important role in determining the spatial–statistical structure of flow and mass transport systems. Consider for instance a subsurface contaminant plume originating from a local source, and convected in a steady groundwater flow field through a heterogeneous porous formation. This situation is schematically illustrated in Fig. 1 (top). As the plume grows and invades larger regions of the porous formation, the transported solute 'responds' to larger and larger heterogeneities. At early times, when the evolving plume is still small, its advance and spreading is strongly influenced by local heterogeneities that appear to be large in comparison with plume size. Later on, the same local heterogeneities become mere fluctuations with respect to the larger size of the plume. In summary, as the plume grows and spreads, there is a gradual change in the threshold size of heterogeneities, such that larger ones affect the global trends of the plume, while smaller ones contribute to random-like dispersion within the plume. According to this scenario, it seems likely that the apparent macrodispersion of contaminant depends on plume size, travel distance, and time scale.

Scale dependence may also play a perturbing role in spatial-statistical analyses of groundwater flow systems. Figure 1 (centre) shows a 200 foot ($\sim$ 60 m) transect of log-conductivity from vertical borehole data in the Mount Simon aquifer (Gelhar, 1976; Bakr, 1976). The straight lines, which are superimposed on the data, illustrate that the apparent mean level and mean trend of the observed variable depends on the size of the domain of interest. Note that the

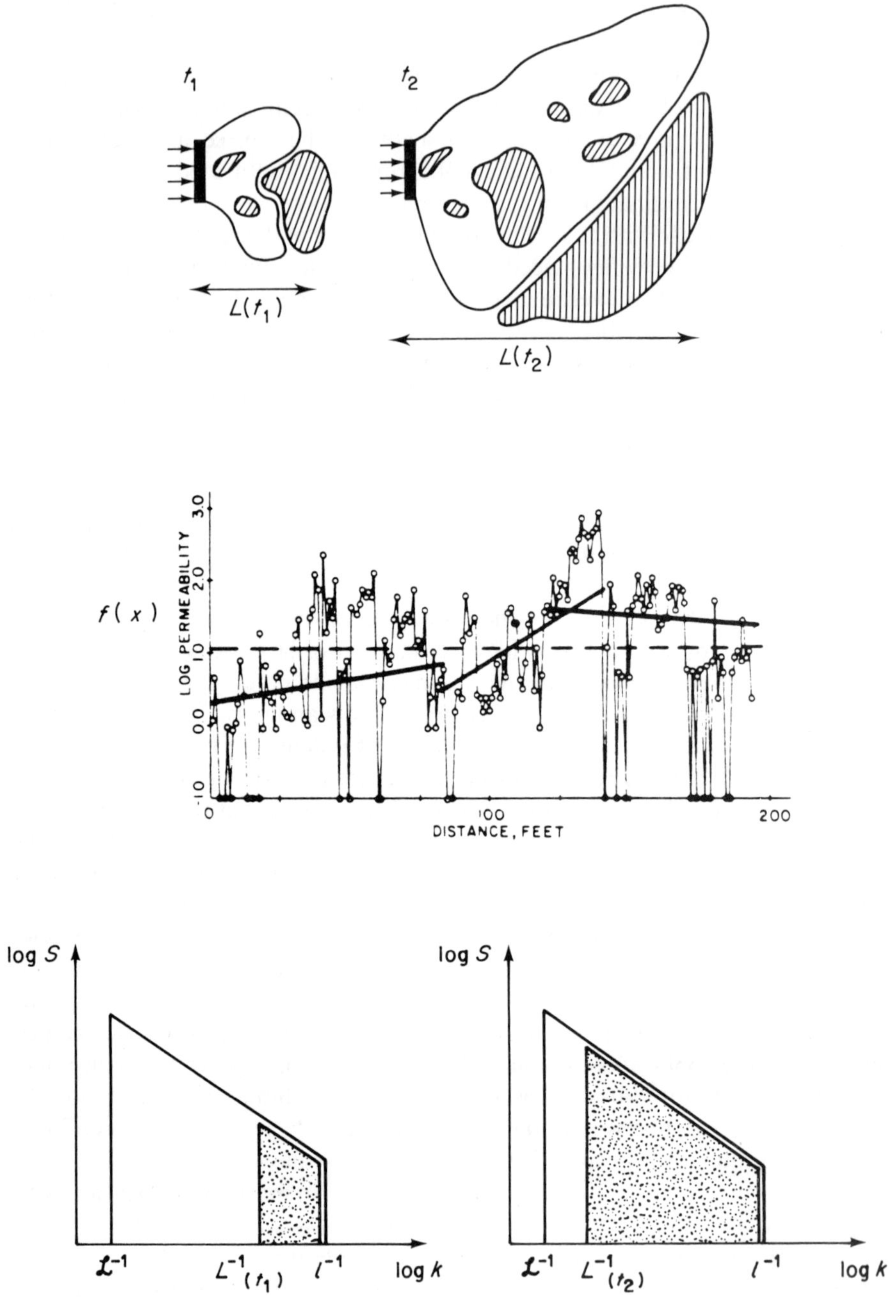

Figure 1. Illustration of finite size effects: contamination plume (top); sample function of log-conductivity process (centre); truncated self-similar spectrum (bottom).

mean log-conductivity appears more or less constant at the scale 200 feet, while there exist pronounced trends in the mean log-conductivity when observed at the scale 75 feet (~ 23 m). Also, the trend varies more or less randomly, depending on the precise location of the 75 foot piece of record examined. These observations—based on field data—suggest that the effective conductivity at a site is scale dependent, and varies at random depending on the location of the site relative to the actual heterogeneities of the porous formation. This, of course, is based on intuitive arguments rather than formal analysis (so far).

Let us now briefly indicate how the standard spectral theory could be manipulated to incorporate finite-size effects. A first step in this direction consists in using a band-pass spectrum with low and high wavenumber cut-offs for representing the spatial structure of the log-conductivity field at the scale of interest. The main idea here is to let the low wavenumber cut-off value be proportional to the inverse size of the domain ($k_{\min} \cong 1/L$). The high wavenumber cut-off $k_{\max} \cong 1/l$ may be related to measurement spacing (or possibly to the scale of conductivity measurements). Strictly speaking, we have $L = L'/2\pi$ where L' is the true domain size, and we may take $l = l'/\pi$ where l' is the measurement spacing, L and l are surrogates for L' and l', and they will be used as such in the sequel. Finally, this band-pass spectral representation of finite-scale phenomena will be considerably simplified by assuming a *self-similar behaviour* of the spectrum within some range of scales.

To illustrate these ideas, we have represented in the bottom part of Fig. 1 a log–log plot of a self-similar spectrum with two distinct low wavenumber cut-offs. The lowest value $k = \mathscr{L}^{-1}$ is the wavenumber below which the spectral content of the log-conductivity becomes negligible, while the value $k = L^{-1}$ is more 'application-specific', being the cut-off wavenumber corresponding to the size of the domain of interest. The main idea is that the spectral region $L^{-1} < k < l^{-1}$, i.e. the dotted region in the bottom part of Fig. 1, describes the 'stationary part' of random spatial variability, which may be analysed using the standard spectral method developed in the previous section. The lower part of the spectrum ($k < L^{-1}$) will be interpreted as uncertainty in the spectral conditioning approach to be developed later.

One-dimensional spectra obtained from log-conductivity data in three vertical boreholes of the Mount Simon aquifer are shown in Figs 2–4, reproduced from Bakr (1976). In Fig. 2, the domain size was $L = 303$ ft and the measurement spacing was $l = 1$ ft. The log-spectral density is plotted against log-frequency, where frequency is expressed in cycles/feet (wavenumber $= 2\pi \times$ frequency). Note that the range of low to high frequencies used in this figure goes approximately from $(L/5)^{-1}$ to $(l/2)^{-1}$. The frequency range is therefore one order of magnitude less than the ratio L/l. In each of Figs 2, 3 and 4 the straight line superimposed on the spectral data depicts a self-similar spectrum of the form:

$$S_{ff}(k) = \frac{S_0}{k^{\alpha}} \tag{3.1}$$

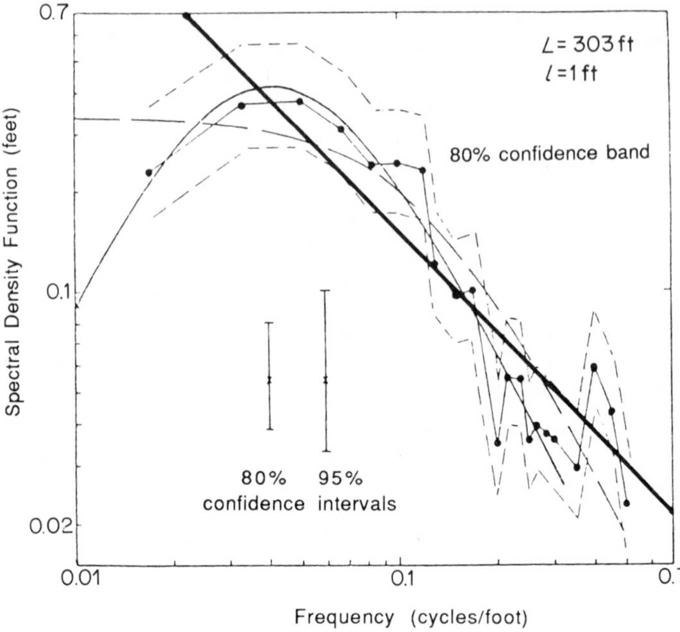

Figure 2. One-dimensional spectral density of measured log-conductivity in a vertical borehole (black circles), Mount Simon aquifer, from Bakr (1976). The straight line represents a self-similar spectrum with exponent $\alpha = 1$.

where the exponent α and the intercept S_0 were fitted visually to the spectral data. Observe that the coefficient α, the slope of the spectral density function in a log–log plot, appears to be close to unity in all three cases ($\alpha \approx 1$). The meaning of this coefficient may be better grasped after a brief review on self-similar randomness.

Self-similarity, or fractal behaviour, is the property of certain physical phenomena to replicate themselves on many scales. This means that certain identifiable patterns, structures, or fluctuations, replicate themselves in almost identical form at many different length scales, via a simple similarity transformation. For instance Fig. 5, reproduced from Mandelbrot (1983), shows a three-dimensional view of a deterministic fractal sponge constructed by dividing the initial cube into 3^3 smaller cubes, replacing some of these by voids, and dividing again each of the remaining cubes into 3^3 smaller cubes, etc, *ad infinitum*. Each part of the cube is similar to the whole, that is, each smaller cube is exactly identical to the whole cube after application of a similarity transformation.

Mandelbrot (1983) developed and expanded both the deterministic and *random* aspects of fractal geometry, and investigated the manifestations of fractal behaviour in nature. In particular, Mandelbrot proposed to use a class of

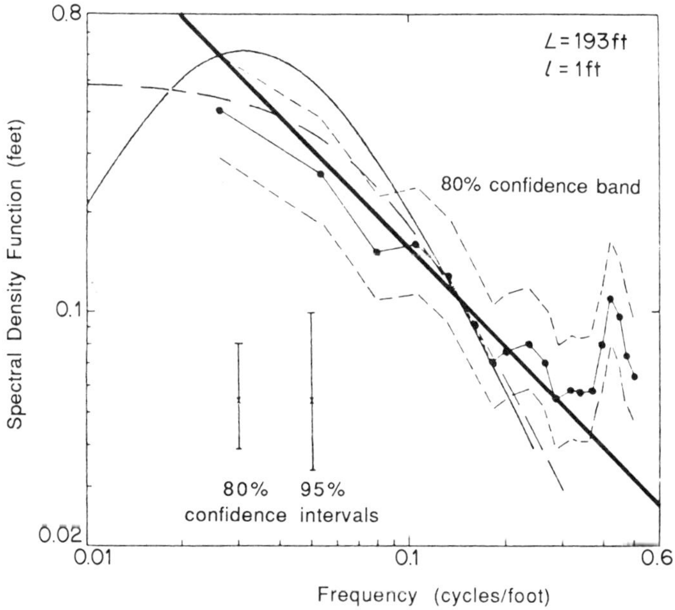

Figure 3. Same as Fig. 2, for another borehole (Bakr, 1976). The black circles represent the spectral density from field data, and the straight line represents a self-similar spectrum with exponent $\alpha = 1$.

statistically self-similar random fields, the fractional Brownian motion (fbm), as a model for generating landscapes. The 'fbm' has also been considered for modelling a number of other natural phenomena, including hydrological time series and geophysical records in space (Mandelbrot and Wallis, 1969). More recent studies along these lines include the modelling of metal fracture surfaces (Mandelbrot *et al.*, 1984), and spatial analysis of soil granulometry data (Burrough, 1981) among others.

Briefly, the one-dimensional fractional Brownian motion is a stationary-increment random function possessing the property of self-similarity, or more appropriately self-affinity. That is, a random function $f(\mathbf{x})$ is said to be 'self-similar' (self-affine) if its increments are statistically invariant under the scaling transformation $\Delta f(\mathbf{x}) \rightarrow \lambda^{-H} \Delta f(\lambda \mathbf{x})$. Here, statistical invariance may be understood in the mean-square sense, although a stronger statement is also possible. By analogy with the deterministic fractal of Fig. 5, any part of the graph of the random function $y = f(\mathbf{x})$ is statistically identical to the whole graph after application of an affinity transformation ($\Delta \mathbf{x} \rightarrow \lambda \Delta \mathbf{x}$ and $\Delta \mathbf{y} \rightarrow \lambda^{H} \Delta \mathbf{y}$).

A more complete review of the properties of one-dimensional fractional Brownian motion can be found in Mandelbrot and Van Ness (1968). The

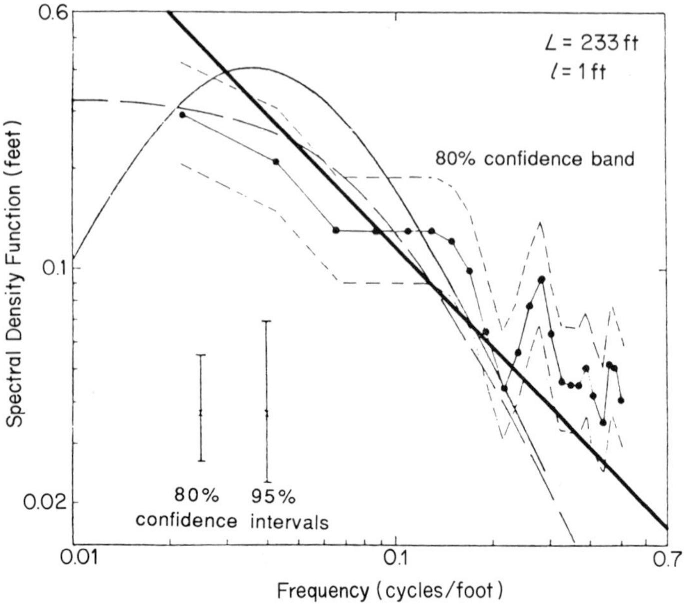

Figure 4. Same as Figs 2 and 3, for another borehole (Bakr, 1976). The black circles represent the spectral density from field data, and the straight line represents a self-similar spectrum with exponent $\alpha = 1$.

parameter H is known as the Hurst coefficient, from Hurst's study of the stages of the Nile river, or Hölder coefficient, as $f(x)$ satisfies a kind of 'Hölder condition'. Note in particular that the case $H = 1/2$ corresponds to the usual Brownian motion, which has independent increments. For $1/2 < H < 1$, the fbm processes are 'persistent', having positively correlated increments, while for $0 < H < 1/2$ they are 'anti-persistent' processed with negatively correlated increments.

Now, the spectral density of the one-dimensional 'fbm' is precisely given by Eq. 3.1 with exponent $\alpha = 2H + 1$. Since $\alpha \approx 1$ for the three log-conductivity records shown in Figs 2, 3, and 4, we conclude that the Hurst coefficient is close to zero for the Mount Simon borehole data. This type of 'fbm' is generally known as '$1/f$ noise', because the spectral density varies as the inverse of frequency. The $1/f$ behaviour indicates a very noisy, anticorrelated structure of the log-conductivity within the available range of scales (1 ft–300 ft). This finding seems in accordance with the survey by Burrough (1981), who obtained a low Hurst coefficient for the sand and clay fractions of soils ($H \simeq 0.2$). It may be worth noting, however, that the Hurst coefficient could be evaluated in a different fashion, not necessarily by spectral analysis. For instance, note that the variance of increments (core function or variogram) of the fractional

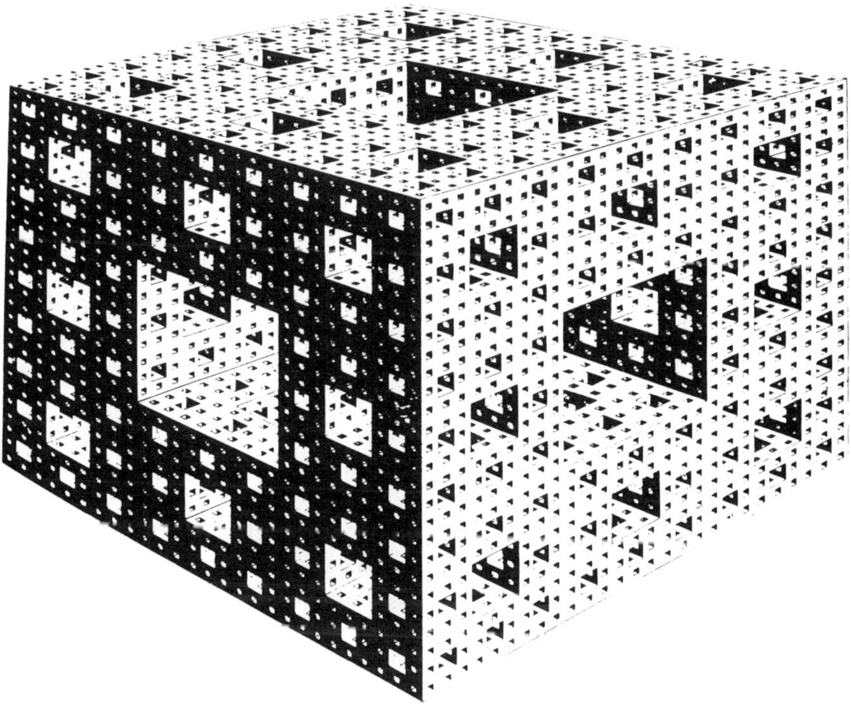

Figure 5. Fractal sponge, a deterministic self-similar set in $\mathbb{R}^3$, which was originally constructed by Menger and is also known as the Menger sponge (from Mandelbrot, 1983).

Brownian motion is of the form:

$$C_H(\xi) \sim |\xi|^{2H} \tag{3.2}$$

where ξ is the lag distance. For the Mount Simon data, the core function would probably be nearly flat since $H \simeq 0$, but this would have to be checked by plotting the variogram directly, without prior detrending of the data. Other methods such as 'R/S analysis' (Mandelbrot, 1983) could also be used.

Finally, note that the above definition of fractional Brownian motion can be extended isotropically to multidimensional space, as will be seen shortly. Many comments and references on multidimensional isotropic Levy–Brown fractional motions can be found in the book by Mandelbrot (1983). The reader may also consult Monin and Yaglom (1965) for applications in the context of statistical turbulence theory, or the more mathematically oriented book by Adler (1981).

In summary, the data shown in Figs 2–4 indicate that the self-similar spectrum, with exponent one in Eq. 3.1, is a reasonable model of randomness within the available range of length scales (1–300 ft). The agreement with spectral data is not too good in the low wavenumber region, but spectral densities estimated at wavenumbers of the order of the inverse domain size or less may not be reliable here. In fact, the estimated confidence intervals themselves—shown as dashed lines in Figs 2–4—are not reliable in the low wavenumber region of the spectrum. They are based on approximate, asymptotic evaluations of the sampling bias and error variance, and may lose their meaning as $k \to 0$, depending on the behaviour of the true spectrum at hand. See Priestley (1981, Section 6.2.4) for a classical exposé of the sample statistics of smoothed spectral estimates.

Based on the previous discussion, we now model the log-conductivity as a band-pass self-similar process with Hurst coefficient zero. Its one-dimensional spectral density is given by:

$$S_{ff}^{(1)}(k_1) = \left| \begin{array}{ll} S_0/|k_1| & \text{for } L^{-1} \leqslant |k_1| \leqslant l^{-1} \\ 0 & \text{otherwise} \end{array} \right. \tag{3.3}$$

where the large scale cut-off L represents the size of the region in which flow or transport takes place (L may evolve in time), while l represents some small scale of the order of measurement spacing or possibly sample size. In a log–log plot, the intercept of the spectrum at the origin (S_0) may be viewed as an intrinsic measure of variability of the porous formation, independent of the particular cut-off scales L and l. Accordingly, we define the 'intrinsic variance' of log-conductivity as:

$$\sigma_0^2 = S_0 \tag{3.4}$$

On the other hand, the 'observed' variance of log-conductivity, within the range of scales (l,L), can be computed by integrating the one-dimensional band-pass spectrum as follows:

$$\sigma_f^2 = 2 \int_{L^{-1}}^{l^{-1}} \frac{\sigma_0^2}{k_1} \, dk_1 = 2\sigma_0^2 \ln\left(\frac{L}{l}\right) \tag{3.5}$$

Note that σ_f^2 is scale-dependent, being a function of the ratio (L/l). This equation specifically shows that the 'observed' variance of log-conductivity increases logarithmically with domain size.

As announced above, we now generalize this approach of randomness to the more realistic case of three-dimensional space, assuming however that the log-conductivity is *statistically isotropic*. First, note that the one-dimensional spectra displayed in Figs 2–4 are essentially the Fourier transforms of the log-conductivity data sampled along a single direction. This procedure yields a one-dimensional marginal spectrum, related to the full three-dimensional spectrum by:

$$S_{ff}^{(1)}(k_1) = \int\int S_{ff}(k_1,k_2,k_3)\,\mathrm{d}k_2\,\mathrm{d}k_3 \tag{3.6}$$

Based on this remark, let us now define a three-dimensional isotropic band-pass self-similar spectrum analogous to its one-dimensional counterpart:

$$S_{ff}(k) = \begin{vmatrix} \dfrac{\sigma_0^2}{2\pi k^3} & \text{for } L^{-1} \leqslant k \leqslant l^{-1} \\ 0 & \text{otherwise} \end{vmatrix} \tag{3.7}$$

By substituting Eq. 3.7 into Eq. 3.6, one obtains the corresponding one-dimensional marginal spectrum. It turns out that this spectrum approximates quite well the one-dimensional self-similar spectrum of Eq. 3.3, at least far enough from the cut-off wavenumbers as shown below:

$$S_1(k_1) = \sigma_0^2\left[\frac{1}{(k_1^2 + L^{-2})^{\frac{1}{2}}} - \frac{1}{(k_1^2 + l^{-2})^{\frac{1}{2}}}\right] \simeq \frac{\sigma_0^2}{|k_1|} \quad \text{for } L^{-1} \ll k_1 \ll l^{-1} \tag{3.8}$$

Since the above spectrum agrees fairly well with the one-dimensional spectral data of Figs 2–4, this justifies our choice of the three-dimensional spectrum Eq. 3.7.

We have assumed that the porous formation is statistically isotropic. Although this may be a serious limitation for applications, the more general case of statistically anisotropic media will not be treated in the present context. For completeness, note that the 'observed' variance of log-conductivity in a finite three-dimensional region of spherical shape can be obtained by integrating the isotropic spectrum Eq. 3.7. This leads to the same relation as in the one-dimensional case, namely:

$$\sigma_f^2 = 2\sigma_0^2 \ln(L/l) \tag{3.9}$$

where L and l are now the *radial cut-off scales* in three-dimensional space. Again, note that σ_f^2 increases logarithmically with radial domain size.

We are now ready to investigate the statistical behaviour of stochastic flow systems with the one-dimensional and three-dimensional truncated self-similar spectra defined above.

4. Stochastic solutions for truncated self-similar spectra

We now develop approximate solutions of stochastic flow that partially take into account the influence of finite domain size, using the standard spectral method of infinite domains previously outlined in Section 2. We assume for instance that the geometric mean conductivity K_G and the mean hydraulic gradient $\mathbf{J} = -\nabla\langle H\rangle$ are spatially constant. Our approach is to introduce finite size effects indirectly, via a low wavenumber cut-off of the conductivity spectrum as explained in the previous section.

By construction, a constant mean random field with a truncated self-similar spectrum of the form Eq. 3.3 or Eq. 3.7 is statistically homogeneous, at least in a formal sense. This implies that the spectral representation theory previously used for infinite domains applies in the present case as well. There may be technical difficulties due to the discontinuity of spectral density at the cut-off wavenumbers, but these can be dealt with by smoothing out the cut-off, and will be ignored here. Admittedly, the above assumptions may be too simplistic to take fully into account the influence of domain size, particularly due to the assumed stationarity in the mean. However, we choose to go along with this simplified analysis, postponing to a later section a more complete treatment of finite size effects via the spectral conditioning approach.

Accordingly, we develop here the stochastic solutions of groundwater flow at a finite scale L by applying the standard spectral method using truncated self-similar spectra for the log-conductivity field (Eqs 3.3 and 3.7). The calculations for the head variance and the effective conductivity are relatively straightforward and need not be reproduced in detail here (see basic equations in Section 2). The truncated self-similar spectrum can also be inserted in the asymptotic expression for the longitudinal macrodispersivity of an advected solute at large times (Gelhar and Axness, 1983). This simplified approach to the transport problem will allow us, in one instance, to examine the possible scale dependence of solute macrodispersivity in the pre-asymptotic regime as well.

For one-dimensional flow, and with the truncated self-similar log-conductivity spectrum Eq. 3.3, the head variance is found to be of the form:

$$\sigma_h^2 = \sigma_0^2 J^2 (L^2 - l^2) \tag{4.1}$$

For three-dimensional flow, with isotropic truncated self-similar log-conductivity spectrum Eq. 3.7, the head variance is similarly:

$$\sigma_h^2 = \tfrac{1}{3}\sigma_0^2 J^2 (L^2 - l^2) \tag{4.2}$$

In comparison, the infinite domain results for a one-dimensional hole-exponential model (Gutjahr and Gelhar, 1981) and for a three-dimensional isotropic Markov spectrum (Bakr *et al.*, 1978) were, respectively:

$$\sigma_h^2 = \sigma_f^2 J^2 \lambda_{3\mathrm{D}}^2 \tag{4.3}$$

$$\sigma_h^2 = \tfrac{1}{3} \sigma_f^2 J^2 \lambda_{1\mathrm{D}}^2 \tag{4.4}$$

Note that the λs do not necessarily represent the same type of correlation length in Eqs 4.3 and 4.4. In comparison with the previous solutions (Eqs 4.1 and 4.2), it is interesting to note that, when $l \ll L$, the results are identical provided that σ_f^2 be replaced by σ_0^2, and λ by L. Since L represents domain size, the head variance obtained with the truncated self-similar spectra appears to depend mainly on large-scale structures, and not on small-scale fluctuations. For instance, assuming $l \ll L$ yields, for the three-dimensional case, the simple result:

$$\sigma_h^2 \simeq \tfrac{1}{3} \sigma_0^2 J^2 L^2 \tag{4.5}$$

Note in particular that σ_0^2 is the 'intrinsic' log-conductivity variance and does not depend on the domain size (L).

For the three-dimensional effective conductivity, the case of isotropic truncated self-similarity yields a result that is formally identical to the solution obtained by Gelhar and Axness (1983) for infinite domains. Namely, the effective conductivity tensor is isotropic and its components are given by:

$$\bar{K}_{11} = K_{\mathrm{G}} \exp(\sigma_f^2/6) \tag{4.6}$$

However, remember that σ_f^2 represents here the observed variance of log-conductivity, or finite domain variance, within the range of scales (l,L). Substituting Eq. 3.9 into Eq. 4.6 shows explicitly the scale dependence of the three-dimensional effective conductivity:

$$\bar{K}_{11} = K_{\mathrm{G}} \left(\frac{L}{l} \right)^{\sigma_0^2/3} \tag{4.7}$$

Let us illustrate numerically the behaviour of effective conductivity from Eq. 4.7. For the data of Fig. 2, the intrinsic log-conductivity variance σ_0^2 is found to be of the order of 0.06. This yields a fairly slow rate of growth of the effective conductivity with domain size, e.g. a + 10% increase of $\bar{K}$ when L increases by two orders of magnitude. Note, however, that this example corresponds to a mildly heterogeneous formation ($\sigma_f^2 = 0.67$, from equation 4.7 with $l = 1$ ft and $L = 300$ ft). For highly variable media, we expect values of σ_0^2 of the order of 0.3–0.6. In such cases the effective conductivity given by Eq. 4.7 could increase by a factor of two or more as the domain size L increases by two orders of magnitude. This increase of the three-dimensional effective conductivity with size can be interpreted as follows. When the size of the domain of interest is allowed to grow, larger heterogeneities are included. The high conductivity zones outweigh the low conductivity zones, due to the relatively high degree of freedom of fluid trajectories in a three-dimensional isotropic medium.

In one dimension however, the situation is reversed. Indeed, it is easily found that the one-dimensional effective conductivity is the harmonic mean:

$$\bar{K}_{11} = K_{\mathrm{H}} = K_{\mathrm{G}} \exp(-\sigma_f^2/2) \tag{4.8}$$

Now, expressing as before the observed variance σ_f^2 in terms of the intrinsic variance σ_0^2, we obtain the explicit scale dependence of the one-dimensional effective conductivity:

$$\bar{K}_{11} = K_{\mathrm{G}} \left(\frac{L}{l} \right)^{-\sigma_0^2} \tag{4.9}$$

For example, if σ_0^2 is of the order of 0.3–0.6, the effective conductivity of a one-dimensional flow system will *decrease* by one order of magnitude when the domain size is increased by two orders of magnitude. This behaviour is due to the low degree of freedom of fluid trajectories in one-dimensional space. As the scale of observation is increased, larger heterogeneities are included in the domain, and low conductivity zones outweigh high conductivity zones (consider what happens if an impervious block is placed in a one-dimensional porous medium).

Finally, the two-dimensional isotropic case appears to be special. The effective conductivity is found to be equal to the geometric mean, and does not depend on domain size or observation scale, unlike the one- and three-dimensional cases. This finding emphasizes the particular symmetry of two-dimensional statistically isotropic media; for infinite domains, it was already known that the effective conductivity is the geometric mean and does not otherwise depend on variance (Matheron, 1967; and Gelhar and Axness, 1983, among others).

The same type of simplified approach can be used to evaluate the scale dependence of the macrodispersivity of a convected plume. For simplicity, we focus here exclusively on the case of a three-dimensional statistically isotropic medium. Following Gelhar (1987), the longitudinal macrodispersity *at large times* and in the absence of local dispersion is given by:

$$A_{11} = \frac{\pi}{\bar{Q}_1^2} \iint_{-\infty}^{+\infty} S_{q_1 q_1}(0, k_2, k_3)\, \mathrm{d}k_2\, \mathrm{d}k_3 \tag{4.10}$$

where $S_{q_1 q_1}(\mathbf{k})$ is the spectral density function of the longitudinal component of flux (analogous to velocity). This spectrum can be obtained using the spectral solution method outlined above (Section 2), and particularly the flux-based approach of Ababou *et al.* (1988, Section 4.3). Accordingly, when the mean flow is aligned with the x_1-axis, the spectral density tensor of flux is given by:

$$S_{q_iq_j}(\mathbf{k}) = \langle Q_1 \rangle^2 \left\{ \delta_{1i} - \frac{k_1 k_i}{k^2} \right\} \left\{ \delta_{1j} - \frac{k_1 k_j}{k^2} \right\} S_{ff}(\mathbf{k}) \tag{4.11}$$

Inserting the three-dimensional isotropic, truncated self-similar spectrum given by Eq. 3.7 in the above expression gives:

$$S_{q_1q_1}(\mathbf{k}) = \langle Q_1 \rangle^2 \frac{\sigma_0^2}{2\pi} \left(1 - \frac{k_1^2}{k^2} \right)^2 \cdot \frac{1}{k^3} \tag{4.12}$$

with the understanding that $S_{q_1q_1}$ vanishes outside the wavenumber range (L^{-1}, l^{-1}). Finally, substituting this spectrum into Eq. 4.10 and integrating gives the following expression for the longitudinal macrodispersivity:

$$A_{11} = \sigma_0^2 (L - l) \tag{4.13}$$

Equation 4.13 should be taken with a grain of salt given the approximations involved here, particularly in the asymptotic dispersion relation (Eq. 4.10). Nevertheless, the result suggests that the pre-asymptotic macrodispersivity of a finite plume grows linearly with the size of the plume. For $l \ll L$, the pre-asymptotic macrodispersivity should increase in time roughly as:

$$A_{11}(t) \simeq \sigma_0^2 L(t) \tag{4.14}$$

where $L(t)$ is some global length scale characteristic of the plume. Intuitively, the growth of $A_{11}(t)$ could be due to the fact that, as the plume travels and spreads, a wider range of conductivity heterogeneities is allowed to contribute to the process of hydrodynamic dispersion. Furthermore, relation 4.14 indicates the possibility of a positive feedback: increasing the macrodispersivity enhances the spread of the plume, which in turn enhances the dispersion process and leads to a large macrodispersivity coefficient.

In summary, the stochastic solutions obtained with truncated self-similar spectra revealed some aspects of finite size effects (physical domain size, scale of observation, etc.). For a given three-dimensional heterogeneous formation, our analysis suggests that the head standard deviation and the longitudinal macrodispersivity both increase linearly with the size of the domain, while the effective conductivity increases more slowly like a mild power function of domain size. In one-dimensional space, though, it is found that the effective conductivity decreases with domain size. This behaviour is thought to be physically realistic, e.g. based on 'degree of freedom' arguments.

In closing, let us recall some of the limitations of the previously obtained results. First and foremost, our analysis of finite scale effects was based on applying the infinite domain spectral theory of stationary flow (Section 2) without modifications. The shortcomings of this kind of analysis will motivate the more general approach of spectral conditioning developed in the next section. Secondly, the important case of statistical anisotropy was ignored, and the scale of the domain was assumed to be the same in all directions. Our analysis of the influence of space dimensionality may be viewed as a preliminary study of the more general case where both the random heterogeneities and the domain's length scales are anisotropic.

5. Spectral conditioning and uncertainty

We now proceed to develop further the finite-domain approach, building on the previous results obtained with the truncated self-similar spectrum. Our purpose is to show how the effects of local trends in the random medium structure could be taken into account by using the idea of 'spectral conditioning'. This will be illustrated shortly for the simple case of one-dimensional saturated flow in a random self-similar medium. However, the range of applications is potentially broader. Ultimately, we have in view the stochastic characterization of groundwater flow and advection-dispersion phenomena in relation to some typical 'observation scale', consistent with the space–time scales of solute contamination.

The spectral conditioning method bears some resemblance to the effective averaging procedure used in renormalization group theory, a powerful method for studying critical phenomena in statistical physics (Wilson and Kogut, 1974; Wilson, 1975, 1979, 1983). Briefly, Wilson's Fourier space version of the renormalization group involves defining some average state (e.g. magnetization) over finite regions of fixed size L. This average state is obtained by truncating a spectral representation so as to include only the low wavenumber range $0 \leqslant k \leqslant L^{-1}$. One may then compute an effective property (effective Hamiltonian) by conditionally averaging the quantity of interest (Hamiltonian) over the local fluctuations or high wavenumbers $k > L^{-1}$. This constitutes only the first step of the whole renormalization procedure, which also involves iterating and rescaling to advance towards a fixed-point solution.

The spectral conditioning approach proposed here does not involve the whole recursion of renormalization group theory, but uses the idea of freezing the low frequencies while solving conditionally for the higher frequencies. The stochastic flow equation is first solved for small-scale fluctuations, up to domain scale, conditioned on larger scale fluctuations. This solution is obtained in Fourier space, with high wavenumbers conditioned on low wavenumbers, hence the name 'spectral conditioning'. Large-scale fluctuations have been kept fixed in this first step. In the second step, the random nature of large-scale fluctuations is finally taken into account. The stochastic flow equation is solved in Fourier space for the 'unfrozen' low wavenumbers. This leads to evaluating the uncertainty of the conditioned spatial statistics that had been calculated in the previous step.

In order to be more specific, let us develop a concrete application of the spectral conditioning approach in a relatively simple case. From now on, we focus on the one-dimensional stochastic equation of steady saturated flow in a random porous medium:

$$\frac{d^2H}{dx^2} + \frac{dF}{dx} \cdot \frac{dH}{dx} = 0 \tag{5.1}$$

where the log-conductivity $F(x)$ is a random field whose spectrum is assumed to be self-similar within some wavenumber range $\mathscr{L}^{-1} \leqslant k \leqslant l^{-1}$, and zero outside.

Suppose we are interested in finding some statistical characteristics of the flow system in a finite domain of size L, comprised between L and $\mathscr{L}$. The situation is illustrated schematically in Fig. 6 (top and bottom). The smallest scale l corresponds, say to measurement spacing or sample size (our precise choice for l is not the main point here). The intermediate scale L represents the size of the domain of interest (to a multiplicative constant), say for instance the expected size of the contaminated zone. And the largest scale $\mathscr{L}$, or regional scale, is such that the log-conductivity is roughly self-similar up to fluctuation scales on the order of $\mathscr{L}$.

If a piece of size L of the sample function $F(x)$ is isolated, as shown in Fig. 6 (bottom), the mean value $\bar{F}(x)$ estimated over the domain of size L will appear different from the regional mean $\langle F \rangle$. Our main purpose is to incorporate this kind of effect in Eq. 5.1 by way of 'spectral conditioning'. The key to this approach consists in writing the spectral representation of $F(x)$—which is formally a constant mean stationary random field—in such a way that the wavenumbers larger and smaller than $k = L^{-1}$ be clearly distinguished. Accordingly, let:

$$F(x) = \langle F \rangle + f(x) = \langle F \rangle + \bar{f}(x) + \tilde{f}(x) \tag{5.2a}$$

where:

$$\begin{aligned} \bar{f}(x) &= \int_{\mathscr{L}^{-1} \leqslant |k| \leqslant L^{-1}} e^{ikx}\, dZ_F(k) \\ \tilde{f}(x) &= \int_{L^{-1} \leqslant |k| \leqslant l^{-1}} e^{ikx}\, dZ_F(k) \end{aligned} \tag{5.2b}$$

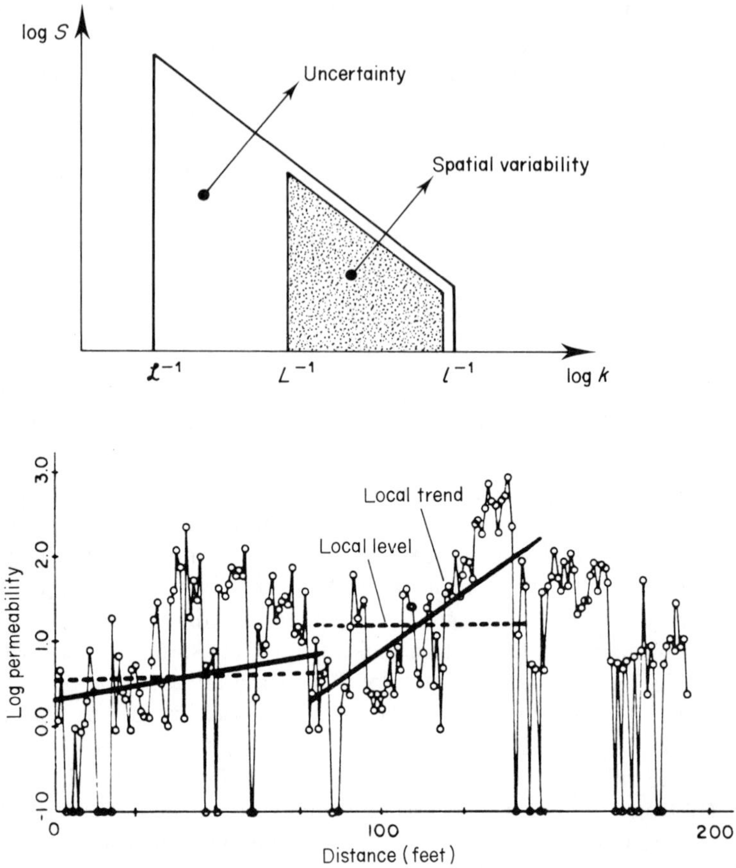

Figure 6. Illustration of the spectral conditioning method: log-spectral density versus log-wavenumber (top); sample function of log-conductivity process (bottom).

and $\langle F \rangle$ represents the ensemble average of log-conductivity, also equivalent to an infinite domain average by the stationarity and ergodicity properties of $F(x)$. In practice, $\langle F \rangle$ corresponds to some large scale regional mean. The perturbation $f(x)$ was split in two parts. The first term, $\bar{f}(x)$, is a smoothly varying zero-mean random function involving only those fluctuations of scale larger than L, and the second term, $\tilde{f}(x)$ is rapidly fluctuating zero-mean random function that involves all the higher wavenumber components (local fluctuations). As the size of the domain L is allowed to increase and reach the value of the regional scale $\mathscr{L}$, the spectral representation of Eq. 5.2 reduces to the usual infinite-domain spectral representation (Eq. 2.21).

Let us now 'freeze' the smooth perturbation $\bar{f}(x)$ by assuming that the Fourier coefficients at low wavenumbers are fixed. This procedure is equivalent to conditioning the log-conductivity field $F(x) = \ln K(x)$ with respect to its low wavenumber fluctuations, and also with respect to the specific location of the domain. Say, for instance, that x_0 defines the spatial location of the geometric centre of the domain $\left(x_0 - \frac{L}{2} \leqslant x \leqslant x_0 + \frac{L}{2}\right)$. Thus, $\bar{f}(x)$ is known when the position x_0 of the domain and all smooth fluctuations on the order of domain scale or larger are known. Now, substituting Eq. 5.2 in Eq. 5.1 while holding $\bar{f}(x)$ fixed, should give a stochastic equation governing the local fluctuations $\tilde{f}(x)$.

In order to implement this idea, we must first split the spectral representation of the hydraulic head process as was done for the log-conductivity field. This gives:

$$H(x) = \langle H(x) \rangle + \bar{h}(x) + \tilde{h}(x) \tag{5.3}$$

where the regional mean $\langle H \rangle$, the smooth perturbation $\bar{h}(x)$, and the local fluctuations $\tilde{h}(x)$, are defined as in Eq. 5.2. Now, substituting Eq. 5.2 and Eq. 5.3 in Eq. 5.1 gives the stochastic flow equation in the desired form:

$$\frac{d^2\langle H \rangle}{dx^2} + \frac{d^2\bar{h}}{dx^2} + \frac{d^2\tilde{h}}{dx^2} + \left(\frac{d\bar{f}}{dx} + \frac{d\tilde{f}}{dx}\right)\left(\frac{d\langle H \rangle}{dx} + \frac{d\bar{h}}{dx} + \frac{d\tilde{h}}{dx}\right) = 0 \tag{5.4}$$

Upon conditional averaging of this equation, with $\bar{f}$ and $\bar{h}$ fixed in the sense defined earlier, the local fluctuations $\tilde{f}$ and $\tilde{h}$ vanish. This leaves a conditional mean equation, of the form:

$$\frac{d^2\langle H \rangle}{dx^2} + \frac{d^2\bar{h}}{dx^2} + \frac{d\bar{f}}{dx} \cdot \frac{d\langle H \rangle}{dx} + \frac{d\bar{f}}{dx} \cdot \frac{d\bar{h}}{dx} = -\overline{\left(\frac{d\tilde{f}}{dx} \cdot \frac{d\tilde{h}}{dx}\right)} \tag{5.5}$$

where the 'bar' sign stands for the conditional spectral average defined in Eq. 5.2.

By subtracting Eq. 5.5 from Eq. 5.4, one also obtains the equation governing the local head fluctuations $\tilde{h}$:

$$\frac{d^2\tilde{h}}{dx^2} + \frac{d\tilde{f}}{dx} \cdot \frac{d\langle H \rangle}{dx} + \frac{d\tilde{f}}{dx} \cdot \frac{d\bar{h}}{dx} + \frac{d\bar{f}}{dx} \cdot \frac{d\tilde{h}}{dx} = -\left\{\frac{d\tilde{f}}{dx} \cdot \frac{d\tilde{h}}{dx} - \overline{\left(\frac{d\tilde{f}}{dx} \cdot \frac{d\tilde{h}}{dx}\right)}\right\} \tag{5.6}$$

Neglecting the second-order perturbation term on the right-hand side by analogy with the formal σ-expansions developed in Section 2, gives finally:

$$\frac{d^2\tilde{h}}{dx^2} + \frac{d\tilde{f}}{dx} \cdot \frac{d\langle H \rangle}{dx} + \frac{d\tilde{f}}{dx} \cdot \frac{d\bar{h}}{dx} + \frac{d\bar{f}}{dx} \cdot \frac{d\tilde{h}}{dx} \simeq 0 \tag{5.7}$$

where the local fluctuations $\tilde{h}$ are conditioned on the smoother fluctuations of $\bar{f}$ and $\bar{h}$ (these may represent subregional scale inhomogeneities).

Before attempting to solve the stochastic equation for $\tilde{h}(x)$, we need some information on the local mean perturbation term $\bar{h}(x)$. This information may be obtained by solving Eq. 5.5 for $\bar{h}(x)$, with $\bar{f}(x)$ considered now as a *random* function rather than a prescribed deterministic function. In other words, this is the unconditional step of the method. The smooth perturbation $\bar{f}(x)$ is allowed to vary randomly over all fluctuation scales larger than L, and over all possible positions of the local flow domain in space: see the local trends shown in Fig. 6, bottom.

Accordingly, we now apply ensemble average operators to Eq. 5.5 in order to obtain, respectively, the *unconditional* ensemble mean equation for $\langle H \rangle$ and the *unconditional* perturbation equation for $\bar{h}$. Note that the sum $\langle H \rangle + \bar{h}(x)$ basically represents the local mean hydraulic head evaluated at the finite scale L. The unconditional mean and perturbation equations are, respectively:

$$\frac{d^2\langle H \rangle}{dx^2} = -\left\langle \frac{d\bar{f}}{dx} \cdot \frac{d\bar{h}}{dx} \right\rangle - \left\langle \overline{\frac{d\tilde{f}}{dx} \cdot \frac{d\tilde{h}}{dx}} \right\rangle \tag{5.8a}$$

$$\begin{aligned} \frac{d^2\bar{h}}{dx^2} + \frac{d\bar{f}}{dx} \cdot \frac{d\langle H \rangle}{dx} &= -\left\{ \frac{d\bar{f}}{dx} \cdot \frac{d\bar{h}}{dx} - \left\langle \frac{d\bar{f}}{dx} \cdot \frac{d\bar{h}}{dx} \right\rangle \right\} \\ &\quad - \overline{\left\{ \frac{d\tilde{f}}{dx} \cdot \frac{d\tilde{h}}{dx} - \left\langle \frac{d\tilde{f}}{dx} \cdot \frac{d\tilde{h}}{dx} \right\rangle \right\}} \end{aligned} \tag{5.8b}$$

As before, the high order perturbation terms appearing in the right-hand sides are assumed to be negligible. Thus, we obtain finally the unconditional mean and perturbation equations:

$$\frac{d^2\langle H \rangle}{dx^2} \simeq 0 \tag{5.9a}$$

$$\frac{d^2\bar{h}}{dx^2} + \frac{d\bar{f}}{dx} \cdot \frac{d\langle H \rangle}{dx} \simeq 0 \tag{5.9b}$$

Equation 5.9a implies in particular that the ensemble mean hydraulic gradient, or 'regional gradient', is a constant which we denote by J:

$$J = -\frac{d\langle H \rangle}{dx} \tag{5.10}$$

Inserting this into Eq. 5.9b yields finally the simple equation:

$$-\frac{\mathrm{d}\bar{h}}{\mathrm{d}x} \simeq -J \cdot \bar{f}(x) \tag{5.11}$$

relating the perturbation of the local mean head gradient $(-\mathrm{d}\bar{h}/\mathrm{d}x)$ to the perturbation of the local mean log-conductivity $(\bar{f})$. This relation appears intuitively correct at first sight, and could have been inferred directly from the simple form taken by the Darcy equation in one-dimensional space: the flux is constant and the hydraulic gradient is inversely proportional to the conductivity.

The *conditional* perturbation Eq. 5.7 can now be solved for the local head fluctuations $\tilde{h}(x)$ by substituting the above relations and holding $\bar{f}(x)$ fixed, non-random. This gives the stochastic equation:

$$\frac{\mathrm{d}^2\tilde{h}}{\mathrm{d}x^2} + \frac{\mathrm{d}\bar{f}}{\mathrm{d}x} \cdot \frac{\mathrm{d}\tilde{h}}{\mathrm{d}x} \simeq J(1 - \bar{f})\frac{\mathrm{d}\tilde{f}}{\mathrm{d}x} \tag{5.12}$$

where randomness is due to the local fluctuation term $\tilde{f}(x)$ on the right-hand side. Recall that $\bar{f}(x)$ is by construction a slowly varying function, compared to the faster oscillations of $\tilde{f}(x)$. This suggests simplifying Eq. 5.12 by assuming that the average slope $\mathrm{d}\bar{f}/\mathrm{d}x$ and the average level $f(x)$, taken separately as two independent variables, are approximately constant within the flow domain of size L. This is illustrated in the bottom part of Fig. 6.

With this provision, Eq. 5.12 becomes a linear stochastic equation, which can be solved conveniently in Fourier space using the standard spectral method of Section 2. This yields the random Fourier–Stieltjes increments of the local head fluctuations $\tilde{h}$, conditioned on the values taken by $\bar{f}$ and $\mathrm{d}\bar{f}/\mathrm{d}x$:

$$\mathrm{d}Z_{\tilde{h}}(k) \simeq -J(1 - \bar{f})\frac{ik}{k^2 - ik\left(\dfrac{\mathrm{d}\bar{f}}{\mathrm{d}x}\right)}\mathrm{d}Z_{\tilde{f}}(k) \tag{5.13}$$

Multiplying the $\mathrm{d}Z_{\tilde{h}}$ term by its complex conjugate and taking the ensemble expectation finally gives the conditional head spectral density:

$$S_{\tilde{h}\tilde{h}}(k) \simeq J^2(1 - \bar{f})^2 \frac{1}{k^2 + \left(\dfrac{\mathrm{d}\bar{f}}{\mathrm{d}x}\right)^2} S_{\tilde{f}\tilde{f}}(k) \tag{5.14}$$

where $S_{\tilde{f}\tilde{f}}(k)$ is the truncated self-similar spectral density, equal to σ_0^2/k within the range $\mathrm{L}^{-1} \leqslant k \leqslant l^{-1}$ and vanishing outside this range. Low wavenumber effects $(k \leqslant L^{-1})$ have been incorporated entirely into the terms $\bar{f}$ and $\mathrm{d}\bar{f}/\mathrm{d}x$. Note that the domain of validity of Eq. 5.14 is limited to moderate departure of the average level of log-conductivity from its ensemble mean, i.e. we must require $|\bar{f}| \ll 1$ *a priori*.

The *conditional* head variance may now be obtained directly from Eq. 5.14 by integrating the spectral density of head over the wavenumber range (L^{-1}, l^{-1}). After some manipulations, and replacing $(1 - \bar{f})$ by $\exp(-\bar{f})$ to extend the range of validity of the first-order approximations, we obtain:

$$\tilde{\sigma}_h^2 \simeq \sigma_0^2 J^2 \frac{e^{-2\bar{f}}}{\left(\dfrac{d\bar{f}}{dx}\right)^2} \ln \left\{ \frac{1 + \left(\dfrac{d\bar{f}}{dx}\right)^2 L^2}{1 + \left(\dfrac{d\bar{f}}{dx}\right)^2 l^2} \right\} \qquad (5.15)$$

It is interesting to note that this expression gives the same head variance as the standard spectral theory when the scale L goes to infinity ($L \gg \mathscr{L}$). In the limit, the local and regional means coincide, so that both $\bar{f}$ and $d\bar{f}/dx$ vanish. And a careful Taylor development of the logarithmic term shows that the right-hand side of Eq. 5.15 coincides exactly with the head variance given by Eq. 4.1 as $\bar{f}$ and $d\bar{f}/dx$ both go to zero. Equation 5.15 may be compared to the *conditional* log-conductivity variance obtained by integrating the log-conductivity spectrum in the range $L^{-1} \leqslant k \leqslant l^{-1}$. We have indeed from Eq. 3.5:

$$\tilde{\sigma}_f^2 = 2\sigma_0^2 \ln\left(\frac{L}{l}\right) \qquad (5.16)$$

where σ_0^2 is the intrinsic variance of log-conductivity, independent of the scale L. It is clear that the conditional head variance Eq. 5.15 may be expressed alternatively in terms of $\tilde{\sigma}_f^2$ rather than σ_0^2.

The last step of the spectral conditioning method consists in recognizing the random nature of the conditional (local) head variance; i.e. the smooth log-conductivity perturbation $\bar{f}(x)$ appearing in Eq. 5.15 may be viewed as a random field rather than a 'frozen' constant. Recall that $\bar{f}(x)$ was defined as the difference between the local mean (L) and the regional mean ($\mathscr{L}$) of the log-conductivity field Eq. 5.2. Therefore, $\bar{f}(x)$ is a random field whose spectral content coincides with that of log-conductivity in the low wavenumber range $\mathscr{L}^{-1} \leqslant k \leqslant L^{-1}$. The conditional head variance $\tilde{\sigma}_h^2$ is viewed as a local variance because it quantifies spatial variability at the finite scale L, given a finite domain whose particular features are embodied in $\bar{f}(x)$. By the same token, $\tilde{\sigma}_h^2$ is a random parameter; its randomness arises from the many possible configurations of the heterogeneous formation at large.

Evaluation of the uncertainty of the $\tilde{\sigma}_h^2$ statistic involves computing the ensemble mean and ensemble variance of Eq. 5.15, which may then be used to obtain a coefficient of variation (note that the ensemble mean of Eq. 5.15 is the *unconditional* head variance). This requires in particular the ensemble variance of the zero-mean random terms $\bar{f}$ and $\mathrm{d}\bar{f}/\mathrm{d}x$:

$$\mathrm{Var}(\bar{f}) = 2\sigma_0^2 \ln\left(\frac{\mathscr{L}}{L}\right) \tag{5.17a}$$

$$\mathrm{Var}\left(\frac{\mathrm{d}\bar{f}}{\mathrm{d}x}\right) = \sigma_0^2\left(\frac{1}{L^2} - \frac{1}{\mathscr{L}^2}\right) \tag{5.17b}$$

Using these intermediate results, it is possible to obtain, after some manipulations, a crude estimate of the coefficient of variation of the local head variance from Eq. 5.15. The result is:

$$CV(\tilde{\sigma}_{\tilde{h}}^2) \approx 1 - \exp\{-2[2\sigma_0^2 \ln(1/\varepsilon)]^{\frac{1}{2}}\}\frac{\ln[1 + \sigma_0^2(1-\varepsilon^2)]}{\sigma_0^2(1-\varepsilon^2)} \tag{5.18}$$

where ε is the length scale ratio:

$$\varepsilon = L/\mathscr{L} \leqslant 1 \tag{5.19}$$

Let us use hypothetical data to mimic the case of a fairly heterogeneous formation. We assume $\sigma_0^2 \simeq 0.5$ and $\varepsilon = 0.1$. This yields a fairly large coefficient of variation of the head variance of the order of 100%. The reason for this large uncertainty can be seen directly from Eq. 5.15; note that the head variance is largest in zones of smallest mean log-conductivity.

Finally, a similar type of analysis can also be carried out for the *effective conductivity*. This is sketched summarily below, again for the simple case of steady one-dimensional groundwater flow. First, the Darcy equation is used to define a *local* effective conductivity $\tilde{K}$ at the scale L, conditioned on larger scale fluctuations. Second, the *uncertainty* in the value of $\tilde{K}$ is compared by averaging the unresolved large-scale fluctuations, i.e. by taking expectations over those terms involving $\bar{f}(x)$.

Using previous results such as Eq. 5.11, the one-dimensional Darcy equation can be expressed in the form:

$$Q = -K_G \exp(\bar{f} + \tilde{f})\left(-J + J\bar{f} + \frac{\mathrm{d}\tilde{h}}{\mathrm{d}x}\right) \tag{5.20}$$

where the flux Q is constant due to one-dimensionality and steady flow assumptions. Now, taking conditional expectations on both sides (with $\bar{f}(x)$ fixed), and linearizing the exponential term, gives:

$$Q \simeq -K_G\left\{J\cdot\left(\frac{\bar{f}^2}{2} - 1\right) + \overline{\tilde{f}\frac{\mathrm{d}\tilde{h}}{\mathrm{d}x}} - J\frac{\overline{(\tilde{f}^2)}}{2}\right\} \tag{5.21}$$

Using again previous definitions and results, the second term on the right-hand side can be computed explicitly by spectral integration, as shown below:

$$\overline{\tilde{f}\frac{\mathrm{d}\tilde{h}}{\mathrm{d}x}} = \int_{L^{-1}\leqslant|k|\leqslant l^{-1}} ik\langle \mathrm{d}Z^*_{\tilde{f}}(k)\cdot \mathrm{d}Z_{\tilde{h}}(k)\rangle \tag{5.22}$$

where $\mathrm{d}Z^*$ denotes the complex conjugate of $\mathrm{d}Z$. For a justification of this relation, see Section 2 on the spectral representation theorem. Substituting in this expression the random increment $\mathrm{d}Z_{\tilde{h}}(k)$ from Eq. 5.13, and using the truncated self-similar spectrum for log-conductivity, finally leads to an effective Darcy equation analogous to Eq. 2.28, of the form:

$$Q = \tilde{K}J \tag{5.23}$$

with local effective conductivity $\tilde{K}$ given by:

$$\tilde{K} \simeq K_{\mathrm{G}}\left\{1 - \frac{\bar{f}^2}{2} + \sigma_0^2 \ln\left(\frac{L}{l}\right) - (1-\bar{f})\cdot 2\sigma_0^2 \ln\left[\frac{L}{l}\left(\frac{1+\bar{a}^2l^2}{1+\bar{a}^2L^2}\right)^{\frac{1}{2}}\right]\right\} \tag{5.24}$$

and where $\bar{a} = \mathrm{d}\bar{f}/\mathrm{d}x$ is the local slope of log-conductivity (see sketch in lower part of Fig. 6). After rearranging the above expression, and replacing as before terms like $(1-\bar{f})$ by exponentials, a more realistic relation is finally obtained for the local effective conductivity:

$$\tilde{K} \simeq K_{\mathrm{G}} \exp\left\{(1-2\mathrm{e}^{-\bar{f}})\sigma_0^2 \ln\left(\frac{L}{l}\right) + \mathrm{e}^{-\bar{f}^2/2} + \mathrm{e}^{-\bar{f}}\sigma_0^2 \ln\left(\frac{1+\bar{a}^2L^2}{1+\bar{a}^2l^2}\right)\right\} \tag{5.25}$$

Equation (5.25) gives the effective conductivity of the flow system in a domain of size L. This 'local' effective conductivity is conditioned on the local geometric mean, $K_{\mathrm{G}}\exp(\bar{f})$, and on the local trend $\bar{a} = \mathrm{d}\bar{f}/\mathrm{d}x$. As the domain size L increases, these parameters converge to their ensemble expectations ($\bar{f} = 0$ and $\bar{a} = 0$). In this case, it is easily seen that $\tilde{K}$ converges as it should to the harmonic mean:

$$K_{\mathrm{H}} = K_{\mathrm{G}} \exp\left[-\sigma_0^2 \ln\left(\frac{\mathscr{L}}{L}\right)\right] = K_{\mathrm{G}}\exp(-\sigma_f^2/2) \tag{5.26}$$

which is the exact solution in the limit of infinite domains. Moreover, the variance of the finite domain effective conductivity $\tilde{K}$ could be computed, in principle, from Eq. 5.25 by letting $\bar{f}(x)$ be random and applying ensemble expectations. Alternatively, a rough approximation of the coefficient of variation, of the type of Eq. 5.18, may also be attempted. This is not pursued here, but the form of Eq. 5.25—with $\bar{f}$ and $\bar{a}$ generally non-null—clearly indicates the significant influence of local inhomogeneities on the actual value of the effective conductivity.

In summary, we have developed in this section a working example of the application of the spectral conditioning approach for one-dimensional stochastic flow. The example was intended to illustrate how scale-dependent parameters of effective variability can be related to high wavenumber fluctuation components, while the remaining uncertainty is expressed in terms of the low wavenumber components of porous medium heterogeneity. Ultimately, the purpose of the method is to quantify finite-scale variability as well as uncertainty in a consistent way. More work will be needed to improve the tractability of the spectral conditioning method in view of treating a broader class of scale-dependent problems of groundwater flow and mass transport.

6. Summary and conclusions

The spectral conditioning approach explicitly shows the influence of domain size on the effective characteristics of flow in a medium having many scales of heterogeneity. The previous section developed in some detail an example of the application of the spectral conditioning approach in combination with a truncated self-similar model of randomness. Closed form expressions were obtained for the scale-dependent variance of hydraulic head and for the effective conductivity in a *finite* flow domain of size L. Moreover, it was shown that these statistical quantities were themselves subject to *uncertainty* due to larger unresolved heterogeneities on the order of the domain's scale or greater.

The particular solutions obtained for the case of one-dimensional flow incorporate such uncertainty in the form of two simple random parameters:

(i) the local mean of conductivity ($K_G \exp(\bar{f})$);
(ii) the local mean slope of log-conductivity ($\bar{a} = d\bar{f}/dx$).

These local mean conductivity parameters were defined by smoothing out the fast fluctuations of $\ln K(x)$, that is by filtering out those fluctuations whose wavenumber is higher than the inverse domain size. Therefore, the spectral content of $\bar{f}$ and $\bar{a}$ is concentrated exclusively in the low wavenumber range, and decreases as domain size increases.

The whole approach leads to a natural distinction between 'spatial variability' and 'uncertainty' for finite scale hydrological phenomena, and seems useful in environments characterized by the presence of many scales of heterogeneity. This type of approach may be particularly relevant in the case of a contamination plume spreading from a local source, a situation where there may be a great uncertainty about the position of the centre of mass and the degree of spreading or macrodispersion of the plume. In particular, it would be interesting to examine how uncertainty changes as the space–time scales of the plume increase. Our preliminary results with the spectral conditioning approach show that the uncertainty of effective flow statistics does decrease with domain size. Extrapolating this argument to contaminant transport seems consistent with the intuitive notion that our uncertainty concerning the spatial moments of the concentration field must eventually decrease as more heterogeneities are being explored by the plume.

However, more work will be needed in order to evaluate fully the potential of the spectral conditioning approach for realistic three-dimensional anisotropic problems of stochastic flow and transport. Some of the results obtained previously—using the simple cut-off approach without spectral conditioning (Section 4)—may indicate the type of dimensionality effects to be expected. In particular, note that the effective conductivity of flow in an isotropic domain of diameter L was found to increase with L in three-dimensional space, remained independent of L in two-dimensional space, and decreased with L in one-dimensional space. An heuristic application of the cut-off approach to advective solute transport also indicated the possibility of a linear scale dependence of the longitudinal macrodispersivity in three-dimensional space. Of course, the latter indications are only approximate and need to be confirmed. Possible extensions of this work towards a consistent scale analysis of multidimensional contaminant transport in heterogeneous media are currently being considered.

Finally, it should be realized that comprehensive sets of conductivity data may be needed in order to ascertain further the validity of our 'truncated self-similar' model of randomness. The latter model seems at least consistent with the borehole data of the Mount Simon aquifer, as shown in Section 3. However, these data are only one-dimensional and cover no more than two orders of magnitudes of length scales. Typically, a useful range of scales for stochastic analysis of subsurface contamination may involve three to four orders of magnitude along each spatial direction. At any rate, given the general lack of data, a parsimonious random field model of the kind used in this work seems an attractive option.

References

Ababou, R. and Gelhar, L.W. (1990). On the Convergence of Infinite Series Solution of Stochastic Flow, *Water Resour. Res.*, submitted.

Ababou, R., McLaughlin, D., Gelhar, L.W. and Tompson, A.F.B. (1985). Numerical simulation of saturated flow fields in randomly heterogeneous porous media. *In* 'The Stochastic Approach to Subsurface Flow', Preprints of the International Symposium, A.I.R.H. and Paris School of Mines, Montvillargennes, France.

Ababou, R., Gelhar, L.W. and McLaughlin, D. (1988). 'Three-Dimensional Flow in Random Porous Media', Parsons Laboratory Technical Report No. 318, Massachusetts Institute of Technology, Cambridge, Massachusetts, 2 vols, 883 pp.

Ababou, R., McLaughlin, D., Gelhar, L.W. and Tompson, A.F.B. (1990). Numerical simulation of saturated flow fields in randomly heterogeneous porous media. *In* 'The Stochastic Approach to Subsurface Flow', *J. Transport Porous Media* (in press).

Adler, R.J. (1981). 'The Geometry of Random Fields', 280 pp. Wiley, New York.

Bakr, A.A. (1976). Stochastic analysis of the effect of spatial variations of hydraulic conductivity on groundwater flow. Ph.D. Thesis, New Mexico Institute of Mining and Technology, Socorro, NM.

Bakr, A.A., Gelhar, L.W., Gutjahr, A.L. and McMillan, R.J. (1978). Stochastic analysis of spatial variability in subsurface flow: 1. Comparison of one and three dimensional flows, *Water Resour. Res.* **14**, 263–272.

Burrough, P.A. (1981). Fractal dimensions of landscapes and other environmental data, *Nature* **294**, 238–242.

Cushman, J.H. (1986). Multiphase transport in the space of stochastic tempered distributions, *IMA J. Appl. Math.* **36**, 159–175.

Cushman, J.H. (1988). Stochastic equations and hierarchy in subsurface transport: assumptions and methodology. *In* 'Proceedings of the International Conference on Validation of Flow and Transport Models for the Unsaturated Zone'. Ruidoso, New Mexico, pp. 50–62.

Dagan, G. (1981). Analysis of flow through heterogeneous random aquifers by the method of embedding matrix: 1. Steady Flow, *Water Resour. Res.* **17**, 107–121.

Dagan, G. (1987). Theory of solute transport by groundwater, *Ann. Rev. Fluid Mech.* **19**, 183–215.

Devary, J.L. and Doctor, P.G. (1982). Pore velocity estimation uncertainties, *Water Resour. Res.* **18**, 1157–1164.

Gelhar, L.W. (1976). Stochastic analysis of flow in aquifers, advances in groundwater hydrology. 'Symposium of the American Water Resources Association', Minneapolis, Minnesota.

Gelhar, L.W. (1986). Stochastic subsurface hydrology from theory to applications, *Water Resour. Res.* **22**, 135–145.

Gelhar, L.W. (1987). Stochastic analysis of solute transport in saturated and unsaturated porous media. *In* 'Fundamentals of Transport in Porous Media' (J. Bear and M.Y. Corapcioglu, eds), pp. 657–700. Martinus Nijhoff, Dordrecht, The Netherlands.

Gelhar, L.W. and Axness, C.L. (1983). Three dimensional stochastic analysis of macrodispersion in aquifers, *Water Resour. Res.* **19**, 161–180.

Graham, W. and McLaughlin, D. (1989). Stochastic analysis of nonstationary subsurface solute transport: 1. Unconditional moments, *Water Resour. Res.* **25**, 215–232.

Gutjhar, A.L. and Gelhar, L.W. (1981). Stochastic models for subsurface flow: infinite versus finite domains and stationarity, *Water Resour. Res.* **17**, 337–350.

Hoeksema, R.J. and Kitanidis, P.K. (1985). Analysis of the spatial structure of properties of selected aquifers, *Water Resour. Res.* **21**, 563–572.

Loève, M. (1978). 'Probability Theory', Vol. II. Springer-Verlag, Berlin.

Ma, F., Wei, M.S. and Mills, W.H. (1987). Correlation structuring and the statistical analysis of steady-state groundwater flow, *SIAM J. Sci. Stat. Comput.* **8**, 848–867.

Mandelbrot, B.B. (1983). 'The Fractal Geometry of Nature', 468 pp. Freeman and Co., New York.

Mandelbrot, B.B., Passoja, D.E. and Paullay, A.J. (1984). Fractal character of fracture surfaces of metals, *Nature*, **308** (No. 5961), 721–722.

Mandelbrot, B.B. and Van Ness, J.W. (1968). Fractional Brownian motions, fractional noises and applications, *SIAM Rev.* **10** (No. 4), 422–437.

Mandelbrot, B.B. and Wallis, J.R. (1969). Some long-run properties of geophysical records, *Water Resour. Res.* **5**, 321–340.

McLaughlin, D. and Wood, E.F. (1988a). A distributed parameter approach for evaluating the accuracy of groundwater models predictions: 1. Theory, *Water Resour. Res.* **24**, 1037–1047.

McLaughlin, D. and Wood, E.F. (1988b). A distributed parameter approach for evaluating the accuracy of groundwater models predictions: 2. Application to groundwater flow, *Water Resour. Res.* **24**, 1048–1060.

Matheron, G. (1967). 'Eléments pour une Théorie des Milieux Poreux', 166 pp. Masson and Cie, Paris.

Monin, A.S. and Yaglom, A.M. (1965). 'Statistical Fluid Mechanics: Mechanics of Turbulence' (J.L. Lumley, ed.), Vol. 2, 874 pp. The MIT Press, Cambridge, MA.

Naff, R.L. and Vecchia, A.V. (1986). Stochastic analysis of three-dimensional flow in a bounded domain, *Water Resour. Res.* **22**, 695–704.

Priestley, M.B. (1981). 'Spectral Analysis and Time Series', 890 pp. Academic Press, New York.

Smith, L. and Freeze, R.A. (1979a). Stochastic analysis of steady state groundwater flow in a bounded domain: 1. One-dimensional simulations, *Water Resour. Res.* **15**, 521–528.

Smith, L. and Freeze, R.A. (1979b). Stochastic analysis of steady state groundwater flow in a bounded domain: 2. Two-dimensional simulations, *Water Resour. Res.* **15**, 1543–1559.

Vomvoris, E. (1986). Concentration variability in transport in heterogeneous aquifers: a stochastic analysis. Ph.D. Thesis, Parsons Laboratory, M.I.T., Cambridge, MA.

Wilson, K.G. (1975). The renormalization group: critical phenomena and the Kondo problem, *Rev. Mod. Phys.* **47**, 773–840.

Wilson, K.G. (1979). Problems in physics with many scales of length, *Scient. Am.* **241** (No. 2), August.

Wilson, K.G. (1983). The renormalization group and critical phenomena, *Rev. Mod. Phys.* **55**, 583.

Wilson, K.G. and Kogut, J. (1974). The renormalization group and the ε expansion, *Phys. Rep.* **12**, 75–200.

Winter, C.L., Newman, C.M. and Newman, S.P. (1984). A perturbation expansion for diffusion in a random velocity field, *SIAM J. Appl. Math.* **24**, 411–424.

Yaglom, A.M. (1962). 'An Introduction to the Theory of Stationary Random Functions' (R.A. Silverman, trans. and ed.). Prentice-Hall, Englewood Cliffs, NJ (Dover, New York, 1973).

Chapter XV

Assessment of the Behaviour of Conservative and Biodegradable Solutes in Heterogeneous Porous Media

E.A. Sudicky, S.L. Schellenberg and K.T.B. MacQuarrie
Waterloo Centre for Groundwater Research, Waterloo, Ontario, Canada, N2L 3G1

1. Introduction

1.1 A question of scale

In recent years, a number of investigators have performed detailed measurements of hydraulic conductivity at several field sites and have concluded that, with rare exception, geological materials are heterogeneous. Because of the uncertainty associated with precisely prescribing the hydraulic conductivity at every point within a groundwater flow system, most of these studies have conceded that it is most appropriate to characterize the heterogeneities in a statistical framework. Sudicky (1986), for example, examined the spatial variability of hydraulic conductivity in the aquifer at Canadian Forces Base Borden, Ontario, by performing many hundreds of closely spaced measurements in the vertical and horizontal directions along two coring transects. He concluded that, although it is possible to make deterministic measurements of hydraulic conductivity over horizontal scales of a few metres to a few tens of metres, it is impractical to make measurements over distal scales that are relevant to the problem of long-term, large-scale contaminant migration. Other recent case studies illustrating the natural heterogeneity of earth materials include efforts by Smith (1980), Pickens and Grisak (1981), Hoeksema and Kitanidis (1985), Molz *et al.* (1985) and Zapico (1987).

The problem which arises is that, when simulating flow and contaminant transport through a geological material, a modelling scale must be chosen that is consistent with the contaminant travel distances of concern and these distances are often large with respect to the scale of the underlying heterogeneity. Because

DYNAMICS OF FLUIDS IN HIERARCHICAL POROUS MEDIA
ISBN 0-12-200260-1

of this dimensional incompatibility, there will generally exist an 'unmodelled' heterogeneity at a scale below the selected modelling scale. The salient question now becomes, how can realistic values for flow and transport parameters be selected that are compatible with the modelling scale and yet provide a reasonable representation of the effects of the unmodelled variability?

1.2 The macroscopic simulation approach

The conventional approach for simulating solute transport in unconsolidated geological materials is a deterministic approach that uses some form of the advection-dispersion equation. Advection is the process by which solutes are transported by the bulk average motion of the flowing groundwater. The spreading of the solute about its mean displacement position, referred to as dispersion, occurs because of the mechanical mixing induced by local fluctuations in the fluid velocity and molecular diffusion. With the conventional development of the advection-dispersion equation, the mechanical mixing is considered to be analogous to a large-scale Fickian diffusion process, where the dispersion parameter, dispersivity, is assumed to be a unique property of the geological medium at the selected scale of continuum description (Bear, 1972).

The traditional way of determining the dispersivity of a medium is by calibration of a deterministic solution of the advection-dispersion equation to observed concentration patterns. In many cases, however, the dispersivity can only be regarded as a curve-fitting parameter because sufficient concentration measurements are often unavailable for adequate calibration and because the source boundary conditions and history are often unknown. A further complication associated with the calibration approach arises because the dispersivity has been shown initially to grow with the solute residence time and displacement distance; the approach to a constant asymptotic value can be slow, if it occurs at all. This phenomenon has been noted in field tracer tests performed in exceptional detail (Dieulin, 1980; Sudicky *et al.*, 1983; Freyberg, 1986; Sudicky, 1986) and has also been demonstrated theoretically (Gelhar *et al.*, 1979; Matheron and de Marsily, 1980; Dagan, 1982, 1984, 1988; Sudicky, 1983; Neuman *et al.*, 1987; Schellenberg, 1987). Thus, use of pre-asymptotic concentration data for calibration purposes can lead to an erroneous estimate of the asymptotic dispersivity, referred to as a macrodispersivity, for application to large-scale field problems of contaminant plume behaviour.

Recent theoretical developments in the area of contaminant dispersion attempt to circumvent the difficulties associated with model calibration by directly relating the magnitude of the macrodispersivity to estimable parameters describing the variations in the hydraulic properties of a geological medium (Sudicky, 1986). When invoking the macroscopic approach to simulate flow and transport within a geological unit, the unmodelled heterogeneities are viewed as being essentially unknowable and the assertion is made that small-scale local variations in the hydraulic conductivity can be treated as a random field. The problem of describing the material variability is then reduced to estimation of the statistical moments of a spatial random variable. The relevant statistical properties include the mean, the variance, and the covariance of the hydraulic conductivity field. The covariance is a convenient measure of the spatial persistence of a random field because it expresses the degree of correlation in the random fluctuations as a function of the separation distance between the points of observation.

By relating the effective macro-scale flow and transport parameters to the statistical properties of the unmodelled heterogeneities, stochastic theories of contaminant dispersion replace the anonymous heterogeneous medium with a macroscopically equivalent homogeneous medium (Dagan, 1982, 1984; Gelhar and Axness, 1983). The functional forms of the relationships describing the hydraulic and dispersive properties of the equivalent aquifer are obtained from the solution of stochastic forms of the groundwater flow and advection-dispersion equations.

It is our aim here to examine the validity of the macroscopic transport approach for dealing with contaminant transport in heterogeneous, sandy-type aquifers. The problem will be approached by performing high-resolution numerical simulations of flow and solute transport in heterogeneous porous materials described at the local scale of an individual bed and then comparing the detailed synthesized plumes to those predicted using macro-scale parameters derived for an equivalent homogeneous medium on the basis of stochastic transport theory. Two types of dissolved chemicals are considered here; one that is unaffected by reaction or transformation processes along the path of migration, and another that undergoes a biologically-mediated transformation under aerobic conditions. The former solute is a non-reactive one such as chloride and the latter represents a chemical such as an aromatic hydrocarbon dissolved in groundwater.

It is of interest to compare the behaviour of conservative and biodegradable plume evolution in heterogeneous aquifers by detailed numerical simulations because stochastic–analytic theories for macrodispersivity estimation have to date been concerned primarily with conservative solutes. Hence, it is not known whether dispersion parameters predicted by stochastic theory for non-reactive solutes can be used as model input for simulating biodegradable chemical transport. Questions have also been raised about the need to develop theory for prediction of macro-scale biodegradation parameters which take into account the effects of aquifer heterogeneity on such factors as the subsurface distribution of oxygen (MacQuarrie and Sudicky, 1989). These issues are of practical importance because contamination of aquifers by organic chemicals is a widespread problem and large-scale models are often utilized in the analysis of such problems.

2. Simulation approach

2.1 Overview

All of the simulations considered here involve flow through a two-dimensional random heterogeneous medium having prescribed geostatistical properties. The flow domain is comprised of a rectangular cross-section in a vertical plane bounded above and below by impermeable boundaries. At the upstream (left) and downstream (right) ends of the domain, the hydraulic head is specified such that the mean velocity $\bar{v}$ is along the principal direction of bedding (horizontal) and equal to a prescribed value of about 0.1 m/day. This value is typical of the velocity in a fine-to-medium-grained sand aquifer in which the hydraulic gradient is moderate to low. As the slug of chemical of interest is transported across the medium, spatial moments of the plume up to and including order four are calculated. From these moments, the mean advective and dispersive properties of the medium can be deduced in a manner consistent with stochastic theories of contaminant dispersion.

The strategy used in this work for studying the development of a contaminant plume differs from the Monte Carlo analysis performed by Smith and Schwartz (1980). Rather than perform Monte Carlo simulations, a single realization of large spatial extent is generated for each set of input parameters. This single realization is then presumed to be an analogue of the real aquifer into which a tracer is introduced. Because the maximum travel distance encountered by the simulated tracer cloud is several to many tens of horizontal correlation scales, thus permitting the plume to encounter numerous heterogeneities, there is reason to expect that the eventual fate of an ever-expanding conservative solute plume will be characteristic of a mean behaviour given by an ensemble average. The applicability of the macroscopic simulation approach is somewhat suspect for a biodegradable plume in heterogeneous aquifers because its dimensions can decrease along its path of migration.

The major assumptions made in this analysis include: (1) the flow field is time invariant; (2) the porosity is spatially constant; (3) the mean flow remains along the principal direction of bedding; (4) the advection-dispersion equation is valid at the local scale of an individual heterogeneity; and (5) dual-Monod kinetics adequately describe the biotransformation of a dissolved hydrocarbon under aerobic conditions by an indigenous bacterial population. The fourth assumption can be justified on the basis that the advection-dispersion equation has been found by numerous experimentalists to provide a reasonable representation of conservative solute transport in laboratory columns packed with homogeneous materials. The applicability of dual-Monod kinetics during subsurface transport is, however, less certain, although it has been employed by several workers (e.g. Borden and Bedient, 1986; Molz *et al.*, 1986; MacQuarrie *et al.*, 1989) and has been shown to yield good results when compared to laboratory column data (MacQuarrie *et al.*, 1989).

2.2 Hydraulic conductivity field

In order to generate random, stationary, two-dimensional hydraulic conductivity fields with prescribed statistical properties, the turning bands method (Mantoglou and Wilson, 1982) was chosen. The mean, $\bar{Y}$, and variance, σ_Y^2, of the log-transformed hydraulic conductivity field, $Y = \ln(K)$, and prescribed values for the horizontal and vertical correlation lengths of Y are used as input in the turning bands model. It is assumed that the spatial persistence of the log-transformed hydraulic conductivity fluctuations is described by an exponentially decaying covariance function given by

$$\mathrm{Cov}(s_L, s_T) = \sigma_Y^2 \exp\{-(s_L^2/\lambda_L^2 + s_T^2/\lambda_T^2)^{\frac{1}{2}}\} \tag{2.1}$$

where s_L and s_T are the longitudinal and transverse components, respectively, of the separation vector, and λ_L and λ_T are the longitudinal and transverse correlation lengths, respectively. An anisotropic exponential covariance is consistent with detailed field measurements performed by Sudicky (1986).

Isotropic point-values of hydraulic conductivity are generated on a rectangular grid at regularly spaced discrete points with horizontal, $\Delta x'$, and vertical, $\Delta z'$, spacings that are much smaller than the correlation lengths in the corresponding directions. This is done in order to preserve the statistical features of the $\ln(K)$ field. Moreover, the fine spatial discretization allows justification of the assumption that the local hydraulic conductivity is isotropic. The ratios of $\lambda_L/\Delta x'$ and $\lambda_T/\Delta z'$ were typically about 10 and 5, respectively. The discretization used for generating the conductivity field was the same as that used for the solution of the flow problem.

2.3 The flow model

The flow in the random hydraulic conductivity field is simulated using the dual-formulation groundwater flow model described by Frind and Matanga (1985). The dual formulation makes use of both potentials (i.e. hydraulic heads), $u = u(x', z')$, and streamfunctions, $\Psi = \Psi(x', z')$. If the hydraulic conductivity varies spatially but is isotropic at the local scale, the streamfunctions are governed by

$$\frac{\partial}{\partial x_i'}\left(\frac{1}{K}\frac{\partial \Psi}{\partial x_i'}\right) = 0 \quad i = x', z' \tag{2.2}$$

and the hydraulic heads by

$$\frac{\partial}{\partial x_i'}\left(K\frac{\partial u}{\partial x_i'}\right) = 0 \quad i = x', z' \tag{2.3}$$

where $K = K(x', z')$ is a local value of the hydraulic conductivity. The potential form of the Darcy equation for average linear velocity v_i is given by

$$v_i = -\frac{K}{n}\frac{\partial u}{\partial x'_i} \quad i = x', z' \tag{2.4}$$

where n is the porosity. The corresponding streamfunction form of the Darcy equation is

$$v_{x'} = \frac{1}{n}\frac{\partial \Psi}{\partial z'} \tag{2.5}$$

$$v_{z'} = -\frac{1}{n}\frac{\partial \Psi}{\partial x'} \tag{2.6}$$

Because all of the detail in the hydraulic conductivity is represented explicitly as $K(x', z')$ in the Darcy equation, the calculated velocities are local-scale values that will in general vary randomly both in direction and magnitude.

A principal direction Galerkin finite-element model (adapted from Frind, 1982) is used for solving the governing flow equations. With this method, the two-dimensional problem is decoupled into two one-dimensional problems by discretizing the flow domain into nodal strips along the z'- and x'-directions. The one-dimensional equations are solved iteratively along the nodal strips until the solution converges to a prescribed level of error tolerance.

The principal direction finite-element technique is chosen over conventional two-dimensional finite-element techniques because the resulting coefficient matrices are only tri-diagonal and thus require much less main computer memory. The solution is also less susceptible to round-off error because only small matrices need to be inverted.

Following the solution for the streamfunction and hydraulic head values, an exceedingly fine flow net is constructed by using linear interpolation between nodal values. The interpolated flow net is then used in the principal direction transport model described briefly below.

2.4 The transport model

In this section, the equations governing the fate and transport of a dissolved hydrocarbon subject to aerobic biodegradation processes will be presented. Because aerobic biodegradation requires dissolved oxygen as an electron acceptor in the transformation reaction, and because this type of reaction is mediated by an indigenous subsurface bacterial population, a total of three coupled transport equations are required: one for the organic substrate, one for the oxygen, and another describing the dynamics of the microbial population. The coupling of the three equations is provided by the biodegradation reaction which is represented here through the use of dual-Monod kinetics (Kissel, 1985; Borden and Bedient, 1986; MacQuarrie *et al.*, 1989). The simulations performed for the case of a conservative solute represent a special situation in which sorption and biodegradation processes are ignored. Thus, the governing equations for oxygen and the bacterial population are not needed for a non-reactive, non-transformable solute.

The equation describing the fate of an organic subject to the processes of advection, dispersion, sorption, and aerobic biodegradation in a saturated porous media is given by (Borden and Bedient, 1986; MacQuarrie *et al.*, 1989)

$$\frac{\partial C}{\partial t} + v'_x \frac{\partial C}{\partial x} - \frac{\partial}{\partial x_i}\left(D'_{ij}\frac{\partial C}{\partial x_j}\right) + M_t k'\left(\frac{C}{K_C + C}\right)\left(\frac{A}{K_A + A}\right) = 0 \quad i = x, z \tag{2.7}$$

where C and A are the organic and oxygen concentrations, respectively, $v'_x = v_x/R_C$ is a retarded groundwater velocity along a curvilinear streamline, R_C is the retardation factor describing sorption of the organic onto the porous medium, $D'_{ij} = D_{ij}/R_C$ are the principal values of the retarded hydrodynamic dispersion tensor, M_t is the total active microbial biomass concentration, $k' = k/R_C$ is a retarded substrate utilization rate, and K_C and K_A are the substrate and oxygen concentrations, respectively, at which the utilization rate is half the maximum rate k. Equation 2.7 presumes that the organic acts as a primary substrate and that the sorbed-phase organic is unavailable for biodegradation (Steen *et al.*, 1980; Ogram *et al.*, 1985). Because Eq. 2.7 and the associated equations describing the oxygen and bacterial population are written in a principal curvilinear coordinate system, x and z in Eq. 2.7 represent coordinates along and orthogonal to the flowlines, respectively.

The principal values of the hydrodynamic dispersion tensor, D_{ii}, are conventionally written in terms of a longitudinal value D_l along the flow direction (i.e. $i = x$) and a transverse value, D_t, that acts transverse to the flow direction (i.e. $i = z$). They are further described as (Bear, 1972)

$$D_l = \alpha_l v_x + D^* \tag{2.8}$$

$$D_t = \alpha_t v_x + D^* \tag{2.9}$$

where α_l and α_t are longitudinal and transverse dispersivities, respectively, and D^* is an effective molecular diffusion coefficient for a porous medium. In general, each of the three different species considered here will have different values for the diffusion coefficient.

Equations analogous to Eq. 2.7 can be written for the oxygen and the microbial population. For the oxygen, we have (Borden and Bedient, 1986; MacQuarrie *et al.*, 1989)

$$\frac{\partial A}{\partial t} + v_x \frac{\partial A}{\partial x_x} - \frac{\partial}{\partial x_i}\left(D_{ij}\frac{\partial A}{\partial x_j}\right) + XM_t k\left(\frac{C}{K_C + C}\right)\left(\frac{A}{K_A + A}\right) = 0 \quad i = x, z \tag{2.10}$$

where X is the ratio of oxygen consumed to substrate degraded as determined from stoichiometry. Equation 2.10 assumes that the electron acceptor does not undergo sorption into the porous medium, which is reasonable for oxygen. The equation describing the growth dynamics of the microbial population together with the transport of free-living bacteria can be expressed as (MacQuarrie *et al.*, 1989)

$$\frac{\partial M_s}{\partial t} + v''_x \frac{\partial M_s}{\partial x} - \frac{\partial}{\partial x_i}\left(D''_{ii}\frac{\partial M_s}{\partial x_i}\right) - BM_s k\left(\frac{C}{K_C + C}\right)\left(\frac{A}{K_A + A}\right) + dM_s = 0$$

$$i = x, z \tag{2.11}$$

where M_s is the microbial mass concentration in solution, $v''_x = v_x/R_m$, $D''_{ij} = D_{ij}/R_m$, R_m is a retardation factor describing microbial attachment onto the porous medium, B is the microbial yield per unit of substrate consumed, and d is a first-order microbial decay coefficient. Most of the microbial population in the subsurface are attached to particle surfaces (Wilson *et al.*, 1983). This situation is handled by using a large value for R_m in Eq. 2.11; however, the major assumption is that organisms can be expressed by a linear equilibrium relationship $M_t = M_s R_m$.

By solving the transport equations presented above using a high-resolution input flow field that reflects the detail of the underlying heterogeneity, the parameters such as the dispersion coefficients appearing in each equation must reflect small-scale local conditions. Thus, for example, input dispersivity values would be typical of values obtained by passing a tracer through a laboratory column packed with a homogeneous sand. Such tests usually yield longitudinal dispersivities in the range of a few millimetres to a few centimetres (Fried, 1975). Transverse local dispersivities are commonly found to be a factor of 10 to 20 smaller than the longitudinal value (Bear, 1972; Fried, 1975).

The coupled system of Eqs 2.7, 2.10 and 2.11 for the case of a biodegradable organic, or Eq. 2.7 only with $R_C = 1$ and $k = 0$ for a conservative solute, require initial and boundary conditions for solution. For the conservative solute simulations, the tracer was released into the medium, assumed free of solute initially, from a source over a 0.5 m depth interval located midway along the left (inflow) boundary. Along the source, the concentration was specified at a unit value for 20 days, after which time it was set to zero. The bottom, top and right (outflow) boundary conditions were of the second type, with $\partial C/\partial x_i = 0$.

The organic contaminant release was represented by an initial condition in which $C = 3$ mg/l in a 2 m × 2 m square patch situated near the inflow boundary and at mid-depth in the aquifer. The remainder of the region was initially free of organic. The initial oxygen concentration in the patch was specified to be 1.0 mg/l, with $A = 3.5$ mg/l over the remainder of the region at time zero. The initial microbial population was assumed to be everywhere given by $M_s = 0.23$ μg/l. By choosing $R_m = 1000$ such that the bulk of the microbes are attached to the porous medium, this yields $M_t = 0.23$ mg per 10^3 g of soil or about 10^5 microbes per gram of solid which is in the range of observed values (Berry-Spark, 1987). Along the inflow boundary, a first-type condition was imposed using $C = 0.0$ mg/l, $A = 3.5$ mg/l, and $M_s = 0.23$ μg/l. The bottom boundary condition for all constituents and the top boundary for the organic and the bacteria were of the second type with $\partial \gamma/\partial z = 0$, where $\gamma = C$, A or M_s. The upper boundary condition for the oxygen was a diffusive boundary where dissolved oxygen at $A = 9.0$ mg/l was allowed to enter across the water table into the saturated zone by molecular diffusion. Details concerning the implementation of this condition can be found in MacQuarrie *et al.* (1989). The condition at the outflow boundary was a free-exit type (Frind, 1988) for all species.

Details related to the numerical solution of the governing transport equations by the Principal Direction (PD) finite-element scheme are provided by MacQuarrie *et al.* (1989) and will not be presented. In order to avoid numerical dispersion or oscillations caused by inadequate grid discretization, the curvilinear PD transport grid interpolated from the flow solution contained from about 30 000 to nearly 300 000 nodes depending on the length and thickness of the domain. The nodal spacing was sufficiently small to ensure that the grid Peclet and Courant numbers were satisfied on average.

3. Evolution of a conservative solute plume

3.1 A single realization example

The statistical properties of the aquifer heterogeneity used in this illustrative example are similar to those estimated by Sudicky (1986) for the Borden aquifer. The geometric mean hydraulic conductivity is 7.1×10^{-3} cm/s, the variance of ln(K) is $\sigma_Y^2 = 0.38$, and the horizontal and vertical correlation lengths are 2.8 m and 0.12 m, respectively. Because the dispersive properties of the Borden aquifer are generally well documented (Sudicky *et al.*, 1983; Freyberg, 1986; Sudicky, 1986), the evolution of the plume in this particular synthesized aquifer will be described in detail. To be consistent with the high-resolution flow solution, a local-scale longitudinal dispersivity value equal to 0.02 m, a transverse value equal to 0.0 m, and an effective diffusion coefficient equal to 1.35×10^{-5} cm^2/s were used as input to the PD transport model. Because the transverse dispersivity is 0.0 in this example, local transverse dispersion will be diffusion-controlled, which is consistent with the experimental findings of Sudicky *et al.* (1983). The local longitudinal dispersivity is larger than values commonly associated with the local scale, but sensitivity analyses performed by Schellenberg (1987) indicate that plume evolution in a random heterogeneous aquifer is relatively insensitive to this parameter. Figure 1 shows a contour plot of the hydraulic conductivity field and the corresponding flow field solution. The flow lines are seen to be tortuous with numerous regions of diverging and converging flow paths. Zones where flow tubes are narrow correspond to higher velocity, more permeable beds, and zones where the flow tubes are wide correspond to lower velocity, less permeable beds.

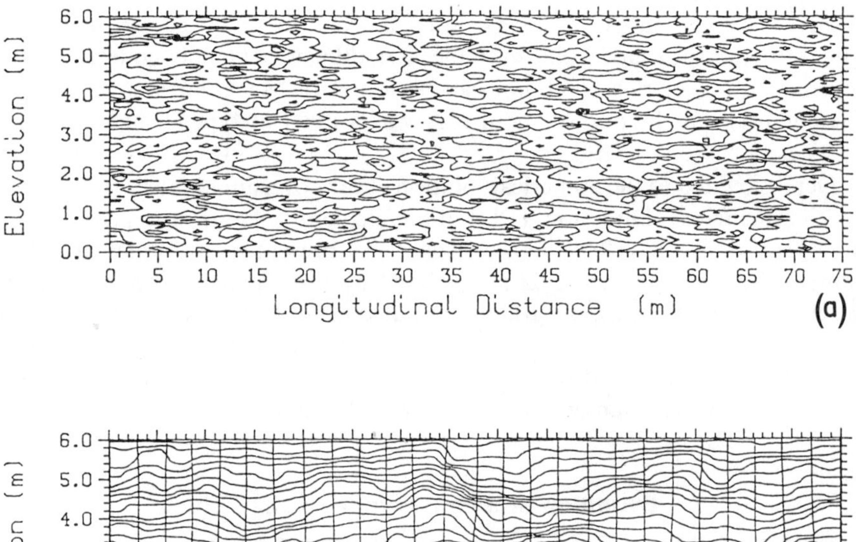

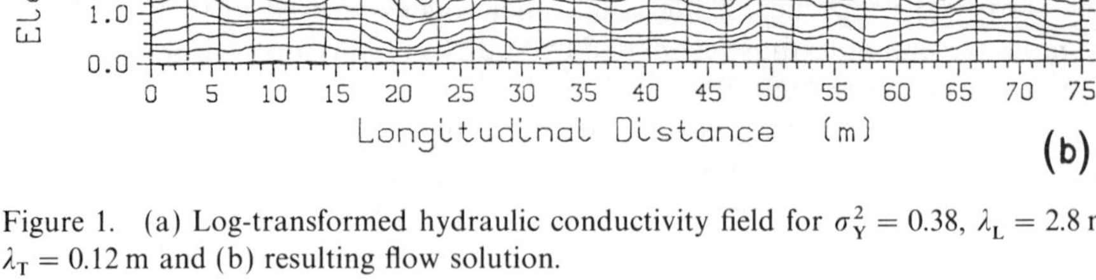

Figure 1. (a) Log-transformed hydraulic conductivity field for $\sigma_Y^2 = 0.38$, $\lambda_L = 2.8$ m, $\lambda_T = 0.12$ m and (b) resulting flow solution.

The tracer was injected for a period of 20 days between elevations 2.75 m and 3.25 m along the left boundary. Figure 2 shows snapshots of the plume at times equal to 40, 185 and 600 days after the commencement of injection. After 40 days, each contour of the plume is irregular with a similar amount of spreading in each direction. After 185 days, the plume is oriented with an upward slant and has developed a narrow trailing tail. By comparing the 185-day location of the plume to the flow net in Fig. 1, it can be seen that the trailing tail is situated in a narrow low-conductivity zone and the remainder of the plume is propagating along upward-slanted streamlines that merge into a higher conductivity zone. Although the low-level concentrations remain erratic after 185 days, the concentration contours in the interior the plume are becoming

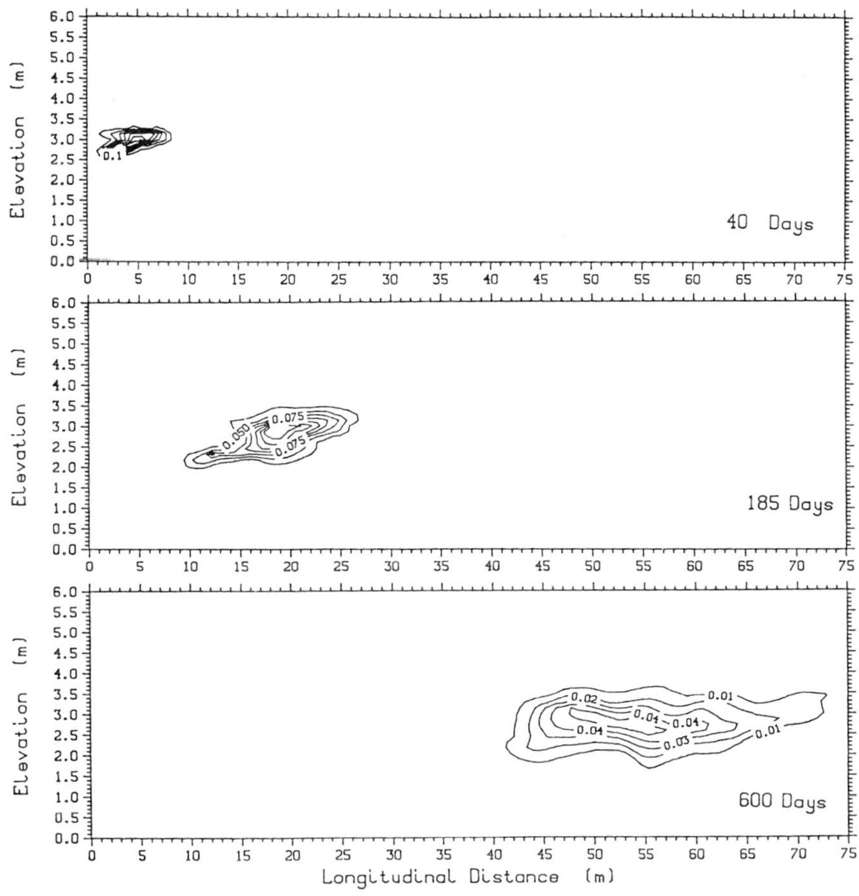

Figure 2. Conservative solute plume after 40, 185, and 600 days for $\sigma_Y^2 = 0.38$, $\lambda_L = 2.8$ m, $\lambda_T = 0.12$ m.

smoother and more regular. The centroid of the plume is advancing at a consistent rate, given by $\bar{v} = 0.09$ m/day, which is identical to the average of the local groundwater velocities. After 600 days of travel time, the plume is very long in the direction of flow but thin in the vertical direction. The concentration contours greater than about 0.01 are seen to be substantially smoothed compared to the irregularity observed at earlier time and are near-Gaussian in shape.

The progression from irregularly shaped plumes at early time to much smoother concentration distributions at later time can be explained by comparing the size of the plume and the distance travelled at different times to the scale of the heterogeneity. At early time, only a small number of heterogeneities have been encountered in the path of migration and, because the solute body is small relative to the size of these heterogeneities, the shape of the plume is controlled by very few zones of contrasting velocity. However, at later times, the plume has traversed a large number of heterogeneities and has spread such that it contains numerous high and low velocity zones. The presence of many contrasting velocity zones in the interior of the plume and the integrated effect of velocity variations previously encountered produces a natural averaging that causes the plume to approach a classical Gaussian shape.

The tendency towards normality in the concentration distributions is demonstrated by the behaviour of the third (skew) and fourth (Kurtosis) spatial moments with time (Fig. 3). The coefficients of skew and Kurtosis oscillate in an irregular fashion at early time and attain their largest values at times prior to about 100 days. After 100 days, the oscillations dampen and the skewness and Kurtosis tend towards 0.0 and 3.0, respectively. These are the values characteristic of a normal distribution. When this behaviour occurs, the plume is behaving in a manner that classical theory of dispersion would predict for a homogeneous medium. It should be noted that the coefficients of skew and Kurtosis are calculated with respect to the x and z axes and therefore the early-time variability in the orientation of the plume is reflected in the erratic shape of the skew and Kurtosis curves.

The principal values of the second moments σ_{11}^2 and σ_{22}^2 which describe the spreading along and perpendicular to the major axis of the plume, respectively, are shown in Fig. 4. It is evident that there is an increase in the slope of the σ_{11}^2 curve at early time prior to about 300 days, but the rate of increase becomes approximately linear for larger values of time. The flattening of the curve beyond 550 days is indicative of a boundary effect caused by arrival of the plume at the exit boundary of the flow system.

The early-time non-linearity in σ_{11}^2 is similar to the behaviour observed by Sudicky *et al.* (1983) and by Freyberg (1986) in the field tracer tests conducted at Borden. It is also similar in nature to the temporal evolution predicted by Sudicky (1983) and Dagan (1982), among others, on the basis of stochastic dispersion theory. Because σ_{11}^2 can be reasonably approximated as a linear function of time (or travel distance) and because the concentration distribution

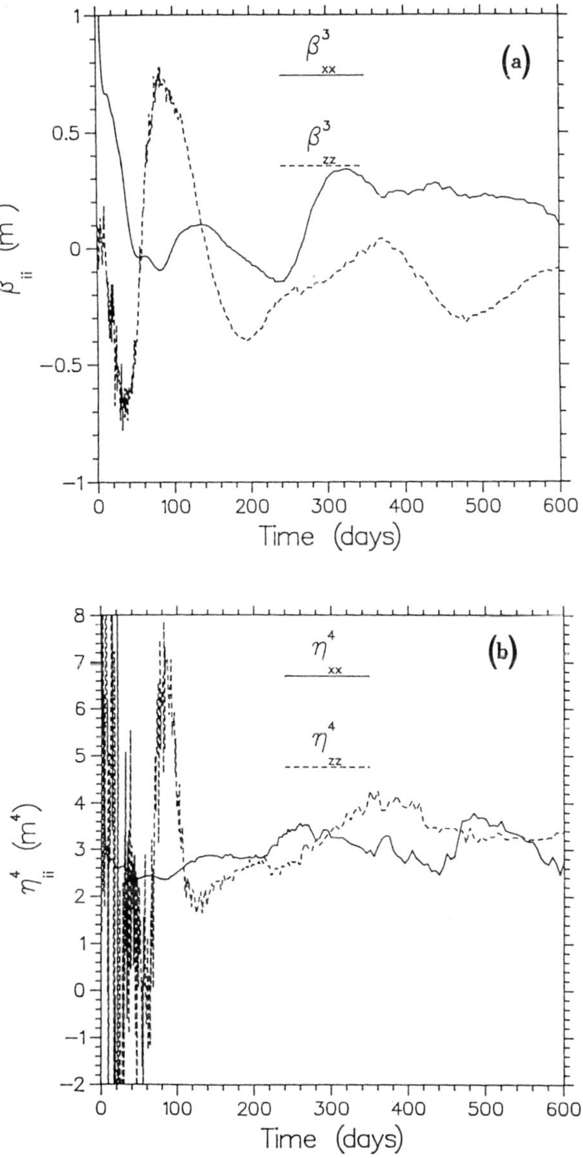

Figure 3. Coefficient of (a) skew and (b) Kurtosis as a function of plume residence time.

is approaching normality, it is possible to determine an asymptotic longitudinal macrodispersivity. Using the slope of the linear portion of the σ_{11}^2 curve between 300 and 500 days and the classical relationships for a macrodispersion

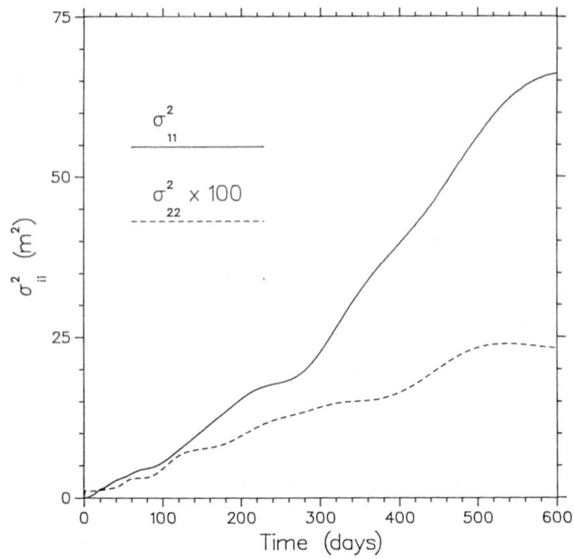

Figure 4. Second moments describing dispersion along ($i = 1$) and transverse ($i = 2$) to the major axis of the plume.

coefficient, $\bar{D}_l$, given by

$$\bar{D}_l = \frac{1}{2}\frac{d\sigma_{11}^2}{dt} \tag{3.1}$$

and a longitudinal macrodispersivity, A_l, determined as $A_l = \bar{D}_l/\bar{v}$, the calculated value for A_l equals about 0.94 m. This value is much larger than the input local dispersivity $\alpha_l = 0.02$ m. Results obtained from the slope of the σ_{22}^2 curve, which is approximately linear over much of the 600 day simulation period, suggest that macro-scale transverse dispersion is essentially controlled by local-scale transverse mixing. This is consistent with stochastic–analytic theory when there is strong statistical anisotropy of the $\ln(K)$ field and when the mean flow direction is along the principal direction of correlation (e.g. Gelhar and Axness, 1983).

It is possible to predict the value of the longitudinal macrodispersivity from theory such as that developed by Dagan (1982), Gelhar and Axness (1983), and Neuman *et al.* (1987). Using the Gelhar and Axness (1983) theory, A_l is given by

$$A_l = \frac{\sigma_Y^2 \lambda_L}{\gamma^2} \tag{3.2}$$

where $\gamma = \exp(\sigma_Y^2(0.5 - g_{11}))$ and $g_{11} = 1/(1 + \lambda_L/\lambda_T)$ for the two-dimensional case. Substituting $\sigma_Y^2 = 0.38$ and $\lambda_L = 2.8$ m into the above yields $A_l = 0.77$ m upon adding in the value of α_l. This value is close to that calculated directly from the numerical simulation and demonstrates that dispersivity values for use in models of large-scale field problems can be obtained without resorting to problematic model calibration provided information is available on the geostatistical structure of an aquifer.

Schellenberg (1987) investigated the robustness of Eq. 3.2 for the prediction of macrodispersivity by performing a number of high-resolution numerical simulations in synthesized aquifers of differing geostatistical properties. In general, it was found that the stochastic–analytic theory provided reliable results except for large $\ln(K)$ variance values.

3.2 Comparison of local-scale and macro-scale simulations

The detailed numerical simulations performed by Schellenberg (1987) have generally shown that contaminant plumes gradually evolve from an irregular shape at early time and for short travel distances to plumes that are much smoother and more highly dispersed at large time and travel distances. Such smoothing suggests that contaminant concentrations become more predictable after the plume traverses a large distance and expands to contain numerous small-scale heterogeneities.

To determine whether or not plume behaviour becomes predictable in a quantitative way, the heterogeneous 'real' aquifers used in a series of detailed numerical simulations will be replaced by their homogeneous counterparts defined on the basis of macroscopic transport parameters (i.e. $\bar{v}$, A_l, A_t). Transport in the equivalent homogeneous aquifer will then be simulated using a two-dimensional analytical solution developed by Sudicky (1985). This two-dimensional analytical solution for a finite-thickness aquifer is based on the classical advection-dispersion equation and assumes a constant, horizontal flow velocity and constant dispersivities. The velocity values used in the analytical model corresponded to the mean horizontal groundwater velocity $\bar{v}$, which was in agreement with the slopes of the first-moment curves. Values for A_l were predicted according to Eq. 3.2 and the macro-scale transverse dispersion parameters were assigned according to their local-scale values because the theory predicts $A_t = 0$ for the case of uniform mean flow along the bedding. A source geometry identical to that used in the numerical simulations was used for consistency.

In addition to providing contour diagrams of the tracer concentrations for a visual comparison between the analytical model results and the detailed numerical simulations, a correlation coefficient describing the degree of similarity was also calculated as a function of travel time. The correlation coefficient, r, is given by Davis (1973) as

$$r = \frac{n\Sigma\langle C\rangle C - \Sigma\langle C\rangle \Sigma C}{\{[n\Sigma\langle C\rangle^2 - (\Sigma\langle C\rangle)^2][n\Sigma C^2 - (\Sigma C)^2]\}^{\frac{1}{2}}} \tag{3.3}$$

where $\langle C\rangle$ represents the macroscopic concentration obtained from the analytical solution and C is a local-scale concentration obtained from the numerical solution. The summations are over the n points where concentrations are compared. When r equals zero, there is no correlation between $\langle C\rangle$ and C and when r is equal to 1.0, the correlation is perfect. A negative value of r indicates an inverse relationship between $\langle C\rangle$ and C.

Figure 5 compares the concentrations predicted by the analytical solution using the predicted macrodispersivities and detailed numerical simulations of transport in an aquifer with geostatistical parameters given by $\sigma_Y^2 = 0.38$, $\lambda_L = 2.8$ m, and $\lambda_T = 0.12$ m. Two different cases were simulated. A value for α_L equal to 0.02 m was used in both cases. Only the local transverse dispersivity was different; a value of 0.003 m was used for case (a) and a value of 0.0 m was used for (b).

It can be seen that the macroscopic approach yields concentration values at 600 days which are close to those obtained from the numerical model which accounts for all of the variability in the flow field. The overall length and thickness of the plumes and the location of the centroid position are in good agreement. The slight irregularities present in the concentration contours for the numerical simulation are not present in the analytical results. This is because the macroscopic approach predicts a mean concentration value at each point which, in effect, is an ensemble average value. It is also evident that use of macrodispersivities in the analytical model provides a reasonable prediction at times prior to 600 days, except for noticeable angular offset of the principal axis of the detailed numerical solution plumes.

The correlation coefficients as functions of time for these two cases are also shown in Fig. 5. At 25 days, the correlation between the analytical and the numerical results is high ($r = 0.90$) but, between 25 and 85 days, the degree of correlation decreases. Because the solute is injected at the same location for both types of simulations and because only a small amount of dispersion has occurred in both the analytical and numerical solutions after 25 days, the high correlation coefficient at early time is to be expected. The correlation decreases until about 100 days because the plumes from the numerical solution must adjust to the ever-changing spatial variations in the flow field. Then, for times greater than about 100 days, the correlation coefficients begin to increase as the plumes from the numerical simulations expand and become smoother, and approach a Gaussian shape. The correlation coefficients between the two types of simulations after 400 days are highly significant. Because of the high degree

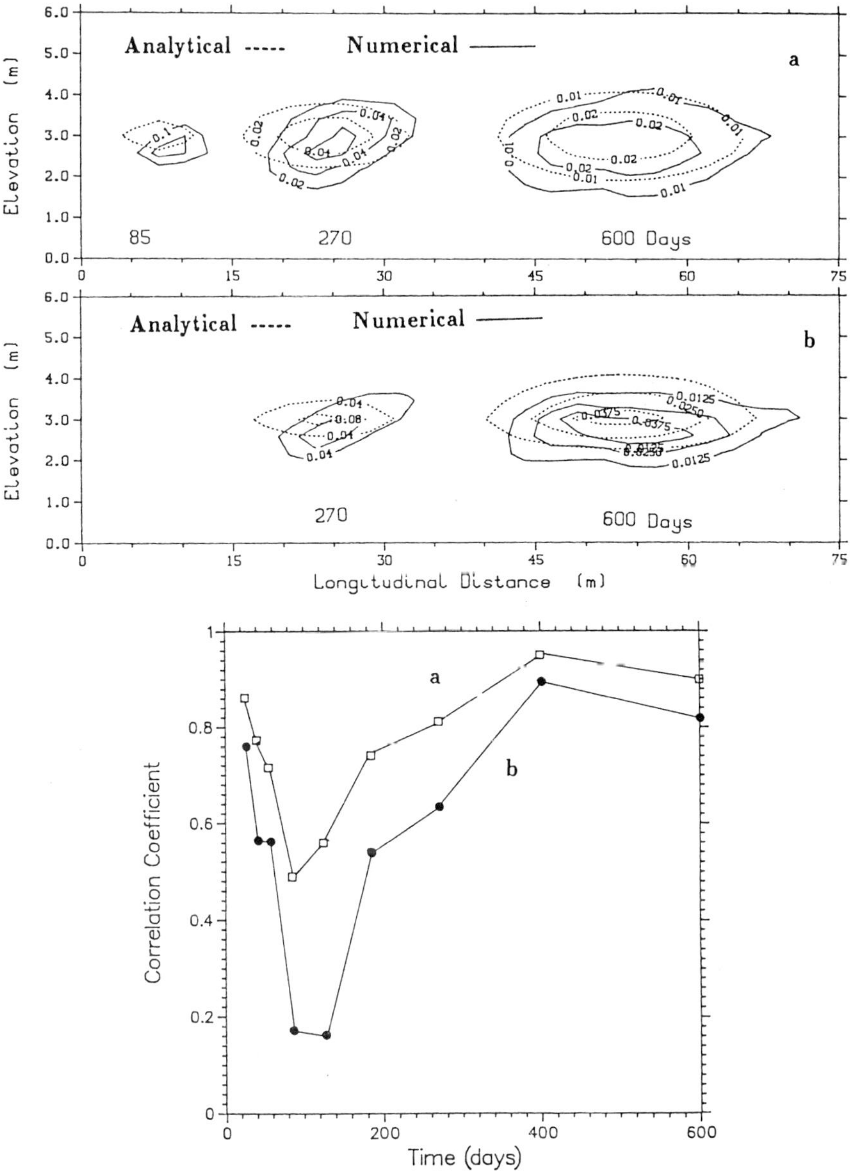

Figure 5. Comparison of numerical and macro-scale solution using predicted macro-dispersivity for different α_T values. Case (a) $\alpha_T = 0.003$ m, and (b) $\alpha_T = 0.0$ m.

of significance, it seems reasonable to accept the macroscopic results as being a likely representation of the actual plumes at large time.

The plume shown as case (b) in Fig. 5 ($\alpha_T = 0.000$ m) is much narrower than the plume for $\alpha_t = 0.003$ m, and the correlation coefficients are noticeably reduced. This is expected because weak transverse local dispersion suppresses vertical expansion of the plume across the bedding and the plume effectively samples fewer velocity variations. Thus, a longer period of transport time will be needed in this case to reach asymptotic mixing conditions.

Figure 6 shows plumes and the correlation coefficients for two simulations in which the correlation lengths of $\ln(K)$ differ. For case (c), the correlation lengths are $\lambda_L = 6.0$ m, $\lambda_T = 0.12$ m, while for case (d) they are $\lambda_L = 0.5$ m, $\lambda_T = 0.5$ m. In both cases, the local dispersivities used in the detailed numerical solution are $\alpha_L = 0.02$ m and $\alpha_T = 0.003$ m. The agreement between the analytical and the numerical solution results for case (c) is poor at early time, but tends to improve with increasing plume residence time. After 600 days, the plumes are similar in size and shape but the plume obtained from the numerical solution is rotated downwards. This angular offset accounts for the lower value of the correlation coefficient at 600 days compared to the previous examples given in Fig. 5. Because the mean flow is horizontal, it is not unreasonable to expect that the offset will tend to become smaller for longer travel distances.

For the isotropic medium of case (d), with $\lambda_L = \lambda_T = 0.5$ m, the plumes from the two types of simulations are located at different elevations. This shift is caused by upward-directed flowlines in the high-resolution flow field near the zone of tracer release. However, by 600 days, the two plumes have similar shapes and peak concentration values. There remains a slight shift in the elevation of the centroid positions although the amount of the shift appears to have diminished slightly. The correlation coefficients for this example steadily increases for times greater than 100 days.

It can be expected that the macroscopic approach will produce realistic plumes even for comparatively small times as the medium becomes more homogeneous. Figure 7 demonstrates this behaviour for an aquifer in which the variance of $\ln(K)$ is equal to 0.18. The correlation lengths are $\lambda_L = 2.8$ m and $\lambda_T = 0.12$ m. The local longitudinal and transverse dispersivities used in the numerical model are 0.02 m and 0.003 m, respectively. The plumes are seen to be similar in shape and size and good correlation exists between the numerical and analytical results over much of the simulation time. After 600 days of transport, the value of r is equal to about 0.95.

The eventual high degree of correlation observed between the detailed, numerically simulated plumes in a heterogeneous aquifer and those obtained using a homogeneous, but hydraulically and dispersively equivalent aquifer, demonstrates conclusively that macroscopic behaviour is reached when a plume expands to a large enough volume such that it encompasses a significant number of heterogeneities. Moreover, the usual arguments about the need to perform some type of spatial averaging of point-to-point concentrations to achieve regularity of a contaminant plume seem unnecessary when a plume becomes large compared to the scale of the heterogeneity. This was effectively

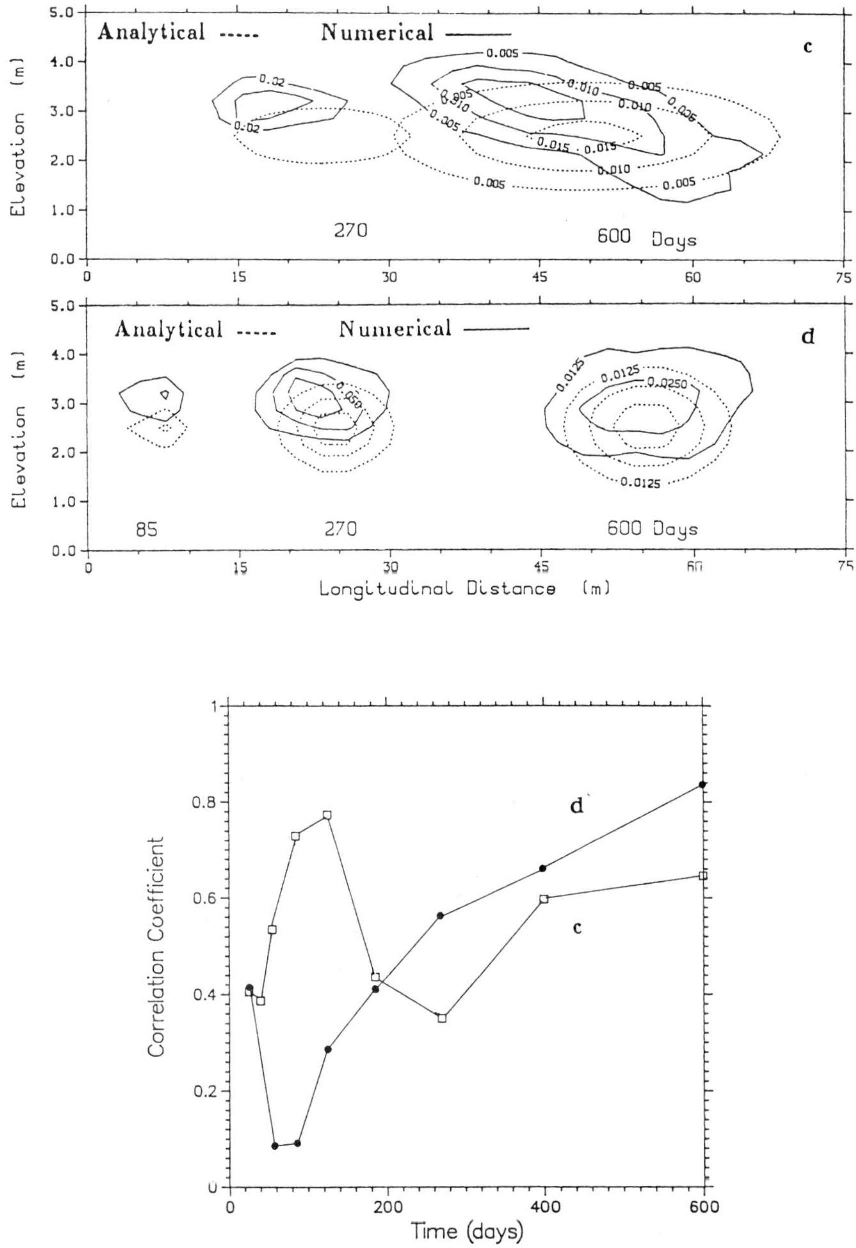

Figure 6. Comparison of numerical and macro-scale analytical solution using predicted macrodispersivity for different λ_L and λ_T values. Case (c) $\lambda_L = 6.0$ m, $\lambda_T = 0.12$ m, and (d) $\lambda_L - 0.5$ m, $\lambda_T = 0.5$ m.

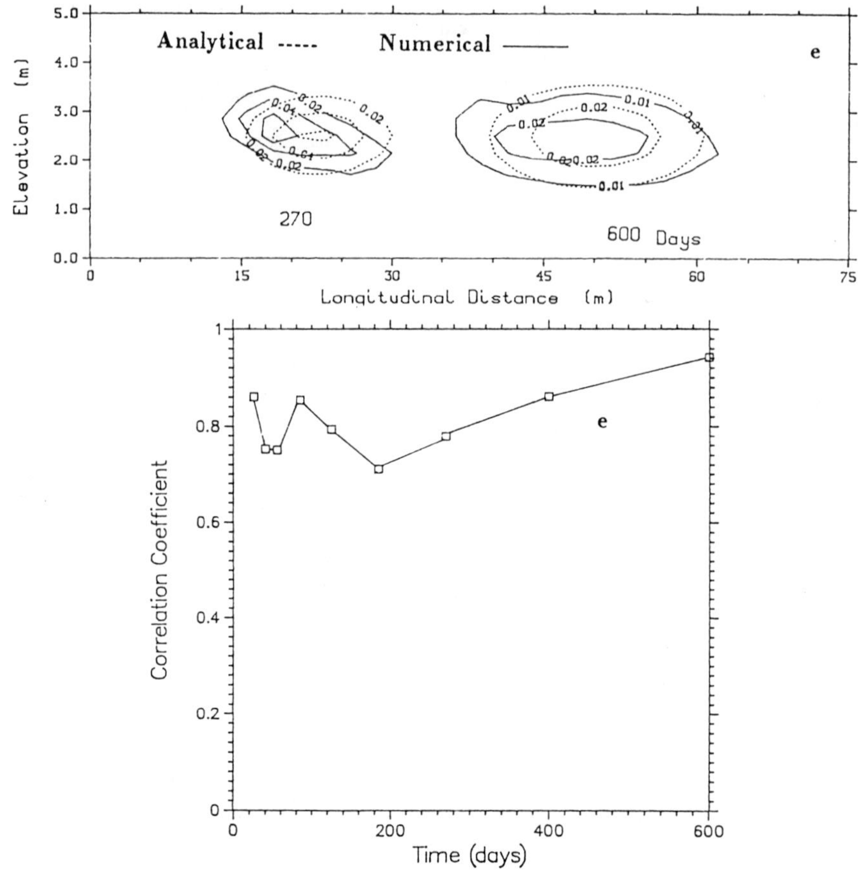

Figure 7. Comparison of numerical and macro-scale analytical solution using predicted macrodispersivity for case (e) $\sigma_Y^2 = 0.18$, $\lambda_L = 2.8$ m, $\lambda_T = 0.12$ m.

demonstrated here because the comparisons were carried out using point values of the local and macro-scale concentrations. The local-scale dispersion process is sufficient to act as a natural self-averaging process.

4. Evolution of a biodegradable organic plume

4.1 A single realization example and the validity of the macro-scale approach

The results presented in the previous section demonstrate that a conservative solute continually spreads with travel time and distance such that it eventually occupies a volume of aquifer that is large compared to the integral scale of the hydraulic conductivity variations. Once this condition is attained, the heterogeneous medium can be replaced with an equivalent homogeneous medium characterized by an effective hydraulic conductivity and a macrodispersivity. It is currently not well known if the migration of organic solute plumes undergoing natural aerobic biodegradation is significantly affected by geological heterogeneities or if an asymptotic mean plume behaviour can be defined. In this section, we will demonstrate some of the features exhibited by organic plumes as they are transported through complex-structured geological materials.

The hydraulic conductivity field used in this example as input to the groundwater flow equation is again assumed to be stationary and log-normally distributed, with point-to-point hydraulic conductivities being locally isotropic. The values of the statistical parameters used to describe the structure of the heterogeneity are the same values as those used in the flow problem for the conservative solution simulation described in detail earlier, except that the variance of $\ln(K)$ is reduced from $\sigma_Y^2 = 0.38$ to 0.27. The length of the flow domain is reduced from 75 m to 45 m due to computer memory limitations when solving the three simultaneous transport equations.

Figure 8 shows the contour plot of the hydraulic conductivity field and the flow net obtained from the dual-formulation flow solution. The medium possesses a lenticular structure due to the statistical anisotropy, with predominantly horizontal discontinuous beds. The mean linear groundwater velocity $\bar{v}$ for the generated flow field is horizontal and equals 0.09 m/day and the maximum and minimum velocities are 0.45 m/day and 0.01 m/day, respectively.

The values of the physical and biochemical transport parameters used in all of the organic transport simulations described in the section are given in Table 1. The organic under consideration is toluene, a hydrocarbon present in petrol, and the electron acceptor is oxygen. The values of the biodegradation kinetic parameters k, K_C, B and d were obtained from the results of a laboratory column biodegradation experiment (MacQuarrie *et al.*, 1989) and the value of the toluene retardation factor, R_C, was estimated from batch-type sorption tests. A value for the utilization ratio $X = 3.13$ was obtained from stoichiometry assuming complete mineralization of the toluene.

Figure 9 shows snapshots of the organic plume and dissolved oxygen distributions at various times after the slug injection. At 50 days, the organic plume is very irregular with two distinct areas of high concentration. The top and middle portions of the plume are observed to be moving forward in high velocity regions. The dissolved oxygen distribution at 50 days shows an area of oxygen depletion below the background value of 3500 μg/l which closely corresponds to the position of the organic plume. This consumption of oxygen within a zone encompassing a dissolved organic plume along with irregularly-shaped organic and oxygen distributions has also been observed in the results

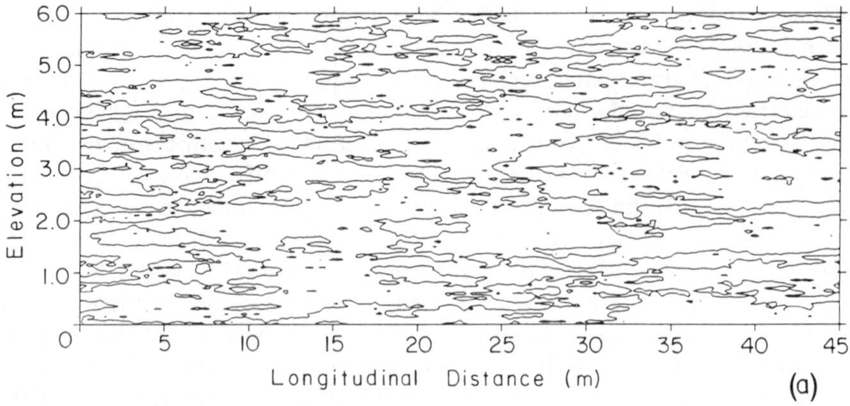

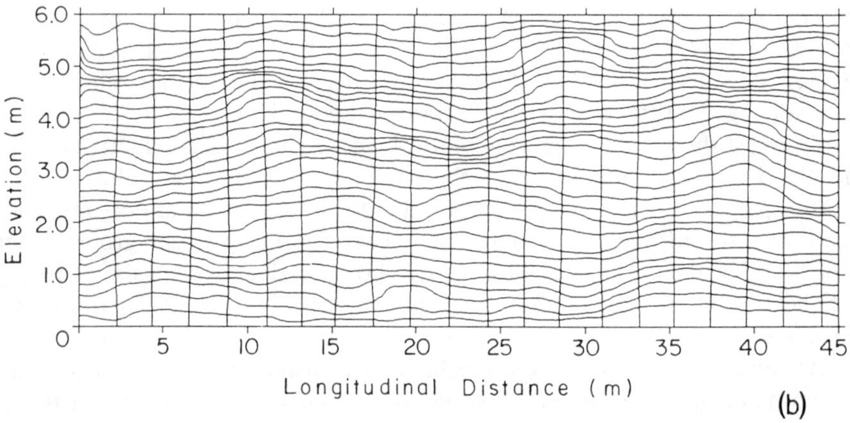

Figure 8. (a) Log-transformed hydraulic conductivity field for $\sigma_Y^2 = 0.27$, $\lambda_L = 2.8$ m, $\lambda_T = 0.12$ m, and (b) resulting flow solutions.

of a highly monitored organic solute tracer experiment performed in the Borden aquifer (Barker *et al.*, 1987). By 250 days, the organic plume has spread longitudinally and narrowed in the vertical direction with peak concentration values decreasing to about 203 μg/l from the initial concentration of 3000 μg/l. By comparing the orientation of the plume at 250 days with the streamlines in the flow net at that position, it can be seen that the plume occupies an area of upward-slanted streamlines. After 300 days, the organic plume has significantly decreased in size and the maximum concentration has been reduced to 5.6 μg/l. The zone of oxygen consumption expands with time in a manner similar to that of a conservative solute such as chloride. At 50 and 150 days, dissolved oxygen is completely consumed by biodegradation in the regions of highest organic concentrations; however, after 300 days the minimum dissolved oxygen

Table 1. Input parameters for organic transport simulations.

Toluene:
Effective diffusion coefficient, $D^* = 6.0 \times 10^{-6}\ cm^2/s$
Retardation factor, $R_C = 1.4$
Maximum utilization rate, $k = 0.493\ 1/day$
Half-utilization constant, $K_C = 654.6\ \mu g/l$

Oxygen:
Effective diffusion coefficient, $D^* = 1.3 \times 10^{-5}\ cm^2/s$
Utilization ratio, $X = 3.13$
Half-utilization constant, $K_A = 100\ \mu g/l$

Microbial population:
Effective diffusion coefficient, $D^* = 0.0$
Retardation factor, $R_m = 1000.0$
Yield coefficient, $B = 0.426$
Decay coefficient, $d = 0.0$

Local dispersivity values:
Longitudinal dispersivity, $\alpha_l = 0.05\ m$
Transverse dispersivity, $\alpha_t = 0.005\ m$

concentration increases to 144 μg/l. Oxygen is able to diffuse into the interior of the small organic plume by 300 days because the organic concentrations are considerably reduced at this time such that a lower amount of dissolved oxygen is required for biodegradation. This is in turn a response to a reduced rate of biomass growth due to limited substrate availability.

Figure 10 shows the total dissolved mass and the position of the centre of mass for the organic plume as functions of time. The mass loss of the plume gradually slows with time, as was observed in the uniform flow simulations. Although the mass loss curve appears to decline exponentially with time, a plot of the log-transformed organic mass versus time curve is non-linear indicating that mass loss cannot be represented by a first-order decay process.

The horizontal velocity calculated from Fig. 10 reveals that, until about 125 days, the centre of mass of the plume is travelling at 0.07 m/day. It then slows at 145 days to 0.03 m/day and remains at this low velocity until 260 days. If a bulk retardation factor is computed using the above mean organic plume velocities and the average groundwater velocity of 0.09 m/day, apparent retardation factor of 1.3 for times up to 125 days and 3.0 for times between 145 and 260 days are obtained.

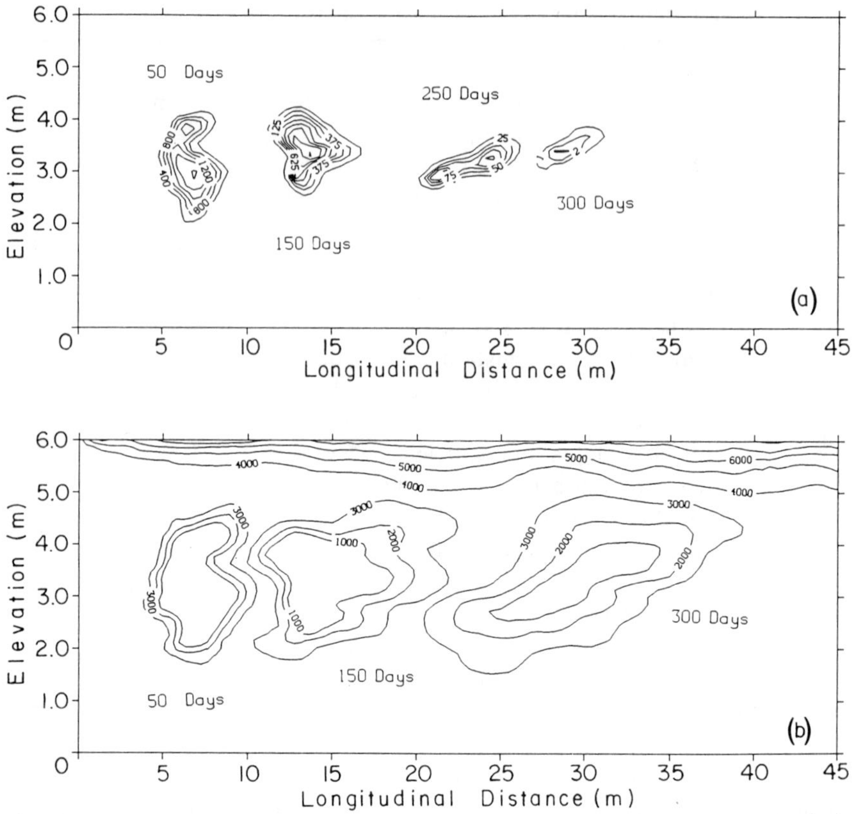

Figure 9. (a) Dissolved organic and (b) dissolved oxygen distributions for $\sigma_Y^2 = 0.27$, $\lambda_L = 2.8$ m, $\lambda_T = 0.12$ m. Concentration isolines are in μg/l. Dissolved oxygen distribution near the water table is for 300 days.

The second-order moments, σ_{11}^2 and σ_{22}^2, are shown in Fig. 11. The plume is decreasing in size in the vertical direction over most of the simulation period, indicating that biodegradation is removing mass at a rate faster than transverse dispersion can spread mass vertically. This is expected because of the typically small value of vertical transverse dispersivity used in the simulation. On the other hand, the plume continues to increase in size in the longitudinal direction at an almost constant rate. Small fluctuations in the spatial variance functions for the longitudinal and transverse directions are probably the result of small-scale velocity fluctuations that cause local distortions in the shape of the plume.

By fitting a straight line to the σ_{11}^2 data shown in Fig. 11 and using Eq. 3.1, we obtain an apparent $\bar{D}_l$ value equal to 0.00325 m^2/day. This calculated value

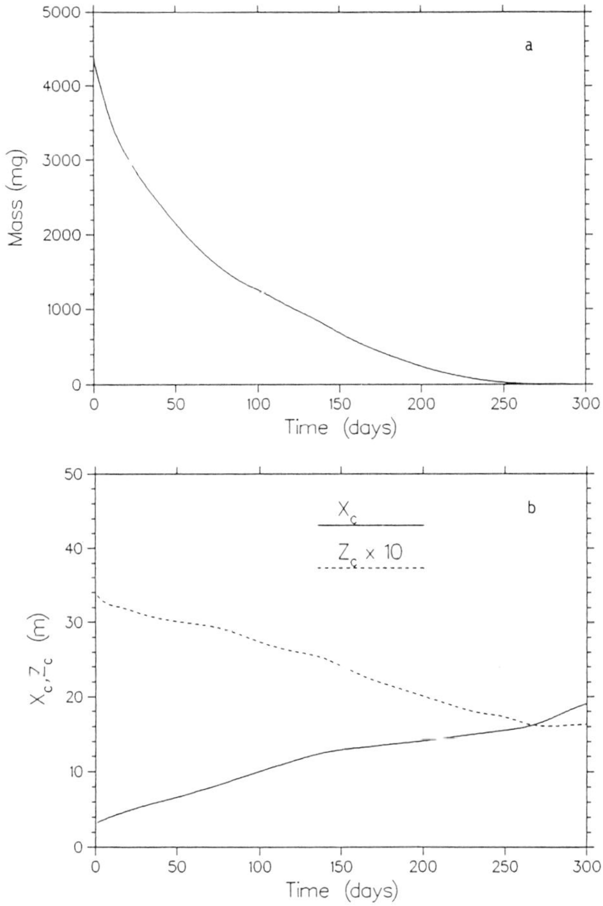

Figure 10. Moment versus time plots for organic plume: (a) mass, (b) centre of mass.

for $\bar{D}_l$ is much less than would be expected for an ever-expanding conservative solute plume on the basis of stochastic macrodispersivity theory. This example calculation illustrates that dispersion theories developed for conservative solute should not be arbitrarily applied to biodegradable chemical plumes and that Eq. 3.1 is probably inappropriate without some modification that takes into account the biodegradation process.

In order to demonstrate the errors that might be committed by using a macro-scale modelling approach based on conservative solute theory, a simulation was performed in which the flow was assumed to be uniform and horizontal with $\bar{v} = 0.09$ m/day and a constant longitudinal macrodispersivity equal to 0.81 m was used as input. Figure 12 shows a comparison between the

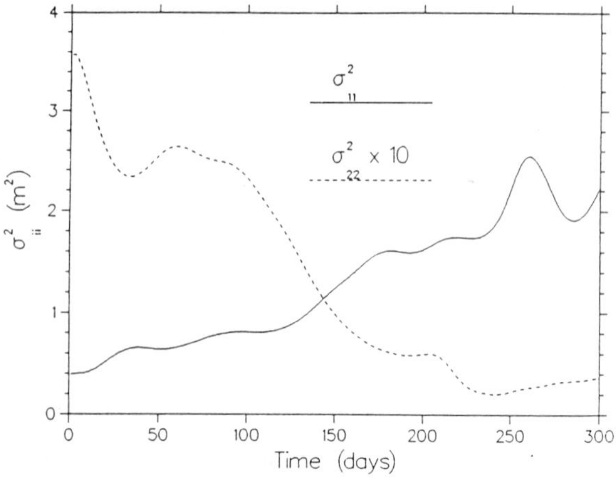

Figure 11. Second moments describing dispersion along ($i = 1$) and transverse ($i = 2$) to the major axis of the organic plume.

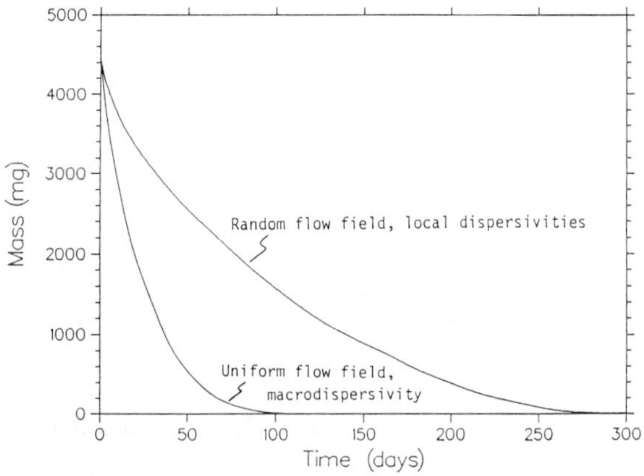

Figure 12. Comparison of organic plume mass loss for different representations of dispersion.

mass loss functions for the organic plume in the random flow field and that in a uniform flow field. The mass loss of the organic plume is greatly overestimated for the simulation where the flow field is taken to be macroscopically uniform and the conservative solute macrodispersivity is used. Using the macro-scale approach for an ever-spreading conservative solute to account for the material heterogeneity allows far too much mixing to occur between the organic plume and the surrounding oxygenated groundwater such that mass loss is artificially enhanced. This observation has significance in practical situations where the distribution of a conservative solute such as chloride is often used to determine dispersion parameters through model calibration. These parameters are then considered as constant, characteristic properties of the geological material. However, as shown in this example, using these parameters to describe the behaviour of a biodegradable organic plume could lead to significant errors in estimating the time required for remediation of the plume. This is of critical importance because the mass loss of the organic will be overpredicted if physical dispersion parameters based on conservative solute distributions are used for organic transport modelling. Because a biodegradable organic plume will always occupy a smaller volume of aquifer compared to that of a conservative solute, especially in the case of a slug-type input, where the volume of the plume will diminish along the path of migration, the ergodic assumption in stochastic transport theories becomes problematic and lacks foundation for its general application to problems involving biodegradable chemicals.

4.2 Influence of aquifer structure on bulk biodegradation rates

In order to investigate how the degree of aquifer heterogeneity affects mass loss of an organic plume undergoing aerobic biodegradation, an aquifer realization was generated with the statistical properties of the previous hydraulic conductivity field, except the variance, σ_Y^2, was increased from 0.27 to 1.04. All of the physical and biochemical transport parameters remain at the values given in Table 1 as used in the previous organic transport simulation. Increasing the variance produces a more heterogeneous medium with larger hydraulic conductivity contrasts between individual lenses. The contoured log-transformed hydraulic conductivity field and the corresponding flow net are given in Fig. 13. Although the aquifer structure is similar to that in Fig. 8, the streamlines are clearly more tortuous than those of the previous case where $\sigma_Y^2 = 0.27$. The bulk average groundwater velocity is again 0.09 m/day while the maximum and minimum groundwater velocities are 1.98 m/day and 0.001 m/day, respectively.

The dissolved organic plume and dissolved oxygen distributions, shown in Fig. 14 are much more erratic than those shown in Fig. 9. At 50 days, the organic plume consists of two distinct lobes, with the top portion moving at a higher velocity. By 150 days, the top portion has completely biodegraded and the remaining organic mass is distributed more elliptically. After 250 days, the plume has split into two smaller plumes, with the lower one being retained in an area of low groundwater velocity and having a maximum concentration of 107 μg/l. The corresponding areas of oxygen consumption, below the background of 3500 μg/l, are likewise very erratic. The minimum oxygen concentration in the oxygen depletion plume is 0.0 μg/l at 50 days and increases to 21 μg/l after 250 days. By 250 days, a portion of the low-dissolved-oxygen plume has moved far ahead of the position of the organic plume because it is non-retarded and because it has entered a high groundwater velocity zone.

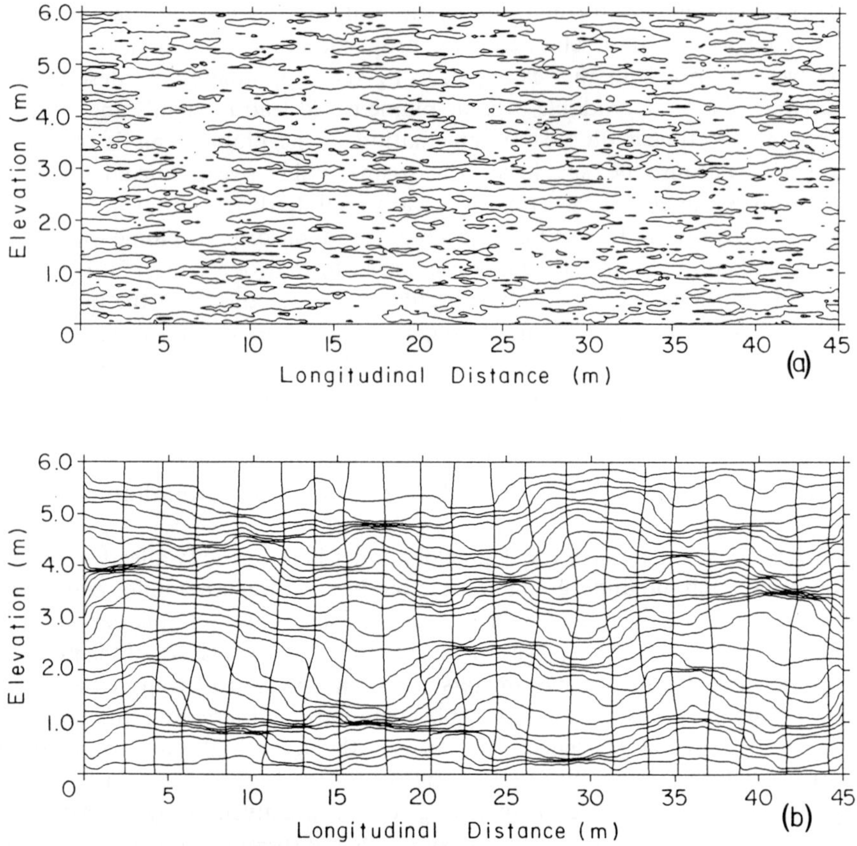

Figure 13. (a) Log-transformed hydraulic conductivity field for $\sigma_Y^2 = 1.04$, $\lambda_L = 2.8$ m, $\lambda_T = 0.12$ m, and (b) resulting flow solution.

Figure 15 shows a comparison of the mass loss curves for the organic plume transported in hydraulic conductivity fields with three different values of σ_Y^2. The case of $\sigma_Y^2 = 0.0$ represents a homogeneous medium. The input dispersivities used in all cases represent the local-scale values of $\alpha_L = 0.05$ m and $\alpha_T = 0.005$ m and the biochemical transport parameters remain unchanged from the values in Table 1. As the variance is increased, the rate of mass loss increases noticeably. Because an organic plume evolves more erratically in a more heterogeneous hydraulic conductivity field, it has a larger surface area over which mixing with oxygenated groundwater can occur. It may also be the case that a high hydraulic conductivity lens may intersect the organic plume and act as an oxygen-supplying conduit. This feature is noted in Fig. 14, where the top portion of the organic plume is quickly biodegraded because it is located in a high groundwater velocity zone.

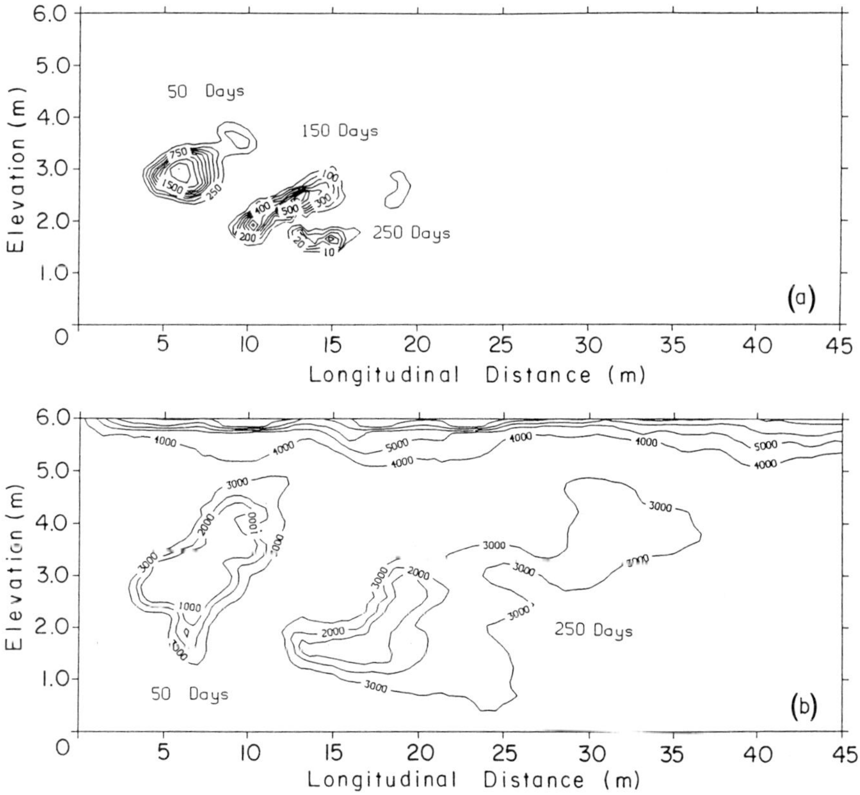

Figure 14. (a) Dissolved organic plume and (b) dissolved oxygen distributions for $\sigma_Y^2 = 1.04$, $\lambda_L = 2.8$ m and $\lambda_T = 0.12$ m. Concentration isolines are in μg/l. Dissolved oxygen distribution near the water table is for 300 days.

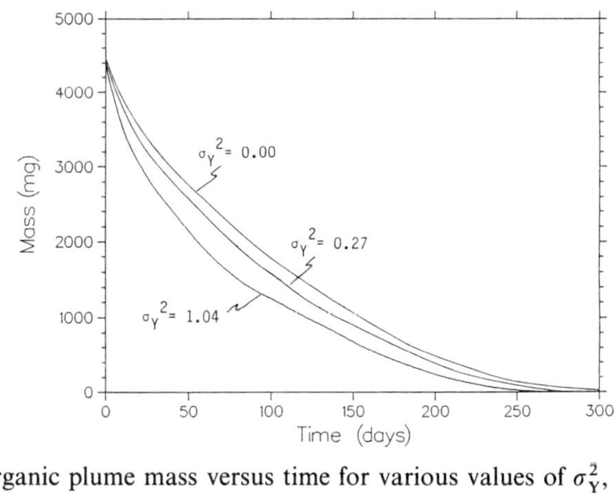

Figure 15. Organic plume mass versus time for various values of σ_Y^2,

It is important to note that there will exist an infinite number of possible realizations of an aquifer, each with the same statistical properties. In order to generalize the results obtained here, a large number of realizations would have to be generated with transport being simulated through each. The average of all the transport simulation results would then provide a better representation of mean plume behaviour along with some measure of the uncertainty in making field-scale predictions. Also, the initial source size relative to the dimensions of the heterogeneities, and whether or not the source is continuous in time, are factors which may produce organic plumes that evolve differently from those simulated here. Nevertheless, for small to moderately-sized spills, our results imply that aquifer structure can in itself influence bulk biodegradation rates and that the behaviour of an organic solute will be much less predictable than that of a conservative solute.

5. Concluding remarks

It has been demonstrated by means of highly detailed numerical simulations of flow and transport in random, synthesized hydraulic conductivity fields that the macro-scale approach for conservative solute plume prediction yields reliable results even without spatial averaging of the local-scale concentrations. This is because an ever-expanding conservative plume gradually smooths itself in response to local-scale dispersion processes and eventually samples a large range of velocity variations. The good agreement between the local- and macro-scale plume simulations suggest that the stochastic–analytic theory provides workable results for prediction of macrodispersivity values from the geostatistical parameters describing the structure of a heterogeneous aquifer. Field studies aimed at characterizing aquifer structures and then relating the statistical parameters to the depositional features of the geological environment would appear to be a logical way of extending the utility of the approach for practical application. It should, nevertheless, be kept in mind that spatial trends in the hydrogeological properties of an aquifer at a scale near or above the modelling scale should be accounted for in a deterministic way.

The simulation of the evolution of a biodegradable organic is, however, more problematic, particularly in the case of a slug-type release of contaminant. This is because a biodegradable plume evolving from a small to moderately-sized spill will continually diminish in volume along its path of migration such that local-scale transport processes, for example advection in an individual bedding feature, eventually control the fate of the plume. Use of a macrodispersivity value characteristic of a conservative solute leads to excessive mixing between the organic plume and oxygenated groundwater on the fringes of the plume and hence will tend to overestimate the actual amount of aerobic degradation. We have also noticed that the degree of heterogeneity of an aquifer can influence bulk biodegradation rates, primarily because the amount of mixing between the oxygenated groundwater and the contaminant plume is controlled by the heterogeneity. This leads us to suspect that macro-scale biodegradation parameters may be needed to account for the heterogeneous nature of the underlying local-scale biochemical transformations.

Acknowledgements

Funding for most of this work was provided by the Natural Sciences and Engineering Research Council of Canada in the form of a University Research Fellowship Grant No. U0508 awarded to E.A. Sudicky and to NSERC post-graduate scholarships held by K.T.B. MacQuarrie and S.L. Schellenberg. The support provided by HydroGeoLogic, Inc., Herndon, Virginia, to the senior author during his stay there is gratefully acknowledged. The typing skills of K.I. Ford are much appreciated.

References

Barker, J.F., Patrick, G.C. and Major, D. (1987). Natural attenuation of aromic hydrocarbons in a shallow sand aquifer, *Ground Water Monitor Rev.*, pp. 64–71, Winter 1987.

Bear, J. (1972). 'Dynamics of Fluids in Porous Media'. Elsevier, New York.

Berry-Spark, K.L. (1987). Gasoline contaminants in groundwater: A field experiment. M.Sc. Thesis, University of Waterloo, Waterloo, Ontario.

Borden, R.C. and Bedient, P.B. (1986). Transport of dissolved hydrocarbons influenced by oxygen-limited biodegradation: 1. Theoretical development, *Water Resour. Res.*, **22**, 1973–1982.

Dagan, G. (1982). Stochastic modeling of groundwater flow by unconditional and conditional probabilities: 2. The solute transport, *Water Resour. Res.* **18**, 835–848.

Dagan, G. (1984). Solute transport in heterogeneous porous formations, *J. Fluid Mech.* **145**, 97–156.

Dagan, G. (1988). Time-dependent macrodispersion for solute transport in anisotropic heterogeneous aquifers, *Water Resour. Res.* **24**, 1491–1500.

Davis, J.C. (1973). 'Statistics and Data Analysis in Geology'. John Wiley and Sons, New York.

Dieulin, A. (1980). Propagation de pollution dans un aquifere alluvial: l'effet de parcours. Thèse de Docteur-Ingenieur, École des Mines de Paris-Université Paris VI.

Freyberg, D.L. (1986). A natural gradient experiment on solute transport in a sand aquifer. 2. Spatial moments and the advection and dispersion of nonreactive tracers, *Water Resour. Res.* **22**, 2031–2046.

Fried, J.J. (1975). 'Groundwater Pollution: Theory, Methodology, Modelling and Practical Rules'. Elsevier, Amsterdam.

Frind, E.O. (1982). The principal direction technique: A new approach to groundwater contaminant transport modeling. 'Proceedings, Fourth International Conference on Finite Elements in Water Resources, Hannover, Germany', **13**, pp. 25–42. Springer-Verlag, New York.

Frind, E.O. (1988). Solution of the advection-dispersion equation with free exit boundary, *Numer. Meth. Part. Diff Equat.* **4**, 301–313.

Frind, E.O. and Matanga, G.B. (1985). The dual-formulation of flow for contaminant transport modeling: 1. Review of theory and accuracy aspects, *Water Resour. Res.* **21**, 159–169.

Gelhar, L.W. and Axness, C.L. (1983). Three-dimensional stochastic analysis of macrodispersion in aquifers, *Water Resour. Res.* **19**, 161–180.

Gelhar, L.W., Gutjahr, A.L. and Naff, R.L. (1979). Stochastic analysis of macrodispersion in a stratified aquifer, *Water Resour. Res.* **15**, 1387–1396.

Hoeksema, R.J. and Kitanidis, P.K. (1985). Analysis of the spatial structure of properties of selected aquifers, *Water Resour. Res.* **21**, 563–572.

Kissel, J.C. (1985). Numerical simulation of mixed-culture biofilms. Ph.D. Thesis, Stanford University, Stanford, CA.

MacQuarrie, K.T.B. and Sudicky, E.A. (1989). Simulation of biodegradable organic contaminants in groundwater: 2. Plume behaviour in uniform and random flow fields, *Water Resour. Res.* (in press).

MacQuarrie, K.T.B., Sudicky, E.A. and Frind, E.O. (1989). Simulation of biodegradable organic contaminants in groundwater: 1. Numerical formulation in principal directions, *Water Resour. Res.* (in press).

Mantoglou, A. and Wilson, J.L. (1982). The turning bands method for simulation of random fields using line generation by a spectral method, *Water Resour. Res.* **18**, 1379–1394.

Matheron, G. and de Marsily, G. (1980). Is transport in porous media always diffusive? A counter example, *Water Resour. Res.* **16**, 901–917.

Moltyaner, G.L. (1985). Stochastic versus deterministic: a case study. 'Proceedings of the International Symposium on the Stochastic Approach to Subsurface Flow', pp. 362–364, Montvillargenne, France, 4–6 June, 1985.

Molz, F.J., Melville, J.G., Güven, O., Crocker, R.D. and Matheson, K.T. (1985). Design and performance of single-well tracer tests at the Mobile site, *Water Resour. Res.* **21**, 1497–1502.

Molz, F.J., Widdowson, M.A. and Benefield, L.D. (1986). Simulation of microbial growth dynamics coupled to nutrient and oxygen transport in porous media, *Water Resour. Res.* **22**, 1207–1216.

Neuman, S.P., Winter, C.L. and Newman, C.M. (1987). Stochastic theory of field-scale Fickian dispersion in anisotropic porous media, *Water Resour. Res.* **23**, 453–466.

Ogram, A.V., Jessup, R.E., Ou, L.T. and Rao, P.S.C. (1985). Effect of sorption on biological degradation rates of (2,4-Dichlorophenoxy)acetic acid in soils, *Appl. Environ. Microbiol.* **49**, 582–587.

Pickens, J.F. and Grisak, G.E. (1981). Scale-dependent dispersion in a stratified granular aquifer, *Water Resour. Res.* **17**, 1191–1211.

Schellenberg, S.L. (1987). Groundwater flow and nonreactive tracer motion in heterogeneous statistically anisotropic porous media. M.Sc. Thesis, University of Waterloo, Waterloo, Ontario.

Smith, L. (1980). Spatial variability of flow parameters in stratified sand, *Int. J. Math. Geol.* **13**, 1–21.

Smith, L. and Schwartz, F.W. (1980). Mass Transport: 1. A stochastic analysis of macroscopic dispersion, *Water Resour. Res.* **16**, 303–313.

Sudicky, E.A. (1983). An advection-diffusion theory of contaminant transport for stratified porous media. Ph.D. Thesis, University of Waterloo, Waterloo, Ontario.

Sudicky, E.A. (1985). A collection of analytical solutions for solute transport in porous and fractured porous media. Unpublished report, Institute for Groundwater Research, University of Waterloo, Waterloo, Ontario.

Sudicky, E.A. (1986). A natural gradient experiment on solute transport in a sand aquifer: Spatial variance of hydraulic conductivity and its role in the dispersion process, *Water Resour. Res.* **22**, 2069–2082.

Sudicky, E.A., Cherry, J.A. and Frind, E.O. (1983). Migration of contaminants in groundwater at a landfill: A case study. 4. A natural-gradient dispersion test, *J. Hydrol.* **63**, 81–108.

Steen, W.C., Paris, D.F. and Buaghman, G.L. (1980). Effects of sediment sorption on microbial degradation of toxic substances. *In* 'Contaminants and Sediments, Vol. 1: Fate and Transport, Case Studies, Modeling, Toxicity' (R.A. Baker, ed.), pp. 477–482. Ann Arbor Science Publishers, Ann Arbor, MI.

Wilson, J.T., MacNabb, J.F., Balkwill, D.L. and Ghiorse, W.C. (1983). Enumeration and characterization of bacteria indigenous to a shallow water-table aquifer, *Ground Water* **21**, 134–142.

Zapico, M.M. (1987). Aquifer evaluation procedure—Core acquisition and an assessment of slug tests. M.Sc. Thesis, University of Waterloo, Waterloo, Ontario.

Chapter XVI

Percolation-based Models for Pore-to-lab Scale Calculations in Multifluid Porous Media

Lin A. Ferrand, Michael A. Celia and Wendy E. Soll
Ralph M. Parsons Laboratory, Department of Civil Engineering, Massachusetts Institute of Technology, Cambridge, MA 02139, USA

1. Introduction

Mathematical models describing the flow of two or more fluids in a porous medium generally consist of continuity equations (one for each mobile phase), physical constraints and a set of constitutive equations. The continuity equations and the physical constraints are essentially material independent; specific material properties are introduced via constitutive relations which are usually determined experimentally. Even for the simplest case of multifluid porous media, involving non-reactive, fully immiscible fluids at constant temperature, the model requires a set of constitutive equations relating fluid pressures, relative saturations and relative permeabilities. Because the relationships between these variables depend upon both fluid and solid phase compositions and upon the microscopic structure of the solid matrix, they must be determined individually for each system. The lack of available constitutive data is frequently cited by both petroleum reservoir and groundwater contaminant transport modellers as a primary hindrance to effective forecasting (Aziz and Settari, 1979; Abriola and Pinder, 1985).

Because laboratory measurements of the necessary constitutive relationships are very costly, an alternative derivation of these relationships which uses simplified laboratory measurements is desirable. One methodology for deriving these constitutive relationships involves modelling fluid–fluid displacement at the pore scale. Physically based displacement models of this type are rooted in network modelling and in percolation theory. With the increasing availability of powerful computers, these models are becoming more feasible. It appears that carefully calibrated percolation-based models may provide physically meaningful prediction of constitutive relationships. In addition, these models

DYNAMICS OF FLUIDS IN HIERARCHICAL POROUS MEDIA
ISBN 0-12-200260-1

provide an important tool in understanding scaling relationships for multiphase porous media systems.

The purpose of this review is to summarize briefly theoretical and computational developments in the use of percolation-based models to describe fluid behaviour in multifluid porous media. In Section 2 constitutive relationships and the procedures used to measure them in the laboratory are presented. Network modelling, percolation theory and the foundations of their application to porous media are introduced in Section 3. In Section 4, we summarize the use of percolation-based models to simulate fluid displacement and associated constitutive data. Finally, in Section 5, we summarize current directions in percolation-based modelling of multiphase flow phenomena in porous media.

2. The physical system

Percolation-based models of pore-scale fluid behaviour have been proposed as an alternative to laboratory measurements to generate laboratory-scale constitutive data. We begin our discussion of these models by examining relevant constitutive relations and the experimental procedures used to measure them.

Most fundamental is the relation between fluid pressures and relative saturations or fluid contents. In porous media containing more than one fluid phase, each fluid will, in general, exhibit a pressure which is different from that of the other fluid(s). Furthermore, when any two fluids are in the presence of a solid surface, as in a porous medium, one of the fluids preferentially wets the solid. Relative wettability with respect to each fluid is determined by the contact angle (θ) between fluid–fluid interface and solid surface. If wetting ($\theta < 90°$) and non-wetting ($\theta > 90°$) fluid pressures are denoted by p_w and p_n, respectively, then the capillary pressure (P_c) is defined as

$$P_c \equiv p_n - p_w \tag{2.1}$$

Existence of non-zero capillary pressures is a result of the ability of a fluid–fluid interface to support a non-zero stress. In a given two-fluid porous medium the relative wetting fluid saturation (S_w), which is equal to the fraction of void space occupied by wetting fluid, is a function of this pressure difference, i.e.

$$S_w = S_w(P_c) \tag{2.2}$$

The typical form of the P_c–S_w relation is shown in Fig. 1A. In a typical two-fluid pressure or suction cell experiment to determine this relation, the sample of porous medium is initially saturated with the wetting fluid ($S_w = 1.0$) and is in contact with reservoirs of both wetting and non-wetting fluids. To generate the main drainage (displacement of wetting fluid by non-wetting fluid)

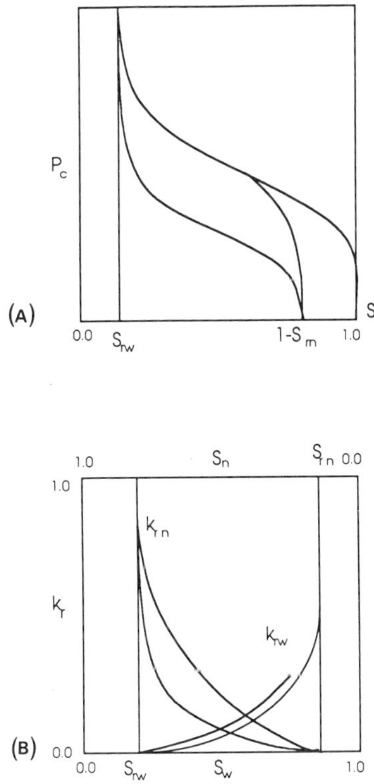

Figure 1. Multiphase constitutive relations. (A) Form of the capillary pressure–saturation relation (after Bear, 1972). (B) Form of the relative permeability–saturation relation (after Bear, 1972).

curve, P_c is increased incrementally and the sample is allowed to equilibrate at each step. After a critical entry value of P_c is reached, successive increases in P_c result in successive decreases in S_w, until wetting fluid exists only as isolated blobs or ganglia. At this point, S_w is essentially irreducible and is referred to as residual wetting fluid saturation (S_{rw}). If the process is reversed, an imbibition (displacement of non-wetting by wetting fluid) branch is generated. This process ends at residual non-wetting fluid saturation (S_{rn}). Reversals in the direction of capillary pressure change at arbitrary locations on either branch produce so-called scanning curves. It is this complex, multivalued relation, including residual saturations, entry pressures and hysteresis between drainage and imbibition branches, that a physically meaningful model must be able to predict.

A mathematical description of multiphase flow in porous media also requires a Darcy-type equation relating phase-velocity ($\mathbf{q}_\alpha$) to a pressure gradient (∇p_α)

for each mobile fluid phase α

$$\mathbf{q}_\alpha = \frac{\mathbf{k}k_{r\alpha}}{\mu_\alpha} \cdot [\nabla p_\alpha - \rho_\alpha g\hat{z}] \quad (\alpha = \mathrm{w,n}) \tag{2.3}$$

where $\mathbf{k}$ is the intrinsic permeability tensor, $k_{r\alpha}$ is relative permeability to the α phase, μ_a and ρ_α are viscosity and density of the α phase, respectively, g is the gravitational acceleration constant and $\hat{\mathbf{z}}$ is a unit vector in the z-direction. Relative permeability depends on relative saturation so that, for the two-fluid case,

$$k_{r\alpha} = k_{r\alpha}(S_w) = k_{r\alpha}(P_c) \quad (\alpha = \mathrm{w,n}) \tag{2.4}$$

Equation 2.4 is the second important constitutive equation which must be included in any multiphase porous medium flow model.

The form of the $k_{r\alpha}$–S_α (α = w,n) relation is shown in Fig. 1B. In a typical two-fluid, steady-state experiment to determine the relative permeability relation, a uniform P_c or S_w is established through the section of porous medium to be tested. Pressures and saturations are held constant in time while flux rates are measured. The relative permeabilities, k_{rw} and k_{rn}, are then determined using Eq. 2.3. Important characteristics of the relation between relative permeabilities and saturations, which must be predicted by a physically meaningful model, include zero relative permeability to each fluid at residual saturation and more significant hysteresis in the non-wetting fluid (k_{rn}–S_n) relation than in the wetting fluid (k_{rw}–S_w) relation (Corey, 1986).

3. Percolation-based models of fluid transport in porous media

3.1 Percolation theory

Percolation theory, which was introduced by Broadbent and Hammersley in 1957, describes the effects of randomness in a medium on the transport of some material or process through the medium. The theory has been used to analyse such widely varying phenomena as the spread of epidemics in a population, the diffusion of hydrogen atoms through disordered solids, and fluid flow in porous media.

In each case, the medium is conceptualized as a geometric lattice of locations called nodes or sites. Each site may be connected to neighbouring sites by bonds or may share a common boundary. Given a lattice, some property which governs movement through the system is assigned, at random, to sites and/or bonds. That is, sites or bonds are randomly available for transport. In addition, an accessibility rule, which defines the conditions under which transport from one available site to another is possible, must be specified.

A very simple example is a site percolation model which predicts the spread of forest fires (Stauffer, 1985). The forest might be represented by a square lattice of sites which represent possible tree locations (Fig. 2). Trees are assigned randomly to sites with some probability p_s. A reasonable accessibility rule might be that fire can spread from one tree to another only if they are nearest neighbours, that is, if their sites share a side. An important question of percolation theory is: at what value of p_s does a percolating cluster, or continuous pathway from one end of the medium to the other, first appear? In terms of our example, we ask at what value of p_s a fire which starts at one side of the forest can spread to the other side. This critical value of p_s is known as the percolation threshold or critical probability, $p_{s(cr)}$. Note that if the accessibility rule is changed such that fire can spread to next nearest neighbours (i.e. across corners), the value of $p_{s(cr)}$ also changes. For the configuration shown in Fig. 2, fire can spread from one side of the forest to the other under the second accessibility rule but not under the first.

For the case of fluid transport in a porous medium, some fraction (p_b) of a network of bonds might be assigned to be 'open' and the remainder ($1 - p_b$) to be 'closed'. The percolation threshold, $(p_b)_{cr}$, for this 'bond percolation' problem is the value of p_b at which we are guaranteed a continuous pathway

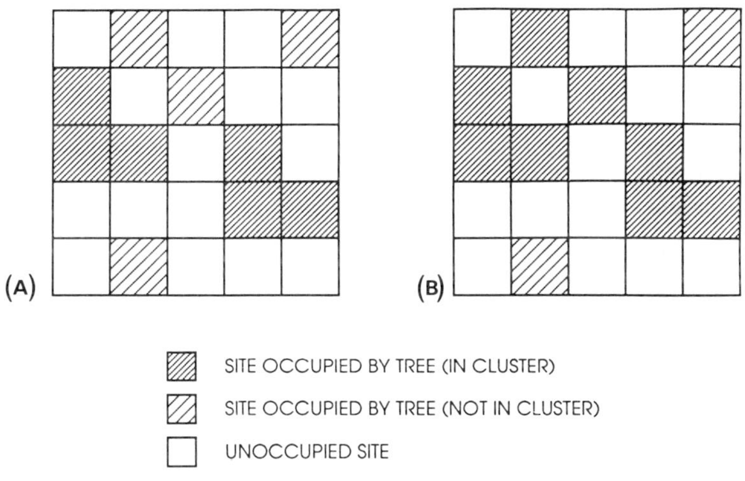

Figure 2. Forest fire model: lattice of possible tree locations. Shaded squares represent sites with trees ($p = 0.4$). (A) Accessibility rule 1: fire can spread to nearest (side-sharing) neighbours. Dark shaded squares are cluster members: no lattice spanning cluster. (B) Accessibility rule 2: fire can spread to nearest and next nearest (corner-sharing) neighbours, lattice spanning cluster exists.

of open bonds across an infinite lattice. Identification of $(p_b)_{cr}$ for a given system is one of the fundamental objectives of percolation theory.

While percolation theory provides a mathematically attractive framework with which to view multiphase displacement in porous media, it is often insufficient to accommodate all of the important physics of the system. Therefore, some modification and enhancements must be introduced. Because the problem of interest is flow in porous media, the discussion which follows will focus exclusively on applications of percolation-based models to these systems. Further discussion of general percolation theory can be found in Broadbent and Hammersley (1957), Frisch and Hammersley (1963), Shante and Kirkpatrick (1971), Essam (1972) and Stauffer (1985).

3.2 Application of percolation concepts to porous media

The movement of fluids in porous solids is an inherently complex process. This is due largely to the complex topology of the pore space through which the fluids move. A fundamentally important property of this topology is the 'connectedness' or 'branchiness' of the flow paths. For the case of multiple fluid phases in porous media, the flow process is further complicated by the existence of fluid–fluid interfaces and the pressure differences that exist across such interfaces.

Representations of pore space

The earliest reported conceptualization of pore space topology was the sphere pack, which dates to the nineteenth century (Hazen, 1892; as reported by Fatt, 1956a). The model represents the solid matrix of the porous medium as spheres of variable radius that are arranged in a regular packing. The resulting pore space topology is of sufficient complexity to render the system essentially mathematically intractable. While some results have been obtained for simple cases of a single fluid flow (dating to Kozeny, 1927, and Carmen, 1938), the void structure appears to be too complex to be useful for analyses of multiphase systems.

The first viable model for multiphase porous media was the 'bundle-of-tubes' model. In this model, the pore space is envisioned as a bundle of independent, parallel tubes, each having a different radius. Some important radius-dependent capillary effects, such as finite entry pressure, are captured. Virtually all flow properties can be written in closed form equations. Pore size distributions can also be calculated directly from drainage and/or imbibition curves. This type of model was apparently first introduced by Blake (1922). It has been used as a pore-space model in many other studies; see, for example, Childs and Collis-George (1948) and Fatt and Dykstra (1951).

The most attractive attribute of the bundle-of-tubes model is its topological simplicity. However, this extreme simplicity precludes model results from matching some observed behaviours of multiple fluids in porous media. In its simplest form, this model cannot predict hysteresis or residual saturations, and is completely anisotropic. While the anisotropy can be eliminated by random orientation of the tubes, and some hysteresis can be produced by allowing tube radii to vary along their lengths, residual saturations cannot be predicted. The most serious deficiency of bundle-of-tubes models is that they ignore the fundamental interconnectedness, or branchiness, of pore space.

In his classic work, Fatt (1956a,b,c) introduced the idea of the network model of porous media. His objective was to combine the best attributes of both the sphere pack model (pore space interconnectedness) and bundle of tubes model (mathematical tractability). He designed a lattice structure in which cylindrical tubes were assumed to represent individual pores, thereby producing a network of interconnected tubes. While not presented in terms of percolation theory, Fatt's model is essentially equivalent to a description of quasi-static immiscible displacement as a bond percolation process.

Typical networks or lattices used by Fatt and later researchers to represent the void space of porous media are shown in Fig. 3. Two important characteristics of lattices are dimensionality (two dimensional versus three-dimensional) and the average number of bonds per site or coordination number (z). Note, however, that two lattices with the same dimensionality and coordination number may have very different properties. For example, both the Bethe tree (Fig. 3A) and the hexagonal network (Fig. 3B) are two-dimensional lattices with coordination number 3, but there is exactly one path connecting any two sites in the Bethe tree while there are many paths connecting any two sites in the hexagonal lattice.

Most early lattice models sought to represent the inherently three-dimensional structure of pore space by a two-dimensional network. The usefulness of these models is very limited however, since use of any two-dimensional network precludes the ability to model fluid bicontinua. In addition, several researchers (Sykes and Essam, 1963; Dean, 1963; Dean and Bird, 1967; Chatzis, 1980), investigating applications of percolation theory to fluid flow in porous media, have shown that two- and three-dimensional networks differ in a fundamental way: the critical probability (p_{cr}) for breakthrough in any two-dimensional model may be approximated as $2/z$, while for a three-dimensional model it is $1.5/z$.

The complexity of early network simulations of fluid displacement in pore space was constrained by the need to perform analyses without recourse to sophisticated computing facilities. These models generally consisted of cylindrical tubes or bonds intersecting in volumeless nodes (Fatt, 1956a; Dodds and Lloyd, 1971/72). In many more recent models (Chatzis and Dullien, 1985; Diaz *et al.*, 1987; Wardlaw *et al.*, 1987; Soll *et al.*, 1988; Celia and Ferrand, 1988; Ferrand and Celia, 1989) pore space is conceptualized as a network of hollow spherical sites, which are connected to neighbouring sites by cylindrical bonds. Specification of site and bond radii, bond lengths, and the pattern of connectedness, serves

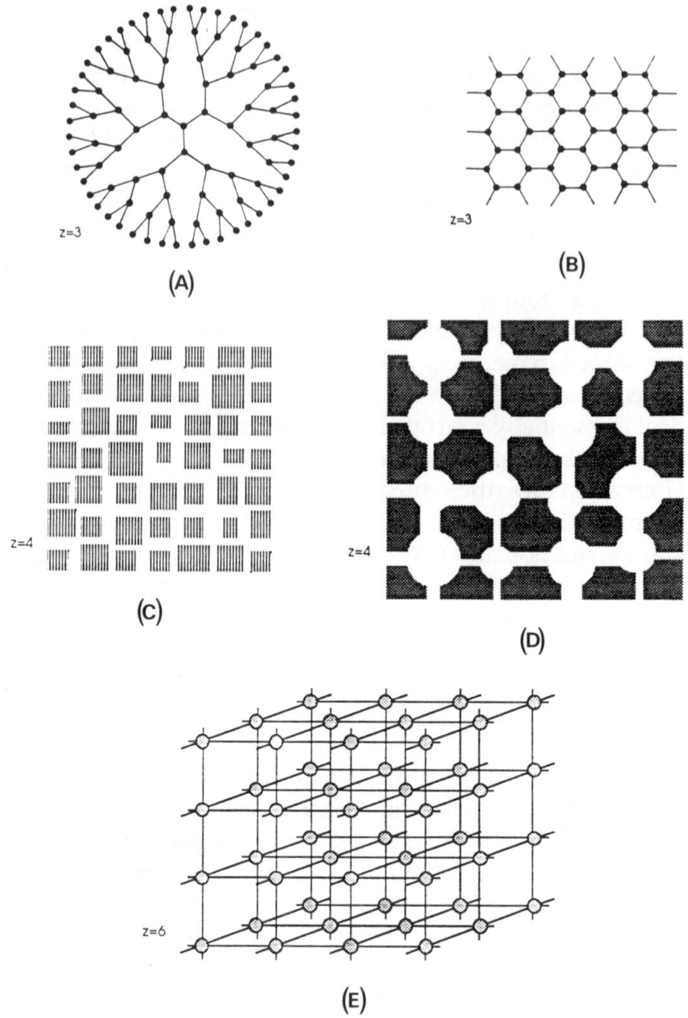

Figure 3. Examples of geometrically regular networks (z is coordination number). Circles represent sites (or nodes), connecting lines represent bonds. (A) Bethe tree; (B) two-dimensional hexagonal lattice; (C) two-dimensional square lattice of bonds intersecting at volumeless sites; (D) two-dimensional square lattice of finite-volume sites and bonds; (E) three-dimensional square lattice of finite volume sites and bonds.

to define the geometry of the pore space. The sizes of the sites and bonds are typically chosen at random from probability distributions that characterize the porous material.

This quantitative characterization of the pore space is not a straightfoward task. Many techniques have been used to try to determine the distribution of pore and throat sizes for various materials. Mercury porosimetry is the most commonly used experimental method (Orr, 1970; Wall and Brown, 1981; Juang and Holtz, 1988). In general, analyses of porosimetry data implicitly assume a bundle-of-tubes model. This does not account for larger pores which may be prevented from filling with mercury until narrower pores surrounding them fill. Therefore, porosimetry results tend to be skewed toward larger numbers of smaller pores for these tests.

One alternative to porosimetry is nuclear magnetic resonance (NMR) analysis. Analyses of NMR data does not require the bundle-of-tubes assumption, but still assumes a simple cylindrical pore shape. The technique gives fairly good results for mid-range distributions, that is, for distributions which are neither very broad nor peaky (Gallegos and Smith, 1988).

A third laboratory technique for characterizing pore space is direct measurement (Yanuka *et al.*, 1986; Gvirtzman *et al.*, 1987; Wardlaw *et al.*, 1987). A sample is saturated with a hardening fluid (e.g. Wood's metal, epoxy) and then thin-sectioned. The sections are analysed visually under a microscope. This method is most often used to quantify solid versus void volumes rather than pore geometries and is extremely time consuming.

A number of theoretical models have also been used to calculate pore-size distributions. One such model, developed by D'Hollander (1979), assumes a log-normal distribution based on the physical characteristics of capillary pressure–saturation curves. The parameters of the distribution are found by fitting to experimental data, making a bundle-of-tubes type assumption. Alternative distributions (e.g. normal, uniform and Weibull) have been investigated by other researchers (Koplik and Lasseter, 1982; Mann *et al.*, 1986; Diaz *et al.*, 1987). Distribution parameters vary widely both from material to material and between researchers, and have significant variations in mean and variance. The applicability of pore-size distributions found using any of these methods to network models of pore space remains unclear. Thus calibration of network parameters remains an active area of research.

Network models of multiphase flow

In a network or percolation-based model, it is assumed that only one fluid resides in a given site or bond at any given time. That is, there is no funicular flow of fluids within a site or bond. This assumption is supported by experimental evidence for very slow displacement (Chatenever and Calhoun, 1952; Craig, 1971; Elsik and Miller, 1988), including photomicrographic observations.

Lenormand and Zarcone (1985a) presented experimental visual evidence of the validity of quasi-static, capillary network models for two phase porous media systems when the capillary number,

$$N_c = \frac{q\mu}{A\sigma} \tag{3.1}$$

which measures the relative importance of viscous versus capillary forces, is sufficiently small. In Eq. 3.1, q is flow rate, μ is viscosity, A is cross-sectional area and σ is interfacial tension. For $N_c \leqslant O(10^{-8})$, the quasi-static capillary description appears to be appropriate. For larger values of N_c, capillary forces lose their dominant role, and viscous effects must be accounted for.

Given a geometric representation of the pore space as a regular lattice with a random distribution of site and/or bond radii, the distribution of the fluid within the porous medium at a given capillary pressure is derived by application of the Young–Laplace equation

$$P_c = \frac{2\sigma}{R} \tag{3.2}$$

where σ is the interfacial tension between fluids and R is the effective radius of curvature of the fluid–fluid interface. Consider the two-fluid problem, with a wetting and a non-wetting fluid. If a cylindrical tube of radius R_T is initially filled with wetting fluid, and is in contact with non-wetting fluid at one end, the non-wetting fluid will not displace the wetting fluid until

$$P_c > \frac{2\sigma \cos\theta}{R_T} \tag{3.3}$$

Non-wetting fluid invades and completely displaces the wetting fluid whenever inequality 3.3 holds. As P_c increases, tubes of largest radius are invaded first, followed by those of successively smaller radii.

In a lattice model of a porous medium with a stable fluid configuration at a fixed capillary pressure, a simulated change in P_c results in movement of some of the fluid–fluid interfaces. An interface moves when it is in contact with a pore opening (radius R_T) for which the critical invasion value of capillary pressure, $2\sigma \cos\theta/R_T$, lies between the original value of P_c and its final (changed) value. When P_c is increased, non-wetting fluid displaces wetting fluid. When P_c is decreased, wetting fluid displaces non-wetting fluid. During drainage, it is the largest pores that are invaded first by the non-wetting fluid. Conversely, during imbibition, the smallest pores are the first to be invaded by the wetting fluid.

Thus, in lattices consisting of both bonds and sites with finite radii, drainage is controlled by bond radii and is modelled as a bond percolation problem; imbibition is controlled by site radii and is modelled as a site percolation problem. Repeated application of the Young–Laplace relation (Eq. 3.3) allows fluid–fluid interfaces to be tracked throughout the network as capillary pressure is varied. Models of this type inherently predict hysteresis between drainage and imbibition processes.

In a simulation of a laboratory procedure to measure the P_c–S_w relation, all sites and/or bonds are assumed to be initially saturated with wetting fluid. Two boundaries of the lattice are assumed to be in contact with fluid (wetting and non-wetting) reservoirs. Incremental changes in P_c are simulated by incremental changes in pressure boundary conditions. As described above, repeated application of the Young–Laplace equation (Eq. 3.3) allows fluid–fluid interfaces to be tracked throughout the network as P_c is varied. The procedure may be reversed to simulate imbibition. At each step the volumes or numbers of wetting fluid-filled sites and bonds are summed to calculate a global wetting fluid-filled saturation.

Percolation-based models of multiphase flow in porous media can also be applied to the dynamic case of creeping flow. This extension allows lattice simulation of the relative permeability–saturation relation. A computational simulation of the $k_{r\alpha}$–S_α ($\alpha = \mathrm{w,n}$) relation mimics the experimental procedure described in Section 2. At the start of a simulation, the permeability of the completely saturated lattice is calculated by assigning resistances to bonds based on their radius. The lattice is drained (or imbibed) incrementally using the P_c–S_w algorithm. At each step, a new value of network permeability to both wetting and non-wetting fluids is calculated based on their distributions in the lattice. Those sites or bonds which are not connected to a lattice spanning cluster are not included in either calculation.

4. Applications of percolation-based models

Network or percolation-based simulations of two-phase constitutive relations have been reported in the porous media literature for more than three decades. As mentioned in Section 3.2, the earliest application was the classic work of Fatt (1956a,b,c) in the context of petroleum reservoir simulation. Fatt (1956a) simulated the drainage branch of the P_c–S_w relation in two-dimensional lattices of bonds in which junctions or sites were assumed to be of zero volume and of negligible resistance to fluid movement. Bond radii were assigned from arbitrarily chosen size distributions.

During a simulation, a given bond becomes available when P_c reached a value sufficient to drain a tube of its radius. Bonds were accessible if they were connected to a bond filled with the invading, non-wetting fluid. Thus, all bonds were eventually allowed to drain and no residual saturations were predicted. Fatt did point out the possibility of including a trapping mechanism but did not attempt any such calculations. Results presented by Fatt included demonstrations of the effects of bond radius distribution and coordination number on resulting capillary pressure curves. A detailed discussion of the calculation of pore size distributions from experimental P_c–S_w curves was also included.

The earliest work on relative permeability in a lattice model of porous media was also reported by Fatt (1956b, c). He noted the analogy between Poiseuille's law for flow in a cylindrical tube

$$q = \frac{\pi r^4}{8\mu l} \Delta p \tag{4.1}$$

and Ohm's law for electrical current flow

$$I = \frac{1}{R} \Delta E \tag{4.2}$$

In Eqs 4.1 and 4.2, q is the volumetric flow rate, μ is fluid viscosity, r and l are tube radius and length, respectively, Δp is the pressure gradient, I is flux or current, R is resistance and ΔE is the voltage gradient. Fatt constructed physical models of porous media based on this comparison between electrical conductivity $(1/R)$ and hydraulic conductivity $(\pi r^4/8\mu l)$. His models consisted of networks of electrical resistors of varying connectivities. Total network resistance was measured using a Wheatstone bridge (Fatt, 1956b). Fatt found that he could qualitatively capture the shape of typical $k_{r\alpha}$–S_α curves by repeatedly measuring resistance as he removed resistors from, or added resistors to, a network. The pioneering work of Fatt presented the general concepts that underlie virtually all later percolation or network models. Other researchers who investigated the use of network models to predict conductivity include Kirkpatrick (1973) and Dullien (1975). Kirkpatrick (1973) calculated the electrical conductivity of regular networks as individual conducting bonds were removed in accordance with bond or site percolation patterns. His model of immiscible displacement gave results for network conductivity as a function of the fraction of open bonds. Dullien (1975) modelled pore space as a network of orthogonally intersecting tubes made up of segments of different diameters. He calculated resistance to flow in fully saturated networks to estimate permeability. Unfortunately, this type of model is not readily adaptable to the partially saturated case.

An important precedent for applications of network modelling to unconsolidated media is the work of Dodds and Lloyd (1971/72). They represented the void structure of a packed bed of known sphere size distribution by a regular two-dimensional network of capillary tubes with void spaces at their junctions. Size distributions of cylindrical bonds and spherical sites were calculated from a tetrahedral model of dense packing for unequal spheres (Wise, 1952). The model differed from earlier work in that it assigned a volume to sites as well as to bonds. The authors reported qualitatively good agreement between fluid retention curves obtained by a stepwise drainage of the model and those generated experimentally for the same sphere-size distribution. However, they noted that pore-size distributions calculated theoretically using the Wise model did not compare well with experimental results.

Percolation theory has been applied extensively to the study of multifluid porous media by Scriven and coworkers (Larson *et al.*, 1981a, b; Heiba *et al.*, 1982, 1983; Sahimi *et al.*, 1986a, b). The pore-space topology assumed in this sequence of papers is the Bethe tree (see Fig. 2A). Due to its peculiar connectedness properties, this is a physically unrealistic model. However, it is often used in theoretical calculations because it allows closed-form solutions and has been found to give surprisingly useful results.

Larson *et al.* (1981a) conceptualized the porous medium as an infinite sheet of cells, some fraction (x) of which are conducting and the rest of which ($1 - x$) are non-conducting. On this lattice, so-called accessibility (fraction of cells accessible via a conducting pathway) and effectiveness (portion of those cells not on a dead-end path) functions were computed. The percolation threshold concept was used by Larson *et al.* (1981b) to calculate residual saturations, given that the lack of a continuous pathway across the sample implies the inability of the fluid to flow through the sample.

Heiba *et al.* (1982, 1983) used the analysis of Larson *et al.* (1981a, b) to calculate two-phase relative permeability relations. Both drainage and imbibition curves were derived. When plotted as a function of saturation, the non-wetting relative permeability curves showed hysteresis while the wetting fluid curves did not. Heiba *et al.* (1982, 1983) reported qualitative agreement with experimental data for petroleum reservoir rocks. Sahimi *et al.* (1986a, b) extended the ideas of the previous publications to look at dispersion phenomena in the context of percolation theory.

Lin and Slattery (1982) derived a two-phase percolation-based model for both the capillary pressure–saturation and the relative permeability–saturation relations. In their model, seven systems parameters, which characterize required probability distributions, were calibrated on the saturated permeability and primary drainage and imbibition curves for the medium of interest. Given these parameters, their model predicts the capillary pressure–saturation relation, including scanning curves, and the relative permeability–saturation curves. Lin and Slattery reported a good match with experimental data for relatively uniform media (e.g. well-sorted sands and mono-sized glass beads), and not for sandstones.

Koplik (1982) reported a model for saturated permeability based on an investigation of flow patterns in a two-dimensional lattice of straight channels and circular junctions. Adopting the electrical resistor analogy proposed by Fatt (1956b), he used techniques developed for electrical circuit analysis to calculate network conductivities. Results found using effective medium theory, in which a random lattice property is replaced by a mean value, compared favorably with computational results.

Dullien and coworkers (Chatzis and Dullien, 1977, 1985; Diaz, 1985; Diaz *et al.*, 1987) have also made significant contributions to percolation-based modelling of constitutive relations in multiphase porous media. Chatzis and Dullien (1977) modelled drainage in two-dimensional networks of bonds of varying connectivity. They found good agreement between their computational results for critical breakthrough probabilities and results predicted by bond percolation theory. They inferred that this agreement would extend to the three-dimensional case and pointed out the importance of developing three-dimensional lattice models to simulate the fluid bicontinua occurring in fluid displacement processes. They also investigated the conversion of number-based saturation values to volume-based values and noted the importance of bond lengths in these calculations. Chatzis and Dullien (1985) extended their analysis to two- and three-dimensional lattices of sites and bonds of finite volume whose sizes were assigned from separate beta distributions. They made comparisons between model results and mercury porosimetry and relative permeability curves for several sandstones. They noted the importance of realistic pore structure data, finding better agreement for angular than for circular pore cross-sections.

In real porous media, blobs or ganglia of a resident fluid may be trapped in individual pores or groups of pores as they are surrounded by the displacing fluid. The ability of early network models to reproduce experimental results was limited by their inability to predict trapping and, thus, residual saturations. For the problem of fluid displacement without trapping (i.e. without the immobilization of a non-continuous incompressible fluid phase) percolation theory applies directly. That is, the specification of a range of available pore diameters ($<$ or $> R_T$) is exactly analogous to the specification of a fraction (p_s or p_b) of open sites or bonds. For the no-trapping case, accessibility is contingent upon direct contact with another site or bond containing the displacing fluid. Computations are simple and the modelling exercise is not computation limited.

For a lattice simulation which includes trapping, a pore can drain (or fill) only if it is available and it is both adjacent to a pore containing displacing fluid and in hydraulic contact with its fluid reservoir. The simulation of physically based trapping within the network was long considered to be algorithmically prohibitive. However, Chandler *et al.* (1982) proposed an area-spanning network search algorithm which resulted in considerable computational savings and made procedures which include trappings more feasible. This is the first reported use of a trapping algorithm in the literature. In this important paper, Chandler *et al.* (1982) also compared results based on percolation theory with computational results of network modelling. They performed careful Monte Carlo simulations including a trapping algorithm on two-dimensional, two-fluid networks. They compared these with results from classical percolation theory, and found both similarities and differences. For example, the network simulations produced threshold percolation values that were very close to those predicted by percolation theory. However, using previous results by Fisher (1971), they found that scaling effects (due to the necessarily finite size of the network model) differed between the network simulation and the percolation theory. They also described the effects of boundary conditions on lattice simulations and provided guidelines for choosing sample sizes.

In the last few years, much of the work reported in the literature has included some form of trapping algorithm. Diaz (1985) and Diaz *et al.* (1987) used a trapping algorithm in which the entire cubic lattice is scanned, during each displacement step, to determine which sites and bonds are trapped. The authors compared simulation results with experimental data for sandstones and found relatively good agreement for drainage branches of the P_c–S_w relation, but poor agreement for the imbibition branches. Bear *et al.* (1987) used an unspecified trapping algorithm in simulations of both P_c–S_w and k_{rw}–S_w relations in anisotropic cubic lattices of bonds. Two-dimensional path-tracking (Soll *et al.*, 1988) and three-dimensional surface-searching (Ferrand and Celia, 1989) algorithms have also been reported. Figure 4 is a sample P_c–S_w curve generated using the path-tracking model. Note that residual saturations of both wetting and non-wetting fluids, as well as hysteresis between drainage and imbibition curves, are predicted.

Based on results from models that include fluid trapping, convincing evidence exists that two-dimensional network models cannot be used to predict accurately capillary pressure–saturation relations. This is because a scaling relationship of the form

$$S_{rw} = S_{rw}(\ln L) \tag{4.3}$$

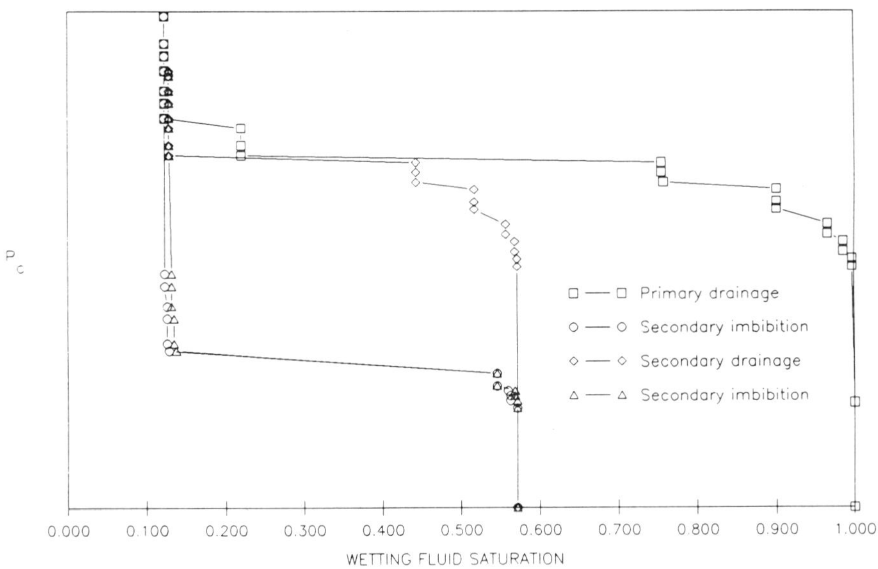

Figure 4. Capillary pressure–saturation relation (including primary drainage, secondary imbibition and secondary drainage branches) generated using two-dimensional (50 × 50, $z = 6$), path-tracking-algorithm model (Soll *et al.*, 1988).

where L is lattice width, exists between residual saturation and lattice size in two dimensions (Chandler *et al.*, 1982). This scaling relation is absent in three-dimensional lattices (Wilkinson and Willemsen, 1983; Ferrand and Celia, 1989). It therefore appears that while many reported applications have used two-dimensional lattices, meaningful predictive models must be three-dimensional.

Figure 5 illustrates typical drainage P_c–S_w curves generated using the model described by Ferrand and Celia (1989). Simulations were randomly assigned from even (curves 2,4) or log-normal (curves 1,3,5,6) distributions. In both cases entry pressure increases as the mean bond radius decreases.

Quasi-static immiscible displacement is typically simulated in standard percolation-based models by imposing a capillary pressure on a sample and allowing the fluid–fluid interfaces to redistribute themselves according to Eq. 3.3. Site or bond radii at the global interface between wetting and non-wetting fluid are examined in arbitrary order. Wilkinson and Willemsen (1983) proposed a new form of percolation theory that they called 'invasion percolation'. In invasion percolation, a specific dynamic model is used that describes fluid–fluid interface movement under a constant flow rate instead of a constant capillary pressure. An advantage of this approach is that any possible ambiguity in the trapping rule is eliminated. Wilkinson (1984) introduced buoyancy effects into invasion percolation models, and demonstrated how residual saturations change

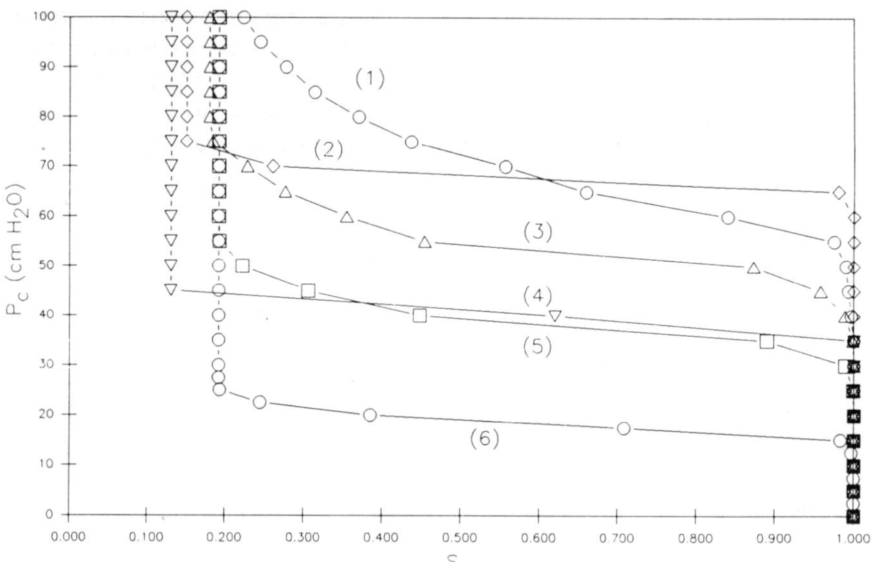

Figure 5. Drainage capillary pressure–saturation curves generated using three-dimensional ($25 \times 25 \times 25$, $z = 4$) lattices with surface-search algorithm (Ferrand and Celia, 1989).

as density effects become more pronounced. Dias and Wilkinson (1986) compared theoretical scaling relations derived for ordinary invasion percolation with those for percolation with trapping.

The development of theoretical and computational models of pore-scale displacement has been aided by the concurrent development of a number of experimental techniques for visualization of pore-scale fluid behaviour. Most important of these techniques is the use of two-dimensional, etched plate micromodels first reported by Mattax and Kyte (1961). A discussion of the technique for etching two-dimensional networks in a clear polymer plate is given by Bonnet and Lenormand (1977) and Lenormand and Zarcone (1985a). These networks are analogous to two-dimensional lattices used in computational network models.

Micromodel studies have provided insight into the physics of quasi-static displacement (Lenormand *et al.*, 1983; Chen, 1986), the influence of flow rate on fluid distribution (Chen and Koplik, 1985), and the occurrence of fingering as a function of viscosity ratio (Chen, 1987). Campbell and Orr (1985) used micromodels to investigate both two- and three-phase flow systems for oil recovery. Lenormand and Zarcone (1985b) experimentally determined a fractal dimension for capillary-dominated displacement in two dimensions. Lenormand and Zarcone (1988) studied the trapping and remobilization of a non-wetting fluid during imbibition. Wardlaw and Li (1988) investigated the effects of fluid topology, i.e. of meniscus position within the individual pores, during imbibition. Conrad *et al.* (1988) looked at the effects of small-scale heterogeneities and varying flow rates on displacements of organic liquids by water. To date, experimental micromodel studies have reported primarily qualitative results.

Thus, since the network concept was introduced by Fatt in 1956, significant progress has been made in theoretical, computational and experimental methods for studying pore-scale multiphase fluid displacement in porous media. To date, nearly all reported computational and physical models have included only two mobile fluids. The simulation of three-fluid systems is an area of research interest to both petroleum reservoir and separate phase contaminant transport modellers. Very recently, development of pore-scale models of three-fluid flow and displacement has begun. Soll and Celia (1988, 1989) reported on the development of a computational, two-dimensional network model for three-fluid displacement including a simple trapping rule. Mason *et al.* (1989) reported on some preliminary three-phase displacement experiments in etched glass micromodels.

5. Research directions

Simulations of multi-fluid behaviour using percolation-based pore-scale models can provide insight into immiscible fluid behaviour as measured in laboratory samples of porous media. These models have been and are being developed by researchers interested in petroleum recovery, water movement in the unsaturated soil zone and contamination of groundwater by separate phase liquids. Ultimately, these models may provide useful predictive tools for multiphase constitutive relations.

Much current work in this area focuses on calibration of network characteristics for more accurate predictions. Attempts to quantify these models have called attention to limitations in our knowledge of pore geometry and of fluid–fluid and fluid–solid interactions at the pore scale. Their development has motivated, and been motivated by, pore-scale flow visualizations in the laboratory. Much work remains to be done in this area, especially in quantifying pore-size distributions, and in understanding physical mechanisms of imbibition, fluid trapping and remobilization.

An obvious direction for the extension of percolation-based models is to the three-fluid case. Three-fluid porous media are highly complex and relatively poorly understood systems. They are, however, of immediate and practical interest to researchers and engineers concerned with enhanced oil recovery and the transport of non-aqueous phase liquid (NAPL) contaminants in the unsaturated soil zone. We are currently witnessing the development of a variety of experimental techniques to measure three-fluid multiphase variables (relative saturations, capillary pressures and relative permeabilities) in laboratory porous medium samples. Simultaneous development of micromodel visualization techniques and percolation-based computational models for the three-fluid case will add significantly to our understanding and ultimately our ability to predict behaviour in these systems.

References

Abriola, L.M. and Pinder, G.F. (1985). A multiphase approach to the modeling of porous media contamination by organic compounds: 1. Equation development, *Water Resour. Res.* **21**, 11–18.

Aziz, K. and Settari, A. (1979). 'Petroleum Reservoir Simulation'. Applied Science, London.

Bear, J. (1972). 'Dynamics of Fluids in Porous Media'. American Elsevier, New York.

Bear, J., Braester, C. and Menier, P.C. (1987). Effective and relative permeabilities of anisotropic porous media, *Transport in Porous Media* **2**, 301–316.

Blake, F.C. (1922). The resistance of packing to fluid flow, *Trans. Am. Inst. Chem. Eng.* **14**, 415–421.

Bonnet, J. and Lenormand, R. (1977). Constructing micromodels for the study of multiphase flow in porous media, *Technical Note, Revue L'Inst. Franc. Petr.*, pp. 477–480, May–June.

Broadbent, S.K. and Hammersley, J.M. (1957). Percolation processes I. Crystals and mazes, *Proc. Camb. Phil. Soc.* **53**, 629–641.

Campbell, B.T. and Orr, F.M. Jr. (1985). Flow visualization for CO_2/crude-oil displacements, *Soc. Petrol. Eng. J.*, pp. 665–678, October.

Carmen, P.C. (1938). Determination of the specific surface of powders, *J. Soc. Chem. Ind.* **57**, 225–234.

Celia, M.A. and Ferrand, L.A. (1988). Parameter estimation for multiphase porous media systems. *In* 'Proceedings of the ASCE National Conference on Hydraulic Engineering' (S. Abt and J. Gessler, eds), pp. 558–563. Am. Soc. Civ. Eng., New York.

Chandler, R., Koplik, J., Lerman, K. and Willemsen, J.F. (1982). Capillary displacement and percolation in porous media, *J. Fluid Mech.* **119**, 249–267.

Chatenevar, A. and Calhoun, J.C. (1952). Visual examination of fluid behavior in porous media, Part I, *Trans. AIME* **195**, 149–156.

Chatzis, I. (1980). A network approach to analyze and model capillary and transport phenomena in porous media. Ph.D. Dissertation, University of Waterloo, Waterloo, Ontario.

Chatzis, I. and Dullien, F.A.L. (1977). Modelling pore structure by 2-D and 3-D networks with application to sandstones, *J. Canad. Petrol. Technol.* **16**, 97–108.

Chatzis, I. and Dullien, F.A.L. (1985). The modeling of mercury porosimetry and the relative permeability of mercury in sandstones using percolation theory, *Int. Chem. Engng.* **25**, 47–66.

Chen, J.-D. (1986). Some mechanisms of immiscible fluid displacement in small networks. *J. Colloid Interface Sci.* **110**, 488–503.

Chen, J.-D. (1987). Pore scale difference between miscible and immiscible viscous fingering in porous media, *AIChE J.* **33**, 307–311.

Chen, J.-D. and Koplik, J. (1985). Immiscible fluid displacement in small networks, *J. Colloid Interface Sci.* **108**, 304.

Chen, J.-D., Koplik, J., Wilkinson, D., Willemsen, J. and Hammond, P. (1987). Oil–water dynamics in porous media, *Science, The Technical Review* **34**, 26–32.

Childs, E.C. and Collis-George, N. (1948). Soil geometry and soil-water equilibria, *Discuss. Faraday Soc.* **3**, 78–85.

Conrad, S.H., Mason, W.B., Peplinski, W., Hagan, E. and Wilson, J.L. (1988). The combined effect of soil heterogeneities and flow rate on the advance and ultimate recovery of spilled organic liquids. Paper presented at annual Fall Meeting, American Geophysical Union, San Francisco, CA.

Corey, A.T. (1986). 'Mechanics of Immiscible Fluids in Porous Media' Water Resources Publ., Littleton, CO.

Craig, F.F. (1971). 'The Reservoir Engineering Aspects of Water Flooding'. H.L. Doherty Memorial Fund of AIME, New York.

Dean, P. (1963). A new Monte Carlo method for percolation problems on a lattice, *Proc. Camb. Phil. Soc.* **59**, 397.

Dean, P. and Bird, N.F. (1967). Monte Carlo estimates of critical percolation probabilities, *Proc. Camb. Phil. Soc.* **63**, 447.

D'Hollander, E.H. (1979). Estimation of pore size distribution from the moisture characteristic, *Water Resour. Res.* **15**, 107–112.

Dias, M.M. and Wilkinson, D. (1986). Percolation with trapping, *J. Phys. A: Math Gen.* **19**, 3131–3146.

Diaz, C.E. (1985). Simulation of fluid distribution and oil–water capillary pressure curves in water wet porous media using network modelling techniques. M.A. Sc. Thesis, University of Waterloo, Waterloo, Ontario.

Diaz, C.E., Chatzis, I. and Dullien, F.A.L. (1987). Simulation of capillary pressure curves using bond correlated site percolation on a simple cubic network, *Transport in Porous Media* **2**, 215–240.

Dodds, J.A. and Lloyd, P.J. (1971/72). A model for the void structure in multi-component sphere packs applied to capillary pressure curves, *Powder Technol.* **5**, 69–76.

Dullien, F.A.L. (1975). New network permeability model of porous media, *AIChE J.* **21**, 299–307.

Elsik, C.M. and Miller, C.A. (1988) Videomicroscopy of two-phase steady cocurrent flow in a model porous medium. *In* 'Surfactant-based Mobility Control' (D.H. Smith, ed.), Am. Chem. Soc. Symposium Series, Washington, DC. No. 373, p. 258.

Essam, J.W. (1972). Percolation and cluster size. *In* 'Phase Transitions and Critical Phenomena', Vol. 2 (C. Domb and M.S. Green, eds). Academic Press, New York.

Fatt, I. (1956a). The network model of porous media, I. Capillary pressure characteristics, *Trans. AIME Petrol. Div.* **207**, 144–159.

Fatt, I. (1956b). The network model of porous media, II. Dynamic properties of a single size tube network, *Trans. AIME Petrol. Div.* **207**, 160–163.

Fatt, I. (1956c). The network model of porous media, III. Dynamic properties of a single size tube network, *Trans. AIME Petrol. Div.* **207**, 164–177.

Fatt, I. and Dykstra, H. (1951). Relative permeability studies, *Trans. AIME Petrol. Div.* **192**, 249–256.

Ferrand, L.A. and Celia, M.A. (1989). Development of a three-dimensional network model for quasi-static immiscible displacement. *In* 'Proceedings of the International Symposium on Contaminant Transport in Groundwater (H.E. Kobus and W. Kinzelbach, eds). A.A. Balkema, Rotterdam.

Fisher, M.E. (1971). *In* 'Critical Phenomena', Proceedings of the International School of Physics Enrico Fermi, Course LI (M.S. Green, ed.). Academic Press.

Frisch, H.L. and Hammersley, J.M. (1963). Percolation processes and related topics, *J. SIAM* **11**, 894–918.

Gallegos, D.P. and Smith, D.M. (1988). A NMR technique for the analysis of pore structure: determination of continuous pore size distributions, *J. Colloid Interface Sci.* **122**, 143–153.

Gvirtzman, H., Magaritz, M., Klein, E. and Nadler, A. (1987). A scanning electron microscopy study of water in soil, *Transport in Porous Media* **2**, 83–93.

Hazen, A. (1892). 24th Annual Report, Massachusetts State Board of Health, Publ. Document 34, p. 539.

Heiba, A.A., Sahimi, M., Scriven, L.E. and Davis, H.T. (1982). Percolation theory of two-phase relative permeability, Soc. Petrol. Eng. Paper No. 11015.

Heiba, A.A., Davis, H.T. and Scriven, L.E. (1983). Effect of wettability on two-phase relative permeabilities and capillary pressure, Soc. Petrol. Eng. Paper No. 12172.

Juang, C.H. and Holtz, R.D. (1988). Fabric, core size distribution and permeability of sandy soils, *J. Geotech. Eng.* **112**, 855–868.

Kirkpatrick, S. (1973). Percolation and conduction, *Rev. Mod. Phys.* **45**, 574–588.

Koplik, J. (1982). Creeping flow in two-dimensional networks, *J. Fluid Mech.* **119**, 219–247.

Koplik, J. and Lassetter, T.J. (1982). Two-phase flow in random network models of porous media. Soc. Petrol. Eng. Paper No. 11014.

Kozeny, J. (1927). Uber kapillare Leitung des Wassers im Boden, *Sitzungsber. Adad. Wiss. Wien* **136**, 271–306.

Larson, R.G., Scriven, L.E. and Davis, H.T. (1981a). Percolation theory of two phase flow in porous media, *Chem. Engng Sci.* **36**, 57–73.

Larson, R.G., Davis, H.T. and Scriven, L.E. (1981b). Displacement of residual non-wetting fluid from porous media, *Chem. Engng Sci.* **36**, 75–81.

Lenormand, R. and Zarcone, C. (1985a). Two-phase flow experiments in a two-dimensional permeable medium, *Phys. Chem. Hydrol.* **6**, 497–506.

Lenormand, R. and Zarcone, C. (1985b). Invasion percolation in an etched network: measurement of a fractal dimension, *Phys. Rev. Lett.* **54**, 2226–2229.

Lenormand, R. and Zarcone, C. (1988). Physics of blob displacement in a two-dimensional porous medium, *S.P.E. Formation Eval.* **3**, 271–275.

Lenormand, R., Zarcone, C. and Saar, A. (1983). Mechanism of the displacement of one fluid by another in a network of capillary ducts, *J. Fluid Mech.* **135**, 337–353.

Lin, C.-Y. and Slattery, J.C. (1982). Three-dimensional, randomized, network model for two-phase flow through porous media, *AIChE J.* **28**, 311–324.

Mann, R., Almeida, J.J. and Mugerwa, M.N. (1986). A random pattern extension to the stochastic network pore model, *Chem. Engng Sci.* **41**, 2663–2671.

Mason, W.R., Peplinski, W.J., Conrad, S.H. and Wilson, J.L. (1989). Visualization of three phase flow on a pore scale. Paper presented at annual Spring meeting, American Geophysical Union, Baltimore, MD.

Mattax, C.C. and Kyte, J.R. (1961). Ever see a water flood? *Oil and Gas J.*, pp. 115–128, October.

Orr, C., Jr. (1970). Application of mercury penetration to materials analysis, *Powder Technol.* **3**, 117–123.

Sahimi, M., Hughes, B.D., Scriven, L.E. and Davis, H.T. (1986a). Dispersion in flow through porous media: I. One-phase flow, *Chem. Engng Sci.* **41**, 2103–2122.

Sahimi, M., Heiba, A.A., Davis, H.T. and Scriven, L.E. (1986b). Dispersion in flow through porous media: II. Two-phase flow, *Chem. Engng Sci.* **41**, 2123–2136.

Shante, V.K.S. and Kirkpatrick, S. (1971). An introduction to percolation theory, *Adv. Phys.* **20**, 325–357.

Soll, W.E. and Celia, M.A. (1988). A pore-scale model of three-phase immiscible fluid flow. Paper presented at annual Fall Meeting, American Geophysical Union, San Francisco, CA.

Soll, W.E. and Celia, M.A. (1989). A pore-scale model of three-phase immiscible fluid flow in porous media. *In* 'Proceedings of the ASME Winter Meeting' (to be published).

Soll, W.E., Ferrand, L.A. and Celia, M.A. (1988). An enhanced percolation model for the capillary pressure–saturation relation. *In* 'Proceedings of the VII International Conference on Computational Methods in Water Resources' (M.A. Celia *et al.*, eds), pp. 165–170. Elsevier, New York.

Stauffer, D. (1985) 'Introduction to Percolation Theory'. Taylor and Francis, London.

Sykes, M.F. and Essam, J.W. (1963). Some exact critical percolation probabilities for bond and site problems in two-dimensions, *Phys. Rev. Lett.* **10**, 3.

Wall, G.C. and Brown, R.J.C. (1981). Determination of pore size distributions from sorption isotherms and mercury penetration of interconnecting pores: the application of percolation theory, *J. Colloid Interface Sci.* **82**, 141–149.

Wardlaw, N.C. and Li, Y. (1988). Fluid topology, pore size and aspect ratio during imbition, *Transport in Porous Media* **3**, 17–34.

Wardlaw, N.C., Li, Y. and Forbes, D. (1987). Pore-throat size correlation from capillary pressure curves. *Transport in Porous Media* **2**, 597–611.

Wilkinson, D. (1984). Percolation model of immiscible displacement in the presence of buoyancy forces, *Phys. Rev. A.* **30**, 520–531.

Wilkinson, D. (1986). Percolation effects in immiscible displacement, *Phys. Rev. A.* **34**, 1380–1391.

Wilkinson, D. and Willemsen, J.F. (1983). Invasion percolation: a new form of percolation theory, *J. Phys. A.* (*Math. Gen.*) **16**, 3365–3376.

Wise, M.E. (1952). Dense random packing of unequal spheres, *Philips Res. Rept.* **7**, 321–343.

Yanuka, M., Dullien, F.A.L. and Elrick, D.E. (1986). Percolation processes in porous media. I. Geometrical and topological model of porous media using a three-dimensional joint pore size distribution, *J. Colloid Interface Sci.* **112**, 24–41.

Chapter XVII

Generalized Hydrodynamics of Microporous Media: Relationships Between the Memory Function and the Scale of Observation

John H. Cushman
Lilly Hall of Life Sciences, Purdue University, West Lafayette, IN 47907, USA

1. Introduction

Other chapters in this text have dealt with scales of motion greater than the pore scale. Various authors have presented theories, numerical techniques, and experimental methods for handling the hierarchy of larger scales of space and time. In this chapter we present a method for looking at information transfer between the molecular scale (of relevance to surface adsorption) and the micropore scale (characteristic of natural clays) (Schoen *et al.*, 1987, 1988) and discuss how the same methodology may be applied to larger scales. The tools which are relied upon in this chapter are normally developed in the fields of molecular hydrodynamics (cf. Boon and Yip, 1980) and in linear filtering theory as presented in Cushman (1986).

The adsorption of chemicals on mineral surfaces or natural soil organic matter, and diffusion to surfaces in micropores are processes with characteristic time scales which are quite often of the order of picoseconds. Yet the time scales of relevance to transport within natural clays and soil organic matter may be much larger. Theories which discuss transport in clays must account for this discrepancy of time scales in some fashion. It is this last point that is the motivation for this chapter.

Molecular hydrodynamics has classically been the study of the pre-continuum nature of fluids, i.e. the study of fluids which retain some detailed molecular information but which are approaching the hydrodynamic limit (lower bound of the classical continuum scale). The basic setting of the theory is that of statistical mechanics. All correlation functions are defined in terms of a Volterra integrodifferential equation which involves a memory function. The entirety of

DYNAMICS OF FLUIDS IN HIERARCHICAL POROUS MEDIA
ISBN 0-12-200260-1

the physics of the transport process is contained in this function. The scale of observation may be introduced through separate linear filters on space and time in the manner of Cushman (1986). The memory function on the observational scale may be expressed in terms of the memory function on the molecular scale and in terms of the time and space filters. Molecular dynamic computer experiments in conjunction with neutron scattering experiments may be used to obtain the scale constraints and filters associated with the measurement of diffusion coefficients in porous media.

The remainder of this chapter is broken down as follows. In Section 2 we present the basic motivation for the underlying theory. Section 3 sets forth the framework for a porous media phase space wherein the porous matrix itself is represented as a random force field. In Section 4 the basic tenets of molecular hydrodynamics are reviewed. Section 5 outlines time and space linear filtering theory as used with porous media. A derivation of the connection between the molecular scale and the measurement scale memory functions is presented in Section 6, and an example involving neutron scattering is given in Section 7. The final section illustrates how the ideas may be applied to larger scales.

2. Motivation

In Bhattacharya and Gupta's earlier chapter (Chapter IV, this volume), a discussion was presented wherein transport in a porous medium was analysed with a Markovian process involving Brownian dynamics. As motivation for the generalized hydrodynamics approach, we summarize part of that discussion here.

Suppose we have a system composed of N particles of mass M, suspended in a fluid of light particles of mass $m \ll M$. The concentration of heavy particles is assumed dilute so that they do not interact with each other. The heavy particles do interact with the light particles which behave as a continuum and satisfy a Langevin equation of the form:

$$M \frac{\mathrm{d}\mathbf{u}}{\mathrm{d}t} = -\xi \mathbf{u}(t) + \mathbf{B}(t) \tag{2.1}$$

where $\mathbf{B}(t)$ is a Standard Brownian motion, $\mathbf{u}$ is the velocity of a heavy particle and ξ is the friction coefficient. It is well known that in the long-time (asymptotic) limit (the Markovian approximation) the friction coefficient satisfies

$$\xi = (3k_\mathrm{B}T)^{-1} \int_0^\infty E[\mathbf{B}(0) \cdot \mathbf{B}(t)]\, \mathrm{d}t \tag{2.2a}$$

$$\xi^{-1} = (3k_\mathrm{B}T)^{-1} \int_0^\infty E[\mathbf{u}(0) \cdot \mathbf{u}(t)]\, \mathrm{d}t \tag{2.2b}$$

$$d = k_B T \xi^{-1} \tag{2.2c}$$

where d is the classical diffusion coefficient, E is the expectation operator, T is temperature, and k_B is the Boltzmann constant. Many other fluctuating quantities, $\mathbf{A}(t)$, may be described by a similar stochastic differential equation of the Langevin type

$$\frac{d\mathbf{A}}{dt} = -\gamma \mathbf{A}(t) + \mathbf{B}(t) \tag{2.3}$$

The reader should note that if we set $\mathbf{A} = M\mathbf{u}$ and $\gamma = \xi M^{-1}$ in Eq. 2.3, then Eq. 2.1 results. If Eq. 2.3 is multiplied by $\mathbf{A}(0)/E[\mathbf{A}(0)]^2$ and expectations are taken (assuming $E[\mathbf{A}(0) \cdot \mathbf{B}(t)] = 0$), we find

$$\frac{dC}{dt} = -\gamma C \tag{2.4}$$

where

$$C = E[\mathbf{A}(0) \cdot \mathbf{A}(t)]/E[\mathbf{A}(0)]^2 \tag{2.5}$$

Equation 2.4 defines the time evolution of the autocorrelation function, e.g. it may define the time evolution of the velocity autocorrelation, which in turn gives us information on d.

In order to describe the dynamics of a homogeneous system without restriction on the time scale, Eq. 2.3 is rewritten as a non-Markovian generalized Langevin equation

$$\frac{d\mathbf{A}}{dt} = -iF\mathbf{A} - \int_0^t \gamma(t')\mathbf{A}(t - t')\, dt' + \mathbf{B}(t) \tag{2.6}$$

where F is real and corresponds to the existence of a propagation process associated with the time evolution of the dynamic variable. The convolution in Eq. 2.6 accounts for memory effects. The Langevin equation for the autocorrelation function in Eq. 2.6 takes the form

$$\frac{dC}{dt} = -\int_0^t \gamma(t')C(t - t')\, dt' - iFC \tag{2.7}$$

If the second term on the right-hand side of Eq. 2.7 is included in the convolution (via a delta function), then Eq. 2.7 is called the memory function equation for the autocorrelation and γ is called the memory function. As we will see, every correlation function satisfies an equation of this form. The entirety of the physics of any transport process is implicitly contained in the memory function $\gamma(t)$ and consequently a knowledge of $\gamma(t)$ allows us to study the pre-asymptotic nature of the transport process, e.g. pre-asymptotic diffusion.

3. Evolution in a microporous phase space

For simplicity in presentation, we will define our porous media phase space somewhat differently from most predecessors (cf. Sposito, 1978; Baveye, 1985). The porous matrix (assumed non-deformable) and external fields (such as gravity) will be represented by a force field $\mathbf{X}(\mathbf{r})$. Conceptually, the porous matrix has no physical content (mass, volume, etc.)—it exists solely as a force field. The 'physical' system consists of N particles (atoms, molecules; including what would be considered as adsorbed atoms) under the influence of the force field $\mathbf{X}$. Intuitively, the physical system would consist of fluids, adsorbed molecules and gases internal to the fictitious porous matrix. Conceptually, the porous-medium fluid is no different from a classical fluid under the influence of an external force field such as gravity. Our approach is in the spirit of the 'mean-field' approximation to solvated molecules. From classical dynamics, we know that the state of the system is completely specified by the $6N$-vector of generalized position and momentum coordinates, $(\mathbf{R}, \mathbf{P}) = (\mathbf{R}_1, \mathbf{R}_2, \ldots, \mathbf{R}_N, \mathbf{P}_1, \ldots, \mathbf{P}_N)$, of the N-particles where

$$\mathbf{R}_i = (R_{1i}, R_{2i}, R_{3i}) \tag{3.1a}$$

$$\mathbf{P}_i = (P_{1i}, P_{2i}, P_{3i}) \tag{3.2b}$$

As the system of particles evolves in time, the phase-space point $(\mathbf{R}, \mathbf{P})$ moves in phase-space according to Hamilton's equations of motion. For several reasons, we can never exactly determine the true molecular state of the system (e.g. due to lack of knowledge of the initial conditions); what we do know (or at least may control), however, is the macroscopic state of the system which is specified by fixing a triad or macroscopic variables, e.g. (N, V, E), the number of particles, volume, and total energy. Clearly, there are a large number of phase-space trajectories that will correspond to fixed (N, V, E). The collection of all such phase-space points, phase-space, is denoted Ω_N. We define a probability density, $f_N(\mathbf{R}, \mathbf{P}; t)$, on phase-space as the probability particle 1 is in $d\mathbf{R}_1\, d\mathbf{P}_1$, at $(\mathbf{R}_1, \mathbf{P}_1)$ at time t, particle 2 is in $d\mathbf{R}_2\, d\mathbf{P}_2$ at $(\mathbf{R}_2, \mathbf{P}_2)$ at time t, etc.

If we assume the number of points in phase-space is constant (e.g. no nuclear reactions) it follows that

$$i\frac{\partial f_N}{\partial t} = L_N f_N \tag{3.2}$$

where the Liouville operator L_N is given by

$$L_N = -i\left[\sum_{j=1}^{N} \frac{\mathbf{P}_j}{m_j}\cdot\nabla_{\mathbf{R}_j} + \sum_{j=1}^{N} \mathbf{F}_j\cdot\nabla_{\mathbf{P}_j}\right] \tag{3.3}$$

$$\mathbf{F}_j = -\nabla_{\mathbf{R}_j} U_j + \mathbf{X}_j \tag{3.4}$$

U_j is the interparticle potential energy of particle j due to all other particles and $\mathbf{X}_j$ represents the force field due to the porous matrix (including gravity, etc.) on particle j. It is easily verified that the time evolution of any dynamic variable, $\mathbf{A}$, which does not depend explicitly on time, is governed by the Liouville equation (McQuarrie, 1976)

$$\frac{\partial \mathbf{A}}{\partial t} = iL_N\mathbf{A} \tag{3.5}$$

or

$$\mathbf{A}(\mathbf{R}, \mathbf{P}; t) = \exp(iL_N t)\mathbf{A}(\mathbf{R}, \mathbf{P}; 0) \tag{3.6}$$

When no confusion can arise we will drop the subscript N above.

The above discussion may be generalized by allowing $\mathbf{X}$ to be random on a second probability space, S. Thus for fixed $s \in S$, $\mathbf{X}$ is a random variable on Ω_N, and for fixed $\mathbf{R}$ and $\mathbf{P}$, $\mathbf{X}$ is a random variable on S. We require that for fixed s, the parameters specifying the macroscopic ensemble remain unchanged (e.g. fixed (N, V, E) for $s_1 \in S$ equals (N, V, E) for $s_2 \in S$). One can imagine fixing $s \in S$ as specifying a specific inter- and intra-particle structure for a porous matrix. As s runs over S, we define a collection of admissible porous media. The elementary events in a specific subset, B of S, might be viewed as the set of all force fields corresponding to different porous media that generate, say, the same 'hydraulic conductivity'. While it is clear that we can include the atoms of the porous matrix in our original ensemble and thus not require the above definitions, we prefer the approach outlined here because fluctuations in the position (e.g. due to deformation) of the particles within the porous matrix itself are in general on a totally different time scale from that of the interstitial fluid. More specifically, one can treat motions of the crystals (porous media matrix particles) as a small perturbation in rigorous treatments of transport processes. In the setting outlined above one might ask questions such as: what is the expected value of the dispersion coefficient at $\mathbf{r}$ given the hydraulic conductivity at $\mathbf{r}$ (i.e. $E[\mathbf{D}(\mathbf{r})|\mathbf{K}]$). The result is of course a random variable on S.

4. Elements of molecular hydrodynamics

In this section we provide a brief review of the memory function formalism of molecular hydrodynamics (Zwanzig, 1961; Harp and Berne, 1970; Boon and Yip, 1980).

We define a projection operator Q_V, ($Q_\mathrm{V}^2 = Q_\mathrm{V}$), such that any square-integrable dynamic variable $\mathbf{A}$ (measurable function on Ω) is projected perpendicular to the square-integrable dynamic variable $\mathbf{V}$ by

$$Q_\mathrm{V}\mathbf{A} = \mathbf{A} - \mathbf{V}E(\mathbf{VA})_0 = \mathbf{A} - \mathbf{V}(\mathbf{V}, \mathbf{A})_0 \tag{4.1}$$

where E denotes expectation with respect to the equilibrium ensemble ($t = 0$) probability density function $f(\mathbf{R}, \mathbf{P}, 0)$ and $(\cdot, \cdot)_0$ denotes inner product on Ω with respect to $f(\mathbf{R}, \mathbf{P}, 0)$. Let $C(\mathbf{r}, t)$ denote the normalized time-correlation function of the space-parameterized dynamic variable $\mathbf{V}(\mathbf{r}, t; \mathbf{R}, \mathbf{P})$,

$$\begin{aligned} C(\mathbf{r}, t) &= (\mathbf{V}^*(\mathbf{r}, t; \mathbf{R}, \mathbf{P}), \mathbf{V}(\mathbf{r}, 0; \mathbf{R}, \mathbf{P}))_0 / \| \mathbf{V}(\mathbf{r}, 0; \mathbf{R}, \mathbf{P}) \|_0^2 \\ &= \int_\Omega \mathbf{V}^*(\mathbf{r}, t; \mathbf{R}, \mathbf{P}) \mathbf{V}(\mathbf{r}, 0; \mathbf{R}, \mathbf{P}) f(\mathbf{R}, \mathbf{P}; 0) \, \mathrm{d}\mathbf{R} \, \mathrm{d}\mathbf{P} \\ &\quad \Big/ \int_\Omega \mathbf{V}^*(\mathbf{r}, 0; \mathbf{R}, \mathbf{P}) \mathbf{V}(\mathbf{r}, 0; \mathbf{R}, \mathbf{P}) f(\mathbf{R}, \mathbf{P}; 0) \, \mathrm{d}\mathbf{R} \, \mathrm{d}\mathbf{P} \end{aligned} \tag{4.2}$$

One readily shows (Boon and Yip, 1980) that C satisfies a Volterra integro-differential equation of the form

$$\frac{\partial C}{\partial t} = -\int_0^t \gamma_\mathrm{V}(\mathbf{r}, t') C(\mathbf{r}, t - t') \, \mathrm{d}t' \tag{4.3}$$

where $\gamma_\mathrm{V}(\mathbf{r}, t)$, called the kernel or memory function of $\mathbf{V}$, is defined by

$$\gamma_\mathrm{V}(\mathbf{r}, t) = E[\dot{\mathbf{V}}^*(\mathbf{r}, 0) Q_\mathrm{V} \exp(itQ_\mathrm{V}L) \dot{\mathbf{V}}(\mathbf{r}, 0)] / \| \mathbf{V}(\mathbf{r}, 0) \|_0^2 - 2\phi(\mathbf{r}, 0)\delta(t) \tag{4.4}$$

The entirety of the physics of any dynamic process involving $\mathbf{V}$ is contained in γ_V. The Laplace transform of γ_V, denoted $\tilde{\gamma}_\mathrm{V}(r, \omega)$, is given by

$$\tilde{\gamma}_\mathrm{V}(\mathbf{r}, \omega) = [\tilde{\Phi}(\mathbf{r}, \omega) - \phi(\mathbf{r}, 0)] / [1 - \omega^{-1}(\tilde{\Phi}(\mathbf{r}, \omega) + \phi(\mathbf{r}, 0))] \tag{4.5}$$

where

$$\tilde{\Phi}(\mathbf{r}, \omega) = \int_0^\infty \mathrm{e}^{-\omega t} \Phi(\mathbf{r}, t) \, \mathrm{d}t \tag{4.6}$$

$$\Phi(\mathbf{r}, t) = C_{\dot{\mathbf{V}}}(\mathbf{r}, t)\,\|\,\dot{\mathbf{V}}(\mathbf{r}, 0)\,\|_0^2/\,\|\,\mathbf{V}(\mathbf{r}, 0)\,\|_0^2 \tag{4.7}$$

and

$$\phi(\mathbf{r}, 0) = (\mathbf{V}(\mathbf{r}, 0), \dot{\mathbf{V}}(\mathbf{r}, 0))_0/\,\|\,\mathbf{V}(\mathbf{r}, 0)\,\|_0^2$$

Clearly,

$$\tilde{C}(\mathbf{r}, \omega) = [\omega + \tilde{\gamma}_{\mathbf{V}}(\mathbf{r}, \omega)]^{-1} \tag{4.8}$$

5. Elements of linear filtering theory

In this section we extend concepts presented in Cushman (1984, 1986, 1987). A measure μ over space is said to be a spatial filter if its Fourier transform exists and

$$\int_{\mathbb{R}^3} \mu(\mathrm{d}\mathbf{r}) = 1 \tag{5.1}$$

or equivalently

$$\hat{\mu}(0) = 1 \tag{5.2}$$

where ˆ denotes Fourier transform, i.e.

$$\hat{\mu}(\mathbf{k}) = \int_{\mathbb{R}^3} \mathrm{e}^{-i\mathbf{r}\cdot\mathbf{k}}\mu\,(\mathrm{d}\mathbf{r}) \tag{5.3}$$

A measure θ over time is said to be a temporal filter if its Laplace transform exists and

$$\int_0^\infty \theta(\mathrm{d}t) = 1 \tag{5.4}$$

or equivalently

$$\tilde{\theta}(0) = 1$$

where ˜ denotes Laplace transform, i.e.

$$\tilde{\theta}(\omega) = \int_0^\infty \mathrm{e}^{-\omega t}\theta(\mathrm{d}t) \tag{5.5}$$

For most applications $\theta(\mathrm{d}t)$ will have bounded support and Eq. 5.4 will be replaced by (assuming t is outside the support of θ)

$$\int_0^t \theta(\mathrm{d}t) = 1 \tag{5.6}$$

(Note: it is sometimes advantageous to allow ω in Eq. 5.5 to take on complex values.)

The act of taking a measurement of a dynamic variable $\mathbf{V}$ defined over $(\mathbb{R}^3 \times [0, t] \times \Omega \times S)$ is defined by

$$\langle \mathbf{V} \rangle = \theta(t) * \mu(\mathbf{r}) * \mathbf{V}(\mathbf{r}, t; \mathbf{R}, \mathbf{P}, s) = [\theta(t) * \mathrm{e}^{iLt}][\mu(\mathbf{r}) * \mathbf{V}(\mathbf{r}, 0; \mathbf{R}, \mathbf{P}, s)] \tag{5.7}$$

In the above $*$ denotes convolution product.

$$\langle \mathbf{V} \rangle = \int_0^t \int_{\mathbb{R}^3} \mathbf{V}(\mathbf{r} - \mathbf{r}', t - t'; \mathbf{R}, \mathbf{P}, s)\theta(\mathrm{d}t')\mu(\mathrm{d}\mathbf{r}') \tag{5.8}$$

The meaning of $\gamma(t) * \mathrm{e}^{iLt}$ is

$$\theta(t) * \mathrm{e}^{iLt} = \sum_{n=0}^{\infty} \int_0^t \frac{(iL)^n}{n!}(t - t')^n \theta(\mathrm{d}t') \tag{5.9}$$

Equation 5.7 is quite useful in that it completely decouples the spatial and temporal evolution of the measured dynamic variable, i.e. given a spatial measurement of $\mathbf{V}$ at time zero and given the temporal measurement of the evolution operator at fixed spatial position, we can compute the space–time evolution of $\mathbf{V}$ as a simple product.

It is also clear that $\langle \mathbf{V} \rangle$ is another dynamic variable on $(\mathbb{R}^3 \times [0, t] \times \Omega \times S)$. The act of measurement is often called coarse graining.

The expected value of a dynamic variable $\langle \mathbf{V} \rangle$ with respect to the probability density for a given equilibrium ($t = 0$) ensemble (specified by macroscopic variables) is

$$E\langle \mathbf{V} \rangle = \int_\Omega \int_0^t \int_{\mathbb{R}^3} \mathbf{V}(\mathbf{r} - \mathbf{r}', t - t'; \mathbf{R}, \mathbf{P}, s)\, f(\mathbf{R}, \mathbf{P}, s; 0)\theta(\mathrm{d}t')\mu(\mathrm{d}\mathbf{r}')\, \mathrm{d}\mathbf{R}\, \mathrm{d}\mathbf{P}$$

$$= \int_\Omega [\mathbf{V}(\mathbf{r}, 0; \mathbf{R}, \mathbf{P}, s) * \mu(\mathrm{d}\mathbf{r})][\mathrm{e}^{-iLt} * \theta(\mathrm{d}t)] f(\mathbf{R}, \mathbf{P}, s; 0)\, \mathrm{d}\mathbf{R}\, \mathrm{d}\mathbf{P} \tag{5.10}$$

Many dynamic variables in porous media are of interest only on scales larger than say $(\lambda_{\mathbf{r}}, \lambda_t)$. That is, spatial and temporal filters are defined such that spatial motions on scales smaller than λ_r and temporal scales smaller than λ_t are considered unwanted noise. Thus, we place constraints on μ and θ such that

$$\hat{\mu}(\mathbf{k}) \downarrow\downarrow 0 \quad \text{for } |k_i| > \lambda_i^{-1} \tag{5.11a}$$

and

$$\tilde{\theta}(\omega) \downarrow\downarrow 0 \quad \text{for } |\omega| > \lambda_t^{-1} \tag{5.11b}$$

where $\mathbf{k} = (k_1, k_2, k_3)$ and $\lambda_r = (\lambda_1, \lambda_2, \lambda_3)$.

It is a straightforward exercise to extend Eq. 4.4 of Cushman (1986) to account for time via the distributional Laplace transform. We thus have

$$\tilde{\hat{C}}_{\langle \mathbf{V} \rangle}(\mathbf{k}, \omega) = |\tilde{\theta}(\omega)|^2 |\hat{\mu}(\mathbf{k})|^2 \tilde{\hat{C}}_{\mathbf{V}}(\mathbf{k}, \omega) \tag{5.12}$$

6. Molecular to measurement scale memory functions

Let $\mathbf{V}$ be a scaler-valued dynamic variable over $(\mathbb{R}^3 \times \mathbb{R} \times \Omega \times S)$. Because $\langle \mathbf{V} \rangle$ is also a dynamic variable, its correlation satisfies

$$\tilde{\hat{C}}_{\langle \mathbf{V} \rangle}(\mathbf{k}, \omega) = \{[\omega + \tilde{\gamma}_{\langle \mathbf{V} \rangle}(\mathbf{r}, \omega)]^{-1}\}\hat{} \tag{6.1}$$

We also have

$$\tilde{\hat{C}}_{\mathbf{V}}(\mathbf{k}, \omega) = \{[\omega + \tilde{\gamma}_{\mathbf{V}}(\mathbf{r}, \omega)]^{-1}\}\hat{} \tag{6.2}$$

Combining Eqs 5.12 and 6.2, we find

$$\tilde{\hat{C}}_{\langle \mathbf{V} \rangle}(\mathbf{k}, \omega) = \{[\omega + \tilde{\gamma}_{\langle \mathbf{U} \rangle}(\mathbf{r}, \omega)]^{-1}\}\hat{}\,|\tilde{\theta}(\omega)|^2 |\hat{\mu}(\mathbf{k})|^2 \tag{6.3}$$

Also, combining Eqs 6.2 and 6.3, we find

$$\{[\omega + \tilde{\gamma}_{\langle \mathbf{V} \rangle}(\mathbf{r}, \omega)]^{-1}\}\hat{} = \{[\omega + \tilde{\gamma}_{\mathbf{V}}(\mathbf{r}, \omega)]^{-1}\}\hat{}\,|\tilde{\theta}(\omega)|^2 |\hat{\mu}(\mathbf{k})|^2 \tag{6.4}$$

This last equation relates the memory function on the measurement scale to the ‘true’ memory function.

There are occasions when Eqs 6.3 and 6.4 are inappropriate. Consider the following example. Let $G(\mathbf{r}, t)$ be the fraction of molecules in an N molecule system that lie in $d\mathbf{r}$ at $\mathbf{r}$ at time t, given they were initially located somewhere else at $t = 0$

$$G(\mathbf{r}, t) = \frac{1}{N} E \sum_{j=1}^{N} \delta[\mathbf{r} - (\mathbf{r}_j(t) - \mathbf{r}_j(0))] \tag{6.5}$$

If all the molecules are identical, then we may set

$$G(\mathbf{r}, t) = E\delta[\mathbf{r} - (\mathbf{r}_j(t) - \mathbf{r}_j(0))] \tag{6.6}$$

The Fourier transform of Eq. 6.6 produces the self-part of the intermediate scattering function (McQuarrie, 1976)

$$\begin{aligned} F_s(\mathbf{k}, t) &= E\{\exp[i\mathbf{k}\cdot\mathbf{r}(t)] \exp[i\mathbf{k}\cdot\mathbf{r}(0)]\} \\ &= C_{\exp[i\mathbf{k}\cdot\mathbf{r}(t)]}(\mathbf{k}, t) \end{aligned} \tag{6.7}$$

When we make actual measurements of these functions Eqs 6.6 and 6.7 are replaced by

$$\langle G \rangle = E\delta[\mathbf{r} - (\mathbf{r}_j(t) - \mathbf{r}_j(0))] * \mu * \theta \tag{6.8}$$

$$\begin{aligned} \langle F_s \rangle &= \langle G \rangle\hat{} \\ &= C_{\exp[i\mathbf{k}\cdot\mathbf{r}(t)]}(\mathbf{k}, t)\hat{\mu}(\mathbf{k}) * \theta(t) \end{aligned} \tag{6.9}$$

We also have

$$\langle F_s \rangle\tilde{} = \tilde{C}_{\exp[i\mathbf{k}\cdot\mathbf{r}(t)]}(\mathbf{k}, \omega)\hat{\mu}(\mathbf{k})\tilde{\theta}(\omega) \tag{6.10}$$

Now because F_s is a correlation function, we have

$$\tilde{F}_s(\mathbf{k}, \omega) = [\omega + \tilde{\gamma}_{\exp[i\mathbf{k}\cdot\mathbf{r}(t)]}(\mathbf{k}, \omega)]^{-1} \tag{6.11a}$$

and consequently

$$\langle F_s \rangle\tilde{} = [\omega + \tilde{\gamma}_{\exp[i\mathbf{k}\cdot\mathbf{r}(t)]}(\mathbf{k}, \omega)]^{-1}\hat{\mu}(\mathbf{k})\tilde{\theta}(\omega) \tag{6.11b}$$

For this specific example, Eq. 6.11b replaces Eq. 6.3. This is because we are not correlating two measured dynamic variables, but rather measuring a single dynamic variable which, when expectations are taken, produces a correlation function in Fourier space.

If we identify $\tilde{\gamma}_{\exp[i\mathbf{k}\cdot\mathbf{r}(t)]}(\mathbf{k}, \omega)$ with $\mathbf{k}\cdot\tilde{\mathbf{D}}(\mathbf{k}, \omega)\cdot\mathbf{k}$, then it may be shown (Schoen *et al.*, 1988, Appendix) that as $\omega \downarrow 0$ and $|\mathbf{k}| \downarrow 0$, $\tilde{\mathbf{D}}(\mathbf{k}, \omega) \to \mathbf{d}$, the classical diffusion tensor. Consequently, it is customary to call $\tilde{\mathbf{D}}(\mathbf{k}, \omega)$ a generalized diffusion tensor which is wave-vector and frequency dependent, and we may write

$$\langle F_s \rangle^{\sim} = [\omega + \mathbf{k} \cdot \tilde{\mathbf{D}}(\mathbf{k}, \omega) \cdot \mathbf{k}]^{-1} \hat{\mu}(\mathbf{k}) \tilde{\theta}(\omega) \tag{6.12}$$

Equation 6.12 gives us information on the porous-medium measured value of the diffusion coefficient, as a function of frequency and wave-vector, in terms of the frequency and wave-vector pore-molecular diffusion tensor and the instruments used to measure $\exp(i\mathbf{k} \cdot \mathbf{r}(t))$ in the porous medium. Moreover, it can be shown that in the asymptotic limit (over $\mathbf{k}$) the power spectra $\tilde{C}_V(\omega)$ of the velocity autocorrelation function is identical to $\tilde{\mathbf{D}}(\mathbf{k}, \omega)$.

7. Computer simulation of inelastic neutron scattering in a microporous medium and comparison with experiment

Neutron scattering techniques are commonly employed to study fluids in microporous media and sorption on surfaces (cf. Dash and Ruvalds, 1980; Lovesay, 1984). By using molecular dynamics in conjunction with statistical mechanics, neutron scattering can be effectively simulated on the computer (Schoen *et al.*, 1988). In this section, we briefly illustrate how molecular dynamics computer simulation, in conjunction with molecular hydrodynamics and linear filtering theory, can be used to examine pre-asymptotic diffusion in micropores and to study the asymptotic limit (Fickian approximation).

We consider a slit-pore containing fluid atoms confined between two parallel square planes of rigidly fixed atoms (walls); the walls have side lengths l and are separated by a distance h. Each wall consists of N_l rigidly fixed rare gas atoms. The structure of each wall corresponds to that of the (1 0 0) plane of the face-centred cubic lattice. The coordinate system is chosen so that the walls are parallel to the $(x-y)$ plane. All atoms interact with one another pairwise via the Lennard–Jones (12, 6) potential

$$U(r_{ij}) = \varepsilon[(\sigma/r_{ij})^{12} - (\sigma/r_{ij})^6] \tag{7.1}$$

where the parameters ε and σ are chosen to represent argon and r_{ij} is the distance between atoms i and j. Summing over the fluid atoms, $i \neq j$ gives U_j of Eq. 3.4, while taking the gradient and summing over the surface atoms gives $\mathbf{X}_j$ of Eq. 3.4. Periodic boundary conditions are used in the $x-y$ plane and we employ the minimum-image convention to compute interatomic distances. Newton's equation (Eq. 3.4) of motion is integrated to obtain trajectories of the atoms. Details of the calculation may be found in Schoen *et al.* (1988).

There was considerable uncertainty as to how wide a pore had to be to have Fick's law of diffusion satisfied in the direction normal to the surfaces. To eliminate this uncertainty, F_s, $\gamma((0,0,k),t)/\gamma((0,0,k),0)$ and $\tilde{D}_{zz}((0,0,k),\omega)$ were computed. First, F_s was computed as a function of $\mathbf{k} = (0,0,\pi/h)$ for several h (Fig. 1). This computed F_s was then used to solve Eq. 4.3 for γ; plots of normalized γ corresponding to F_s in Fig. 1 are presented in Fig. 2. By Laplace-transforming γ, $\tilde{D}_{zz}(\omega)$ is obtained and plotted in Fig. 3. Also shown in Fig. 3 are the corresponding power spectra, $\tilde{C}_w$, for w (the z-component of the velocity). Only for the narrowest pore ($h = 2.9\sigma$) do we observe a significant difference between $\tilde{C}_w$ and $\tilde{D}_{zz}$. Thus Fick's law of diffusion is valid for pores greater than 3σ in width. Moreover, by use of these results for $\tilde{D}(\mathbf{k},\omega)$ in Eq. 6.12 and by use of actual neutron scattering experiments which obtain $\langle F_s \rangle^{\tilde{}}$, μ and θ (the spatial and temporal filters of the experimental device) may be determined.

8. Generalizations

In this chapter we have discussed a technique for examining pre-asymptotic diffusion in microporous media. The overriding concept is that pre-asymptotic

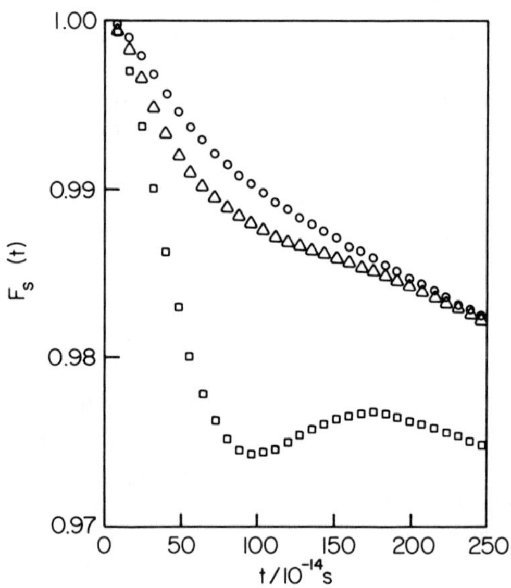

Figure 1. Self part of the intermediate scattering function $k = \pi/h$. (□) $h = 2.9\sigma$, (△) $h = 3.7\sigma$, (○) $h = 4.8\sigma$. (Schoen *et al.*, 1988.)

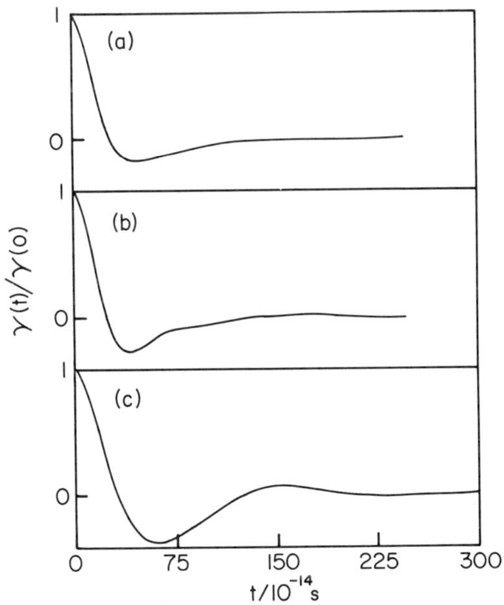

Figure 2. Normalized memory function corresponding to the normalized intermediate scattering functions of Fig. 1. (a) $h = 4.8\sigma$, (b) $h = 3.7\sigma$, (c) $h = 2.9\sigma$. (Schoen *et al.*, 1988.)

transport processes are inherently non-local in either time or space. As was pointed out in Cushman (1987), whenever a system displays long-range order, thermotransport coefficients will be wave-vector and frequency dependent, i.e. they will result from non-local theories. Two methods have been used to examine such phenomena in the past: non-local continuum theories (cf. Eringen, 1976, Vol. 4) and molecular hydrodynamics (Boon and Yip, 1980). We have illustrated the use of the latter theory at the pore scale. There are several routes one might follow to develop a 'generalized hydrodynamics' on larger scales. The simplest would be to extend the *ad hoc* theory of Matheron and de Marsily (1980). This theory envisions a ficticious 'particle' of fluid on a scale which is large relative to the pore scale. The theory can be extended to account for particles which are non-locally defined, i.e. the properties of which are a function of all of space. This proposed extension of particle mechanics would be much in the spirit of non-local continuum theories. In these theories, 'localization residuals' are introduced to localize global properties, i.e. the axiom of locality (Eringen, 1980) is not enforced and additional (non-local) source terms must be added to the standard transport equations; this is in addition to the development of non-local constitutive equations.

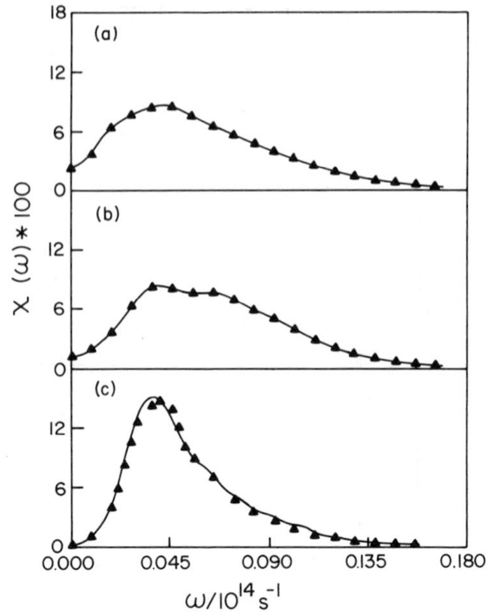

Figure 3. The normalized z-component of the Laplace transform of the generalized diffusion tensor $\tilde{D}_{zz}(\bar{k}, \omega)$ and the power spectra of the velocity autocorrelation $\tilde{C}_w(\omega)$ for the data of Fig. 1. $\chi(\omega) = \tilde{D}_{zz}((0, 0, k), \omega)$ (——); $\chi(\omega) = \tilde{C}_w(\omega)$ (▲). (a) $h = 4.86\sigma$, (b) $h = 3.76\sigma$, (c) $h = 2.96\sigma$. (Schoen *et al.*, 1988.)

Acknowledgements

This research was supported in part from a Department of Energy, Office of Health and Environmental Research grant DE-FG02/85ER 60310. The author wishes to thank Professors Dennis J. Diestler and Martin Schoen for providing stimulating discussions.

References

Baveye, P. (1985). The nonequilibrium statistical mechanics of simultaneous steady-state transport of water and solutes in soils. Unpublished Ph.D. Thesis, University of California at Riverside, CA.

Boon, J.P. and Yip, S. (1980). 'Molecular Hydrodynamics'. McGraw-Hill, New York.

Cushman, J.H. (1987). Development of stochastic partial differential equations for subsurface hydrology, *Stoch. Hyd. Hydrol.* **1**, 241–242.

Cushman, J.H. (1984). On unifying the concepts of scale, instrumentation, and stochastics in the development of multiphase transport theory, *Water Resour. Res.* **20**, 1668–1676.

Cushman, J.H. (1986). Multiphase transport in the space of stochastic tempered distributions, *I.M.A. J. Appl. Math.* **36**, 159–175.

Dash, J.G. and Ruvalds, J. (eds) (1980). 'Phase Transitions in Surface Films'. Plenum Press, London.

Eringen, A.C. (1976). 'Continuum Physics', Vol. 4. Academic Press, New York.

Eringen, A.C. (1980). 'Mechanics of Continua'. Krieger, New York.

Harp, G.D. and Berne, B.J. (1970). Time-correlation functions, memory functions, and molecular dynamics, *Phys. Rev. A.* **2**, 975–996.

Lovesay, S.W. (1984). 'Theory of Neutron Scattering from Condensed Matter', Vols 1 and 2. Clarendon, Oxford.

Matheron, G. and de Marsily, G. (1980). Is transport in porous media always diffusive? A counterexample, *Water Resour. Res.* **16**, 901–917.

McQuarrie, D.A. (1976). 'Statistical Mechanics'. Harper and Row, New York.

Schoen, M., Diestler, D.J. and Cushman, J.H. (1987). Fluids in micropores, I. Structure of a simple classical fluid in a slit pore, *J. Chem. Phys.* **87**, 5464–5476.

Schoen, M., Cushman, J.H., Diestler, D.J. and Rhykerd, Jr., C.L. (1988). Fluids in micropores, II. Self-diffusion of a simple classical fluid in a slit-pore, *J. Chem. Phys.* **88**, 1394–1406.

Sposito, G. (1978). The statistical mechanical theory of water transport through unsaturated soil, 1. Conservation laws, *Water Resour. Res.* **14**, 474–478.

Zwanzig, R. (1961). 'Lectures in Theoretical Physics' (W.B. Britling, W.B. Downs and J. Downs, eds), Vol. III, p. 106. Interscience, New York.

Index